Dietary Polyphenols

IFT Press

The *IFT Press* series reflects the mission of the Institute of Food Technologists—to advance the science of food contributing to healthier people everywhere. Developed in partnership with Wiley-Blackwell, *IFT Press* books serve as leading-edge handbooks for industrial application and reference and as essential texts for academic programs. Crafted through rigorous peer review and meticulous research, *IFT Press* publications represent the latest, most significant resources available to food scientists and related agriculture professionals worldwide. Founded in 1939, the Institute of Food Technologists is a nonprofit scientific society with 22 000 individual members working in food science, food technology, and related professions in industry, academia, and government. IFT serves as a conduit for multidisciplinary science thought leadership, championing the use of sound science across the food value chain through knowledge sharing, education, and advocacy.

WILEY Blackwell

Dietary Polyphenols

Metabolism and Health Effects

Edited by

Francisco A. Tomás-Barberán
Department of Food Science and Technology
CEBAS-CSIC, Murcia, Spain

Antonio González-Sarrías
Department of Food Science and Technology
CEBAS-CSIC, Murcia, Spain

Rocío García-Villalba
Department of Food Science and Technology
CEBAS-CSIC, Murcia, Spain

WILEY Blackwell

This edition first published 2021

Registered Office
John Wiley & Sons, Inc., 111 River Street, Hoboken, NJ 07030, USA

Editorial Office
The Atrium, Southern Gate, Chichester, West Sussex, PO19 8SQ, UK

For details of our global editorial offices, customer services, and more information about Wiley products visit us at www.wiley.com.

Wiley also publishes its books in a variety of electronic formats and by print-on-demand. Some content that appears in standard print versions of this book may not be available in other formats.

Library of Congress Cataloging-in-Publication Data

Names: Tomás-Barberán, F. A. (Francisco A.), editor. | González-Sarrías, Antonio, editor. | García-Villalba, Rocío, editor.
Title: Dietary polyphenols : metabolism and health effects / edited by Francisco A. Tomás-Barberán, Antonio González-Sarrías, Rocío García-Villalba.
Other titles: IFT press series.
Description: Hoboken, NJ : Wiley-Blackwell, 2021. | Series: IFT press series | Includes bibliographical references and index.
Identifiers: LCCN 2020019599 (print) | LCCN 2020019600 (ebook) | ISBN 9781119563723 (cloth) | ISBN 9781119563716 (adobe pdf) | ISBN 9781119563747 (epub)
Subjects: MESH: Polyphenols–metabolism | Polyphenols–pharmacology | Nutritive Value
Classification: LCC QK898.P764 (print) | LCC QK898.P764 (ebook) | NLM QV 223 | DDC 613.2/86–dc23
LC record available at https://lccn.loc.gov/2020019599
LC ebook record available at https://lccn.loc.gov/2020019600

Cover Design: Wiley
Cover Images: Pomegranate © October 22 / Getty Images, Chemical bonding structures courtesy of Francisco A. Tomás-Barberán, Human figure © Christos Georghiou / Shutterstock

Set in 10/12pt WarnockPro by SPi Global, Chennai, India

MIX
Paper from responsible sources
FSC® C013604

10 9 8 7 6 5 4 3 2 1

Titles in the IFT Press series

- Accelerating New Food Product Design and Development, 2nd Edition (Jacqueline H. Beckley, Leslie J. Herzog, M. Michele Foley)
- Anti-Ageing Nutrients: Evidence-Based Prevention of Age-Associated Diseases (Deliminda Neves)
- Bioactive Compounds from Marine Foods: Plants and Animal Sources (Blanca Hernandez-Ledesma, and Miguel Herrero)
- Biofilms in the Food Environment, 2nd Edition (Anthont L. Pometto III, and Ali Demirci)
- Bitterness: Perception, Chemistry and Food Processing (Michel Aliani, and Michael N.A. Eskin)
- Essential Oils in Food Processing: Chemistry, Safery and Applications (Seyed Mohammed Bagher Hashemi, Amin Mousavi Khaneghah, and Anderson de Souza Sant'Ana)
- Flavor, Satiety and Food Intake (Beverly Tepper, and Martin Yeomans)
- Food Carotenoids: Chemistry, Biology and Technology (Delia B. Rodriguez-Amaya)
- Food Oligosaccharides: Production, Analysis and Bioactivity (F. Javier Moreno, and Maria Luz Sanz)
- Food Safety Design, Technology and Innovation (Helmut Traitler, Birgit Coleman, and Karen Hofmann)
- Food Texture Design and Optimization (Yadunandan Lal Dar, Joseph M. Light)
- Functional Foods and Beverages: In vitro Assessment of Nutritional, Sensory, and Safety Properties (Nicolas Bordenave, and Mario G. Ferruzzi)
- Mathematical and Statistical Methods in Food Science and Technology (Daniel Granato, and Gaston Ares)
- Membrane Processing for Dairy Ingredient Separation (Kang Hu, and James Dickson)
- Microbial Safety of Fresh Produce (Xuetong Fan, Brendan A. Niemira, Christopher J. Doona, Florence E. Feeherry, and Robert B. Gravani)
- Microbiology and Technology of Fermented Foods, 2nd Edition (Robert W. Hutkins)
- Microbiology in Dairy Processing: Challenges and Opportunities (Palmiro Poltronieri)
- Nanotechnology and Functional Foods: Effective Delivery of Bioactive Ingredients (Cristina Sabilov, Hongda Chen, and Rickey Yada)
- Natural Food Flavors and Colorants, 2nd Edition (Mathew Attokaran)
- Packaging for Nonthermal Processing of Food, 2nd Edition (Melvin A. Pascall, and Jung H. Han)
- Processing and Nutrition of Fats and Oils (Ernesto M. Hernandez, and Afaf Kamal-Eldin)
- Resistant Starch: Sources, Applications and Health Benefits (Yong-Cheng Shi, Clodualdo C. Maningat)
- Spray Drying Techniques for Food Ingredient Encapsulation (C. Anandharamakrishnan, and Padma Ishwarya S.)
- Trait-Modified Oils in Foods (Frank T. Orthoefer, and Gary R. List)
- Water Activity in Foods: Fundamentals and Applications, 2nd Edition (Gustavo V. Barbosa-Ćanovas, Anthony J. Fontana Jr., Shelly J. Schmidt, and Theodore P. Labuza)

WILEY Blackwell

Contents

List of Contributors

Aadil Bajoub
Department of Basic Sciences, National School of Agriculture, km 10, Haj Kaddour Road, B.P. S/40, Meknès, Morocco.

Adrian Cortés-Martin
Laboratory of Food and Health, Research Group on Quality, Safety and Bioactivity of Plant Foods,
Dept. Food Science and Technology, CEBAS-CSIC, Murcia, Spain.

Alan Crozier
Department of Nutrition, University of California, Davis, California, USA.

and

School of Medicine, Dentistry and Nursing, University of Glasgow, Glasgow, UK.
Email: alan.crozier44@gmail.com

Alba Macià
Departament de Tecnologia dels Aliments-Àrea Nutrició, Universitat de Lleida, Lleida, Spain.

Alegría Carrasco-Pancorbo
Department of Analytical Chemistry, Faculty of Science, University of Granada,
Ave. Fuentenueva s/n, 18071, Granada, Spain.
Email: alegriac@ugr.es

Alexandre Foito
The James Hutton Institute, Invergowrie, Dundee, Scotland UK.

Alfredo Fernández-Quintela
Nutrition and Obesity group, Department of Nutrition and Food Sciences,
Faculty of Pharmacy, University of the Basque Country (UPV/EHU) and Lucio Lascaray Research Centre, Vitoria, Spain.

and

CIBER Physiopathology of Obesity and Nutrition (CIBERobn), Institute of Health Carlos III, Spain.

Ana Faria
Nutrition and Metabolism, Faculdade de Ciências Médicas|NOVA Medical Schoo|, Universidade NOVA de Lisboa, Lisboa, Portugal.

and

CINTESIS, Center for Health Technology Services Research, Porto, Portugal.

and

Comprehensive Health Research Centre, Universidade NOVA de Lisboa, Lisboa, Portugal.

Anne K. Bolvig
Aarhus University, Department of Animal Science, DK-8830 Tjele, Denmark

Antonio González-Sarrías
Laboratory of Food and Health, Research Group on Quality, Safety and Bioactivity of Plant Foods,
Dept. Food Science and Technology, CEBAS-CSIC, Murcia, Spain.
Email: agsarrias@cebas.csic.es

Cara L. Frankenfeld
Department of Global and Community Health, George Mason University, Fairfax, VA, USA.
Email: cfranken@gmu.edu

Claudia Favari
Human Nutrition Unit, Department of Food & Drug, University of Parma, Parma, Italy.

Cláudia Marques
Nutrition and Metabolism, Faculdade de Ciências Médicas|NOVA Medical Schoo|, Universidade NOVA de Lisboa, Lisboa, Portugal.

and

CINTESIS, Center for Health Technology Services Research, Porto, Portugal.

Claudia Nunes dos Santos
CEDOC, NOVA Medical School, Faculdade de Ciências Médicas, Universidade NOVA de Lisboa, Lisboa, Portugal.
Email: claudia.nunes.santos@nms.unl.pt

Claudio Curti
Department of Food & Drug, University of Parma, Parma, Italy.

Colin D. Kay
Food Bioprocessing and Nutritional Sciences, Plants for Human Health Institute, North Carolina State University, Kannapolis, North Carolina, USA.

Conceição Calhau
Nutrição e Metabolismo, NOVA Medical School, Faculdade de Ciências Médicas, Universidade Nova de Lisboa, Lisboa, Portugal.

and

CINTESIS - Center for Research in Health Technologies and Information Systems, Porto, Portugal.

Daniele Del Rio
Human Nutrition Unit,
Department of Veterinary
Science, University of Parma,
Parma, Italy.

Derek Stewart
The James Hutton Institute,
Invergowrie, Dundee, Scotland
UK.

and

School of Engineering and
Physical Sciences,
Heriot Watt University,
Edinburgh, Scotland.

Diogo Carregosa
CEDOC, NOVA Medical School,
Faculdade de Ciências Médicas,
Universidade NOVA de Lisboa,
Lisboa, Portugal.

Donato Angelino
Human Nutrition Unit,
Department of Veterinary
Science, University of Parma,
Parma, Italy.

Francisco A. Tomás-Barberán
Laboratory of Food and Health,
Research Group on Quality,
Safety and Bioactivity of Plant
Foods,
Dept. Food Science and
Technology, CEBAS-CSIC,
Murcia, Spain.
Email: fatomas@cebas.csic.es

Gema Pereira-Caro
Department of Food Science and
Health, Andalusian Institute of
Agricultural and Fishery
Research and Training, Alameda
del Obispo, Córdoba, Spain.

Gordon McDougall
The James Hutton Institute,
Invergowrie, Dundee, Scotland
UK.

Hélder Oliveira
REQUIMTE/LAQV, Department
of Chemistry and Biochemistry,
Faculty of Sciences,
University of Porto, Porto,
Portugal.

Helle Nygaard Lærke
Aarhus University, Department
of Animal Science, DK-8830
Tjele, Denmark

Iñaki Milton-Laskibar
Nutrition and Obesity group,
Department of Nutrition and
Food Sciences,
Faculty of Pharmacy, University
of the Basque Country
(UPV/EHU) and Lucio Lascaray
Research Centre, Vitoria, Spain.
and

CIBER Physiopathology of
Obesity and Nutrition
(CIBERobn), Institute of Health
Carlos III, Spain.

Itziar Eseberri
Nutrition and Obesity group,
Department of Nutrition and
Food Sciences, Faculty of
Pharmacy, University of the
Basque Country (UPV/EHU) and
Lucio Lascaray Research Centre,
Vitoria, Spain.

and

CIBER Physiopathology of
Obesity and Nutrition
(CIBERobn), Institute of Health
Carlos III, Spain.

Iva Fernandes
REQUIMTE/LAQV, Department of Chemistry and Biochemistry, Faculty of Sciences, University of Porto, Porto, Portugal.

Iziar A. Ludwig
Departament de Tecnologia dels Aliments-Àrea Nutrició, Universitat de Lleida, Lleida, Spain.
Email: iludwig@tecal.udl.cat

Jara Pérez-jiménez
Department of Metabolism and Nutrition, Institute of Food Science, Technology and Nutrition (ICTAN-CSIC), José Antonio Novais 10, 28040, Madrid, Spain.
Email: jara.perez@ictan.csic.es

Juan A. Giménez-Bastida
Laboratory of Food and Health, Research Group on Quality, Safety and Bioactivity of Plant Foods, Dept. Food Science and Technology, CEBAS-CSIC, Murcia, Spain.

Juan C. Espín
Laboratory of Food and Health, Research Group on Quality, Safety and Bioactivity of Plant Foods, Dept. Food Science and Technology, CEBAS-CSIC, Murcia, Spain.

Katerina Valentova
Laboratory of Biotransformation, Institute of Microbiology of the Czech Academy of Sciences, Vídeňská 1083, 142 20 Prague, Czech Republic.

Knud E. Bach Knudsen
Aarhus University, Department of Animal Science, DK-8830 Tjele, Denmark.
Email: knuderik.bachknudsen @anis.au.dk

Laura Rubió
Departament de Tecnologia dels Aliments-Àrea Nutrició, Universitat de Lleida, Lleida, Spain.

Lucía Olmo-García
Department of Analytical Chemistry, Faculty of Science, University of Granada, Ave. Fuentenueva s/n, 18071, Granada, Spain.

María A. Ávila-Gálvez
Laboratory of Food and Health, Research Group on Quality, Safety and Bioactivity of Plant Foods,
Dept. Food Science and Technology, CEBAS-CSIC, Murcia, Spain.

Maria P. Portillo
Nutrition and Obesity group, Department of Nutrition and Food Sciences, Faculty of Pharmacy, University of the Basque Country (UPV/EHU) and Lucio Lascaray Research Centre, Vitoria, Spain.
CIBER Physiopathology of Obesity and Nutrition (CIBERobn), Institute of Health Carlos III, Spain.
Email:
mariapuy.portillo@ehu.eus

Maria P. Romero
Departament de Tecnologia dels Aliments-Àrea Nutrició, Universitat de Lleida, Lleida, Spain.

Maria Romo-Vaquero
Laboratory of Food and Health, Research Group on Quality, Safety and Bioactivity of Plant Foods, Dept. Food Science and Technology, CEBAS-CSIC, Murcia, Spain.

Maria V. Selma
Laboratory of Food and Health, Research Group on Quality, Safety and Bioactivity of Plant Foods, Dept. Food Science and Technology, CEBAS-CSIC, Murcia, Spain.
Email: mvselma@cebas.csic.es

Mette Skou Hedemann
Aarhus University, Department of Animal Science, DK-8830 Tjele, Denmark

Michael N. Clifford
School of Biosciences and Medicine, University of Surrey, Guildford, UK.

Natalja Nørskov
Aarhus University, Department of Animal Science, DK-8830 Tjele, Denmark

Nuno Mateus
REQUIMTE/LAQV, Department of Chemistry and Biochemistry, Faculty of Sciences, University of Porto, Porto, Portugal.

Pedro Mena
Human Nutrition Unit, Department of Food & Drug, University of Parma, Parma, Italy.
Email: pedromiguel.menaparreno@unipr.it

Regina Menezes
CEDOC, NOVA Medical School, Faculdade de Ciências Médicas, Universidade NOVA de Lisboa, Lisboa, Portugal.

Rocío García-Villalba
Laboratory of Food and Health, Research Group on Quality, Safety and Bioactivity of Plant Foods, Dept. Food Science and Technology, CEBAS-CSIC, Murcia, Spain.
Email: rgvillalba@cebas.csic.es

Romina P. Monasterio
Instituto de Biología Agrícola de Mendoza (IBAM), UNCuyo, CONICET. Alt. Brown 500, Chacras de Coria, Mendoza, Argentina.

Rosalía Reynoso-Camacho
Research and Graduate Studies in Food Science, Facultad de Química, Universidad Autónoma de Querétaro, Cerro de las campanas s/n, 76010 Querétaro, Qro., México.

Saioa Gómez-Zorita
Nutrition and Obesity group, Department of Nutrition and Food Sciences, Faculty of Pharmacy, University of the Basque Country (UPV/EHU) and Lucio Lascaray Research Centre, Vitoria, Spain.

and

CIBER Physiopathology of Obesity and Nutrition (CIBERobn), Institute of Health Carlos III, Spain.

Victor de Freitas
REQUIMTE/LAQV, Department of Chemistry and Biochemistry, Faculty of Sciences, University of Porto, Porto, Portugal.
Email: vfreitas@fc.up.pt

Yuridia Martínez-Meza
Research and Graduat Studies in Food Science, Facultad de Química, Universidad Autónoma de Querétaro, Cerro de las campanas s/n, 76010 Querétaro, Qro., México.

1

Structural Diversity of Polyphenols and Distribution in Foods

Antonio González-Sarrías, Francisco A. Tomás-Barberán, and Rocío García-Villalba*

Laboratory of Food and Health, Research Group on Quality, Safety, and Bioactivity of Plant Foods, Department Food Science and Technology, CEBAS-CSIC, Murcia, Spain

1.1 Introduction

(Poly)Phenolic compounds or polyphenols are the most common and ubiquitous groups of secondary metabolites widely distributed in the Plant Kingdom. These metabolites are involved in important roles in plants, such as pigmentation, growth and reproduction functions, protection against ultraviolet (UV) radiation, resistance to pathogens and herbivores, and many other functions. They also contribute substantially to the organoleptic characteristics of flowers, leaves, fruits, and vegetables such as bitterness, astringency, color, and flavor (Bravo, 1998; Lattanzio *et al.*, 2008; Pandey and Rizvi, 2009; Tomás-Barberán and Espín, 2001). Apart from beneficial effects on plants, many of these nonnutrient metabolites have been attributed as the molecules potentially responsible for the health effects in humans. Vegetable- and fruit-rich diets exhibit a wide spectrum of potential biological activities related to the prevention of many of the major chronic diseases such as cardiovascular, neurodegenerative, and cancer diseases (D'Archivio *et al.*, 2007; Espín *et al.*, 2017; Rothwell *et al.*, 2017). In this book, the most recent studies about metabolism and the current evidence on the health effects of the different group of

*Corresponding author.

Dietary Polyphenols: Metabolism and Health Effects, First Edition.
Edited by Francisco A. Tomás-Barberán, Antonio González-Sarrías, and Rocío García-Villalba.

polyphenols, as well as their bioavailable metabolites, will be reviewed and discussed.

1.2 Classification and Chemistry of Polyphenols

The structure of phenolic compounds varies extensively but presents as a common feature the presence of one (simple phenolics) or more (polyphenols) hydroxyl substituents attached directly to one or more aromatic or benzene rings. Therefore, they have been classified into different groups or classes according to the pattern of their basic skeleton, from relatively simple, such as phenolic acids, to polymerized molecules of relatively high molecular mass, such as hydrolyzable and condensed tannins (Manach *et al.*, 2004; Pereira *et al.*, 2010). In general, the phenolic compounds are found in plants in the conjugated form rather than as free compounds, with one or more sugar residues linked by β-glycosidic bonds to a hydroxyl group (*O*-glycosides) or a carbon atom of the aromatic ring (*C*-glycosides). The associated sugars can be monosaccharides, disaccharides or even oligosaccharides, glucose being the most common followed by others such as galactose, rhamnose, xylose, arabinose, etc. (Manach *et al.*, 2004).

Moreover, the wide structural diversity in phenolic compounds encompasses over 8000 compounds described in nature that traditionally are divided into two main groups based on their basic structure, flavonoids and nonflavonoids, that are subdivided into different subgroups according to the number of aromatic or phenol rings and the structural elements that bind these rings to one another (Bravo, 1998; D'Archivio *et al.*, 2007; Del Rio *et al.*, 2013; Waterhouse, 2002).

1.2.1 Flavonoids

Flavonoids are the largest group of phenolic compounds, accounting for more than 5000 different compounds present in dietary plant foods, although they usually occur as glycosides rather than aglycones, mostly linked to glucose, rhamnose, xylose or galactose (Harbone and Williams, 2000; Tsao, 2010).

Figure 1.1 The basic structure of flavonoids.

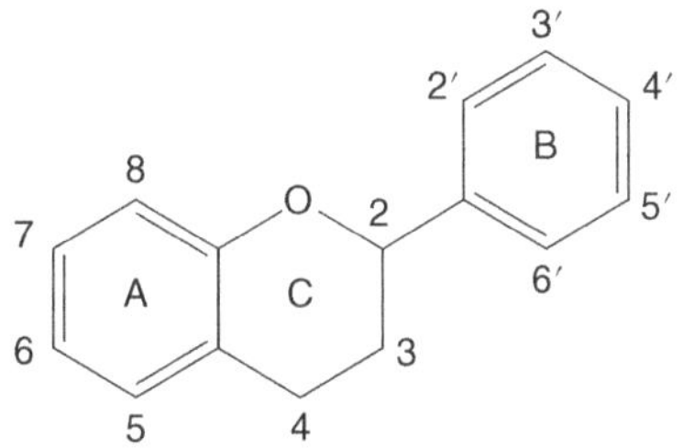

The basic flavonoid structure is composed of two phenol rings (A and B) linked through a linear three-carbon chain that forms a heterocyclic pyran ring (C) containing one oxygen atom (Figure 1.1).

Based on the degree of oxidation, saturation, and hydroxylation of the central pyran ring, flavonoids can be divided into different subgroups as flavan-3-ols (catechins and proanthocyanidins), flavones, flavonols, flavanones, isoflavones, and anthocyanidins (Table 1.1) (Bravo, 1998). The diversity of each group of flavonoids depends on the different patterns of substitution of the hydroxyl groups in the basic flavonoid skeleton, mainly the conjugation with various mono- and disaccharides creating highly complex structures (Bravo, 1998). In addition to these major flavonoid groups, there are other minor ones such as chalcones, dihydrochalcones, dihydroflavonols, and flavan-3,4-diols. In Table 1.1, the most common examples of different flavonoid subgroups found in plant foods are listed.

Flavan-3-ols or flavanols are structurally characterized by the presence of a hydroxyl group in the heterocyclic ring C. Unlike other flavonoid subgroups, they cannot occur as glycosides in food sources, but exist as simple monomers such as catechin and epicatechin, to the oligomeric and polymeric condensed tannins, which are also known as proanthocyanidins. Proanthocyanidins are highly complex chemical structures formed by oligomerization or polymerization of up to 50 subunits of monomeric flavanols joined by one (type B proanthocyanidins) or two (type A proanthocyanidins) oxidative couplings between two monomers. Proanthocyanidins containing only catechin/epicatechin units are known as procyanidins, which are the most common in nature, while those formed by gallocatechin/epigallocatechin units are called prodelphinidins,

Table 1.1 Main flavonoid groups and distribution in foods

Structure	Main compounds	Food sources
Flavan-3-ols 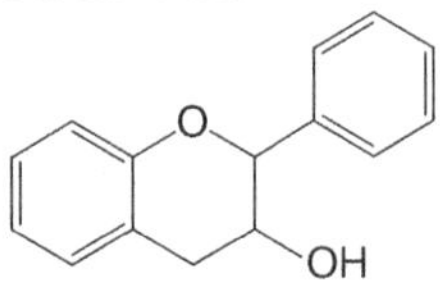	Catechin Epicatechin Epicatechin gallate Gallocatechin	Apple, apricot, peach, grape, berries, cereals, chocolate, red wine, nuts, black and green tea
Proanthocyanidins 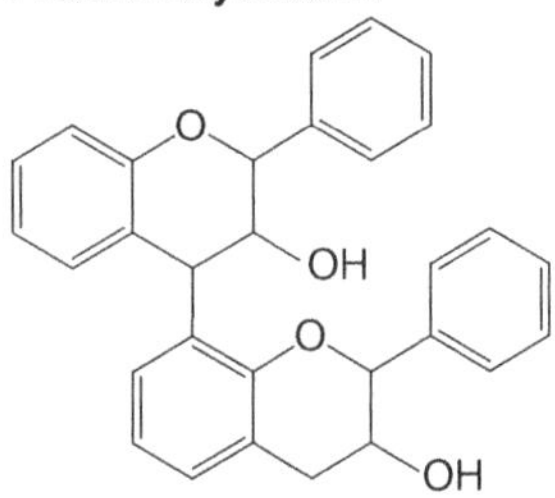	Procyanidin B1 Procyanidin B2	Red wine, beer, cider, apple, pear, grape, chocolate
Flavones 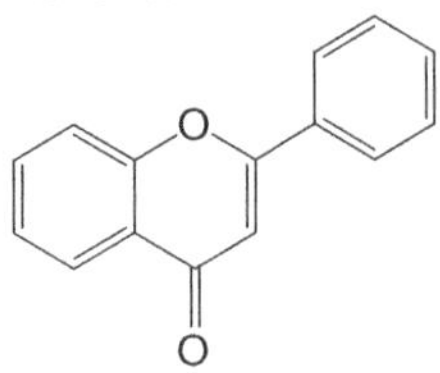	Apigenin Luteolin Chrysin	Parsley, celery, lettuce, artichoke, herbs (rosemary, thyme, oregano, etc.), citrus fruits, cereal grains, sweet peppers
Flavonols 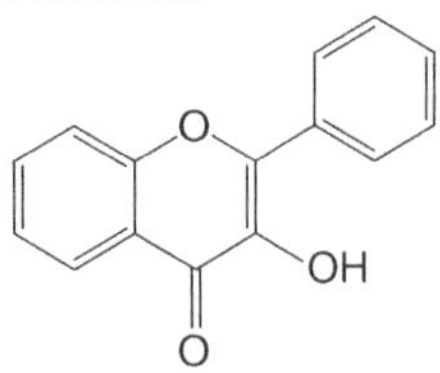	Quercetin Kaempferol Myricetin Isorhamnetin	Yellow and red onion, caper, lettuce, parsley, berries, green and black tea, mango, carrot, pumpkin, kale, cabbage, broccoli, garlic
Flavanones 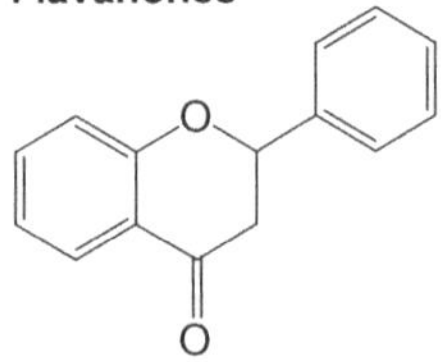	Naringenin Hesperetin	Orange, grapefruit, lemon, lime

Table 1.1 (Continued)

Structure	Main compounds	Food sources
Isoflavones 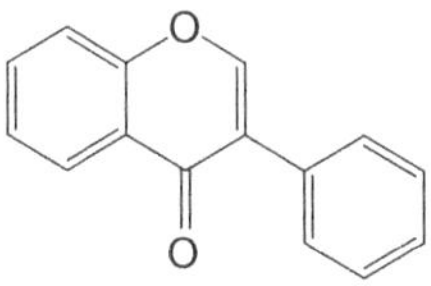	Daidzein Genistein Glycitein	Soybean, tofu, green bean, lentil, chickpea, pea, mung bean, broad bean, medicinal herbs
Anthocyanidins 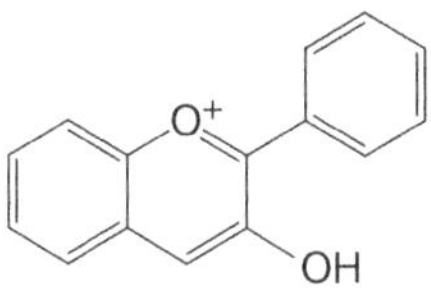	Cyanidin Delphinidin Pelargonidin Peonidin	Berries, currant, grape, aronia, cherries, plum, pomegranate, red wine, red cabbage, eggplant, red onion, radish, hazelnut, pistachio nut, black and red bean, medicinal herbs

and those with afzelechin/epiafzelechin units are known as propelargonidins (Smeriglio *et al.*, 2017; Spranger *et al.*, 2008).

Flavones are structurally characterized by a double bond and an oxygen atom in the heterocyclic ring C of the flavonoid skeleton. Flavones, such as apigenin and luteolin, can be found in plants showing a wide range of substitutions, including methylations, hydroxylations, acylations, and glycosylations leading mainly to *O*- or *C*-glycosides (Hostetler *et al.*, 2017).

Flavonols contain a similar structure to flavones but with the presence of a hydroxyl group at carbon 3 of the flavone nucleus (3-hydroxyflavones). Flavonols are one of the most abundant flavonoid subgroups widely found in plants; they are commonly found as glycosides and the most common one is quercetin (Leo and Woodman, 2015).

Flavanones and dihydroflavonols contain a similar structure to that of flavones and flavonols in which the double bond in the heterocyclic ring C has been reduced (hydrogenated). Flavanones are one of the main flavonoid subgroups and are mostly found in the form of glycosylated derivatives through the formation of an *O*-glycosidic linkage usually with a rutinosyl (rhamnosyl 1-6 glucosyl-) or a neohesperidosyl (rhamnosyl 1-2 glucosyl-) moiety to the aglycone hydroxyl groups, the most common being glycosylation of the hydroxyl at C-7 of ring A (Barreca *et al.*, 2017).

Isoflavones or isoflavonoids differ from the other flavonoid subgroups because the ring B is bound to the heterocyclic ring C at C-3 position instead of C-2. Unlike other flavonoid subgroups, the occurrence of isoflavones in plants is limited, almost exclusively, to leguminous plants, mainly found in the form of β-glucosides and their acetyl- or malonyl-derivatives. However, there is a large structural variation of isoflavones according to the oxidation level of their skeleton. Isoflavones, like lignans, and stilbenes are also classified as phytoestrogens due to their structural similarities to estrogens and, therefore, their capacity to bind to estrogen receptors (Heinonen *et al.*, 2002). The dietary glucosylated isoflavones, such as daidzin or genistin, are poorly absorbed after consumption. However, they are cleaved to their aglycones, daidzein and genistein, which are readily absorbed into the circulatory system and/or further metabolized in the colon by the action of the intestinal microbiota to other bioactive metabolites such as equol, *O*-desmethylangolensin (ODMA), and dihydrogenistein (Frankenfeld *et al.*, 2014; Heinonen *et al.*, 2002; Zaheer and Humayoun Akhtar, 2017). Thus, it is well established that interindividual differences in the conversion of the isoflavone daidzein to equol and ODMA are associated with the heterogeneity of individual biological responsiveness to the consumption of isoflavones-containing products (Frankenfeld *et al.*, 2014; Heinonen *et al.*, 2002).

Anthocyanidins are water-soluble pigments responsible for the red, blue, and purple-colored plant organs, mainly flowers, fruits, and leaves, depending on the light, pH, and temperature (Khoo *et al.*, 2017; Laleh *et al.*, 2006). They differ from other flavonoid subgroups because they have a positive charge at the oxygen atom of the heterocyclic ring C of the basic flavonoid structure, also called the flavylium (2-phenylchromenylium) cation. They lead to a wide variety of pigments in plants and are commonly found as glycosides, called anthocyanins, which are bonded to various sugar residues mainly attached to the hydroxyl at C-3 on the heterocyclic ring C or attached to the hydroxyl groups of the ring A at C-5 and C-7 position. Among monosaccharides, such as glucose, xylose or galactose, and disaccharides, such as rutinose or neohesperidose, glucose is the most common glycosyl unit found in anthocyanins. These sugar moieties can also be acylated with different aromatic (*p*-coumaric, ferulic, caffeic, sinapic) or aliphatic acids (malonic, acetic) (D'Archivio *et al.*, 2007; Khoo *et al.*, 2017; Krga and Milenkovic, 2019; Wallace and Giusti, 2015).

1.2.2 Nonflavonoids

Nonflavonoids are the other principal group of phenolic compounds with dietary importance which generally have both a simpler chemical structure than that of the flavonoids as well as large and complex polyphenols. The main nonflavonoid phenolics include the simple phenolic acids (hydroxycinnamic and hydroxybenzoic acids), the hydrolyzable tannins (ellagitannins and gallotannins), stilbenes, coumarins, and lignans (Bravo, 1998). Table 1.2 shows the most common examples of nonflavonoid phenolics found in plant foods.

Phenolic acids are simple phenols that contain a carboxyl group and occur mainly as hydroxybenzoic (C6-C1 skeleton) and hydroxycinnamic acids (C6-C3 skeleton) which derive from benzoic or cinnamic acid, respectively. They can occur in plant foods either in their free or conjugated form attached to different functional groups or esterified to organic acids (Razzaghi-Asl *et al.*, 2013; Robbins, 2003).

The hydrolyzable tannins have a high molecular weight and are formed by a carbohydrate moiety, usually glucose, partially or totally esterified with phenolic residues such as gallic acid in the case of gallotannins or hexahydroxydiphenic acid (precursor of ellagic acid after hydrolysis) for ellagitannins. Unlike the flavonoid-derived condensed tannins, they are readily hydrolyzed under acid hydrolysis (Okuda *et al.*, 1995; Smeriglio *et al.*, 2017; Tomás-Barberán *et al.*, 2008). It is well documented that ellagitannins and ellagic acid have limited bioavailability. Indeed, when ellagic acid, either released from ellagitannins or free ellagic acid occurring naturally in foods, reaches the distal part of the gastrointestinal tract, it is further hydrolyzed

Table 1.2 Nonflavonoid (poly)phenols and main dietary sources

Structure	Main compounds	Food sources
Phenolic acids Cinnamic acid 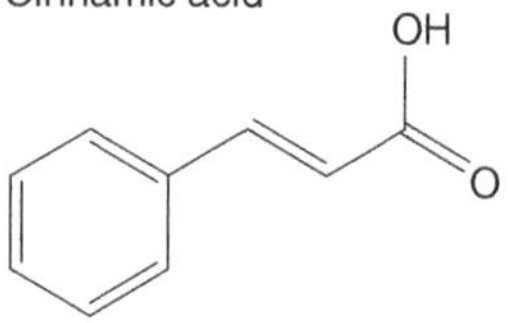	*p*-Coumaric acid Caffeic acid Ferulic acid Sinapic acid	Coffee, potato, broccoli, spinach, lettuce, cabbage, apple, pear, cherries, apricot, peach, blackcurrant, blueberry, asparagus, wine, rye bread
Benzoicacids 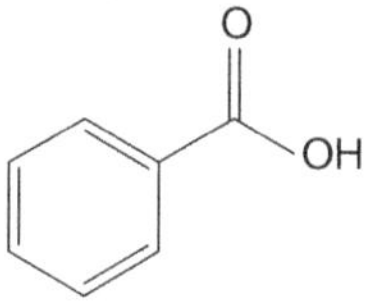	Gallic acid Protocatechuic acid Syringic acid Vanillic acid	Cloudberry, raspberry, red cabbage, chestnut, tea
Hydrolyzable tannins Ellagitannins 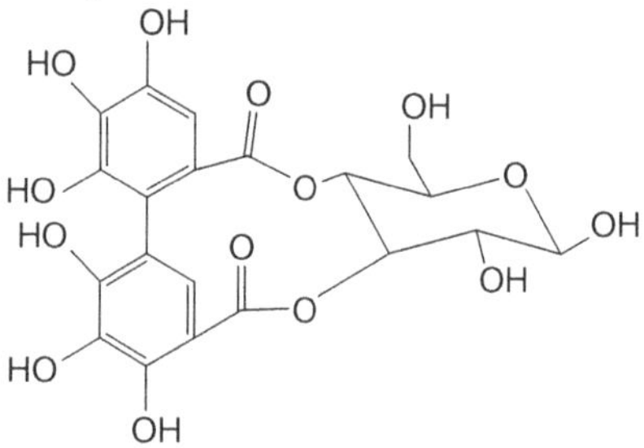	Sanguiin H6 Punicalagin Pedunculagin	Strawberry, raspberry, blackberry, pomegranate, walnut, chestnut, hazelnut, mango, green and black tea, oak-aged beverages
Gallotannins 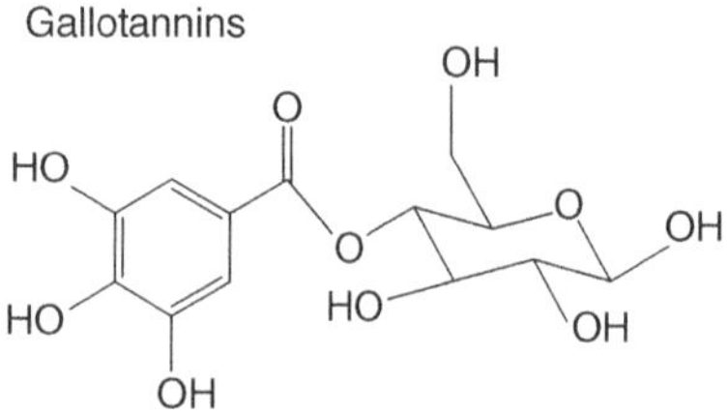	Galloyl-hexoside Digalloyl-hexoside	Mango, chestnut, red sword bean

Table 1.2 (Continued)

Structure	Main compounds	Food sources
Stilbenes 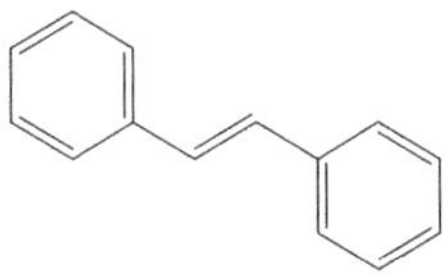	Resveratrol	Red wine, grape
Coumarins 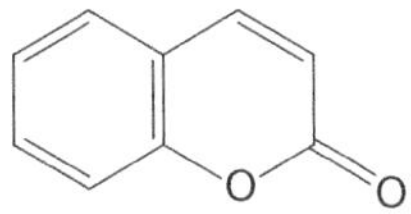	Umbelliferona Esculetina Scoparone	Citrus, parsley, celery, medicinal herbs
Lignans 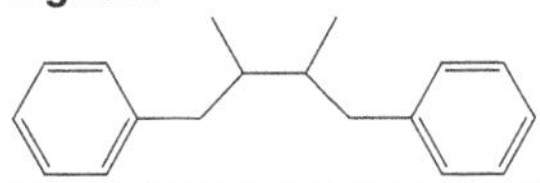	Secoisolariciresinol Matairesinol Pinoresinol Lariciresinol	Flaxseed, sesame seeds

and/or metabolized by the colonizing microbiota into a family of dibenzo[*b,d*]pyran-6-one derivatives known as urolithins that can reach systemic tissues (Cerdá *et al.*, 2004; Tomás-Barberán *et al.*, 2017). Urolithins are bioavailable microbial metabolites characterized by a nucleus of a dibenzo [*b,d*]pyran6-one with different hydroxylation patterns. In recent years, three different ellagitannin-metabolizing metabotypes have been described in humans associated with interindividual variability in urolithin production, which depends on gut microbiota composition (Tomás-Barberán *et al.*, 2014).

Stilbenes are structurally characterized by the presence of two phenyl moieties connected by a two-carbon methylene bridge (C6-C2-C6). They can be found as both monomeric and oligomeric forms that are produced by oxidative coupling between monomeric stilbenes such as *trans*-resveratrol (Rivière *et al.*, 2012; Shen *et al.*, 2009). Since there are more than 400 natural stilbenes in the plant kingdon, low quantities of stilbenes are present in the human diet, resveratrol being the most representative which occurs in both *cis* and *trans* isomers as well as in glycosylated forms such as its glucoside, piceid (D'Archivio *et al.*, 2007; Del Rio *et al.*, 2013; Shen *et al.*, 2009).

Coumarins are a family of benzopyrones derived from hydroxycinnamic acids (C6-C3) by lactonization. The most common are coumarins, isocoumarins, furanocoumarins, and benzocoumarins. They are highly bioactive, and even toxic, compounds that are seldom found in foods (Matos *et al.*, 2015).

Finally, lignans are nonflavonoid phytoestrogens whose structure derives from oxidative dimerization of two phenylpropanoid units (C6-C3) linked at the central carbon (C8-C8′). Lignans are generally found in free forms, although to a lesser extent they can be coupled to sugars as glycosidic derivatives. It is well established that dietary lignans are metabolized by intestinal microbiota to the bioactive mammalian lignans or enterolignans, enterodiol and enterolactone, that contain a structure with only two phenolic hydroxyl groups, at the metaposition of each aromatic ring (D'Archivio *et al.*, 2007; Raffaelli *et al.*, 2002; Saleem *et al.*, 2005).

1.3 Dietary Intake and Food Sources of Polyphenols

As indicated above, phenolic compounds or polyphenols are nonnutrient secondary metabolites widely spread throughout the plant kingdom as constituents of almost all vegetables, fruits, cereals, beverages such as tea, coffee, and red wine, and other plant-derived foods, and therefore, they represent an important source of bioactive compounds in the human diet (Pérez-Jiménez *et al.*, 2010a; Scalbert and Williamson, 2000). Moreover, polyphenols are involved, both positively and negatively, in the sensory and organoleptic properties of fruits and vegetables such as color, flavor, and astringency (Ignat *et al.*, 2011; Tomás-Barberán and Espín, 2001).

According to several observational studies conducted in different cohorts, the estimated mean total daily intake of polyphenols can reach over 1 g, becoming the most abundant micronutrients present in a regular diet (Manach *et al.*, 2004; Miranda *et al.*, 2016; Ovaskainen *et al.*, 2008; Pérez-Jiménez *et al.*, 2011; Pinto and Santos, 2017; Tresserra-Rimbau *et al.*,

2013; Zamora-Ros *et al.*, 2016). Over 500 different polyphenols are found in low or high amounts in most of the over 400 plant species regularly consumed in the human diet. One-third of dietary polyphenols is dominated by phenolic acids and the remaining two-thirds by the largest subgroup of flavonoids (Gupta *et al.*, 2013; Pérez-Jiménez *et al.*, 2010b). It is well known that fruit and beverages such as tea, coffee, and red wine are the most relevant from their content in the diet, but vegetables, cereals, and leguminous plants are also important sources. However, their polyphenol content may significantly differ among different varieties of a specific plant food based on genotype and ecophysiological factors as well as environmental and agronomic conditions (high or low temperature, UV exposure, insect attack, postharvest handling, water supply) and food processing-related factors (type of storage, culinary preparation, type of processing) (D'Archivio *et al.*, 2007; Manach *et al.*, 2004; Schreiner, 2005).

Most plant foods contain complex mixtures of polyphenols. Some of them, however, are mainly present in particular foods such as flavanones in citrus fruit, isoflavones in legumes (soybean and derived foods), dihydrochalcones (phloridzin) in apples, or flavones in celery and parsley. Other polyphenols, such as quercetin or catechin, are, however, found in many food products (fruit, vegetables, cereals, tea, wine). In Tables 1.1 and 1.2, the most common sources of each phenolic subgroup are presented.

1.3.1 Flavonoids

Flavonoids (see Table 1.1) are extensively found in most foodstuffs of plant origin but mainly in fruits such as apples, berries, and citrus fruits, vegetables such as onions and parsley, together with red wine, green and black tea, cocoa, nuts and certain spices (Beecher, 2003; Crozier *et al.*, 2009; Manach *et al.*, 2004; Marzocchella *et al.*, 2011).

Regarding flavanols, mainly catechin and epicatechin, the main representative sources are fruits such as apples, apricots, peaches, grapes, and some berries, cereals, chocolate, red wine, and nuts, whereas flavanols such as epigallocatechin gallate, gallocatechin or epigallocatechin are found especially in

Camellia sinensis teas (black, green, etc.) (Arts *et al.*, 2000a,b; Manach *et al.*, 2004). On the other hand, the most abundant type of proanthocyanidins found in plant foods is the dimeric procyanidins B1, B2, B3, and B4, that consist exclusively of epicatechin/catechin units. These procyanidins have been reported to be responsible for the astringent character of beverages (red wine, beer, and cider) and fruits (apples, pears, grapes, etc.) and the bitterness of chocolate (D'Archivio *et al.*, 2007; Crozier *et al.*, 2009; Gu *et al.*, 2004; Rasmussen *et al.*, 2005).

Flavones are the least common flavonoids in food. They occur in relatively high amounts in parsley and celery (apigenin and luteolin). They are also present as *O*-glycosides and *C*-glycosides in many different food products of the family Asteraceae (lettuce and artichoke) and Lamiaceae (herbs such as rosemary, thyme, oregano, mint, sage, etc.). They also occur in citrus fruits (vicenin-2, orientin), and cereal grains, (tricetin, tricin, luteolin, and apigenin *C*- and di-*C*-glycosides) as well as in sweet peppers (D'Archivio *et al.*, 2007; Del Rio *et al.*, 2013; Hostetler *et al.*, 2017). In addition, low amounts of methylated flavones such as diosmetin-, acacetin-, and chrysoeriol-*C*-glycosides are also found in citrus juices, mainly in mandarin orange, orange, citron, and bergamot juices, as well as low quantities of luteolin and apigenin in red and white wine (Del Rio *et al.*, 2013; Hostetler *et al.*, 2017). Finally, lesser amounts are found in other food sources such as blue fruits, pumpkin, chicory, kumquat, olive oil, honey, etc. as well as in some cereals and legumes such as wheat grain, black rice, fava bean, chickpea, etc. (Hostetler *et al.*, 2017).

Flavonols constitute the most ubiquitous flavonoid subgroup in our diet, with quercetin as the most consumed type of flavonols, typically found as glycosides. The main food sources of quercetin are yellow and red onions, capers, lettuce, parsley, and some types of berries, and in lesser amounts also found in apples, figs, Brussels sprouts, and buckwheat (Bhagwat *et al.*, 2011; D'Archivio *et al.*, 2007; Del Rio *et al.*, 2013). Other dietary flavonols also commonly found as *O*-glycosides are kaempferol and myricetin, found in green and black tea as well as in fruits and vegetables such as mango, carrot or pumpkin, and Brassicaceae such as kale, cabbage, and broccoli or Alliaceae such as garlic (Crozier *et al.*, 2009; Miean and Mohamed, 2001).

Flavanones are found in high concentrations mainly in citrus fruits and their juices (orange, grapefruit, lemon, lime, kumquat, etc.) where they account for approximately 95% of the flavonoids in the *Citrus* genus (Bhagwat *et al.*, 2011; Peterson *et al.*, 2006). They are also found in artichokes, tomatoes, and certain aromatic plants such as oregano (Bhagwat *et al.*, 2011; Crozier *et al.*, 2009; Ignat *et al.*, 2011). The main flavanone glycosides (rhamnosyl-glucosides) are hesperidin (hesperetin-7-*O*-rutinoside) found in oranges, naringin (naringenin-7-*O*-neohesperidoside) found in grapefruit, neohesperidin (hesperetin-7-*O*-neohesperidoside) found in bitter oranges, and eriocitrin (eriodictyol-7-*O*-rutinoside) found in lemons (D'Archivio *et al.*, 2007; Peterson *et al.*, 2006).

Isoflavones occur almost exclusively in legumes (Leguminosae), with the highest amount found in the cultivated soybean (*Glycine max* (L.)). Thus, soybean, also referred to as soy or soya, and its processed products including soy flour, soy flakes, miso, tempeh, natto, tofu, and soy milk, represent the main source of isoflavones (Danciu *et al.*, 2018; D'Archivio *et al.*, 2007; Zaheer and Humayoun Akhtar, 2017). The main isoflavones, referred to as phytoestrogens to indicate their estrogenic properties, are genistein, daidzein, and glycitein that occur as aglycones, mainly in processed soy products, or more often as glycosidic forms, mainly in grains, that are less well absorbed (Danciu *et al.*, 2018; Mazur *et al.*, 1998; Zaheer and Humayoun Akhtar, 2017). In addition to soybean, other legumes also contain significant amount of isoflavones such as green beans, lentils, chickpeas, peas, mung beans, and broad beans, as well as several medicinal plants including red clover, lucerne, and sohphlang flax (Danciu *et al.*, 2018; Ko, 2014; Zaheer and Humayoun Akhtar, 2017).

Finally, anthocyanins represent one of the most important components of flavonoids in the human diet. Anthocyanins are widely distributed in vegetables and fruits and are responsible for the blue, purple, and red pigments found in flowers, fruits, leaves, and roots. They are also increasing being used as colorants for the food industry (D'Archivio *et al.*, 2007; Khoo *et al.*, 2017; Krga and Milenkovic, 2019). The most common types of anthocyanidins widespread in fruits and vegetables are cyanidin (responsible for reddish-purple pigment), delphinidin (responsible for blue-reddish or purple pigment), pelargonidin (responsible for red and orange pigment), peonidin (responsible for reddish-purple pigment), malvidin (responsible for purple-blue and red pigment), and petunidin (responsible for dark red or purple pigment) (Bąkowska-Barczak, 2005; Katsumoto *et al.*, 2007; Khoo *et al.*, 2017). Among colored fruits, the main dietary sources are berries (elderberries, bilberries, blueberries, blackberries, strawberries, raspberries), currants, grapes, aronia, cherries, plums, pomegranates, some tropical fruits, and fruit-derived products (red wine, fruit juices, and jams). Among dark-colored vegetables and cereals, anthocyanins are found in red cabbage, eggplant, red onions, radishes, hazelnuts, pistachio nut, and black and red beans, as well as certain varieties of herbal medicinal plants including red clover, red hibiscus, and purple passion flower (Bhagwat *et al.*, 2011; D'Archivio *et al.*, 2007; Khoo *et al.*, 2017).

1.3.2 Nonflavonoids

The nonflavonoid polyphenols group (see Table 1.2) is rich and diverse. It includes the phenolic acids, commonly found in many foods such as coffee and many types of fruits, the hydrolyzable tannins found in pomegranate, berries, nuts, tropical fruits, the stilbenes such as resveratrol found mostly in red wine, and the lignans found in flaxseed, sesame, and many grains and fruits (D'Archivio *et al.*, 2007; Del Rio *et al.*, 2013; Manach *et al.*, 2004; Robbins, 2003).

Phenolic acids are abundant in the human diet, being present in all plant food groups. Phenolic acids can be distinguished in two main classes: cinnamic acid derivatives (hydroxycinnamic acids) and benzoic acid derivatives (hydroxybenzoic acids). Hydroxycinnamic acids such as *p*-coumaric, caffeic, ferulic, and sinapic acids are more abundant in plant foods and are commonly found as glycosides, esters of glucose, and esters of quinic acid. They are rich in coffee and some vegetables, fruits, and cereals, particularly in potatoes, broccoli, spinach, lettuce, cabbage, apples, pears, cherries, apricots, peaches, blackcurrants, blueberries, asparagus, wine, and rye bread (Bravo, 1998; D'Archivio *et al.*, 2007; Del Rio *et al.*, 2013; El Gharras, 2009;

Manach *et al.*, 2004). Regarding hydroxybenzoic acids, in particular, gallic, protocatechuic, syringic and vanillic acids, they are found in very few edible plant foods, mainly in some berries such as cloudberry or raspberry, red cabbage, chestnut, and tea (D'Archivio *et al.*, 2007; El Gharras, 2009; Manach *et al.*, 2004). Hydroxybenzoic acids, such as gallic or hexahydroxydiphenic acids, are constituents of hydrolyzable tannins such as the gallotannins found in mango and the ellagitannins of various types of fruit, such as strawberries, raspberries, blackberries and pomegranate, and in nuts (Manach *et al.*, 2004).

The hydrolyzable tannins are divided into two classes: those composed of ellagic and gallic acid esters of glucose or related sugars (ellagitannins and galotannins, respectively) (Okuda *et al.*, 1995; Tomás-Barberán *et al.*, 2008). Ellagitannins such as sanguiin H6, punicalagin or pedunculagin are found in significant amounts in many berries, in particular strawberries and raspberries, as well as in other fruits and nuts such as pomegranate, muscadine grapes, walnuts, chestnuts, hazelnuts, mango, green and black tea, and are also present in oak-aged beverages (wine, whiskey, etc.) (Crozier *et al.*, 2009; Tomás-Barberán *et al.*, 2008, 2016). Gallotannins, unlike ellagitannins, are rarely found in plant foods, and occur almost exclusively in mango, chestnuts, and red sword bean (Gan *et al.*, 2018; Luo *et al.*, 2014; Smeriglio *et al.*, 2017).

Stilbenes are present in low quantities in the human diet, resveratrol being the most characteristic, mainly found in grapes and red wine, mostly in glycosylated forms rather than its *cis*/*trans* isomers, as well as oligomers containing different resveratrol units such as δ-viniferin and ε-viniferin (Burns *et al.*, 2002; D'Archivio *et al.*, 2007; Vitrac *et al.*, 2005). Other dietary sources with lesser amounts of stilbenes, mainly resveratrol, are peanuts, pistachios, some berries, red cabbage, and spinach (Crozier *et al.*, 2009; D'Archivio *et al.*, 2007; Rivière *et al.*, 2012), as well as certain medicinal remedies such as *Polygonum cuspidatum* that contains very high levels of resveratrol and its glucoside piceid (Vastano *et al.*, 2000).

Coumarins and derivatives are common in members of the Rutaceae (citrus), and Apiaceae (parsley, celery, etc.) families. They are also found in several species belonging to different botanical families used as herbal medicinal remedies such as

Aesculus hippocastanum (horse chestnut), *Passiflora incarnata* (passion flower), and *Hypericum perforatum* (St John's wort) (Matos *et al.*, 2015).

Lignans have been found in many plant foods, commonly as glycosides. They are receiving growing attention as precursors of the enterolignans (enterolactone and enterodiol), microbial metabolites that exert potential biological effects (Aehle *et al.*, 2011; Raffaelli *et al.*, 2002). The richest dietary source of lignans, mainly of secoisolariciresinol diglucoside and matairesinol, in lesser amount, is flaxseed (also called linseed). Relatively high amounts of other lignans, such as pinoresinol and lariciresinol, are found in sesame seeds (Milder *et al.*, 2005). Other minor sources include several cereals (triticale and wheat), legumes (soybeans and lentils), vegetables (garlic, asparagus, broccoli, carrots), and fruits (pears, prunes, strawberries, lingonberries, blackcurrants) (Aehle *et al.*, 2011; Gerstenmeyer *et al.*, 2013; Mazur *et al.*, 2000; Raffaelli *et al.*, 2002; Smeds *et al.*, 2007).

1.4 Databases Used to Assess Dietary Exposure to Polyphenols

One of the main tasks in nutritional studies with plant food products is determination of the dietary intake of specific food phytochemicals, and particularly polyphenols, due to the large structural diversity and variability among food products, and the changes occurring with processing, storage, and culinary practices. To help with this task, there are several free access internet-based databases, two of which are described here.

- *Phenol Explorer* (http://phenol-explorer.eu/). This database is freely accessible and describes the polyphenol content of food, with more than 500 polyphenols and their occurrence in 400 foods. The database also contains a comprehensive description of polyphenol metabolism, including pharmacokinetics data, as well as a description of the effect of food processing and cooking on these metabolites.
- *Phytohub* (http://phytohub.eu/). Phytohub is a comprehensive database of food phytochemicals. It is linked with FooDB

(http://foodb.ca/) and ITIS (Integrated Taxonomic Information System) (www.itis.gov//) which provides interesting and relevant information regarding the origin of plant foods and food sources. It is the first inventory of all phytochemicals present in foods (>350) commonly ingested in human diets and also includes >560 human and animal metabolites. Phytohub includes a connection with a mass search and structure application, as well as a food search link, which is very useful for metabolomic studies, and identification of discriminant metabolites in nutri-metabolomic studies.

1.5 Bioavailability, Metabolism, and Bioactivity of Dietary Polyphenols

Over recent decades, spectacular advances have been made through in our understanding of the possible role of dietary polyphenols in preventive nutrition. In fact, for decades, epidemiological and observational studies have been pointing out that dietary polyphenols present in a fruit- and vegetable-rich diet exert protective effects against several chronic degenerative diseases including cardiovascular diseases, diabetes, cancer, and neurodegenerative conditions such as Alzheimer's and Parkinson's diseases, mainly in the context of regular or long-term intake (Bravo, 1998; Cory *et al.*, 2018; Espín and Tomás-Barberán, 2005; Fraga *et al.*, 2019; Tresserra-Rimbau *et al.*, 2014). Moreover, in parallel, extensive preclinical research in animal and cell models has described a wide spectrum of biological activities for many dietary polyphenols beyond the antioxidant properties classically attributed to plants, including antimicrobial, antiinflammatory, and anticarcinogenic properties, displayed in both the digestive tract and systemic tissues (Bravo, 1998; Espín and Tomás-Barberán, 2005; Del Rio *et al.*, 2013; Fraga *et al.*, 2019; Quideau *et al.*, 2011; Vauzour *et al.*, 2010).

According to these findings and with the aim of responding to consumer demands, many dietary polyphenols are being used in so-called functional foods as well as nutraceuticals according to their presumably favorable effect on health and to contribute to the prevention of certain chronic diseases (González-Sarrías *et al.*, 2013; Liu, 2003). However, it is important to note that there is still not enough convincing evidence from human studies to fully support the link between dietary polyphenols intake and health benefits, mainly due to inaccurate *in vitro–in vivo* extrapolation (Ávila-Gálvez *et al.*, 2018a; Espín *et al.*, 2017; Hollman, 2014). Thus, to date, unlike vitamins and minerals, only cocoa flavanols and olive oil hydroxytyrosol have gained approval from the European Food Safety Authority (EFSA) for their cardioprotective action (EFSA, 2011, 2014).

Moreover, despite outstanding progress in our understanding of the association between polyphenol-rich food intake and the potential effects on chronic diseases and general health, it is important to note that the beneficial effects depend on bioavailability and metabolism. Thus, the identification and measurement of physiologically relevant polyphenol metabolites, that might exert higher or lower beneficial biological activities than their precursors, represent a crucial requisite for the understanding of the role of dietary polyphenols in human health (D'Archivio *et al.*, 2007; Espín *et al.*, 2017; González-Sarrías *et al.*, 2017; Manach *et al.*, 2004).

In this regard, most dietary polyphenols occur as glycosides and as complex oligomeric structures in plant foods, that are poorly bioavailable and cannot reach systemic tissues in their native form. Once they reach the distal gastrointestinal tract, they are further hydrolyzed and metabolized by either the intestinal enzymes or the gut microbiota. The resulting polyphenol metabolites are then absorbed, and rapidly undergo extensive phase II metabolism by glucuronyl transferases, sulfate transferases, and catechol methyl transferases, yielding sulfate, glucuronide, and methyl conjugates that appear in the circulatory system and can be detected in urine up to 3–4 days after intake. Significant concentrations of the metabolites, in the range from nM to low μM, are present in plasma and these metabolites can target systemic tissues, and therefore, they may trigger beneficial effects attributed to dietary polyphenols (Del Rio *et al.*, 2013; Espín *et al.*, 2017; González-Sarrías *et al.*, 2017; Selma *et al.*, 2009). In recent years, it has been reported that once the polyphenol-derived metabolites are absorbed, they are conjugated by phase II metabolism enzymes to favor excretion. These metabolites do not always demonstrate relevant health effects as the conjugates often show lower bioactivity in different models than that of their deconjugated counterparts (Aires *et al.*, 2013; Ávila-Gálvez *et al.*, 2018b; Giménez-Bastida *et al.*, 2016; González-Sarrías *et al.*, 2014).

In relation to this, it is important to consider several factors that may directly or indirectly affect the bioavailability of dietary polyphenols. These include the food matrix, that involves bonds with food constituents as proteins or lipids, as well as food processing-related factors such as storage and cooking that can significantly limit the release needed to be potentially bioavailable and bioactive (D'Archivio *et al.*, 2007). Growing evidence also suggests that interindividual variability in the absorption, distribution, metabolism, and excretion (ADME) of bioactive compounds may be one of the main reasons behind the heterogeneity found in individual biological responsiveness after intake. This variability is determined by genetic and nongenetic factors, such as age, gender, (epi)genotype, and gut microbiota composition (Manach *et al.*, 2017; Milenkovic *et al.*, 2017).

Regarding microbiota, it is important to point out the two-way interaction existing with dietary polyphenols. Thus, dietary polyphenols not only undergo transformations by specific bacteria, but they can also modulate the gut microbiota composition and even their activity, acting as a prebiotic conferring a potential beneficial effect (Espín *et al.*, 2017; Selma *et al.*, 2009). In this regard, recently, Tomás-Barberán *et al.* (2018) critically suggested that polyphenol gut microbiota metabolites could be considered either as well-known bioactives or as biomarkers regarding gut microbiota composition and functionality, which might be related to the potential response to dietary interventions.

In conclusion, it is becoming more and more important to understand the bioavailability, metabolism, and tissue distribution of dietary polyphenols to establish an association of their metabolites with their biological effects, rather than with the original dietary polyphenol counterparts.

Acknowledgments

This research was supported by the Projects Fundación Séneca de la Región de Murcia, Ayudas a Grupos de Excelencia 19900/GERM/15, AGL-2015-73106-EXP (MINEICO, Spain), AGL2015-73744-JIN (MINECO, Spain), 201870E014 and 201870I028 (CSIC, Spain).

References

Aehle, E., Müller, U., Eklund, P.C., Willför, S.M., Sippl, W., and Dräger, B. (2011) Lignans as food constituents with estrogen and antiestrogen activity. *Phytochemistry*, **72**, 2396–405.

Aires, V., Limagne, E., Cotte, A.K., Latruffe, N., Ghiringhelli, F., and Delmas, D. (2013) Resveratrol metabolites inhibit human metastatic colon cancer cells progression and synergize with chemotherapeutic drugs to induce cell death. *Molecular Nutrition and Food Research*, **57**, 1170–81.

Arts, I.C., van de Putte, B., and Hollman, P.C. (2000a) Catechin contents of foods commonly consumed in The Netherlands. 1. Fruits, vegetables, staple foods, and processed foods. *Journal of Agricultural and Food Chemistry*, **48**, 1746–51.

Arts, I.C., van de Putte, B., and Hollman, P.C. (2000b) Catechin contents of foods commonly consumed in The Netherlands. 2. Tea, wine, fruit juices, and chocolate milk. *Journal of Agricultural and Food Chemistry*, **48**, 1752–7.

Ávila-Gálvez, M.Á., González-Sarrías, A., and Espín, J.C. (2018a) In vitro research on dietary polyphenols and health: A call of caution and a guide on how to proceed. *Journal of Agricultural and Food Chemistry*, **66**, 7857–8.

Ávila-Gálvez, M.Á., Espín, J.C., and González-Sarrías, A. (2018b) Physiological relevance of the antiproliferative and estrogenic effects of dietary polyphenol aglycones versus their phase-II metabolites on breast cancer cells: a call of caution. *Journal of Agricultural and Food Chemistry*, **66**, 8547–55.

Bąkowska-Barczak, A. (2005) Acylated anthocyanins as stable, natural food colorants – a review. *Polish Journal of Food and Nutrition Sciences*, **14**, 107–16.

Barreca, D., Gattuso, G., Bellocco, E., *et al.* (2017) Flavanones: citrus phytochemical with health-promoting properties. *Biofactors*, **43**, 495–506.

Beecher, G.R. (2003) Overview of dietary flavonoids: nomenclature, occurrence and intake. *Journal of Nutrition*, **133**, 3248S–54S.

Bhagwat, S., Haytowits, D.B., and Holden, J.M. (2011) USDA database for the flavonoid content of selected foods. Nutrient Data Laboratory, US Departament of Agriculture.

Bravo, L. (1998) Polyphenols: chemistry, dietary sources, metabolism, and nutritional significance. *Nutrition Reviews*, **56**, 317–33.

Burns, J., Yokota, T., Ashihara, H., Lean, M.E., and Crozier, A. (2002) Plant foods and herbal sources of resveratrol. *Journal of Agricultural and Food Chemistry*, **50**, 3337–40.

Cerdá, B., Espín, J.C., Parra, S., Martínez, P., and Tomás-Barberán, F.A. (2004) The potent in vitro antioxidant ellagitannins from pomegranate juice are metabolized into bioavailable but poor antioxidant hydroxy-6H-dibenzopyran-6-one derivatives by the colonic microflora in healthy humans. *European Journal of Nutrition*, **43**, 205–20.

Cory, H., Passarelli, S., Szeto, J., Tamez, M., and Mattei, J. (2018) The role of polyphenols in human health and food systems: a mini-review. *Frontiers in Nutrition*, **5**, 87.

Crozier, A., Jaganath, I.B., and Clifford, M.N. (2009) Dietary phenolics: chemistry, bioavailability and effects on health. *Natural Product Reports*, **26**, 1001–43.

Danciu, C., Avram, S., Pavel, I.Z., *et al.* (2018) Main isoflavones found in dietary sources as natural anti-inflammatory agents. *Current Drug Targets*, **19**, 841–53.

D'Archivio, M., Filesi, C., di Benedetto, R., Gargiulo, R., Giovannini, C., and Masella, R. (2007) Polyphenols, dietary sources and bioavailability. *Annali dell'Istituto Superiore di Sanità*, **43**, 348–61.

Del Rio, D., Rodriguez-Mateos, A., Spencer, J.P., Tognolini, M., Borges, G., and Crozier, A. (2013) Dietary (poly)phenolics in human health: structures, bioavailability, and evidence of protective effects against chronic diseases. *Antioxidants & Redox Signaling*, **18**, 1818–92.

El Gharras, H. (2009) Polyphenols: food sources, properties and applications – a review. *International Journal of Food Science and Technology*, **44**, 2512–18.

Espín, J.C. and Tomás-Barberán, F.A. (2005) *Constituyentes Bioactivos No-Nutricionales de Alimentos Vegetales y Su Aplicación en Alimentos Funcionales* (pp. 101–65). Madrid: Fundación Española para la Ciencia y la Tecnología.

Espín, J.C., González-Sarrías, A., and Tomás-Barberán, F.A. (2017) The gut microbiota: a key factor in the therapeutic effects of (poly)phenols. *Biochemical Pharmacology*, **139**, 82–93.

European Food Safety Authority (EFSA). (2011) Scientific Opinion on the substantiation of health claims related to polyphenols in olive and protection of LDL particles from oxidative damage (ID 1333, 1638, 1639, 1696, 2865), maintenance of normal blood HDL-cholesterol concentrations (ID 1639), maintenance of normal blood pressure (ID 3781), "anti-inflammatory properties" (ID 1882), "contributes to the upper respiratory tract health" (ID 3468), "can help to maintain a normal function of gastrointestinal tract" (3779), and "contributes to body defences against external agents" (ID 3467) pursuant to Article 13(1) of Regulation (EC) No 1924/2006. *EFSA Journal*, **9**, 2033.

European Food Safety Authority (EFSA). (2014) Scientific Opinion on the modification of the authorisation of a health claim related to cocoa flavanols and maintenance of normal endothelium-dependent vasodilation pursuant to Article 13(5) of Regulation (EC) No 1924/20061 following a request in accordance with Article 19 of Regulation (EC) No 1924/2006. *EFSA Journal*, **12**, 3654.

Fraga, C.G., Croft, K.D., Kennedy, D.O., and Tomás-Barberán, F.A. (2019) The effects of polyphenols and other bioactives on human health. *Food & Function*, **10**, 514–28.

Frankenfeld, C.L., Atkinson, C., Wähälä, K., and Lampe, J. W. (2014) Obesity prevalence in relation to gut microbial environments capable of producing equol or O-desmethylangolensin from the isoflavone daidzein. *European Journal of Clinical Nutrition*, **68**, 526–30.

Gan, R.Y., Kong, K.W., Li, H.B., *et al.* (2018) Separation, identification, and bioactivities of the main gallotannins of red sword bean (*Canavalia gladiata*) coats. *Frontiers in Chemistry*, **6**, 39.

Gerstenmeyer, E., Reimer, S., Berghofer, E., Schwartz, H., and Sontag, G. (2013) Effect of thermal heating on some lignans in flax seeds, sesame seeds and rye. *Food Chemistry*, **138**, 1847–55.

Giménez-Bastida, J.A., González-Sarrías, A., Vallejo, F., Espín, J.C., and Tomás-Barberán, F.A. (2016) Hesperetin and its sulfate and glucuronide metabolites inhibit TNF-α induced human aortic endothelial cell migration and decrease plasminogen activator inhibitor-1 (PAI-1) levels. *Food & Function*, **7**, 118–26.

González-Sarrías, A., Larrosa, M., García-Conesa, M.T., Tomás-Barberán, F.A., and Espín, J.C. (2013) Nutraceuticals for older people: facts, fictions and gaps in knowledge. *Maturitas*, **75**, 313–34.

González-Sarrías, A., Giménez-Bastida, J.A., Núñez-Sánchez, M.A., *et al.* (2014) Phase-II metabolism limits the antiproliferative activity of urolithins in human colon cancer cells. *European Journal of Nutrition*, **53**, 853–64.

González-Sarrías, A., Espín, J.C., and Tomás-Barberán, F.A. (2017) Non-extractable polyphenols produce gut microbiota metabolites that persist in circulation and show anti-inflammatory and free radical-scavenging effects. *Trends in Food Science & Technology*, **69**, 281–8.

Gu, L., Kelm, M.A., Hammerstone, J.F., *et al.* (2004) Concentrations of proanthocyanidins in common foods and estimations of normal consumption. *Journal of Nutrition*, **134**, 613–17.

Gupta, A., Kagliwal, L.D., and Singhal, R.S. (2013) Biotransformation of polyphenols for improved bioavailability and processing stability. *Advances in Food and Nutrition Research*, **69**, 183–217.

Harbone, J.B. and Williams, C.A. (2000) Advances in flavonoid research since 1992. *Phytochemistry*, **55**, 481–504.

Heinonen, S.M., Wähälä, K., and Adlercreutz. H. (2002) Metabolism of isoflavones in human subjects. *Phytochemistry Reviews*, **1**, 175–82.

Hollman, P.C.H. (2014) Unravelling of the health effects of polyphenols is a complex puzzle complicated by metabolism. *Archives of Biochemistry and Biophysics*, **559**, 100–5.

Hostetler, G.L., Ralston, R.A., and Schwartz, S.J. (2017) Flavones: food sources, bioavailability, metabolism, and bioactivity. *Advances in Nutrition*, **8**, 423–35.

Ignat, I., Volf, I., and Popa, V. (2011) A critical review of methods for characterisation of polyphenolic compounds in fruits and vegetables. *Food Chemistry*, **126**, 1821–35.

Katsumoto, Y., Fukuchi-Mizutani, M., Fukui, Y., *et al.* (2007) Engineering of the rose flavonoid biosynthetic pathway successfully generated blue-hued flowers accumulating delphinidin. *Plant and Cell Physiology*, **48**, 1589–600.

Khoo, H.E., Azlan, A., Tang, S.T., and Lim, S.M. (2017) Anthocyanidins and anthocyanins: colored pigments as food, pharmaceutical ingredients, and the potential health benefits. *Food & Nutrition Research*, **61**, 1361779.

Ko, K.P. (2014) Isoflavones: chemistry, analysis, functions and effects on health and cancer. *Asian Pacific Journal of Cancer Prevention*, **15**, 7001–10.

Krga, I. and Milenkovic, D. (2019) Anthocyanins: from sources and bioavailability to cardiovascular-health benefits and molecular mechanisms of action. *Journal of Agricultural and Food Chemistry*, **67**, 1771–83.

Laleh, G.H., Frydoonfar, H., Heidary, R., Jameei, R., and Zare, S. (2006) The effect of light, temperature, pH and species on stability of anthocyanin pigments in four *Berberis* species. *Pakistan Journal of Nutrition*, **5**, 90–2.

Lattanzio, V., Kroon, P., Quideau, S., and Treutter, D. (2008) *Plant Phenolics – Secondary Metabolites with Diverse Functions.* Oxford: Wiley-Blackwell, Oxford.

Leo, C.H. and Woodman, O.L. (2015) Flavonols in the prevention of diabetes-induced vascular dysfunction. *Journal of Cardiovascular Pharmacology*, **65**, 532–44.

Liu, R.H. (2003) Health benefits of fruit and vegetables are from additive and synergistic combinations of phytochemicals. *American Journal of Clinical Nutrition*, **78**, 517S–20S.

Luo, F., Fu, Y., Xiang, Y., *et al.* (2014) Identification and quantification of gallotannins in mango (*Mangifera indica* L.) kernel and peel and their antiproliferative activities. *Journal of Functional Foods*, **8**, 282–91.

Manach, C., Scalbert, A., Morand, C., Rémésy, C., and Jiménez, L. (2004) Polyphenols: food sources and bioavailability. *American Journal of Clinical Nutrition*, **79**, 727–47.

Manach, C., Milenkovic, D., van de Wiele, T., *et al.* (2017) Addressing the inter-individual variation in response to

consumption of plant food bioactives: towards a better understanding of their role in healthy aging and cardiometabolic risk reduction. *Molecular Nutrition and Food Research*, **61**(6), doi: 10.1002/mnfr.201600557.
Marzocchella, L., Fantini, M., Benvenuto, M., *et al.* (2011) Dietary flavonoids: molecular mechanisms of action as anti-inflammatory agents. *Recent Patents on Inflammation & Allergy Drug Discovery*, **5**, 200–20.
Matos, M.J., Santana, L., Uriarte, E., Abreu, O.A., Molina, E., and Yordi, E.G. (2015) *Coumarins – An Important Class of Phytochemicals* (pp. 113–40). London: InTech.
Mazur, W.M., Duke, J.A., Wahala, K., Rasku, S., and Adlercreutz, H. (1998) Isoflavones and lignans in legumes: nutritional and health aspects in human. *Journal of Nutritional Biochemistry*, **9**, 193–200.
Mazur, W.M., Uehara, M., Wähälä, K., and Adlercreutz, H. (2000) Phyto-oestrogen content of berries, and plasma concentrations and urinary excretion of enterolactone after a single strawberry-meal in human subjects. *British Journal of Nutrition*, **83**, 381–7.
Miean, K.H. and Mohamed, S. (2001) Flavonoid (myricetin, quercetin, kaempferol, luteolin, and apigenin) content of edible tropical plants. *Journal of Agricultural and Food Chemistry*, **49**, 3106–12.
Milder, I.E., Arts, I.C., van de Putte, B., Venema, D.P., and Hollman, P.C. (2005) Lignan contents of Dutch plant foods: a database including lariciresinol, pinoresinol, secoisolariciresinol and matairesinol. *British Journal of Nutrition*, **93**, 393–402.
Milenkovic, D., Morand, C., Cassidy, A., *et al.* (2017) Interindividual variability in biomarkers of cardiometabolic health after consumption of major plant-food bioactive compounds and the determinants involved. *Advances in Nutrition*, **8**, 558–70.
Miranda, A.M., Steluti, J., Fisberg, R.M., and Marchioni, D.M. (2016) Dietary intake and food contributors of polyphenols in adults and elderly adults of Sao Paulo: a population-based study. *British Journal of Nutrition*, **115**, 1061–70.
Okuda, T., Yoshida, T., Hatano, T. (1995) Hydrolyzable tannins and related polyphenols. *Fortschritte der Chemie organischer Naturstoffe*, **66**, 1–117.

Ovaskainen, M.L., Törrönen, R., Koponen, J.M., *et al.* (2008) Dietary intake and major food sources of polyphenols in Finnish adults. *Journal of Nutrition*, **138**, 562–6.

Pandey, K.B. and Rizvi, S.I. (2009) Plant polyphenols as dietary antioxidants in human health and disease. *Oxidative Medicine and Cellular Longevity*, **2**, 270–8.

Pereira, D.M., Valentão, P., Pereira, J.A., and Andrade P.B. (2010) Phenolics: from chemistry to biology. *Molecules*, **14**, 2202–11.

Pérez-Jiménez, J., Neveu, V., Vos, F., and Scalbert, A. (2010a) Identification of the 100 richest dietary sources of polyphenols: an application of the Phenol-Explorer database. *European Journal of Clinical Nutrition*, **64**, S112–20.

Pérez-Jiménez, J., Neveu, V., Vos, F., and Scalbert, A. (2010b) Systematic analysis of the content of 502 polyphenols in 452 foods and beverages: an application of the Phenol-Explorer database. *Journal of Agricultural and Food Chemistry*, **58**, 4959–69.

Pérez-Jiménez, J., Fezeu, L., Touvier, M., *et al.* (2011) Dietary intake of 337 polyphenols in French adults. *American Journal of Clinical Nutrition*, **93**, 1220–8.

Peterson, J.J., Dwyer, J.T., Beecher, G.R., *et al.* (2006) Flavanones in oranges, tangerines (mandarins), tangors and tangelos: a compilation and review of the data from the analytical literature. *Journal of Food Composition and Analysis*, **19**, S66–S73.

Pinto, P. and Santos, C.N. (2017) Worldwide (poly)phenol intake: assessment methods and identified gaps. *European Journal of Clinical Nutrition*, **56**, 1393–408.

Quideau, S., Deffieux, D., Douat-Casassus, C., and Pouységu, L. (2011) Plant polyphenols: chemical properties, biological activities, and synthesis. *Angewandte Chemie (International Edition in English)*, **50**, 586–621.

Raffaelli, B., Hoikkala, A., Leppälä, E., and Wähälä, K. (2002) Enterolignans. *Journal of Chromatography B: Analytical Technologies in the Biomedical and Life Sciences*, **777**, 29–43.

Rasmussen, S.E., Frederiksen, H., Struntze Krogholm, K., and Poulsen, L. (2005) Dietary proanthocyanidins: occurrence, dietary intake, bioavailability, and protection against cardiovascular disease. *Molecular Nutrition and Food Research*, **49**, 159–74.

Razzaghi-Asl, N., Garrido, J., Khazraei, H., Borges, F., and Firuzi, O. (2013) Antioxidant properties of hydroxycinnamic acids: a review of structure–activity relationships. *Current Medicinal Chemistry*, **20**, 4436–50.

Rivière, C., Pawlus, A.D., and Mérillon, J.M. (2012) Natural stilbenoids: distribution in the plant kingdom and chemotaxonomic interest in Vitaceae. *Natural Product Reports*, **29**, 1317–33.

Robbins, R.J. (2003) Phenolic acids in foods: an overview of analytical methodology. *Journal of Agricultural and Food Chemistry*, **51**, 2866–87.

Rothwell, J.A., Knaze, V., and Zamora-Ros, R. (2017) Polyphenols: dietary assessment and role in the prevention of cancers. *Current Opinion in Clinical Nutrition & Metabolic Care*, **20**, 512–21.

Saleem, M., Kim, H.J., Ali, M.S., and Lee, Y.S. (2005) An update on bioactive plant lignans. *Natural Product Reports*, **22**, 696–716.

Scalbert, A. and Williamson, G. (2000) Dietary intake and bioavailability of polyphenols. *Journal of Nutrition*, **130**, 2073S–85S.

Schreiner, M. (2005) Vegetable crop management strategies to increase the quantity of phytochemicals. *European Journal of Nutrition*, **44**, 85–94.

Selma, M.V., Espín, J.C., and Tomás-Barberán, F.A. (2009) Interaction between phenolics and gut microbiota: role in human health. *Journal of Agricultural and Food Chemistry*, **57**, 6485–501.

Shen, T., Wang, X.N., and Lou, H.X. (2009) Natural stilbenes: an overview. *Natural Product Reports*, **26**, 916–35.

Smeds, A.I., Willför, S.M., Pietarinen, S.P., Peltonen-Sainio, P., and Reunanen, M.H. (2007) Occurrence of "mammalian" lignans in plant and water sources. *Planta*, **226**, 639–46.

Smeriglio, A., Barreca, D., Bellocco, E., and Trombetta, D. (2017) Proanthocyanidins and hydrolysable tannins: occurrence, dietary intake and pharmacological effects. *British Journal of Pharmacology*, **174**, 1244–62.

Spranger, I., Sun, B., Mateus, A.M., de Freitas, V., and Ricardo-da-Silva, J.M. (2008) Chemical characterization and antioxidant activities of oligomeric and polymeric procyanidin fractions from grape seeds. *Food Chemistry*, **108**, 519–32.

Tomás-Barberán, F.A. and Espín, J.C. (2001) Phenolic compounds and related enzymes as determinants of quality in fruits and vegetables. *Journal of the Science of Food and Agriculture*, **81**, 853–76.

Tomás-Barberán, F.A., García-Conesa, M.T., Larrosa, M., *et al.* (2008) *Bioavailability, Metabolism, and Bioactivity of Food Ellagic Acid and Related Polyphenols* (pp. 264–77). Oxford: Blackwell Publishing.

Tomás-Barberán, F.A., García-Villalba, R., González-Sarrías, A., Selma, M.V., and Espín, J.C. (2014) Ellagic acid metabolism by human gut microbiota: consistent observation of three urolithin phenotypes in intervention trials, independent of food source, age, and health status. *Journal of Agricultural and Food Chemistry*, **62**, 6535–8.

Tomás-Barberán, F.A., García-Villalba, R., González-Sarrías, A., *et al.* (2016) *Gut Microbiota Metabolism, Bioavailability and Health Benefits of Pomegranate Ellagitannin-Derived Urolithins* (pp. 129–48). New York: Nova Publishers, New York.

Tomás-Barberán, F.A., González-Sarrías, A., García-Villalba, R., *et al.* (2017) Urolithins, the rescue of "old" metabolites to understand a "new" concept: metabotypes as a nexus among phenolic metabolism, microbiota dysbiosis, and host health status. *Molecular Nutrition and Food Research*, **61**(1) doi: 10.1002/mnfr.201500901.

Tomás-Barberán, F.A., Selma, M.V., and Espín, J.C. (2018) Polyphenols' gut microbiota metabolites: bioactives or biomarkers? *Journal of Agricultural and Food Chemistry*, **66**, 3593–4.

Tresserra-Rimbau, A., Medina-Remón, A., Pérez-Jiménez, J., *et al.* (2013) Dietary intake and major food sources of polyphenols in a Spanish population at high cardiovascular risk: the PREDIMED study. *Nutrition, Metabolism & Cardiovascular Diseases*, **23**, 53–9.

Tresserra-Rimbau, A., Rimm, E.B., Medina-Remón, A., *et al.* (2014) Polyphenol intake and mortality risk: a re-analysis of the PREDIMED trial. *BMC Medicine*, **12**, 77.

Tsao, R. (2010) Chemistry and biochemistry of dietary polyphenols. *Nutrients*, **2**, 1231–46.

Vastano, B.C., Chen, Y., Zhu, N., Ho, C.T., Zhou, Z., and Rosen, R.T. (2000) Isolation and identification of stilbenes in two varieties of *Polygonum cuspidatum*. *Journal of Agricultural and Food Chemistry*, **48**, 253–6.

Vauzour, D., Rodriguez-Mateo, A., Corona, G., Oruna-Concha, M.J., and Spencer, J.P.E. (2010) Polyphenols and human health: prevention of disease and mechanisms of action. *Nutrients*, **2**, 1106–31.

Vitrac, X., Bornet, A., Vanderlinde, R., Valls, J., *et al*. (2005) Determination of stilbenes (delta-viniferin, trans-astringin, trans-piceid, cis- and trans-resveratrol, epsilon-viniferin) in Brazilian wines. *Journal of Agricultural and Food Chemistry*, **53**, 5664–9.

Wallace, T.C. and Giusti, M.M. (2015) Anthocyanins. *Advances in Nutrition*, **6**, 620–2.

Waterhouse, A.L. (2002) Wine polyphenolics. *Annals of the New York Academy of Sciences*, **957**, 21–36.

Zaheer, K. and Humayoun Akhtar, M. (2017) An updated review of dietary isoflavones: nutrition, processing, bioavailability and impacts on human health. *Critical Reviews in Food Science and Nutrition*, **57**, 1280–93.

Zamora-Ros, R., Knaze, V., Rothwell, J.A., *et al*. (2016) Dietary polyphenol intake in Europe: the European Prospective Investigation into Cancer and Nutrition (EPIC) study. *European Journal of Nutrition*, **55**, 1359–75.

2

Nonextractable Polyphenols: A Relevant Group with Health Effects

Yuridia Martínez-Meza[1], Rosalía Reynoso-Camacho[1], and Jara Pérez-Jiménez[2]*

[1] Research and Graduate Studies in Food Science, Facultad de Química, Universidad Autónoma de Querétaro, Querétaro, México
[2] Department of Metabolism and Nutrition, Institute of Food Science, Technology and Nutrition (ICTAN-CSIC), Madrid, Spain

2.1 Introduction: The Concept of Nonextractable Polyphenols (NEPP)

Research on food polyphenols has experienced a huge development during the last century. Thus, from initial studies more focused on phytochemical aspects and consideration of polyphenols as antinutrients, etc., research evolved towards describing the relevance of polyphenols to improve several health markers by a combination of mechanisms. These advances used many kinds of studies (*in vitro*, preclinical, clinical or epidemiological) and several approaches (mechanistic, bioavailability, health effects). Nevertheless, in most of these studies, an important aspect of polyphenols was, and still is, neglected: the nonextractable polyphenols (NEPP). This partial approach to these dietary bioactive constituents hampers our understanding of the whole relevance of dietary polyphenols. For instance, if they are not systematically considered when evaluating food polyphenol content when a clinical trial is designed based on a certain polyphenol intake, only a fraction (extractable polyphenols, EPP) is actually considered. Thus,

*Corresponding author.

Dietary Polyphenols: Metabolism and Health Effects, First Edition.
Edited by Francisco A. Tomás-Barberán, Antonio González-Sarrías, and Rocío García-Villalba.

although information on the outcomes of the intervention may be achieved, that on the input will remain incomplete.

But what does the term NEPP mean and why they are not considered in most polyphenol studies? When food polyphenols started to be studied, different strategies for their release from food matrix were explored, until the best combinations of solvents were found for obtaining an overall polyphenol profile or for extracting specific polyphenol classes. For instance, it is known that acidified acetone:water (70:30) is the most efficient combination for extracting proanthocyanidins (Hümmer and Schreier, 2008). This approach, common for studying food constituents, was based on the consideration that, although these food extractions provided a residue, the polyphenol content in it could be considered negligible. However, further research showed that a fraction of food polyphenols remained in those residues, and they could even constitute a major fraction of total polyphenols in certain food items (Saura-Calixto, 2018). It should be highlighted that this fraction remains, whatever solvents are used or even applying new extraction techniques, such as ultrasound, subcritical water extraction, etc. (Sanz-Pintos *et al.*, 2017; Wu *et al.*, 2018). The reason for this is how NEPP are present in foods.

Briefly, NEPP may be divided into two categories: hydrolyzable polyphenols (HPP) and nonextractable proanthocyanidins (NEPA). HPP are small phenolic compounds which are associated with food macromolecules – proteins but especially dietary fiber – by a combination of weak and strong bonds (Pérez-Jiménez and Torres, 2011). They have been mostly studied in cereals (where they are also known as insoluble or bound phenolics), but they have been reported in other food groups (Arranz *et al.*, 2010). NEPA are high molecular weight proanthocyanidins, i.e., a polymeric chain of flavanols which may reach several dozens of units which also limits their release from food matrix (Saura-Calixto, 1988). Both HPP and NEPA have a common characteristic, in that they are present in foods forming macromolecules (NEPA are macromolecules *per se*, while HPP are associated with macromolecules). For this reason, in recent years, the term "macromolecular antioxidants" has been suggested to describe NEPP; besides, this term has higher application potential in a nutritional and medical context (Saura-Calixto, 2018).

Thus, from a chemical point of view, HPP and NEPA do not constitute specific chemical entities but they are present in the food matrix in a different way (either by molecular weight or by association with food matrix) from the EPP, i.e., low molecular weight phenolic compounds easily released from the food matrix. This, as will be further discussed, will give place to specific metabolic features associated with NEPP and not with EPP and, ultimately, it may have consequences for the health effects of NEPP. Therefore, the relevance of considering NEPP is not only a chemical or theoretical discussion but also has practical implications for the health effects associated with plant food consumption. Indeed, independently of the practical problems that still face researchers attempting a comprehensive characterization of polyphenol profile, the fact is that when consuming food, we are consuming both EPP and NEPP; this is something that should not be disregarded.

2.2 Contribution of NEPP to Total Polyphenol Content and Intake

A first step towards unraveling the relevance of NEPP for health and nutrition is to determine their content in common foods, as well as their quantitative contribution to total polyphenol content, i.e., their proportion compared to that of EPP. These data may be used for different purposes, such as identifying NEPP-rich foods, designing clinical trials focused on NEPP supplementation, or evaluating NEPP intake in different populations. Indeed, some preliminary studies have already made estimations of NEPP intake compared to EPP intake (Arranz *et al.*, 2010; Hervert-Hernández *et al.*, 2011; Saura-Calixto *et al.*, 2007). Nevertheless, in performing all these determinations, a critical step is to have a routine methodology for the analysis of NEPP in foods, which is still far from being achieved (Pérez-Jiménez and Torres, 2011).

2.2.1 Strategies for the Extraction and Analysis of NEPP

For the isolation and analysis of NEPP, several steps are needed. The first, common to the analysis of EPP, is generally a particle size reduction since this diminution allows an increase in the contact surface between matrix and solvent (Pérez-Jiménez *et al.*, 2008). Also, for fat-rich samples, such as nuts, a defatting step is recommended since the fat may interfere with further determinations (Arranz *et al.*, 2008).

The second step is the removal of EPP and other soluble substances like soluble sugars, vitamins, organic acids, and others. This step is usually performed by combining successive solid-liquid extractions, such as acid-methanol/water (50:50) for extraction of the most polar polyphenols, e.g., phenolic acids, followed by extraction with acetone/water (70:30) or acetone/water/acetic acid (70:29.5:0.5) to extract more apolar polyphenols such as proanthocyanidins. After centrifugation, the residue is treated for the release of NEPP, which include, as described above, HPP and NEPA.

For both HPP and NEPA release, the most common procedures include the use of acids at high temperatures, with different purposes. Thus, in the case of HPP, such treatments allow the release of polyphenols attached to the plant matrix, releasing small phenolic structures that may be further determined by spectrophotometry (Folin assay) or by HPLC or HPLC-MS procedures (Arranz *et al.*, 2009; Pérez-Jiménez *et al.*, 2008). Thus, a standard procedure for the release of HPP from the EPP residue is treatment with methanol/H_2SO_4 (90:10, v/v) at 85 °C for 20 hours with agitation (Hartzfeld *et al.*, 2002). The derived hydrolyzate, exhibiting strong acidity, must be partially neutralized before determining polyphenol content by some of the techniques mentioned above. Indeed, in the case of chromatography analysis, an additional solid phase extraction procedure is needed to remove the high concentration of salts released during the acid treatment (Pérez-Jiménez and Saura-Calixto, 2015).

It should be indicated that, for HPP analysis, alkaline treatments in the residues of extraction have been more commonly used than acid treatment, particularly in the case of cereals. However, although this may be more useful for the specific case of ferulic acid (a major constituent in cereal samples), for most phenolic acids attached to the plant cell wall, acid treatment is the most efficient way to release HPP (Arranz and Saura-Calixto, 2010).

With regard to NEPA release from the EPP residue, the aim of the butanol/HCl/$FeCl_3$ procedure (based on the Porter method; Porter, 1988) is the depolymerization of the long polymeric structures, releasing the corresponding colored anthocyanidins. Then, absorbance is measured at both 450 and 555 nm to calculate the concentration in relation to a known standard (Zurita *et al.*, 2012). An alternative procedure for release of NEPA is nucleophilic depolymerization, for which several reagents may be used. This procedure was applied for determining NEPA content in a wide number of samples (Hellström and Mattila, 2008), but it seems to be less efficient than the butanolysis procedure (Pérez-Jiménez and Torres, 2011).

One of the problems arising from the use of strong acid treatments on the EPP residues is that, in the case of HPP, either significant losses can occur due to degradation of some polyphenols (usually hydroxybenzoic and hydroxycinnamic acids) or artifacts may be generated (Pérez-Jiménez and Saura-Calixto, 2015). This means that the structures finally measured in HPP hydrolyzates may not be exactly those originally present in foods. In the same way, the depolymerization of NEPA creates structures different from those ingested, which complicates the application of *in vitro* procedures to determine their biological activities. Nevertheless, such strong conditions are currently the only valid ones for a comprehensive characterization of NEPP since other procedures, such as enzymatic treatments, are much less effective (Pérez-Jiménez *et al.*, 2009a).

Another important aspect for the characterization of NEPP, since a fraction of them is strongly associated with food matrix, is the understanding of the linkages involved in this association. For instance, it was recently reported that the association between NEPP and the cell wall increases during pear overripening (Brahem *et al.*, 2019).

Overall, the most important problem for the analysis of NEPP is that there is not yet an accepted standard methodology (Chen *et al.*, 2015; de Camargo *et al.*, 2016; de Mira *et al.*, 2009; Domínguez-Rodríguez *et al.*, 2017; García-Villalba *et al.*, 2015;

Table 2.1 Strategies for obtaining extractable and nonextractable polyphenols

Sample	Pretreatment of sample	Procedure for EPP removal	Procedure for NEPP release	Reference
Apples (Golden delicious), peach (Royal) and nectarine (Royal)	Freeze-dried and milled to a particle size of <0.5 mm of edible part from each fruit	Extraction of EPP and EPA with methanol/water (50:50, v/v; pH 2) acidified with 2 N HCl and acetone/water (70:30, v/v)	Release of HPP with methanol/H_2SO_4 90:10 (v/v), 85 °C for 20 h Release of NEPA with 10 mL of butanol/HCl (97.5:2.5 v/v) and 0.7 g of $FeCl_3$ at 100 °C for 1 h	Arranz *et al.*, 2009.
Apple (Golden delicious), banana (Cavendish), cooked pinto beans and red grape pomace (Cencibel)	Freeze-dried and milled to a particle size of <0.5 mm of edible part from each fruit	Extraction of EPP with methanol/water (50:50, v/v; pH 2) and acetone/water (70:30, v/v)	Release of NEPA with HCl/butanol (5:95, v/v) containing 0.7 g of $FeCl_3$/l at 100 °C for 1 h	Zurita *et al.*, 2012.
Mixtures of five types of plant foods: cereals, vegetables, legumes, nuts, and fruits	Freeze-dried and milled to a particle size of 0.5 mm	Extraction of EPP with methanol/water (50:50 v/v at pH 2) followed by acetone/water extraction (70:30, v/v).	Release of HPP with methanol/H_2SO_4 90:10 v/v, 85 °C during 20 h Release of NEPA with butanol/HCl 97.5:2.5 v/v, 100 °C during 1 h	Arranz *et al.*, 2010.

Pomegranate peel	Freeze-dried	Extraction of EPP with methanol/DMSO/H_2O (40:40:20, v/v/v) containing 0.1% HCl	Release of ellagitannins with 4 M HCl in water at 90 °C for 24 h followed by extraction with dimethyl sulfoxide/methanol (50:50, v/v)	García-Villalba *et al.*, 2015.
Cranberry beans (*Phaseolus vulgaris* L.)	Milled	Extraction of EPP with 70% methanol + 1% HCl (v/v) + sonication for 15 min, rotational mixing at room temperature for 2 h	Release of HPP with 2 N NaOH followed by sonication for 15 min, rotational mixing at room temperature for 16 h	Chen *et al.*, 2015.
Rice (*Oryza sativa* L.)	Milled	Extraction of EPP with 80% methanol and hexane	Release of HPP with 4 M NaOH acidified with 6 M HCl at room temperature for 4 h and release with ethyl acetate	de Mira *et al.*, 2009.
Winemaking by-products	Freeze-dried and defatted with hexane	Extraction of EPP with 70% (v/v) acetone at 30 °C for 20 min	Release of HPP with Viscozyme solution (2% in 0.1 M phosphate buffer, pH 4) at 37 °C for 12 h or pronase solution (1 mg/mL in 0.1 M phosphate buffer, pH 8) and stirred for 1 h	de Camargo *et al.*, 2016.

Wu *et al.*, 2018). As shown in Table 2.1, many differences may be found between the procedures used in the literature, first for EPP removal and then for NEPP release. Such standard methodology would help to increase the number of laboratories routinely measuring NEPP in their phenolic compound determinations. And this would increase the data available on polyphenol content in foods, needed, for instance, for proper estimations of NEPP intake, as discussed below.

2.2.2 NEPP Content in Common Foods

The available data on NEPP content in common foods, although scarcer than those for EPP, have shown the wide distribution of NEPP in different groups of plant foods (fruits, vegetables, legumes, cereals, and nuts). When mixtures of individual foods corresponding to the mean intake in Spain of each food group were prepared, and NEPP content was determined, it was found that the richest NEPP food group is fruits (880 mg/g dry weight) followed by legumes (568 mg/100 g), vegetables (326 mg/100 g), nuts (333 mg/100 g), and cereals (210 mg/100 g) (Arranz *et al.*, 2010).

Indeed, NEPP are not only found in most common foods but are present at similar or even higher concentration than EPP. Figure 2.1 shows the proportion of EPP, HPP, and NEPA for a selection of common fruits and vegetables. It can be seen

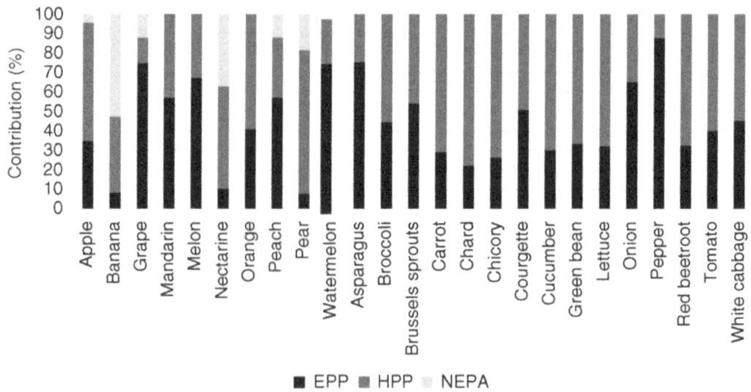

Figure 2.1 Contribution of nonextractable polyphenols to total polyphenol content in different common fruits and vegetables. EPP, extractable polyphenols; HPP, hydrolyzable polyphenols; NEPA, nonextractable proanthocyanidins.

that all samples contained NEPP as HPP and, in the case of fruits, some of them also contained NEPA (these were not detected in any of the samples of vegetables). And in 40% of the samples NEPP constituted at least half of total polyphenol content; indeed, in samples such as banana, nectarine or pear, this fraction was above 80% of total polyphenol content (Pérez-Jiménez and Saura-Calixto, 2015). Cereals are also a food group where a very high proportion of polyphenols are present as NEPP; for instance, in wheat bran, the content of HPP has been reported to be up to 10-fold that of EPP (Arranz and Saura-Calixto, 2010). Overall, the specific contribution of NEPP to total polyphenol content in different food items may be due to variations in polyphenol profile and subcellular location in different foodstuffs, as well as to the various food processes and experimental methodologies used (Pérez-Jiménez and Torres, 2011).

NEPP have also been reported in other less common food items, such as the tropical fruits acerola, açai or cashew apple (Maria do Socorro *et al.*, 2010, 2011). And they were found to represent about 40% of total polyphenol content in edible seaweeds (Sanz-Pintos *et al.*, 2017). Interestingly, cocoa, which is widely studied for its high EPP content, also presents a relevant proportion of NEPP in the form of NEPA (Tabernero *et al.*, 2006). Still, more data on NEPP content in common foods are needed. For instance, although a detailed list was published for fruits and vegetables (Pérez-Jiménez and Saura-Calixto, 2015), similar data have not been provided for common legumes or nuts. And, in the case of cereals, since this analysis was commonly performed by using alkaline conditions which, as discussed above, may not be as efficient as acid conditions, more data on NEPP content in cereals and derived products by using this procedure should be produced. Also, it should be remembered that beverages also contain a small proportion of NEPP, being associated with dietary fiber (Díaz-Rubio and Saura-Calixto, 2011), which also has to be considered in the characterization of these food items.

Another important source of NEPP is food wastes. Many fruit peels have a significant content of these compounds, even higher than that of EPP, as is the case for banana, kiwi, melon, orange, pear and watermelon peels (Pérez-Jiménez and

Saura-Calixto, 2018). Recently, prickly pear peels of different tonalities (yellow-orange, red, and green) were reported to present a high content of NEPP, with gallic acid 3-O-gallate, cinnamic acid, hesperidin, and myricetin 3-O-rhamnoside being the main contributors (Amaya-Cruz *et al.*, 2019). Other by-products such as grape pomace from wine processing or the by-product of peach juice production also present a very high NEPP content (Pérez-Jiménez *et al.*, 2009a) and detailed information by HPLC-MS analysis has been provided for them (Pérez-Ramírez *et al.*, 2018; Rodríguez-Gonzáles *et al.*, 2018). These results show that such by-products should be further studied for their potential as new functional ingredients as well as for the economic and environmental advantage of integral use of vegetal materials.

2.2.3 Estimation of NEPP Intake in Different Populations

The existing data on NEPP content in foods have been used to estimate NEPP intake in some populations. The most detailed study was performed in Spain, based on the reported data from mean food consumption in this country (Table 2.2). It was found that the daily intake of total polyphenols from solid foods was 1207 mg/day/person where 949 mg was NEPP and 258 mg was EPP. Adding to this value the intake of phenolic compounds provided by consumption of beverages and oils in Spain (Saura-Calixto *et al.*, 2007), the total polyphenol intake increased to a value of about 1840 mg/day, of which 898 mg would be EPP and 949 mg NEPP. Although legumes and cereals were the food groups showing the highest NEPP content, the major contributors to NEPP intake were fruits (46%) and cereals (30%), followed by vegetables (13%) (Arranz *et al.*, 2010).

Another study evaluated the intake of NEPP from fruits and vegetables in four European countries (Spain, France, Germany, and the Netherlands). It was found that intake was higher in the Netherlands and Spain than in France and Germany (about 500 mg/day/person vs 400 mg/day/person, respectively). Among the countries with the highest consumption, fruits

Table 2.2 Contribution of EPP and NEPP to total polyphenol intake in the Spanish diet*

Food group		EPP		NEPP		Total
	Consumption[a)]	Content[b)]	Intake[c)]	Content[b)]	Intake[c)]	Intake[c)]
Fruits	192.8	58.4	110.4	168.8	439	549.5
Vegetables	230.1	32.1	73.6	53.4	122.9	196.5
Cereals	139.8	43.8	61.1	171.1	291.7	352.8
Nuts	4.3	125.6	6.8	314.4	28.8	35.3
Legumes	12.1	51.7	6.2	497.7	66.3	72.5
Total	579.1	311.6	258.1	1205.4	948.7	1206.6

EPP, extractable polyphenols; NEPP, nonextractable polyphenols; fw, fresh weight.
a) g fresh edible portion/day/person.
b) mg/100 g fw.
c) mg/day/person.
*Data from Arranz *et al.* (2010).

were the main contributors in Spain, while vegetables were the main contributors in the Netherlands (Pérez-Jiménez and Saura-Calixto, 2015). The intake of NEPP from fruit and vegetables was also estimated in a Mexican rural population; in this case, NEPP intake was about 640 mg/day/person, with NEPA being provided only by fruits and a similar contribution of fruit and vegetables for HPP intake. Interestingly, NEPP intake was three-fold that of EPP (Hervert-Hernández *et al.*, 2011).

Currently, no other study has estimated NEPP intake in other populations, so there is a need to update data. For this purpose, a specific database on NEPP content in foods similar to the existing ones for EPP content (Neveu *et al.*, 2010) would be very useful. Finally, data on polyphenol intake should be used in observational studies in order to evaluate potential association with health outcomes, as has been done for EPP in many nutritional cohorts (Karam *et al.*, 2018; Zamora-Ros *et al.*, 2016).

2.3 Metabolic Fate of NEPP: A Key Process for Their Health Effects

To exert their health effects, polyphenols must be available to a certain extent in the target tissue. Therefore, their biological properties, as with those of other dietary bioactive compounds, are closely linked to the different steps of their metabolic fate: release from the food matrix by the action of digestive enzymes (small intestine) or bacterial microbiota (large intestine), thus becoming bioaccessible, followed by partial absorption and further transformations. The metabolic transformations of EPP have been widely studied and their key aspects are already known. Only a fraction is absorbed in the small intestine, while many of them are absorbed in the colon, after microbiota transformation; the metabolites absorbed may be transported through the portal vein to the liver where other transformations take place, giving rise to phase II metabolites. After extensive circulation in the human body, the absorbed metabolites are excreted in urine, while nonabsorbed metabolites or unfermented substrates are excreted through the feces (Rodríguez-Mateos *et al.*, 2014).

NEPP transformations through the gastrointestinal tract must be studied to understand their potential biological effects. In particular, the following aspects should be considered: solubilization in gastrointestinal fluids, which may take place in both the small and large intestine; transformation by the colonic microbiota, since the colon is the key organ for their transformation; absorption of the resulting metabolites (Pérez-Jiménez *et al.*, 2013).

2.3.1 Current Evidence of the Metabolic Transformation of NEPP

As already explained for other aspects of NEPP research, studies on their metabolic fate have been scarce compared to those regarding EPP. Nevertheless, there is a certain level of evidence suggesting that NEPP are at least partially transformed and further absorbed after intake, which was recently reviewed (Ludwig *et al.*, 2018). Based on this evidence, it is possible to suggest a metabolic fate for NEPP, as depicted in Figure 2.2. After intake, there may be partial solubilization of NEPP either in the stomach or in the small intestine, which would affect a minor fraction of these compounds. This may take place by the action of intestinal esterases (Andreasen *et al.*, 2001a); this would mean that some of the HPP (small phenolic compounds linked to the cell wall by covalent linkages) were released from the food matrix. Other possible methods of NEPP solubilization throughout the digestive tube are still controversial, since *in vitro* and *in vivo* studies have provided contradictory results regarding the depolymerization of NEPA (Mateos-Martín *et al.*, 2012; Tsang *et al.*, 2005; Williamson and Clifford, 2017). Therefore, further research on the transformation of polymeric proanthocyanidins (either as extractable proanthocyanidins or as NEPA) to effectively answer this question remains to be performed. However, although solubilization of NEPP in the upper part of the digestive tract may be followed by absorption of these compounds, this will still be a minor fraction of the metabolic fate of NEPP.

These processes for the release of NEPP from food matrix correspond to the metabolic pathway 1 of NEPP transformation: release from food matrix without transformation of chemical structures. This pathway takes place in both the small and large intestine since esterases are also present there (Andreasen *et al.*, 2001b).

At the same time, another process takes place in the colon: the metabolic pathway 2 of NEPP transformation (see Figure 2.2), which corresponds to the transformation of NEPP structures by microbiota action. These transformations may affect both HPP and NEPA.

In the case of HPP, there is evidence on the colonic transformation of phenolic compounds – whether they were originally free in the food matrix or they were released from it (Rodríguez-Mateos *et al.*, 2014), although new metabolites are still being reported (Gómez-Juaristi *et al.*, 2018). Moreover, how the overall food matrix affects the colonic transformation of HPP still needs more research. For instance, a preclinical study with durum wheat showed that the particle size of the aleurone fraction tested greatly affected the transformation and final bioavailability of ferulic acid linked to the cell wall (Calani *et al.*, 2014).

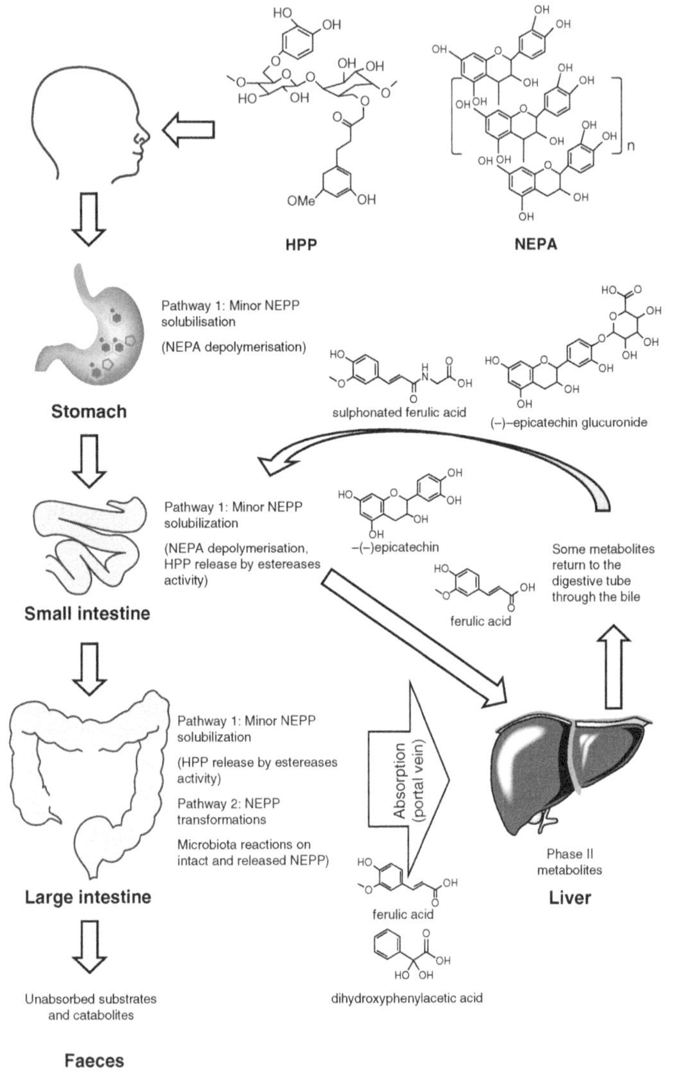

Figure 2.2 Metabolic fate of nonextractable polyphenols (NEPP). HPP, hydrolyzable polyphenols; NEPA, nonextractable proanthocyanidins; NEPP, nonextractable polyphenols.

With regard to the metabolic pathway 2, it must be considered that the constituents of NEPA are indeed the same as those of extractable proanthocyanidins, the only difference being in the degree of polymerization. Therefore, those compounds described as the most common microbial metabolites of extractable proanthocyanidins, i.e., (epi)catechin and several phenolic acids, would be expected to be also formed from NEPA (Monagas *et al.*, 2010). This was first confirmed in an *in vitro* study where the colonic fermentation of a grape product and a concentrate of proanthocyanidins from *Ceratonia siliqua* L. was performed (Saura-Calixto *et al.*, 2010). It was found that the major metabolites were hydroxyphenylacetic acid and hydroxyphenylvaleric acid typically formed during colonic transformation of flavanols. The colonic transformation of NEPA was later confirmed in an *in vivo* study, where rats were fed with a diet supplemented with a grape by-product devoid of extractable proanthocyanidins (Mateos-Martín *et al.*, 2012). It was found that NEPA are a precursor source of bioavailable compounds (phenolic acids and (epi)catechin) for at least 24 hours after intake. In particular, 20 metabolites from different families were identified in urine. Interestingly, some of these metabolites were conjugated, indicating that, after absorption, these compounds were further transformed by hepatic enzymes. Additionally, two phenolic metabolites derived from microbial transformation were detected in feces: 4-hydroxyphenylpropionic acid and 3,4-dihydroxyphenylpropionic acid. As discussed later, even if these compounds are finally excreted, their contact with colonic epithelium during a prolonged period may also have biological implications. Indeed, another preclinical study performed with a grape product, where most proanthocyanidins were present as NEPA, detected microbial metabolites in cecal content (the fluid previous to feces formation) 24 h after supplementation (Touriño *et al.*, 2011). This means that more than one day after NEPA intake, there are still microbial-derived metabolites in contact with colonocytes.

The relevance of pathway 2 (colonic transformation of NEPP) should be explored for the specific case of nonextractable ellagitannins. It is known that ellagitannins are transformed by colonic microbiota, yielding specific metabolites, the urolithins (García-Villalba *et al.*, 2013). At the same time, it has been reported that ellagitannins are distributed in foods as extractable and nonextractable ellagitannins (García-Villalba *et al.*, 2015). Therefore, it may be hypothesized that nonextractable ellagitannins contribute to the colonic production of urolithins, although this aspect has not been specifically explored.

Finally, it should not be forgotten that colonic transformation is a two-way process, i.e., the formation of microbial metabolites implies effects on the growth or activity of specific bacterial species. This aspect and its relation to potential biological effects of NEPP will be later discussed.

2.3.2 Specific Features of the Metabolic Fate of NEPP

As far as we know, the metabolic transformations of EPP and NEPP produce, as shown above, the same chemical structures. This is logical, since the structures of the starting compounds are similar, changing only the molecular size or the association with dietary fiber. However, these specific facts of NEPP are the origin of differential features of their metabolic fate, compared to that of EPP: delayed absorption and interactions with dietary fiber (Pérez-Jiménez *et al.*, 2013).

First, the delayed absorption of NEPP may be expected, considering other known facts on the metabolic fate of EPP: proanthocyanidin oligomers need to reach the colon to be transformed, producing longer circulation times of their metabolites than monomers or dimers (Monagas *et al.*, 2010), or glycosides may be less available than their corresponding aglycone since an additional hydrolysis step is needed (Steensma *et al.*, 2006). Thus, since NEPP have either high molecular weight or associations with food matrix, this implies additional steps in their metabolic transformation. This was demonstrated in a preclinical study where rats were fed a diet supplemented with either ferulic acid or wheat bran, where a major fraction of ferulic acid is associated with the cell wall, i.e., present as HPP (Rondini *et al.*, 2004). It should be remembered that although phenolic acids are the major constituents of the HPP fraction, other compounds such as ellagitannins or different flavonoid classes have also been reported in several foods. Indeed, it has been observed that, in food matrixes rich in pectin (a soluble dietary fiber), there is a delay in the release of some phenolic compounds due to their retention in the gel formed by this substance (Mosele *et al.*, 2016). Overall, in the case of HPP, even if they may not be covalently linked to dietary fiber but exhibit weaker associations, aspects such as physical entrapment, increased viscosity or volume will contribute to this delayed release (Bohn, 2014).

Regarding NEPA, acute supplementation of healthy volunteers with a grape product where most polyphenols were present as NEPA produced a significant increase in plasma antioxidant capacity eight hours after intake (Pérez-Jiménez *et al.*, 2009b), longer than the 1–2 hours described for the peak in polyphenol concentration when supplementing with extractable proanthocyanidins (Rodríguez-Mateos *et al.*, 2014). It has been suggested that another aspect that may be involved in the delayed absorption of NEPP, besides their intrinsic structural aspects (high molecular weight or association with dietary fiber), is the potential interactions that these compounds may establish with intestinal tissues or mucin, since this avoids the access of gut microbiota to NEPP (González-Sarrías *et al.*, 2017). From a practical point of view, this delayed absorption of NEPP will give way to sustained circulation of potentially active metabolites as a particular characteristic of NEPP metabolic fate.

The other specific feature of the metabolic fate of NEPP is derived from their close association with dietary fiber. Thus, at the same time that phenolic compounds are transformed by the microbiota, dietary fiber is simultaneously fermented, and mutual enhancements may take place between both processes. With regard to the effect of this interaction on NEPP transformation, it was observed in an *in vitro* model that the colonic transformation of NEPA (adjusting the results per g of these compounds) was higher in the presence of a combination of soluble and insoluble dietary fiber than in the presence of only insoluble dietary fiber (Saura-Calixto *et al.*, 2010). Since it is known that soluble dietary fiber is much more fermented than the insoluble fraction, it seems that the transformation of this dietary fiber also contributed to the transformation of NEPA.

Similarly, a study in rats supplemented with resistant starch (a fermentable insoluble dietary fiber) showed higher levels of plasma phenolic metabolites in supplemented rats than in the control group, finding significant associations between the levels of some phenolic metabolites and those of short-chain fatty acids derived from dietary fiber fermentation (Vitaglione *et al.*, 2019). Moreover, the association of NEPP with dietary fiber would affect not only the rate of transformation of these compounds but also their transport and absorption; thus, it was recently reported that sustained exposure to short-chain fatty acids derived from dietary fiber fermentation increased the transport of ferulic acid in a monolayer model of epithelial cells, as well as the formation of hesperetin conjugates and their efflux towards the basolateral compartment (van Rymenant *et al.*, 2017). This agrees with previous results indicating that butyric acid increased the transepithelial transport of ferulic acid through upregulation of specific transporters (Ziegler *et al.*, 2016). Conversely, NEPP may also enhance the fermentation of dietary fiber (Aprikian *et al.*, 2003), which would be an additional biological effect of these compounds, as will be discussed in the next section.

2.4 How NEPP may Exhibit Health Effects

The potential health effects of NEPP have been attributed to four different mechanisms, some of these specific to them and others shared with EPP: antioxidant effect, modulation of colonic microbiota, biological activities of metabolites released during NEPP catabolism, and synergistic effects with other macromolecules such as dietary fiber. These should not be considered as independent effects, but as combined processes that will take place simultaneously after the consumption of NEPP-rich foods.

2.4.1 Antioxidant Effects

The most widely known property of dietary polyphenols is their antioxidant action. Although the determination of antioxidant capacity in foods or biological samples cannot be automatically associated with health effects, a number of studies have found significant associations between the antioxidant capacity of a diet and several health outcomes such as inflammation management (Brighenti *et al.*, 2005) or their association with chronic diseases (Nascimento-Souza *et al.*, 2018). Nevertheless, these studies have been mostly based on evaluation of antioxidant capacity derived from EPP, ignoring that corresponding to NEPP.

Studies evaluating the antioxidant capacity of NEPP in common foods have found that they make an important contribution to total food antioxidant capacity, equivalent to that of EPP. One study found that, considering overall consumption of foods and beverages in Spain (Pérez-Jiménez *et al.*, 2015), the antioxidant intake from soluble antioxidants was equivalent to that from macromolecular antioxidants. Beverages (53%) and fruits (16%) were the main contributors to antioxidant capacity intake from soluble antioxidants, while cereals (43%) and fruits (23%) were the main contributors to antioxidant capacity derived from macromolecular antioxidants (Figure 2.3).

Other studies evaluated the antioxidant capacity of NEPP in specific foods, providing additional information such as the high antioxidant capacity values associated with NEPP in tropical fruits (Maria do Socorro *et al.*, 2010; Páez-Peñuñuri

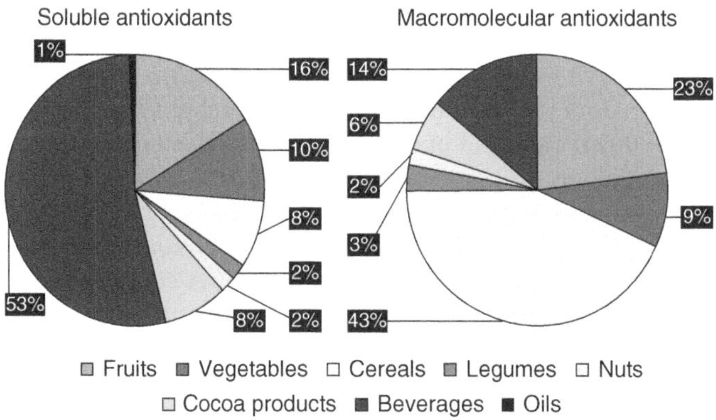

Figure 2.3 Percentage of daily antioxidant capacity per capita intake in the Spanish diet by FRAP assay (μmolTrolox), expressed per food group. FRAP, ferric-reducing antioxidant power assay.

et al., 2016) or in fruit peels (Liu *et al.*, 2018; Pérez-Jiménez and Saura-Calixto, 2018). Another important aspect regarding the antioxidant capacity associated with NEPP in foods is that processing may affect these values, sometimes increasing them and sometimes decreasing them, as shown in different studies (Pérez-Jiménez, 2018). It is remarkable that when antioxidant capacity derived from EPP and NEPP in several Italian dishes, as commonly prepared, was determined, in order to calculate the contribution of each fraction to total antioxidant capacity, there was a major contribution of NEPP (Durazzo *et al.*, 2017).

The above data may be used as the basis for studies evaluating potential associations between the antioxidant capacity of diets derived from NEPP and several health outcomes, as done for EEP. But, obviously, in order to obtain the strongest evidence of the antioxidant effects of NEPP, this must be measured by *in vivo* studies based on supplementation with different NEPP-rich products. In this case, the antioxidant capacity of intact NEPP in the upper part of the digestive tract will be similar to that measured by *in vitro* techniques but after colonic transformation, NEPP-derived antioxidant capacity should be estimated from that of their metabolites, lower than that of their parent compounds (Dueñas *et al.*, 2011).

Studies evaluating *in vivo* antioxidant capacity after supplementation with NEPP-rich products are scarce, but they provide interesting results. First, regarding local modifications of antioxidant capacity, a study in rats supplemented with a NEPP-rich grape pomace found a five-fold increase in cecum antioxidant capacity (Goñi and Serrano, 2005). Additional preclinical studies found that this kind of supplementation also enhanced the endogenous antioxidant system and decreased lipid oxidation in colonic mucosa (López-Oliva *et al.*, 2010, 2013). Interestingly, a recent study found significant increases in the antioxidant capacity of all the sections of the digestive tract after supplementing rats with high molecular weight melanoidins (Patrignani *et al.*, 2019); since it has been shown that these complexes also contain a fraction of NEPP (Pérez-Jiménez *et al.*, 2014), it may be hypothesized that these compounds partially contributed to the observed effect.

It may be expected that NEPP, once metabolized, contribute to increases in plasma antioxidant capacity. Supplementation of rats with NEPP from persimmon significantly increased this parameter (Matsumura *et al.*, 2016) and supplementation with a grape product rich in NEPP caused a sustained increase in plasma antioxidant capacity in healthy volunteers compared to a nonsupplemented control group (Pérez-Jiménez *et al.*, 2009b). More *in vivo* studies should be done with NEPP-derived products to deeply understand their effect on oxidative status, either locally in the digestive tract or in the bloodstream after absorption.

2.4.2 Microbiota Modulation

In recent decades, clear evidence has shown the importance of microbiota regarding both the development of obesity and associated chronic diseases; this affects both the need to avoid dysbiosis and to preserve gut microbes functionality (Khan *et al.*, 2016; Weiss and Hennet, 2017). Therefore, modulation of microbiota is one of the main mechanisms by which polyphenols can produce their beneficial health effects. There are two important aspects to take into account when evaluating the interaction between microbiota and NEPP: first, the importance of identifying bacteria responsible for the transformation and, second, modulation of the microbiota by NEPP.

Currently, information on the gut bacteria involved in colonic transformation of polyphenols is still limited, although there is a certain level of knowledge of the modifications caused by these dietary constituents (Espín *et al.*, 2017) and even some specific bacterial species involved in the transformation of ellagitannins have been reported (Gaya *et al.*, 2018; Selma *et al.*, 2014, 2017). Since, the chemical structures of EPP and NEPP reaching the colon are similar (differing in chain length or association with macromolecules), it may be assumed that they will be transformed by the same species. Thus, NEPP would contribute to the overall modifications in gut microbiota originated by all dietary polyphenols. Nevertheless, some studies have explored specific characteristics of the metabolic fate of whole-grain cereals (rich in NEPP), which deserve some comment.

When whole-grain cereals reach the colon, the first step needed is the release of HPP from the cell wall, which requires the direct action of feruloyl esterase with a potential combination of xylanase activity. These enzymes have been reported in bacterial species belonging to the genera *Lactobacillus* and *Roseburia* (*Firmicutes*), *Bifidobacterium* (*Actinobacteria*), *Bacteroides*, and *Prevotella* (*Bacteroidetes*) (Chassard *et al.*, 2007; Couteau *et al.*, 2001; Dodd *et al.*, 2011; Flint *et al.*, 2012). For instance, it was observed that fermentation of whole-grain barley or oat groats in the presence of three lactic acid bacteria (*L. acidophilus, L. johnsonii,* and *L. reuteri*) increased the production of free phenolic acids (Hole *et al.*, 2012). Additionally, when evaluating the bacterial species involved in the metabolic transformation of HPP, this should be considered as a whole, also evaluating those directly modified by the presence of dietary fiber. A recent preclinical study where rats were supplemented with a whole-grain flour rich in resistant starch and HPP found that, although the common genera involved in the transformation of HPP were not modified, there was an increase in members of *Ruminococcaeae/Ruminococcus*, known to be involved in the transformation of resistant starch, and this finally caused an increase in the levels of dihydroferulic acid, a colonic metabolite of ferulic acid (Vitaglione *et al.*, 2019).

As regards the other relevant aspect in the interaction between NEPP and microbiota, i.e., how NEPP affect the growth of different bacterial species, a clinical trial with overweight/obese subjects supplemented with whole grains (rich in HPP) observed sustained increases in levels of *Firmicutes* and *Bacteroidetes*, accompanied by a decrease in *Clostridia* (Vitaglione *et al.*, 2015). And regarding the effects of NEPA on colonic microbiota, an *in vitro* study evaluated the effect of whole grape antioxidant dietary fiber (GADF) (contanining 80% of polyphenols as NEPP), the EPP fraction, and the NEPP fraction on the growth of different beneficial microbes. It was found that NEPP specifically enhanced the growth of *L. reuteri* but the most interesting result was that it was the whole GADF and not the separated fractions (EPP and NEPP) which had the highest effect on the growth of *L. acidophilus*, suggesting a synergy between both fractions (Pozuelo *et al.*, 2012). Another study isolated NEPP from corn and tested their effect on the growth of several bacterial species, finding no significant differences in the effect on *L. helveticus* and *B. longum*, compared to a control (Galvez-Ranilla *et al.*, 2017). Besides the differences in the microbes tested, since NEPP in corn are mostly HPP while in GADF they are mostly NEPA, the potential role of the different fractions of NEPP on colonic microbiota deserves further research.

Finally, it should be recalled that, besides stimulating the growth of certain bacterial species, polyphenols (including NEPP) may contribute to shaping the gut microbiota by other mechanisms: antimicrobial effects which may be of health relevance, inhibition of microbial enzymes, alterations of mucus viscosity or disruption of quorum sensing (Havlik and Edwards, 2018).

Much more information is still needed regarding the gut microbes involved in the transformation of NEPP as well as on the effects that NEPP have in the overall microbiota profile. Moreover, certain limitations of previous studies should be considered. For instance, common microdilution methods originally developed for the study of antibiotics may not be as useful in the context of colon microbiota, since they were developed for compounds with better solubility and slower metabolism, they use isolated microbial strains, and there are many differences in the methodologies used; for these reasons, more clinical studies evaluating microbiota after the supplementation with NEPP-rich derived products should be performed. Finally, it is relevant to ensure that *in vitro* models use concentrations similar to those circulating in the human body after polyphenol intake, since many studies have been based on much higher doses, limiting the relevance of the reported results (Havlik and Edwards, 2018).

2.4.3 Biological Activities of Microbial Metabolites

The metabolites formed from NEPP by the action of colonic microbiota have been reported to be at least partially absorbed, as shown by their detection first in the colonic tissue and later in urine (Touriño *et al.*, 2009, 2011). Once absorbed, these metabolites show relevant biological effects (Dal-Pan *et al.*, 2017; di Gesso *et al.*, 2015; Esteban-Fernández *et al.*, 2017; Giménez-Bastida *et al.*, 2012; González de Llano *et al.*, 2019; Houghton *et al.*, 2018; Krga *et al.*, 2016; Lee *et al.*, 2017; Pourová *et al.*, 2018; Russell *et al.*, 2008; Sala *et al.*, 2015; Singh *et al.*,

2019; Verzelloni *et al.*, 2011; Waldecker *et al.*, 2008; Wang *et al.*, 2019). The chemical nature of these metabolites and, therefore, their biological activities will be similar to those derived from EPP, but it may be expected that they will be absorbed and transferred to the bloodstream at delayed periods compared to EPP-derived metabolites.

Several studies have evaluated the biological effects of polyphenol metabolites, considering colonic transformations as well as hepatic conjugations. These modifications may either increase or decrease the biological activity of the parent compound (Lodi *et al.*, 2009). In the specific case of conjugated metabolites, a deconjugation process before cell entry has been reported (Menendez *et al.*, 2011), so it might happen that these compounds ultimately exert their potential biological effects as deconjugated structures (Galindo *et al.*, 2012). A summary of the biological activities reported for microbial phenolic metabolites is shown in Table 2.3. Briefly, these studies have reported antiinflammatory, anticancer, antiglycative, neuroprotective, inhibition of lipid synthesis or insulin modulating activities, by a combination of mechanisms of action; of course, these effects should be further confirmed in *in vivo* studies. Also, these effects may be relevant in different target tissues but also locally, since some of them are beneficial for the colonic environment; for instance, urolithin A has been shown to enhance gut barrier integrity (Singh *et al.*, 2019). Since these studies were mostly developed in a context of evaluating mechanistic aspects, they were performed in *in vitro* models; nevertheless, it is important that these models reflect the actual concentrations of phenolic metabolites circulating in the body after absorption (Álvarez-Cilleros *et al.*, 2018; Kroon *et al.*, 2004). Interestingly, as shown in Table 2.3, some *in vivo* studies have found significant associations between circulating levels of some of these metabolites and certain health outcomes.

It should also be considered when evaluating the biological activities of NEPP how their interactions with food matrix take place. For instance, it was recently shown that NEPA linked to the cell wall by noncovalent bonds showed an antiinflammatory activity nearly equivalent to extractable proanthocyanidins, while this activity was reduced in NEPA associated with the cell wall by covalent linkages (Le Bourvellec *et al.*, 2019).

Table 2.3 Summary of biological activities reported for phenolic microbial metabolites

Compound	Effect	Model	Reference
Gluc-urolithin A	Inhibition of monocytes adhesion and endothelial cell migration	HAEC stimulated with TNF-α	Giménez-Bastida *et al.*, 2012
Urolithin B, gluc-urolithin B	Reduction of MCP-1, fractalkine and VEGF expresion	Ventricular cardiomyocytes from neonatal rats exposed to high glucose concentration	Sala *et al.*, 2015
Urolithin A and its analogue	Enhancement of gut barrier integrity and decreased systemic inflammation	C57BL/6, AhR−/−, Nrf2−/− mice and human colon epithelial carcinoma cell lines, HT29 mouse bone marrow-derived macrophages	Singh *et al.*, 2019
4-Hydroxybenzoic acid	Reduction of TNF-α concentration and IL-1β	THP-1 stimulated with LPS	di Gesso *et al.*, 2015
Isovanillic acid, gluc-isovanillic acid, gluc-vanillic acid, sulf-protocatechuic acid-3 and sulf-benzoic acid	Reduction of TNF-α secretion	THP-1 stimulated with LPS	di Gesso *et al.*, 2015
Protocatechuic, vanillic, ferulic and hippuric acid,	Reduction of the adhesion of monocytes to endothelial cells	THP-1 human monocytic cells to HUVEC stimulated with TNF-α	Krga *et al.*, 2016
5-(3′,4′-Dihydroxyphenyl-γ-valerolactone)	Reduction of the adhesion of monocytes to endothelial cells	THP-1 human monocytic cells to HUVEC stimulated with TNF-α	Lee *et al.*, 2017

(*continued*)

Table 2.3 (Continued)

Compound	Effect	Model	Reference
Hippuric acid, α-hydroxyhippuric acid and sulf-3-(3,4-dihydroxyphenyl) propionic acid	Decrease of adhesive activity against uropathogenic bacteria	Uropathogenic bacteria (*E. coli* ATCC 53503, *E. coli* DSM 10791, and *E. faecalis* 04-1) and T24 bladder cells (ATCC HTB4)	González de Llano *et al.*, 2019
Dihydrocaffeic	Decrease of oxidative stress	HepG2 and human vascular endothelium cells (EA.hy926) stimulated with TNF-α	Wang *et al.*, 2019
Dihydroferulic acid	Decrease of ENOS concentrations	HepG2 and human vascular endothelium cells (EA. hy926) stimulated with TNF-α	Wang *et al.*, 2019
3,4-Dihydroxyphenylacetic acid and 4-methylcatechol	Vasorelaxant effects on mesenteric artery	Isolated rat thoracic aorta and mesenteric artery	Pourová *et al.*, 2018
	Hypotensive effect	Normotensive Wistar: Han rats and spontaneously hypertensive rats	
Sulf-isovanillic acid	Stimulation of glucose transport	Human skeletal muscle myoblast line, LHCN-M2	Houghton *et al.*, 2018
p-Coumaric acid, 3-(4-OH-phenyl)-propionate and caffeic acid	Inhibition of HDAC activity	HT-29 human colon carcinoma cells	Waldecker *et al.*, 2008

3-(3,4-Dihydroxyphenyl) propionic acid, 3-(4-hydroxy-3-methoxyphenyl) propionic acid and 3-(3-hydroxyphenyl) propionic acid	Inhibition of prostanoid production	Fibroblast (CCD-Co18) cells stimulated with IL-1β	Russell *et al.*, 2008
3,4-Dihydroxyphenylacetic, 3-hydroxyphenylacetic acids and gluc-salicylic acid	Protection against neuronal death	SH-SY5Y neuroblastoma cells induced with SIN-1	Esteban-Fernández *et al.*, 2017
Dihydroxyphenyl valerolactone	Prevention of memory impairment	Non-Tg mice	Dal-Pan *et al.*, 2017
Urolithin A and B, 3-methoxy-4-hydroxyphenylacetic acid, pyrogallol, dihydrocaffeic acid, feruloylglycine, 3-hydroxyphenylacetic acid, 3,4-dihydroxyphenylacetic acid, 3-methoxy-4-hydroxyphenylacetic acid	Reduction of AGEs formation Protection of oxidative stress	Albumin glycation *in vitro* Human neuroblastoma SK-N-MC clonal cell line	Verzelloni *et al.*, 2011

AGE, advanced glycation end-products; gluc, glucuronide; HAEC, human aortic endothelial cells; HDAC, histone deacetylase; HUVEC, human umbilical vein endothelial cell; LPS, lipopolysaccharide; TNF, tumor necrosis factor; sulf, sufate

A final aspect to be considered is that, although the circulating phenolic metabolites are the compounds ultimately responsible for the action of either EPP or NEPP, they exert their activities in combination with the other mechanisms of action of these compounds, including modulation of dysbiosis. Thus, equol (a colonic metabolite of isoflavones) supplementation to nonequol producers did not promote the same health effects as those observed when the subjects were able to produce this metabolite; similarly, several discrepancies were found between the *in vitro* biological activities of certain ellagitannin metabolites and the health status of subjects with the highest production of those metabolites (Tomas-Barberan *et al.*, 2018). Therefore, these metabolites should be considered in the overall context of the combination of mechanisms of action that finally produce the biological effects of NEPP.

2.4.4 Synergy with Dietary Fiber

While the previously described mechanisms of action of NEPP are mostly shared with extractable polyphenols, there is a specific effect derived from the close association between NEPP and dietary fiber, i.e., the synergy that may take place between both food constituents.

When describing NEPP metabolic fate, it was explained that the presence of dietary fiber might stimulate the formation of microbial-derived metabolites from NEPP. Indeed, it has been suggested that an additional physiological effect of dietary fiber would be to behave as a carrier of phenolic compounds until the short intestine, allowing such transformations to take place (Saura-Calixto, 2011).

It may be expected that this interaction between dietary fiber and phenolic compounds does not only affect the generation of phenolic metabolites, but also that of short-chain fatty acids (SCFAs) from dietary fiber. Some studies have suggested that NEPP may stimulate the fermentation of dietary fiber. Thus, an improvement in the total production of SCFAs was observed after the *in vitro* incubation of chlorogenic acid, rutin, caffeic acid, quercetin, and olive oil polyphenols (Parkar *et al.*, 2013; Zampa *et al.*, 2006). A similar tendency was observed when rats were supplemented with pectin and freeze-dried apple or only with pectin (Aprikian *et al.*, 2003). And when several flavonoids were added to maize cell walls artificially lignified, forming new linkages with the cell wall, this led to an increase in the fermentability of the cell wall (Grabber *et al.*, 2012).

Moreover, a recent study found that proanthocyanidins not bound to cell wall produced, after an *in vitro* fermentation, fewer SCFAs, particularly butyrate, than those associated with the cell wall either by covalent or noncovalent linkages (and, at the same time, led to a higher production of phenolic microbial-derived metabolites, showing the complex interactions with the microbiota) (Le Bourvellec *et al.*, 2019). In contrast, other studies found that the presence of proanthocyanidins inhibited the formation of SCFAs (Bazzocco *et al.*, 2008) or that the co-occurrence of tea polyphenols and β-glucans did not stimulate the microbial transformation of soluble dietary fiber (Jalil *et al.*, 2019). At the same time, several *in vivo* studies have found synergystic effects between polyphenols and dietary fiber regarding cardiometabolic markers (Vitaglione *et al.*, 2015); this would imply that there was no inhibition of the fermentation of dietary fiber since its colonic metabolites are the main agents in the cardiometabolic effects of dietary fiber.

These divergences show that the potential interactions between NEPP and dietary fiber should be explored further, considering the different proportions of both substrates that may be found in a real situation. Many *in vitro* studies on colonic transformations studied either phenolic compounds or dietary fiber, i.e., providing a single source of carbon; however, this situation is different from what occurs after consuming foods, which means that these results may not be completely translated to real situations (Edwards *et al.*, 2017). Thus, the simultaneous effects that NEPP and dietary fiber have on microbiota should be evaluated, considering not only its profile but also its activity, since polyphenols may also affect the activity of bacterial enzymes (Kosmala *et al.*, 2014). Additionally, it would be interesting to determine not only the total formation of SCFAs in the presence of NEPP but also the section of the short intestine where they are formed; for instance, it was hypothesized that a potential retardation of this fermentation in the proximal colon by the action of polyphenols, with a higher production in the distal colon, would avoid the action of proteolytic bacterial species present in that section of the colon and associated with harmful metabolites (Tuohy *et al.*, 2012).

Finally, and showing the complexity of this matter, although several beneficial effects of SCFAs have been reported, an excess of them may increase caloric intake so there is a narrow equilibrium between a deficit and an excess of these compounds (Mosele *et al.*, 2015). In this sense, the concept of hierarchy in the preference of gut bacteria by one or another substrate (Hamaker and Tuncil, 2014) becomes relevant, since depending on the overall food intake, the same bacterial species might show a higher preference for the transformation of dietary fiber or of polyphenols, particularly when they are associated in a single complex, as in the case of NEPP.

2.5 Studies on the Health Effects of NEPP

2.5.1 Local vs Systemic Effects

Based on the metabolic fate and mechanisms of action of NEPP previously reported, their potential health effects will take place mainly in two different ways (Pérez-Jiménez *et al.*, 2013), summarized in Figure 2.4.

- *Local effects in the digestive tract.* These effects may be due either to the completely intact NEPP structures (since, for instance, they exhibit antioxidant capacity) or to the metabolites produced after the action of colonic microbiota. While intact NEPP may exert an effect throughout the whole digestive tract, the derived metabolites will only be able to exert an effect in the colon. Nevertheless, most studies have focused on colon health.
- *Systemic effects after absorption.* After NEPP transformation, a fraction of the generated metabolites is absorbed and transferred into the bloodstream. These metabolites may originate several systemic effects during circulation through the body (and eventually enter into the cell or accumulate into target tissues). Most studies have focused on cardiometabolic makers.

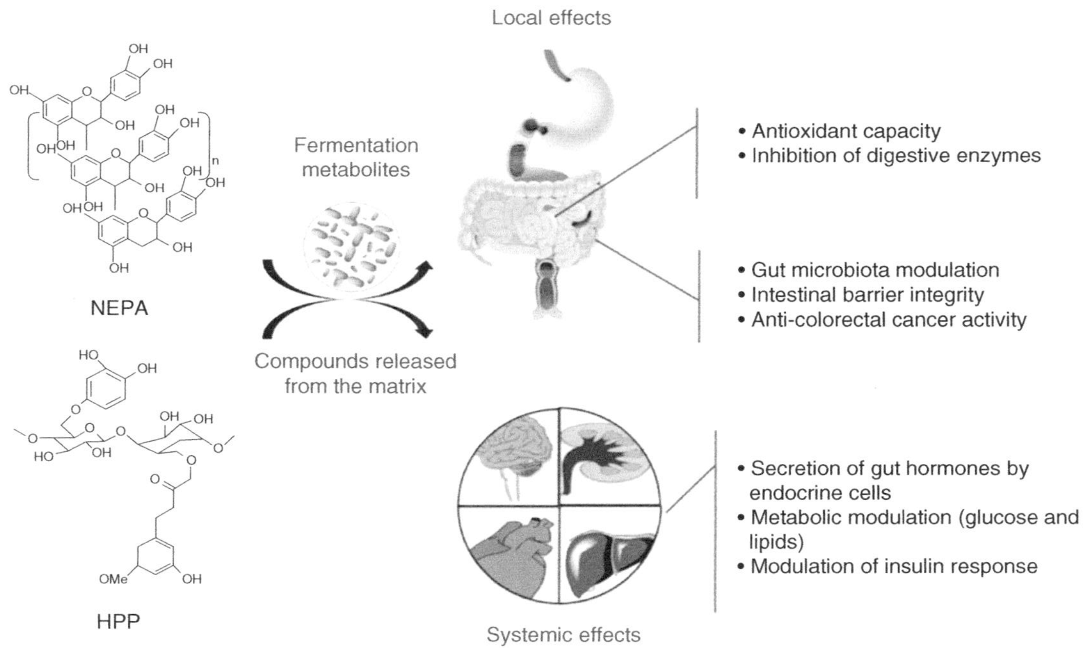

Figure 2.4 Local and systemic effects of NEPP. HPP, hydrolyzable polyphenols; NEPA, nonextractable proanthocyanidins.

Current evidence on both methods of action, based on preclinical and clinical studies, will be summarized in the following sections. It should be recalled that, due to the specific characteristics of NEPP regarding their close association with dietary fiber, it is quite difficult to perform studies with NEPP isolates. Indeed, in the case of HPP, if they were isolated, they would lose their specific macromolecular characteristics, i.e., their association with dietary fiber. For this reason, most studies are performed with NEPP-rich products, but which also present an important proportion of dietary fiber.

Although the scope of this chapter is restricted to the relevance of NEPP for human health, it should not be forgotten that some studies have also reported a promising role for NEPP in animal nutrition. Thus, from the traditional view of polyphenols as compounds that should be removed from animal studies, some experiments performed during recent years have found that supplementing chicken broilers with grape pomace (where most polyphenols are present as NEPP) has beneficial effects in intestinal microbiota and gut morphology (among other parameters) without compromising growth and performance (Brenes *et al.*, 2016; Viveros *et al.*, 2011)' significantly increasing the levels of circulating polyphenol-derived metabolites (Muñoz-González *et al.*, 2019). Therefore, this is a complementary aspect of the biological relevance of NEPP.

2.5.2 Effects on Gastrointestinal Health

Once foods are consumed, they will stay in contact with the digestive tract for hours. While a fraction of EPP may be absorbed in the small intestine, almost all NEPP will reach the colon unaltered. This means that for hours after intake, the original structures will be available to exert radical-scavenging effects. In this way, several *in vitro* digestion studies have shown that NEPP contribute to antioxidant capacity in the different stages of the digestive system (Fogliano *et al.*, 2011; Saura-Calixto *et al.*, 2007; Serrano *et al.*, 2007; Solari-Godiño *et al.*, 2017). NEPP may also exert inhibitory effects on certain digestive tract enzymes. This effect has been widely studied in recent years for dietary polyphenols, but, to the best of our knowledge, only one study specifically explored it in a NEPP concentrate (Yan *et al.*, 2018).

Regarding gastrointestinal health, the main target for NEPP is the colon, where they will remain for longer periods. It is worth highlighting that an epidemiological study exploring the association between (extractable) proanthocyanidin intake and cancer colon risk observed that the highest decrease in cancer risk was found for polymeric structures with more than 10 units (Rossi *et al.*, 2010). This seems logical since they will arrive in a higher proportion at the colon and will stay longer; although this study did not include NEPA, it should be recalled that they are also high molecular weight proanthocyanidins reaching the colon in an intact state.

Studies on the effects of NEPP in the colon have mostly used grape-derived products, e.g., grape pomace or GADF, where NEPP constitute the major fraction of phenolic compounds reaching the colon after the intake of these products. One of the most explored aspects was their effect on colonic oxidative status; this is a relevant outcome since, for instance, it has been reported that colonic cancer patients exhibit impaired colon oxidative status compared to healthy subjects (Chang *et al.*, 2008). Thus, supplementation of rats with grape seeds significantly improved the cecal antioxidant capacity compared with the control group, with NEPP being responsible for 50% of the antioxidant capacity found in these samples (Goñi and Serrano, 2005). In the same way, supplementation of healthy rats with GADF significantly decreased levels of lipid oxidation and increased the endogenous antioxidant defenses (López-Oliva *et al.*, 2010, 2013). Additionally, the previously explained effects of NEPP on colonic microbiota are also another way to improve gastrointestinal health.

But the reported local effects of NEPP go further, and have been directly related to the modulation of certain diseases. Supplementation with GADF to healthy mice significantly modified several biochemical routes in the colonic mucosal in a beneficial way, as measured by gene expression analysis: it upregulated genes involved in the metabolism of xenobiotic compounds, regulation of plasma glucose and tumoral suppression, while it downregulated genes involved in inflammation and lipid biosynthesis as well as protooncogenes (Lizarraga *et al.*, 2011).

But the most relevant result of the effect of NEPP in gastrointestinal health was observed in a study with APCMin/+

(the validated animal model for the study of colon cancer). The authors reported a significant decrease (more than 50%) in total polyps after GADF supplementation compared to a control group (Sánchez-Tena *et al.*, 2013). Additionally, another animal study explored other potential mechanisms of action of NEPP in the modulation of gastrointestinal health, and particularly cancer, reporting a reduction of cleaved caspase-3 and a minor release of cytochrome c from mitochondria to the cytosol (López-Oliva *et al.*, 2013). Overall, it may be expected that the action of NEPP in the colon, together with other agents such as nonabsorbed EPP and dietary fiber, contributes to providing a flow of substrates needed to downregulate cancer pathways (Ricciardiello *et al.*, 2011).

Additionally, it may be that NEPP contribute to the effects observed in some studies providing whole materials rich in both EPP and NEPP, although the studies did not specifically measure NEPP content. For instance, a recent animal experiment found that incorporating the red seaweed *Pyropia columbina* to the diet of Wistar rats improved the endogenous antioxidant system in the colon as well as the intestinal mucosal barrier function (Cian *et al.*, 2018). Since edible seaweeds have been reported to contain relevant amounts of NEPP (Sanz-Pintos *et al.*, 2017) and, based on the results with NEPP-rich materials, it may be hypothesized that these constituents contributed to the observed effects.

2.5.3 Effects on Cardiometabolic Health

Although research on the effects of NEPP on cardiometabolic regulation is still scarce, the results have been rather promising. Again, grape products, particularly grape pomace and GADF, have been the most studied products for the above reasons. Thus, supplementation with GADF (7.5 g/day) to normo- and hypercholesterolemic subjects for 16 weeks did not have any effect on the lipid profile of normocholesterolemic subjects (nor in the control group), but it caused a significant decrease in total cholesterol, LD cholesterol, and triglycerides in the hypercholesterolemic subjects. Moreover, a significant decrease in blood pressure was observed in the supplemented group as a whole (Pérez-Jiménez and Saura-Calixto, 2008). Interestingly, the authors reported that the observed effects were higher than those previously reported separately for dietary fiber or polyphenols, suggesting a synergy among all the beneficial compounds of the product (both EPP and NEPP in the case of phenolic compounds). Indeed, some authors have highlighted that certain physiological effects attributed to insoluble dietary fiber may be partially due to associated phenolic compounds (Macho-González *et al.*, 2017, 2018).

A potential mechanism for the hypocholesterolemic effect of NEPP is their bile acid-binding capacity. This effect, widely known for dietary fiber, was reported specifically for NEPP in an *in vitro* experiment where the alcohol-insoluble solids (including removal of EPP) were prepared from different fruits. Interestingly, although cellulose, pectin, or lignin alone also provided this effect, it was much higher when NEPP were also present (Hamauzu and Mizuno, 2011).

Some clinical studies with grape pomace have been published in recent years. Supplementation of males with at least one factor of metabolic syndrome with 20 g/day of grape pomace significantly decreased systolic and diastolic blood pressure, being the antihypertensive effect related to the role of polyphenols in the activation of endothelial nitric oxide synthase (eNOS) (Urquiaga *et al.*, 2015). But the most interesting aspect of this study was a significant decrease in plasma fasting glucose, which agrees with a study in healthy rats where one of the metabolic pathways observed to be modified in colonic mucosal was that of gene G6PC2, related to the maintenance of plasma blood glucose as well as with the inhibitory effects of NEPP on digestive enzymes (Yan *et al.*, 2018). Therefore, there may be connections between local effects in the colon and systemic outcomes. Another recent study also reported the potential of grape pomace in the regulation of carbohydrate metabolism, since a significant decrease in fasting plasma insulin and an improvement in insulin sensitivity indices were observed after supplementing subjects with at least two factors of metabolic syndrome with grape pomace (8 g/day) for six weeks (Martínez-Maqueda *et al.*, 2018).

Nevertheless, due to differences in subject characteristics, study design, product doses, etc., these results should be confirmed in further clinical trials. At the same time, to better understand the potential systemic effects of NEPP, these clinical trials should be concomitant with preclinical studies that may provide very useful information, such as the potential accumulation in target tissues of NEPP metabolites. For instance, the accumulation of metabolites in brown adipose tissue after supplementation with extractable proanthocyanidins has been reported (Serra *et al.*, 2013); similar research on NEPP would help to better understand the potential biological effects of these compounds.

All the studies mentioned were mostly focused on the role of NEPA but HPP may also play an important role in cardiometabolic regulation. Indeed, the widely known beneficial effects of whole-grain cereals are due to their high dietary fiber content compared to that of refined cereals, but it should be remembered that they also contain a very important fraction of HPP associated with dietary fiber. A study in obese subjects supplemented with whole-grain wheat where a significant decrease in several inflammation markers was observed also reported a significant association between modifications in gut microbiota populations and the circulating levels of ferulic acid, mostly present in wholegrain wheat as HPP (Vitaglione *et al.*, 2015).

2.6 Perspectives

This chapter has summarized current knowledge on NEPP or macromolecular antioxidants. Interest in this fraction of dietary phytochemicals has slowly increased during the last two decades, and at the moment there is some level of evidence that provides certain knowledge of NEPP: they are relevant contributors to total polyphenol content, in many cases equivalent to EPP; in consequence, they also present a high contribution of total daily polyphenol intake; they are at least partially bioavailable after intake, with specific features compared to the metabolic fate of EPP (longer circulation times of derived metabolites, interactions with dietary fiber); they may exhibit either local effects through the digestive tract or systemic effects after absorption of derived metabolites, with several health implications.

Nevertheless, at the same time, the information on NEPP is still limited in many aspects, with some important gaps that need to be filled. Thus, standard methodologies that allow the analysis of NEPP and, if possible, without affecting the original structures in food samples should be adopted. Maybe two parallel methodologies should be considered: one for those laboratories not completely familiar with the topic but which may run some of these analyses in a routine way, and another for a much more detailed comprehensive characterization. Increasing available data on a wide number of food items is a key aspect for improving the estimations performed up to this moment on NEPP intake in different populations. This kind of estimation should be extended to more countries or populations, as the first step to finally establish potential associations with health outcomes. At the same time, studies on biological aspects of NEPP (including both metabolic fate and *in vivo* effects) are essential to finally understand the potential relevance of these compounds. For this, clinical trials should be chosen whenever possible to obtain results that may be translated into common clinical/nutritional practice.

In this context, it is especially relevant to keep in mind that nutritional doses, i.e., doses that may be achieved through a common diet, are tested. The role of NEPP as food constituents and not as drugs should never be disregarded; they will not have drastic effects on the measured markers or by a specific mechanistic action, but it will be the combination of different molecules which may ultimately have a health effect, as suggested for EPP (de Pascual-Teresa and Clifford, 2017).

Finally, it should be borne in mind that since foods are consumed as a whole, and due to the strong association between NEPP and dietary fiber, both food constituents will behave as a whole complex, and their isolated effects can be only partially studied. So this complex as a whole will be ultimately responsible for the potential observed health effects and even in combination with other constituents associated with it in certain food products, such as melanoidins in maillardized dietary fibers, or the recently reported insoluble methylxanthines in coca products (Mudenuti *et al.*, 2018). In the same way, a whole approach to total food polyphenols should be considered, because both classes of phenolic compounds, with specific characteristics, may finally act together in improving the overall health status of a subject; moreover, studies in model systems have reported synergies between both classes of phenolic compounds (Çelik *et al.*, 2015). Thus, in the same way as the suggested concept of the "three Ps" (probiotics, prebiotics, and polyphenols) (Marchesi *et al.*, 2016) for microbiota regulation, where mutual interactions take place among them, the same perspective should be used for approaching total dietary polyphenols, as the combination and interaction between EPP and NEPP.

References

Álvarez-Cilleros, D., Martín, M.Á., and Ramos, S. (2018) (-)Epicatechin and the colonic 2, 3-dihydroxybenzoic acid metabolite regulate glucose uptake, glucose production, and improve insulin signaling in renal NRK-52E cells. *Molecular Nutrition & Food Research*, **62**(4), 1700470.

Amaya-Cruz, D.M., Pérez-Ramírez, I. F., Delgado-García, J., Mondragón-Jacobo, C., Dector-Espinoza, A., and Reynoso-Camacho, R. (2019) An integral profile of bioactive compounds and functional properties of prickly pear (Opuntia ficus indica L.) peel with different tonalities. *Food Chemistry*, **278**, 568–78.

Andreasen, M.F., Kroon, P.A., Williamson, G., and Garcia-Conesa, M.T. (2001a) Esterase activity able to hydrolyze dietary antioxidant hydroxycinnamates is distributed along the intestine of mammals. *Journal of Agricultural and Food Chemistry*, **49**(11), 5679–84.

Andreasen, M.F., Kroon, P.A., Williamson, G., and Garcia-Conesa, M.T. (2001b) Intestinal release and uptake of phenolic antioxidant diferulic acids. *Free Radical Biology and Medicine*, **31**(3), 304–14.

Aprikian, O., Duclos, V., Guyot, S., *et al.* (2003) Apple pectin and a polyphenol-rich apple concentrate are more effective together than separately on cecal fermentations and plasma lipids in rats. *Journal of Nutrition*, **133**(6), 1860–5.

Arranz, S. and Saura-Calixto, F. (2010) Analysis of polyphenols in cereals may be improved performing acidic hydrolysis: a study in wheat flour and wheat bran and cereals of the diet. *Journal of Cereal Science*, **51**(3), 313–18.

Arranz, S., Pérez-Jiménez, J., and Saura-Calixto, F. (2008) Antioxidant capacity of walnut (Juglans regia L.): contribution of oil and defatted matter. *European Food Research and Technology*, **227**(2), 425–31.

Arranz, S., Saura-Calixto, F., Shaha, S., and Kroon, P.A. (2009) High contents of nonextractable polyphenols in fruits suggest that polyphenol contents of plant foods have been underestimated. *Journal of Agricultural and Food Chemistry*, **57**(16), 7298–303.

Arranz, S., Silván, J.M., and Saura-Calixto, F. (2010) Nonextractable polyphenols, usually ignored, are the major part of dietary polyphenols: a study on the Spanish diet. *Molecular Nutrition & Food Research*, **54**(11), 1646–58.

Bazzocco, S., Mattila, I., Guyot, S., Renard, C.M., and Aura, A.M. (2008) Factors affecting the conversion of apple polyphenols to phenolic acids and fruit matrix to short-chain fatty acids by human faecal microbiota in vitro. *European Journal of Nutrition*, **47**(8), 442–52.

Bohn, T. (2014) Dietary factors affecting polyphenol bioavailability. *Nutrition Reviews*, **72**(7), 429–52.

Brahem, M., Renard, C.M., Bureau, S., Watrelot, A.A., and Le Bourvellec, C. (2019) Pear ripeness and tissue type impact procyanidin-cell wall interactions. *Food Chemistry*, **275**, 754–62.

Brenes, A., Viveros, A., Chamorro, S., and Arija, I. (2016) Use of polyphenol-rich grape by-products in monogastric nutrition. A review. *Animal Feed Science and Technology*, **211**, 1–17.

Brighenti, F., Valtuena, S., Pellegrini, N., *et al.* (2005) Total antioxidant capacity of the diet is inversely and independently related to plasma concentration of high-sensitivity C-reactive protein in adult Italian subjects. *British Journal of Nutrition*, **93**(5), 619–25.

Calani, L., Ounnas, F., Salen, P., *et al.* (2014) Bioavailability and metabolism of hydroxycinnamates in rats fed with durum wheat aleurone fractions. *Food & Function*, **5**(8), 1738–46.

Çelik, E. E., Gökmen, V., and Skibsted, L.H. (2015) Synergism between soluble and dietary fiber bound antioxidants. *Journal of Agricultural and Food Chemistry*, **63**(8), 2338–43.

Chang, D., Wang, F.A.N., Zhao, Y.S., and Pan, H.Z. (2008) Evaluation of oxidative stress in colorectal cancer patients. *Biomedical and Environmental Sciences*, **21**(4), 286–9.

Chassard, C., Goumy, V., Leclerc, M., Del'homme, C., and Bernalier-Donadille, A. (2007) Characterization of the xylan-degrading microbial community from human faeces. *FEMS Microbiology Ecology*, **61**(1), 121–31.

Chen, P.X., Tang, Y., Marcone, M.F., *et al.* (2015) Characterization of free, conjugated and bound phenolics and lipophilic antioxidants in regular-and non-darkening cranberry beans (Phaseolus vulgaris L.). *Food Chemistry*, **185**, 298–308.

Cian, R.E., Llopart, E., Muñoz, M.E.L.O., Weisstaub, A.R., Zuleta, Á., and Drago, S.R. (2018) Low level of red seaweed Pyropia columbina added to extruded maize products promotes colonic and systemic antioxidant environment in growing Wistar rats. *Journal of Applied Phycology*, **30**(1), 637–48.

Couteau, D., McCartney, A.L., Gibson, G.R., Williamson, G., and Faulds, C.B. (2001) Isolation and characterization of human colonic bacteria able to hydrolyse chlorogenic acid. *Journal of Applied Microbiology*, **90**(6), 873–81.

Dal-Pan, A., Dudonne, S., Bourassa, P., *et al.* (2017) Cognitive-enhancing effects of a polyphenols-rich extract from fruits without changes in neuropathology in an animal model of Alzheimer's disease. *Journal of Alzheimer's Disease*, **55**(1), 115–35.

de Camargo, A.C., Regitano-d'Arce, M.A.B., Biasoto, A.C.T., and Shahidi, F. (2016) Enzyme-assisted extraction of phenolics from winemaking by-products: antioxidant potential and inhibition of alpha-glucosidase and lipase activities. *Food Chemistry*, **212**, 395–402.

de Mira, N.V.M., Massaretto, I.L., Pascual, C.D.S.C.I., and Marquez, U.M.L. (2009) Comparative study of phenolic compounds in different Brazilian rice (Oryza sativa L.) genotypes. *Journal of Food Composition and Analysis*, **22**(5), 405–9.

de Pascual-Teresa, S. and Clifford, M.N. (2017) Advances in Polyphenol Research: A *Journal of Agricultural and Food Chemistry* Virtual Issue. *Journal of Agricultural and Food Chemistry*, **65** (37), 8093–5.

di Gesso, J.L., Kerr, J.S., Zhang, Q., *et al.* (2015) Flavonoid metabolites reduce tumor necrosis factor-α secretion to a greater extent than their precursor compounds in human

THP-1 monocytes. *Molecular Nutrition & Food Research*, **59**(6), 1143–54.

Díaz-Rubio, M.E. and Saura-Calixto, F. (2011) Beverages have an appreciable contribution to the intake of soluble dietary fibre: a study in the Spanish diet. *International Journal of Food Sciences and Nutrition*, **62**(7), 715–18.

Dodd, D., Mackie, R.I., and Cann, I.K. (2011) Xylan degradation, a metabolic property shared by rumen and human colonic Bacteroidetes. *Molecular Microbiology*, **79**(2), 292–304.

Domínguez-Rodríguez, G., Marina, M.L., and Plaza, M. (2017) Strategies for the extraction and analysis of non-extractable polyphenols from plants. *Journal of Chromatography A*, **1514**, 1–15.

Dueñas, M., Surco-Laos, F., González-Manzano, S., González-Paramás, A.M., and Santos-Buelga, C. (2011) Antioxidant properties of major metabolites of quercetin. *European Food Research and Technology*, **232**(1), 103–11.

Durazzo, A., Lisciani, S., Camilli, E., *et al.* (2017) Nutritional composition and antioxidant properties of traditional Italian dishes. *Food Chemistry*, **218**, 70–7.

Edwards, C.A., Havlik, J., Cong, W., *et al.* (2017) Polyphenols and health: interactions between fibre, plant polyphenols and the gut microbiota. *Nutrition Bulletin*, **42**(4), 356–60.

Espín, J.C., González-Sarrías, A., and Tomás-Barberán, F.A. (2017) The gut microbiota: a key factor in the therapeutic effects of (poly) phenols. *Biochemical Pharmacology*, **139**, 82–93.

Esteban-Fernández, A., Rendeiro, C., Spencer, J.P.E., *et al.* (2017) Neuroprotective effects of selected microbial-derived phenolic metabolites and aroma compounds from wine in human SH-SY5Y neuroblastoma cells and their putative mechanisms of action. *Frontiers in Nutrition*, **4**, 3.

Flint, H.J., Scott, K.P., Duncan, S.H., Louis, P., and Forano, E. (2012) Microbial degradation of complex carbohydrates in the gut. *Gut Microbes*, **3**(4), 289–306.

Fogliano, V., Corollaro, M.L., Vitaglione, P., *et al.* (2011) In vitro bioaccessibility and gut biotransformation of polyphenols present in the water-insoluble cocoa fraction. *Molecular Nutrition & Food Research*, **55**(S1), S44–S55.

Galindo, P., Rodriguez-Gómez, I., González-Manzano, S., *et al.* (2012) Glucuronidated quercetin lowers blood pressure in spontaneously hypertensive rats via deconjugation. *PLoS One*, **7**(3), e32673.

Galvez-Ranilla, L., Christopher, A., Sarkar, D., Shetty, K., Chirinos, R., and Campos, D. (2017) Phenolic composition and evaluation of the antimicrobial activity of free and bound phenolic fractions from a Peruvian purple corn (Zea mays L.) accession. *Journal of Food Science*, **82**(12), 2968–76.

García-Villalba, R., Beltrán, D., Espín, J.C., Selma, M.V., and Tomás-Barberán, F.A. (2013) Time course production of urolithins from ellagic acid by human gut microbiota. *Journal of Agricultural and Food Chemistry*, **61**(37), 8797–806.

García-Villalba, R., Espín, J.C., Aaby, K., *et al.* (2015) Validated method for the characterization and quantification of extractable and nonextractable ellagitannins after acid hydrolysis in pomegranate fruits, juices, and extracts. *Journal of Agricultural and Food Chemistry*, **63**(29), 6555–66.

Gaya, P., Peirotén, Á., Medina, M., Álvarez, I., and Landete, J.M. (2018) Bifidobacterium pseudocatenulatum INIA P815: the first bacterium able to produce urolithins A and B from ellagic acid. *Journal of Functional Foods*, **45**, 95–9.

Giménez-Bastida, J.A., González-Sarrías, A., Larrosa, M., Tomás-Barberán, F., Espín, J.C., and García-Conesa, M.T. (2012) Ellagitannin metabolites, urolithin A glucuronide and its aglycone urolithin A, ameliorate TNF-α-induced inflammation and associated molecular markers in human aortic endothelial cells. *Molecular Nutrition & Food Research*, **56**(5), 784–96.

Gómez-Juaristi, M., Martínez-López, S., Sarria, B., Bravo, L., and Mateos, R. (2018) Bioavailability of hydroxycinnamates in an instant green/roasted coffee blend in humans. Identification of novel colonic metabolites. *Food & Function*, **9**(1), 331–43.

Goñi, I. and Serrano, J. (2005) The intake of dietary fiber from grape seeds modifies the antioxidant status in rat cecum. *Journal of the Science of Food and Agriculture*, **85**(11), 1877–81.

González de Llano, D., Liu, H., Khoo, C., Moreno-Arribas, M.V., and Bartolomé, B. (2019. Some new findings regarding the antiadhesive activity of cranberry phenolic compounds and their microbial-derived metabolites against uropathogenic bacteria. *Journal of Agricultural and Food Chemistry*, **67**, 2166–74.

González-Sarrías, A., Espín, J.C., and Tomás-Barberán, F.A. (2017) Non-extractable polyphenols produce gut microbiota metabolites that persist in circulation and show anti-inflammatory and free radical-scavenging effects. *Trends in Food Science & Technology*, **69**, 281–8.

Grabber, J.H., Ress, D., and Ralph, J. (2012) Identifying new lignin bioengineering targets: impact of epicatechin, quercetin glycoside, and gallate derivatives on the lignification and fermentation of maize cell walls. *Journal of Agricultural and Food Chemistry*, **60**(20), 5152–60.

Hamaker, B.R. and Tuncil, Y.E. (2014) A perspective on the complexity of dietary fiber structures and their potential effect on the gut microbiota. *Journal of Molecular Biology*, **426**(23), 3838–50.

Hamauzu, Y. and Mizuno, Y. (2011) Non-extractable procyanidins and lignin are important factors in the bile acid binding and radical scavenging properties of cell wall material in some fruits. *Plant Foods for Human Nutrition*, **66**(1), 70–7.

Hartzfeld, P.W., Forkner, R., Hunter, M.D., and Hagerman, A.E. (2002) Determination of hydrolyzable tannins (gallotannins and ellagitannins) after reaction with potassium iodate. *Journal of Agricultural and Food Chemistry*, **50**(7), 1785–90.

Havlik, J. and Edwards, C.A. (2018. Non-extractable polyphenols and the gut microbiome. In F. Saura-Calixto, F. and J. Pérez-Jiménez (Eds.), *Non-extractable Polyphenols and Carotenoids* (pp. 241–62). London: Royal Society of Chemistry.

Hellström, J.K. and Mattila, P.H. (2008) HPLC determination of extractable and unextractable proanthocyanidins in plant materials. *Journal of Agricultural and Food Chemistry*, **56**(17), 7617–24.

Hervert-Hernández, D., García, O.P., Rosado, J.L., and Goñi, I. (2011) The contribution of fruits and vegetables to dietary intake of polyphenols and antioxidant capacity in a Mexican rural diet: Importance of fruit and vegetable variety. *Food Research International*, **44**(5), 1182–9.

Hole, A.S., Rud, I., Grimmer, S., Sigl, S., Narvhus, J., and Sahlstrøm, S. (2012) Improved bioavailability of dietary phenolic acids in whole grain barley and oat groat following fermentation with probiotic Lactobacillus acidophilus, Lactobacillus johnsonii, and Lactobacillus reuteri. *Journal of Agricultural and Food Chemistry*, **60**(25), 6369–75.

Houghton, M.J., Kerimi, A., Mouly, V., Tumova, S., and Williamson, G. (2018) Gut microbiome catabolites as novel modulators of muscle cell glucose metabolism. *FASEB Journal*, **33**(2), 1887–98.

Hümmer, W. and Schreier, P. (2008) Analysis of proanthocyanidins. *Molecular Nutrition & Food Research*, **52**(12), 1381–98.

Jalil, M., Abbe, M., Combet, E., Edwards, C.A., and Garcia, A.L. (2019) Effect of β-glucan and black tea in a functional bread on short chain fatty acid production by the gut microbiota in a gut digestion/fermentation model. *International Journal of Environmental Research and Public Health*, **16**(2), 227.

Karam, J., del Mar Bibiloni, M., and Tur, J. A. (2018) Polyphenol estimated intake and dietary sources among older adults from Mallorca Island. *PLoS One*, **13**(1), e0191573.

Khan, M.J., Gerasimidis, K., Edwards, C.A., and Shaikh, M.G. (2016) Role of gut microbiota in the aetiology of obesity: proposed mechanisms and review of the literature. *Journal of Obesity*, **2016**, 7353642.

Kosmala, M., Zduńczyk, Z., Karlińska, E., and Juśkiewicz, J. (2014) The effects of strawberry, black currant, and chokeberry extracts in a grain dietary fiber matrix on intestinal fermentation in rats. *Food Research International*, **64**, 752–61.

Krga, I., Monfoulet, L.E., Konic-Ristic, A., *et al.* (2016) Anthocyanins and their gut metabolites reduce the adhesion of monocyte to TNFα-activated endothelial cells at physiologically relevant concentrations. *Archives of Biochemistry and Biophysics*, **599**, 51–9.

Kroon, P.A., Clifford, M.N., Crozier, A., *et al.* (2004) How should we assess the effects of exposure to dietary polyphenols in vitro? *American Journal of Clinical Nutrition*, **80**(1), 15–21.

Le Bourvellec, C., Bagano Vilas Boas, P., Lepercq, P., *et al.* (2019) Procyanidin–cell wall interactions within apple matrices decrease the metabolization of procyanidins by the human gut microbiota and the anti-inflammatory effect of the resulting microbial metabolome in vitro. *Nutrients*, **11**(3), 664.

Lee, C.C., Kim, J.H., Kim, J.S., *et al.* (2017) 5-(3′,4′-Dihydroxyphenyl-γ-valerolactone), a major microbial metabolite of proanthocyanidin, attenuates THP-1

monocyte-endothelial adhesion. *International Journal of Molecular Sciences*, **18**(7), 1363.
Liu, H., Jiang, W., Cao, J., and Ma, L. (2018) Evaluation of antioxidant properties of extractable and nonextractable polyphenols in peel and flesh tissue of different peach varieties. *Journal of Food Processing and Preservation*, **42**(6), e13624.
Lizarraga, D., Vinardell, M.P., Noé, V., *et al.* (2011) A lyophilized red grape pomace containing proanthocyanidin-rich dietary fiber induces genetic and metabolic alterations in colon mucosa of female C57BL/6J mice. *Journal of Nutrition*, **141**(9), 1597–604.
Lodi, F., Jimenez, R., Moreno, L., *et al.* (2009) Glucuronidated and sulfated metabolites of the flavonoid quercetin prevent endothelial dysfunction but lack direct vasorelaxant effects in rat aorta. *Atherosclerosis*, **204**(1), 34–9.
López-Oliva, M.E., Agis-Torres, A., Goni, I., and Muñoz-Martínez, E. (2010) Grape antioxidant dietary fibre reduced apoptosis and induced a pro-reducing shift in the glutathione redox state of the rat proximal colonic mucosa. *British Journal of Nutrition*, **103**(8), 1110–17.
López-Oliva, M.E., Pozuelo, M.J., Rotger, R., Muñoz-Martínez, E., and Goni, I. (2013) Grape antioxidant dietary fibre prevents mitochondrial apoptotic pathways by enhancing Bcl-2 and Bcl-x L expression and minimising oxidative stress in rat distal colonic mucosa. *British Journal of Nutrition*, **109**(1), 4–16.
Ludwig, I.A., Rubió, L., Mosele, J.I., and Motilva, M.J. (2018) Metabolic fate of extractable and non-extractable polyphenols. In F. Saura-Calixto and J. Pérez-Jiménez (Eds.), *Non-extractable Polyphenols and Carotenoids* (pp. 220–40). London: Royal Society of Chemistry.
Macho-González, A., Garcimartín, A., López-Oliva, M.E., *et al.* (2017) Fiber purified extracts of carob fruit decrease carbohydrate absorption. *Food & Function*, **8**(6), 2258–65.
Macho-González, A., Garcimartín, A., Naes, F., *et al.* (2018) Effects of fiber purified extract of carob fruit on fat digestion and postprandial lipemia in healthy rats. *Journal of Agricultural and Food Chemistry*, **66**(26), 6734–41.
Marchesi, J.R., Adams, D.H., Fava, F., *et al.* (2016) The gut microbiota and host health: a new clinical frontier. *Gut*, **65**(2), 330–9.

Maria do Socorro, M.R., Pérez-Jiménez, J., Tabernero, M., Alves, R.E., de Brito, E.S., Saura-Calixto, F. (2010) Acerola and cashew apple as sources of antioxidants and dietary fibre. *International Journal of Food Science & Technology*, **45**(11), 2227–33.

Maria do Socorro, M.R., Pérez-Jiménez, J., Arranz, S., *et al.* (2011) Açaí (Euterpe oleraceae)'BRS Pará': a tropical fruit source of antioxidant dietary fiber and high antioxidant capacity oil. *Food Research International*, **44**(7), 2100–106.

Martínez-Maqueda, D., Zapatera, B., Gallego-Narbón, A., Vaquero, M.P., Saura-Calixto, F., Pérez-Jiménez, J. (2018) A 6-week supplementation with grape pomace to subjects at cardiometabolic risk ameliorates insulin sensitivity, without affecting other metabolic syndrome markers. *Food & Function*, **9**(11), 6010–19.

Mateos-Martín, M.L., Pérez-Jiménez, J., Fuguet, E., and Torres, J.L. (2012) Non-extractable proanthocyanidins from grapes are a source of bioavailable (epi) catechin and derived metabolites in rats. *British Journal of Nutrition*, **108**(2), 290–7.

Matsumura, Y., Ito, T., Yano, H., *et al.* (2016) Antioxidant potential in non-extractable fractions of dried persimmon (Diospyros kaki Thunb.). *Food Chemistry*, **202**, 99–103.

Menendez, C., Dueñas, M., Galindo, P., *et al.* (2011) Vascular deconjugation of quercetin glucuronide: the flavonoid paradox revealed? *Molecular Nutrition & Food Research*, **55**(12), 1780–90.

Monagas, M., Urpi-Sarda, M., Sánchez-Patán, F., *et al.* (2010) Insights into the metabolism and microbial biotransformation of dietary flavan-3-ols and the bioactivity of their metabolites. *Food & Function*, **1**(3), 233–53.

Mosele, J., Macià, A., and Motilva, M.J. (2015) Metabolic and microbial modulation of the large intestine ecosystem by non-absorbed diet phenolic compounds: a review. *Molecules*, **20**(9), 17429–68.

Mosele, J.I., Macià, A., Romero, M.P., and Motilva, M.J. (2016) Stability and metabolism of Arbutus unedo bioactive compounds (phenolics and antioxidants) under in vitro digestion and colonic fermentation. *Food Chemistry*, **201**, 120–30.

Mudenuti, N.V. de R., de Camargo, A.C., Shahidi, F., Madeira, T.B., Hirooka, E.Y., Grossmann, M.V.E. (2018) Soluble and insoluble-bound fractions of phenolics and alkaloids and their antioxidant activities in raw and traditional chocolate: a comparative study. *Journal of Functional Foods*, **50**, 164–71.

Muñoz-González, I., Chamorro, S., Pérez-Jiménez, J., *et al.* (2019) Phenolic metabolites in plasma and thigh meat of chickens supplemented with grape byproducts. *Journal of Agricultural and Food Chemistry*, **67**(16), 4463–71.

Nascimento-Souza, M.A., Paiva, P.G., Martino, H.S.D., and Ribeiro, A.Q. (2018) Dietary total antioxidant capacity as a tool in health outcomes in middle-aged and older adults: a systematic review. *Critical Reviews in Food Science and Nutrition*, **58**(6), 905–12.

Neveu, V., Perez-Jiménez, J., Vos, F., *et al.* (2010) Phenol-Explorer: an online comprehensive database on polyphenol contents in foods. *Database*, **2010**, bap024.

Páez-Peñuñuri, M.E., Mercado-Mercado, G., Blancas-Benitez, F.J., Villegas-González, R.B., and Sáyago-Ayerdi, S.G. (2016) Bioactive compounds and health properties of tamarindo (Tamarindus indica L). *Biotecnia*, **18**(1), 10–21.

Parkar, S.G., Trower, T.M., and Stevenson, D.E. (2013) Fecal microbial metabolism of polyphenols and its effects on human gut microbiota. *Anaerobe*, **23**, 12–19.

Patrignani, M., Rinaldi, G.J., Rufián-Henares, J.Á., and Lupano, C.E. (2019) Antioxidant capacity of Maillard reaction products in the digestive tract: an in vitro and in vitro study. *Food Chemistry*, **276**, 443–50.

Pérez-Jiménez, J. (2018) Antioxidant capacity of non-extractable polyphenols as a key dietary parameter. . In F. Saura-Calixto and J. Pérez-Jiménez (Eds.), *Non-extractable Polyphenols and Carotenoids* (pp. 263–83). London: Royal Society of Chemistry.

Pérez-Jiménez, J. and Saura-Calixto, F. (2008) Grape products and cardiovascular disease risk factors. *Nutrition Research Reviews*, **21**(2), 158–73.

Pérez-Jiménez, J. and Saura-Calixto, F. (2015) Macromolecular antioxidants or non-extractable polyphenols in fruit and vegetables: intake in four European countries. *Food Research International*, **74**, 315–23.

Pérez-Jiménez, J. and Saura-Calixto, F. (2018) Fruit peels as sources of non-extractable polyphenols or macromolecular antioxidants: analysis and nutritional implications. *Food Research International*, **111**, 148–52.

Pérez-Jiménez, J. and Torres, J.L. (2011) Analysis of nonextractable phenolic compounds in foods: the current state of the art. *Journal of Agricultural and Food Chemistry*, **59**(24), 12713–24.

Pérez-Jiménez, J., Arranz, S., Tabernero, M., *et al.* (2008) Updated methodology to determine antioxidant capacity in plant foods, oils and beverages: extraction, measurement and expression of results. *Food Research International*, **41**(3), 274–85.

Pérez-Jiménez, J., Arranz, S., and Saura-Calixto, F. (2009a) Proanthocyanidin content in foods is largely underestimated in the literature data: an approach to quantification of the missing proanthocyanidins. *Food Research International*, **42**(10), 1381–8.

Pérez-Jiménez, J., Serrano, J., Tabernero, M., *et al.* (2009b) Bioavailability of phenolic antioxidants associated with dietary fiber: plasma antioxidant capacity after acute and long-term intake in humans. *Plant Foods for Human Nutrition*, **64**(2), 102–7.

Pérez-Jiménez, J., Díaz-Rubio, M.E., and Saura-Calixto, F. (2013) Non-extractable polyphenols, a major dietary antioxidant: occurrence, metabolic fate and health effects. *Nutrition Research Reviews*, **26**(2), 118–29.

Pérez-Jiménez, J., Díaz-Rubio, M.E., and Saura-Calixto, F. (2014) Non-extractable polyphenols in plant foods: nature, isolation, and analysis. In R. Watson (Ed.), *Polyphenols in Plants* (pp. 203–18). New York: Academic Press.

Pérez-Jiménez, J., Díaz-Rubio, M.E., and Saura-Calixto, F. (2015) Contribution of macromolecular antioxidants to dietary antioxidant capacity: a study in the Spanish Mediterranean diet. *Plant Foods for Human Nutrition*, **70**(4), 365–70.

Pérez-Ramírez, I.F., Reynoso-Camacho, R., Saura-Calixto, F., and Pérez-Jiménez, J. (2018) Comprehensive characterization of extractable and nonextractable phenolic compounds by high-performance liquid chromatography–electrospray ionization–quadrupole time-of-flight of a grape/pomegranate pomace dietary supplement. *Journal of Agricultural and Food Chemistry*, **66**(3), 661–73.

Porter, L.J. (1988) Flavans and proanthocyanidins. In J.B. Harborne (Ed.), *The Flavonoids* (pp. 21–62). New York: Springer.

Pourová, J., Najmanová, I., Vopršalová, M., *et al.* (2018) Two flavonoid metabolites, 3, 4-dihydroxyphenylacetic acid and

4-methylcatechol, relax arteries ex vivo and decrease blood pressure in vitro. *Vascular Pharmacology*, **111**, 36–43.

Pozuelo, M.J., Agis-Torres, A., Hervert-Hernández, D., *et al.* (2012) Grape antioxidant dietary fiber stimulates Lactobacillus growth in rat cecum. *Journal of Food Science*, **77**(2), H59–H62.

Ricciardiello, L., Bazzoli, F., and Fogliano, V. (2011) Phytochemicals and colorectal cancer prevention – myth or reality? *Nature Reviews Gastroenterology & Hepatology*, **8**(10), 592.

Rodríguez-González, S., Pérez-Ramírez, I.F., Castaño-Tostado, E., Amaya-Llano, S., Rodríguez-García, M.E., and Reynoso-Camacho, R. (2018) Improvement of physico-chemical properties and phenolic compounds bioavailability by concentrating dietary fiber of peach (Prunus persica) juice by-product. *Journal of the Science of Food and Agriculture*, **98**(8), 3109–18.

Rodríguez-Mateos, A., Vauzour, D., Krueger, C.G., *et al.* (2014) Bioavailability, bioactivity and impact on health of dietary flavonoids and related compounds: an update. *Archives of Toxicology*, **88**(10), 1803–53.

Rondini, L., Peyrat-Maillard, M.N., Marsset-Baglieri, A., *et al.* (2004) Bound ferulic acid from bran is more bioavailable than the free compound in rat. *Journal of Agricultural and Food Chemistry*, **52**(13), 4338–43.

Rossi, M., Bosetti, C., Negri, E., Lagiou, P., and Vecchia, C.L. (2010. Flavonoids, proanthocyanidins, and cancer risk: a network of case-control studies from Italy. *Nutrition and Cancer*, **62**(7), 871–7.

Russell, W.R., Scobbie, L., Chesson, A., *et al.* (2008) Anti-inflammatory implications of the microbial transformation of dietary phenolic compounds. *Nutrition and Cancer*, **60**(5), 636–42.

Sala, R., Mena, P., Savi, M., *et al.* (2015) Urolithins at physiological concentrations affect the levels of pro-inflammatory cytokines and growth factor in cultured cardiac cells in hyperglucidic conditions. *Journal of Functional Foods*, **15**, 97–105.

Sánchez-Tena, S., Lizárraga, D., Miranda, A., *et al.* (2013) Grape antioxidant dietary fiber inhibits intestinal polyposis in Apc Min/+ mice: relation to cell cycle and immune response. *Carcinogenesis*, **34**(8), 1881–8.

Sanz-Pintos, N., Pérez-Jiménez, J., Buschmann, A.H., Vergara-Salinas, J.R., Pérez-Correa, J.R., and Saura-Calixto, F. (2017) Macromolecular antioxidants and dietary fiber in edible seaweeds. *Journal of Food Science*, **82**(2), 289–95.

Saura-Calixto, F. (1988) Effect of condensed tannins in the analysis of dietary fiber in carob pods. *Journal of Food Science*, **53**(6), 1769–71.

Saura-Calixto, F. (2011) Dietary fiber as a carrier of dietary antioxidants: an essential physiological function. *Journal of Agricultural and Food Chemistry*, **59**(1), 43–9.

Saura-Calixto, F. (2018) The story of the introduction of non-extractable polyphenols into polyphenol research: origin, development and perspectives. . In F. Saura-Calixto and J. Pérez-Jiménez (Eds.), *Non-extractable Polyphenols and Carotenoids* (pp. 1–16). London: Royal Society of Chemistry.

Saura-Calixto, F., Serrano, J., and Goni, I. (2007) Intake and bioaccessibility of total polyphenols in a whole diet. *Food Chemistry*, **101**(2), 492–501.

Saura-Calixto, F., Pérez-Jiménez, J., Touriño, S., *et al.* (2010) Proanthocyanidin metabolites associated with dietary fibre from in vitro colonic fermentation and proanthocyanidin metabolites in human plasma. *Molecular Nutrition & Food Research*, **54**(7), 939–46.

Selma, M.V., Beltrán, D., García-Villalba, R., Espín, J.C., and Tomás-Barberán, F.A. (2014) Description of urolithin production capacity from ellagic acid of two human intestinal Gordonibacter species. *Food & Function*, **5**(8), 1779–84.

Selma, M.V., Beltrán, D., Luna, M.C., *et al.* (2017) Isolation of human intestinal bacteria capable of producing the bioactive metabolite isourolithin a from ellagic acid. *Frontiers in Microbiology*, **8**, 1521.

Serra, A., Bladé, C., Arola, L., Macia, A., and Motilva, M.J. (2013) Flavanol metabolites distribute in visceral adipose depots after a long-term intake of grape seed proanthocyanidin extract in rats. *British Journal of Nutrition*, **110**(8), 1411–20.

Serrano, J., Goñi, I., and Saura-Calixto, F. (2007) Food antioxidant capacity determined by chemical methods may underestimate the physiological antioxidant capacity. *Food Research International*, **40**(1), 15–21.

Singh, R., Chandrashekharappa, S., Bodduluri, S.R., *et al.* (2019) Enhancement of the gut barrier integrity by a microbial metabolite through the Nrf2 pathway. *Nature Communications,* **10**(1), 89.

Solari-Godiño, A., Pérez-Jiménez, J., Saura-Calixto, F., Borderías, A.J., and Moreno, H.M. (2017) Anchovy mince (Engraulis ringens) enriched with polyphenol-rich grape pomace dietary fibre: in vitro polyphenols bioaccessibility, antioxidant and physico-chemical properties. *Food Research International,* **102**, 639–46.

Steensma, A., Faassen-Peters, M.A., Noteborn, H.P., and Rietjens, I.M. (2006) Bioavailability of genistein and its glycoside genistin as measured in the portal vein of freely moving unanesthetized rats. *Journal of Agricultural and Food Chemistry,* **54**(21), 8006–12.

Tabernero, M., Serrano, J., and Saura-Calixto, F. (2006) The antioxidant capacity of cocoa products: contribution to the Spanish diet. *International Journal of Food Science & Technology,* **41**, 28–32.

Tomas-Barberan, F.A., Selma, M.V., and Espín, J.C. (2018) Polyphenols' gut microbiota metabolites: bioactives or biomarkers? *Journal of Agricultural and Food Chemistry,* **66**(14), 3593–4.

Touriño, S., Fuguet, E., Vinardell, M.P., Cascante, M., and Torres, J.L. (2009) Phenolic metabolites of grape antioxidant dietary fiber in rat urine. *Journal of Agricultural and Food Chemistry,* **57**(23), 11418–26.

Touriño, S., Pérez-Jiménez, J., Mateos-Martín, M.L., *et al.* (2011) Metabolites in contact with the rat digestive tract after ingestion of a phenolic-rich dietary fiber matrix. *Journal of Agricultural and Food Chemistry,* **59**(11), 5955–63.

Tsang, C., Auger, C., Mullen, W., *et al.* (2005) The absorption, metabolism and excretion of flavan-3-ols and procyanidins following the ingestion of a grape seed extract by rats. *British Journal of Nutrition,* **94**(2), 170–81.

Tuohy, K.M., Conterno, L., Gasperotti, M., and Viola, R. (2012) Up-regulating the human intestinal microbiome using whole plant foods, polyphenols, and/or fiber. *Journal of Agricultural and Food Chemistry,* **60**(36), 8776–82.

Urquiaga, I., d'Acuña, S., Pérez, D., *et al.* (2015) Wine grape pomace flour improves blood pressure, fasting glucose and protein damage in humans: a randomized controlled trial. *Biological Research*, **48**(1), 49.

Van Rymenant, E., Abrankó, L., Tumova, S., *et al.* (2017) Chronic exposure to short-chain fatty acids modulates transport and metabolism of microbiome-derived phenolics in human intestinal cells. *Journal of Nutritional Biochemistry*, **39**, 156–68.

Verzelloni, E., Pellacani, C., Tagliazucchi, D., *et al.* (2011) Antiglycative and neuroprotective activity of colon-derived polyphenol catabolites. *Molecular Nutrition & Food Research*, **55**(S1), S35–S43.

Vitaglione, P., Mennella, I., Ferracane, R., *et al.* (2015) Whole-grain wheat consumption reduces inflammation in a randomized controlled trial on overweight and obese subjects with unhealthy dietary and lifestyle behaviors: role of polyphenols bound to cereal dietary fiber. *American Journal of Clinical Nutrition*, **101**(2), 251–61.

Vitaglione, P., Mennella, I., Ferracane, R., *et al.* (2019) Gut fermentation induced by a resistant starch rich whole grain diet explains serum concentration of dihydroferulic acid and hippuric acid in a model of ZDF rats. *Journal of Functional Foods*, **53**, 286–91.

Viveros, A., Chamorro, S., Pizarro, M., Arija, I., Centeno, C., and Brenes, A. (2011) Effects of dietary polyphenol-rich grape products on intestinal microflora and gut morphology in broiler chicks. *Poultry Science*, **90**(3), 566–78.

Waldecker, M., Kautenburger, T., Daumann, H., Busch, C., and Schrenk, D. (2008) Inhibition of histone-deacetylase activity by short-chain fatty acids and some polyphenol metabolites formed in the colon. *Journal of Nutritional Biochemistry*, **19**(9), 587–93.

Wang, S., Sarriá, B., Mateos, R., Goya, L., and Bravo-Clemente, L. (2019) TNF-α-induced oxidative stress and endothelial dysfunction in EA. hy926 cells is prevented by mate and green coffee extracts, 5-caffeoylquinic acid and its microbial metabolite, dihydrocaffeic acid. *International Journal of Food Sciences and Nutrition*, **70**(3), 267–84.

Weiss, G.A. and Hennet, T. (2017) Mechanisms and consequences of intestinal dysbiosis. *Cellular and Molecular Life Sciences*, **74**(16), 2959–77.

Williamson, G. and Clifford, M.N. (2017) Role of the small intestine, colon and microbiota in determining the metabolic fate of polyphenols. *Biochemical Pharmacology*, **139**, 24–39.

Wu, T., Phuong, N.N.M., van Camp, J., Smagghe, G., Le, T.T., and Raes, K. (2018) Analysis of non-extractable polyphenols (NEPP). In F. Saura-Calixto and J. Pérez-Jiménez (Eds.), *Non-extractable Polyphenols and Carotenoids* (pp. 46–67). London: Royal Society of Chemistry.

Yan, S., Shao, H., Zhou, Z., Wang, Q., Zhao, L., and Yang, X. (2018) Non-extractable polyphenols of green tea and their antioxidant, anti-α-glucosidase capacity, and release during in vitro digestion. *Journal of Functional Foods*, **42**, 129–36.

Zamora-Ros, R., Knaze, V., Rothwell, J.A., *et al.* (2016) Dietary polyphenol intake in Europe: the European Prospective Investigation into Cancer and Nutrition (EPIC) study. *European Journal of Nutrition*, **55**(4), 1359–75.

Zampa, A., Silvi, S., Servili, M., Montedoro, G., Orpianesi, C., and Cresci, A. (2006) In vitro modulatory effects of colonic microflora by olive oil iridoids. *Microbial Ecology in Health and Disease*, **18**(3-4), 147–153.

Ziegler, K., Kerimi, A., Poquet, L., and Williamson, G. (2016) Butyric acid increases transepithelial transport of ferulic acid through upregulation of the monocarboxylate transporters SLC16A1 (MCT1) and SLC16A3 (MCT4). *Archives of Biochemistry and Biophysics*, **599**, 3–12.

Zurita, J., Díaz-Rubio, M.E., and Saura-Calixto, F. (2012) Improved procedure to determine non-extractable polymeric proanthocyanidins in plant foods. *International Journal of Food Sciences and Nutrition*, **63**(8), 936–9.

3

Analytical Strategies for Determining Polyphenols in Foods and Biological Samples

Lucía Olmo-García[1], Romina P. Monasterio[2], Aadil Bajoub[3], and Alegría Carrasco-Pancorbo[1]*

[1] Department of Analytical Chemistry, Faculty of Science, University of Granada, Granada, Spain
[2] Instituto de Biología Agrícola de Mendoza (IBAM), UNCuyo, Mendoza, Argentina
[3] Department of Basic Sciences, National School of Agriculture, Meknès, Morocco

3.1 Introduction: Importance of the Determination of Polyphenols

Polyphenols are among the most relevant micronutrients linked to the health beneficial effects of our diet. They are widely found in vegetables, fruits, beverages, nuts, seeds, and also in manufactured foods (as components of their natural ingredients). Olive oil, apples, tea, berries, honey, coffee, wine, chocolate, etc. are good examples of widely consumed sources of polyphenols in the human diet (Kyle and Duthie, 2005; Naczk and Shahidi, 2006). Food manufacturers, consumers, and researchers are nowadays aware that polyphenol consumption may play a crucial role in health through the regulation of metabolism, weight, chronic disease, and cell proliferation (Cory *et al.*, 2018). Their health effects will logically depend on the amount consumed and their bioavailability.

Polyphenols are aromatic secondary metabolites ubiquitously spread through the plant kingdom comprising several thousands of substances with highly diverse structures. Different classifications can be found in the literature (Andrés-Lacueva *et al.*,

*Corresponding author.

Dietary Polyphenols: Metabolism and Health Effects, First Edition.
Edited by Francisco A. Tomás-Barberán, Antonio González-Sarrías, and Rocío García-Villalba.

2009; Antolovich *et al.*, 2000; Bravo, 1998; Crozier *et al.*, 2009; Manach *et al.*, 2004; Motilva *et al.*, 2013), but in general terms, all the possible categorizations classify the polyphenols into two main groups, based on their basic structure: flavonoids and nonflavonoids. The flavonoids may themselves be divided into six subclasses, depending on the type of heterocycle involved: flavonols, flavones, isoflavones, flavanones, anthocyanidins, and flavanols (catechins and proanthocyanidins). The main nonflavonoid phenolic compounds include phenolic acids, hydrolyzable tannins, stilbenes, coumarins and lignans (more information regarding this point can be found in Chapter 1). In addition to this diversity, polyphenols can be found in their plant sources as free molecules (so-called aglycones), in conjugated forms (recurrently as glycosides) or even as oligomers or polymers. In other words, they constitute a very heterogeneous group of substances which includes a molecular mass that ranges from small molecules (<100 Da) such as phenolic acids to quite big analytes (>30 000 Da) of highly polymerized compounds. In addition, polyphenols can differ drastically in terms of polarity (and other physicochemical characteristics) and can be found at remarkably variable concentration levels.

Therefore, determining them in a reliable manner is a very challenging analytical task. Many efforts have been made to provide highly sensitive, selective, fast, and robust analytical methods for the determination of polyphenols from different matrices. Given the complexity, their quantitative analysis from plant tissues and food (or biological fluids) implies a series of stages or steps: extraction or isolation, structural elucidation (and chromatographic or electrophoretic separation, if needed), and further quantification.

The importance of polyphenols is undeniable. For instance, in plants, phenolic substances may act as antifeedants, antioxidants, attractants for pollinators, phytoalexins, protective agents against UV light, contributors to the plant pigmentation, etc. (Shahidi and Naczk, 2004). In food, phenolic compounds may contribute to the astringency, color, bitterness, flavor, odor, and oxidative stability (Shahidi and Naczk, 2004). These factors, together with their great abundance in our diet and their role in the prevention of various diseases associated with oxidative stress, make them the focus of uncountable studies all over

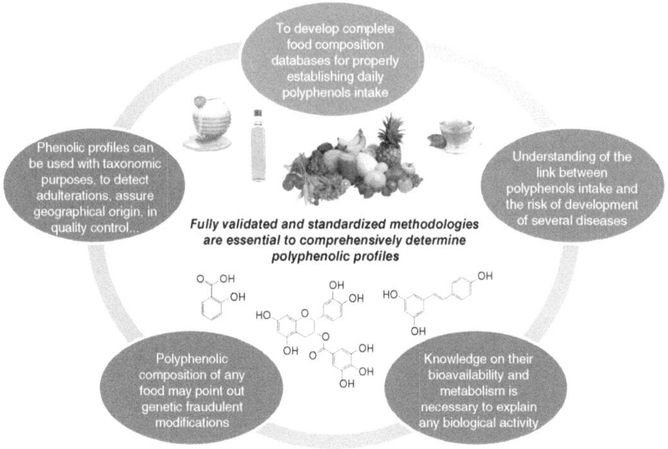

Figure 3.1 Schema representing the importance of having fully validated and standardized methodologies to comprehensively determine polyphenolic profiles.

the world. Figure 3.1 illustrates the relevance of polyphenol characterization in diverse areas. For example, it is essential to develop fully validated and standardized methodologies to comprehensively determine the nature and distribution of these compounds in our diet. Such knowledge could be used with taxonomic purposes, to detect adulterations and/or to establish daily polyphenol intake and carry out epidemiological studies providing an understanding of the link between their intake and the risk of development of several diseases. In this regard, it is important to take into account that not all polyphenols are absorbed with equal efficacy; extending our knowledge of the bioavailability and metabolism of these substances is absolutely necessary to evaluate their biological activity within target tissues. Bioavailability is a very complex process involving several different stages, such as liberation, absorption, distribution, metabolism, and elimination.

Figure 3.2 shows the different steps of any analytical procedure to achieve determination of the polyphenols present in a plant or food material, including options to carry out the sample preparation, analytical determination, and data treatment.

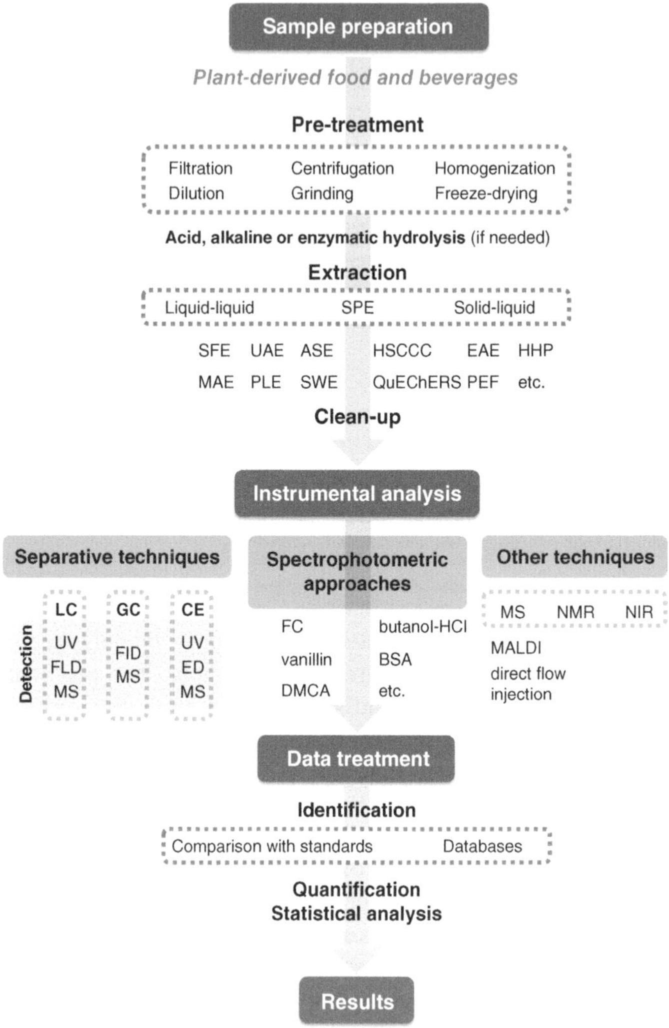

Figure 3.2 Different steps of a complete analytical procedure to achieve the determination of the phenolic compounds present in plant-derived foods and beverages, including sample preparation, instrumental analysis, and data treatment.

Notice that the terms *polyphenols* and *phenolic compounds* will be used interchangeably throughout this chapter despite the fact that they are not strictly synonyms.

3.2 Most Widely Used Extraction Systems and New Trends

Advances in electrophoretic and chromatographic separations and mass spectrometry detection systems have provided excellent separation efficiency, sensitivity and resolution; however, sample preparation is still one of the most essential parts of the whole analytical procedure. It is a critical step that determines the accuracy of the analytical results, since partial extractions and degradations during the procedure have a great impact in analytes recovery.

This stage encompasses a number of steps in the overall analytical scheme, such as the extraction itself, sample clean-up, fractionation, and compound purification. The selected extraction approach will depend on the general goal of the analyst, the nature of the sample, and the chemical characteristics of the polyphenols to be extracted (structure, polarity, concentration, position of functional groups, etc.). Very interesting reports covering sample preparation aspects have been published (Aires, 2017; Ajila *et al.*, 2011; Azmir *et al.*, 2013; Belwal *et al.*, 2018; da Silva *et al.*, 2016; Garcia-Salas *et al.*, 2010; Giacometti *et al.*, 2018; Ignat *et al.*, 2011; Kelly *et al.*, 2019; Khoddami *et al.*, 2013; La Barbera *et al.*, 2017; Ligor *et al.*, 2018; Lucci *et al.*, 2017; Moreira *et al.*, 2019; Motilva *et al.*, 2013; Naczk and Shahidi, 2004; Oroian and Escriche, 2015; Xu *et al.*, 2017); some of them are specific reviews on extraction procedures and others are more general reviews which contain thorough sections regarding sample treatment.

The ultimate goal of any sample preparation protocol is to achieve the preparation of an extract uniformly enriched in all components of interest (and ideally free from interfering matrix components) (Tura and Robards, 2002). Among the most widely used pretreatment operations are filtration, dilution, centrifugation, grinding, homogenization, freeze-drying, etc.

The unequal distribution in various structural forms of the phenolic compounds from fruits has to be considered when dealing with the isolation of polyphenols from this kind of sample; accumulation of soluble phenols is greater in the outer tissues of the fruit (epidermal and subepidermal layers) than in the inner tissues (mesocarp and pulp). Solid samples are usually first homogenized, which may be preceded by (freeze-)drying or freezing with liquid nitrogen. Afterwards, the analytical isolation of phenolic compounds is tackled. Liquid matrices are normally filtered and/or centrifuged; subsequently, the sample is either directly submitted to the measurement systems (best case scenario) or, if necessary, it undergoes further extraction procedures.

Even though conventional extraction protocols such as heating, boiling, refluxing, etc. have been frequently applied in the past, liquid-liquid extraction (LLE) and solid phase extraction (SPE) are the most used techniques for the fractionation/purification step. In LLE, solvents such as methanol, ethanol, propanol, acetone, ethyl acetate, diethyl ether, and their combinations, have been used for extraction of phenolic substances (eventually with different proportions of water) (Aires, 2017; Ajila *et al.*, 2011; Garcia-Salas *et al.*, 2010; Khoddami *et al.*, 2013). In this regard, in 2011, Ignat and co-workers published a compelling review that includes a table presenting the solvents usually applied to the extraction of diverse families of phenolic compounds. Extraction time, number of extraction stages, ratio of solvent to sample, sample particle size, temperature, etc., are some of the variables to be optimized and carefully controlled to guarantee appropriate recoveries. SPE is an efficient and versatile option that can be easily automated. For the extraction and preconcentration of polyphenols, the use of silica gels (mainly based on C18) modified with hydrophobic alkyl- or aryl-bonded silica and polymeric sorbents is prevalent (Ignat *et al.*, 2011; Khoddami *et al.*, 2013; Xu *et al.*, 2017). Diol cartridges and mixed-mode cartridges (providing dual retention modes, both reversed-phases and ion-exchange retention modes) have been used in other applications (Lucci *et al.*, 2017; Motilva *et al.*, 2013).

Besides LLE and SPE, stimulating approaches based on the application of techniques such as pressurized liquid extraction

(PLE) or accelerated solvent extraction (ASE) (Pereira *et al.*, 2019), microwave-assisted extraction (MAE) (Jesus *et al.*, 2019), ultrasound-assisted extraction (UAE) (Manzoor *et al.*, 2019), subcritical water extraction (SWE) (Gabaston *et al.*, 2018), supercritical fluid extraction (SFE) (da Silva *et al.*, 2016), high-speed countercurrent chromatography (HSCCC) (Li *et al.*, 2017), membrane extraction (Kelly *et al.*, 2019), surfactant cloud point extraction, QuEChERS (Quick, Easy, Cheap, Effective, Rugged and Safe) and dispersive-SPE (Galarce-Bustos *et al.*, 2019), immunosorbents and molecularly imprinted polymers (MIPs) (Euterpio *et al.*, 2013; Gomes *et al.*, 2017), enzyme assisted extraction (EAE) (Santana and Macedo, 2019), microwave-assisted enzymatic extraction (MAEE) (Zhang *et al.*, 2013), comprehensive enzyme-assisted supercritical fluid extraction (EASCFE), pulsed electric field (PEF) (Manzoor *et al.*, 2019), high hydrostatic pressure extraction (HHP), etc. can be also found in literature (Aires, 2017; Ajila *et al.*, 2011; Azmir *et al.*, 2013; Belwal *et al.*, 2018; da Silva *et al.*, 2016; Garcia-Salas *et al.*, 2010; Giacometti *et al.*, 2018; Ignat *et al.*, 2011; Kelly *et al.*, 2019; Ligor *et al.*, 2018; Lucci *et al.*, 2017; Moreira *et al.*, 2019; Motilva *et al.*, 2013; Xu *et al.*, 2017). The development of new rapid, economical, robust and reproducible polyphenols extraction techniques, overcoming the common drawbacks of traditional protocols, is a hot topic in the field. Moreover, great efforts are being made in the investigation of green extraction methods, which pursue the reduction of operational costs and the environmental impact of the sample preparation stage by optimizing time, energy, and solvent consumption (Ameer *et al.*, 2017; Panja, 2017).

Another factor to be considered is that there is a possibility of interaction between phenols and other plant components, such as carbohydrates and proteins; that interaction will surely have some influence on the extraction yield. In some instances, a hydrolysis step has to be included, mainly as an aid to structural elucidation and characterization of bound polyphenols. Three types of hydrolytic treatment can be chosen by the analyst: acidic, enzymatic, and alkaline. Hydrolysis could also be used to minimize possible interferences in further stages of the analytical determination.

An in-depth discussion about the diverse extraction systems which can be used to carry out the sample preparation is beyond the scope of this chapter (please check the comprehensive reviews and research articles cited within this section).

3.3 Determination of the Phenolic Compounds in Foods

The scientific community has shown interest in phenolic compounds, mainly due to their therapeutic value, since the end of the nineteenth century (Sadtler, 1889). The development of highly robust, sensitive, selective, rapid and reliable analytical methodologies for the determination of polyphenols makes possible the continuous progress in this area. Despite the great number of investigations, the quantitative determination of these substances remains intricate. The difficulty will depend on the main aim of the particular study; we can find targeted or nontargeted researches, pursuing the determination of total polyphenols (or a specific class of them) or individual comprehensive profiling. The lack of pure standards, the heterogeneity of this group of analytes, the very variable concentration levels they exhibit in different samples, etc. are some of the prevailing problems that the analyst has to deal with.

This section will cover applications from classic spectrophotometric assays to modern liquid chromatography, gas chromatography and capillary electrophoresis coupled to mass spectrometry (LC-MS, GC-MS, CE-MS respectively) or bidimensional platforms, mentioning some other analytical approaches as well. As stated before, this chapter does not aim to present a comprehensive review but a brief general overview. We recommend reading the excellent reviews previously published (Aires, 2017; Ajila *et al.*, 2004; Ignat *et al.*, 2011; Khoddami *et al.*, 2013; Naczk and Shahidi, 2006; Xu *et al.*, 2017).

3.3.1 Classic Methods For Polyphenols Determination: Spectrophotometric Assays

Spectrophotometric assays evaluating total phenol and flavonoid content in plant and food samples are cheaper and faster and more accessible than assays based on chromatographic techniques. The clear disadvantage of using them is that they determine compound categories rather than individual compounds; also, their chemically unspecific nature is one of their main limitations.

Among all the existing methods to determine total phenolic content, the Folin–Ciocalteu (FC) assay has been the most widely used in past decades owing to its simplicity and acceptable reproducibility (AOAC, 1980; Singleton and Rossi, 1965). The FC method is based on the reduction (electron-donating) of a phosphotungstates-phosphomolybdates complex by reductants (or antioxidants) to a blue chromogen. Therefore, it is expected to find good correlation between total phenolic content obtained by the FC method and the antioxidant activity determined by any assay based on an electron transfer mechanism (e.g., ferric reduction antioxidant assay). Numerous modifications of the FC method have been made to eliminate interference and improve its performance (Sánchez-Rangel *et al.*, 2013).

Total flavonoids, proanthocyanidins (condensed tannin), and hydrolyzable tannins can also be estimated by colorimetric methods. Extracts of plant phenols (methanolic or ethanolic extracts) mixed with $AlCl_3$ allow measurement of total flavonoids in the range 410–423 nm (Khoddami *et al.*, 2013). Vanillin and dimethylaminocinnamaldehyde (DMCA) assays are used to determine the level of proanthocyanidins (Ajila *et al.*, 2011), providing information about the degree of polymerization and the hydroxylation pattern and stereochemistry of flavan-3-ol subunits. The butanol-HCl and bovine serum albumin (BSA) methods are other options when proanthocyanidin determination is intended (Shahidi and Naczk, 2004).

Anthocyanins constitute the other main class of phenolic compounds which can be measured by spectrophotometry. Giusti and Wrolstad (2003) reviewed the most relevant spectrophotometric methods to determine anthocyanins (which is usually performed in weak acidic media in the wavelength range 490–550 nm).

Further information about the use of spectrophotometric assays to determine total polyphenols or specific classes of them can be found in the following sources (Ajila *et al.*, 2011; Khoddami *et al.*, 2013; Ignat *et al.*, 2011; Shahidi and Naczk, 2004).

3.3.2 Evolution of the Traditional Methods to Characterize the Polyphenolic Fraction of Foods: Chromatographic and Electrophoretic Separation and Subsequent Detection

LC-MS (coupled by means of different interfaces and analyzers) is by far the technique of choice for the structural characterization and determination of polyphenols in food samples. The evolution of the LC separations (either using sub-2 μm particle packed columns or porous-shell columns with sub-3 μm superficially porous particles) together with the continuous improvements in MS technology (providing more robust high-resolution MS systems) has opened up great possibilities in this field (Cavaliere *et al.*, 2018; Cortés-Herrera *et al.*, 2018; Kaufmann, 2014; La Barbera *et al.*, 2017; Lucci *et al.*, 2017; Motilva *et al.*, 2013).

For determination of phenolic compounds in food samples, different analytical columns have been used. Different types of reverse phase (RP) C18 stationary phases have been mainly utilized, although applications based on the use of other stationary phases such as C8 or high-strength silica (HSS) T3 can be found in the literature (Lucci *et al.*, 2017; Khoddami *et al.*, 2013; Motilva *et al.*, 2013). All these columns work on RP mode; however, the performance of other chromatographic separation modes such as hydrophilic interaction chromatography (HILIC) with amide-bonded stationary phases or pentafluorophenyl (PFP) columns has also been tested (Marrubini *et al.*, 2018). As far as mobile phases are concerned, acidified water (with small proportions of formic, acetic or trifluoroacetic acids) and organic solvents (methanol or acetonitrile with small amounts of acidic modifiers) are principally chosen in RP-LC (Motilva *et al.*, 2013; Saurina and Sentellas, 2015). Some LC methods with elution in isocratic mode have been described, but gradient elution is commonly preferred since it provides more efficient separations. Flow rate and sample volume injection can change a lot from one application to another, logically depending on the LC system and the kind and dimension of the analytical columns.

Phenolic substances have been frequently determined using UV-VIS and photodiode array (PDA) detectors at different wavelengths (Aires, 2017; Ajila *et al.*, 2011; Khoddami *et al.*,

Table 3.1 UV absorption maxima of the different phenolic categories

Compound class	Absorption maxima	
	Band I	Band II
Simple phenols	266–295	
Phenolic acids	235–305	
Hydroxycinnamic acids	227–245, 310–332	
Hydroxycoumarins	≈ 210, 250–260, 280–303	312–351
Flavonoids		
Flavonols	250–280	350–385
Flavones, biflavones	250–280	310–350
Isoflavones	245–275	310–330
Flavanones	275–295	310–330
Chalcones	240–260	365–390
Aurones	240–270	390–430
Anthocyanins	265–275	465–560
Flavanols	270–280	
Stilbenes	220–236, 280–321	
Lignans	227–240, 267–292	310–350

Usual solvent is methanol with the exception of anthocyanins (methanol containing 0.01% HCl), stilbenes (ethanol), and lignans (not specified).

2013; Robards and Antolovich, 1997). Table 3.1 shows UV characteristic absorption bands of the main families of phenolic compounds (adapted from Ayres and Loike, 1990; Beale and Roe, 1953; Harborne, 1984; Robards and Antolovich, 1997; Shahidi and Naczk, 2006). Fluorimetric (FLD) and electrochemical (ED) are other broadly employed detection systems for LC (Cabrera-Benegil *et al.*, 2017; Rodrigues *et al.*, 2019).

The importance of MS as detector for elucidating the chemical structures of polyphenols and also for quantifying them is indisputable. With regard to ionization, electrospray (ESI) is the most extensively used interface for polyphenol determination (preferably in negative ionization mode). Atmospheric pressure chemical ionization (APCI) and, to a lesser extent, atmospheric pressure photoionization (APPI) have also been described for the determination of phenolic substances. Concerning MS analyzers, triple quadrupole (QqQ) and ion-trap (IT)-based analyzers are very common as low-resolution methods; the first stands out for its remarkable sensitivity and the latter for their excellent performance when elucidation and structural characterization are intended by means of MS/MS. Time-of-flight (TOF) and Orbitrap (OT) mass analyzers are outstanding examples of high-resolution MS systems for the study of polyphenols in food matrices; their accuracy expedites the unambiguous determination of the elemental composition of food constituents of complex matrices.

Interesting applications can be found in the compendious reviews cited in the current section; moreover, Table 3.2 includes some interesting examples of methods for the determination of phenolic analytes from diverse matrices using the techniques that we examine in this part of the chapter.

Over the last decade, comprehensive two-dimensional LC (2D LC or LCxLC) has been proved to determine polyphenols in diverse matrices (Cacciola *et al.*, 2017); the improved separation that bidimensional approaches can achieve is essentially grounded in the combination of two analytical columns with different separation mechanisms.

Gas chromatography is another technique that has been employed for separation, identification, and quantification of phenolic compounds in a wide range of food and plant matrices (Nolvachai and Marriott, 2013). GC methods for determining polyphenols generally require a previous derivatization step (mainly conversion to trimethylsilyl derivatives, but also methylation and trifluoroacetylation, among others, can be applied) to assure the volatility and thermal stability of the substances under study prior to analysis (Ajila *et al.*, 2011; Ignat *et al.*, 2011; Khoddami *et al.*, 2013).

Fused silica capillary columns of 30 m length, with internal diameters within the approximate range of 0.25–0.32 mm and stationary phase particle size of 0.25 μm are the most accepted (Khoddami *et al.*, 2013). A flame ionization detector (FID) is still routinely employed (in particular, for its utility to properly quantify in a wide linear concentration range), but the prominence of MS is nowadays unquestionable. Electron impact (EI) is the ionization source that covers a wider number of applications (Deslaurien, 2000; Hurtado-Fernandez *et al.*, 2015a), but more

Table 3.2 Selected examples of determination of polyphenols in widely studied food matrices.

Analytical technique	Extraction protocol	Conditions of analysis	Number of analytes (classes)	Reference
		Avocado		
LC-ESI-IT MS	Solid-liquid extraction: 0.5 g of freeze-dried and homogenized sample + 40 mL methanol (vortex); + 20 mL methanol (2nd step); dryness → 1 mL of methanol	– Zorbax C18 analytical column (4.6 × 150 mm, 1.8 μm particle size) – Water with 0.5 % acetic acid (phase A) and acetonitrile (phase B); gradient: 0 min, 95% A; 15 min, 50% A; 16 min, 0% A; at 17.5 min, initial conditions. – 10 μL of injection volume – 0.8 mL/min flow rate – MS in negative ion mode (+3200 V); 50–1000 m/z	6 relevant metabolites (*p*-coumaric, ferulic, chlorogenic, pantothenic and abscisic acids and epicatechin)	(Hurtado-Fernandez *et al.*, 2016)
GC-FID and GC-APCI-TOF MS	Solid-liquid extraction: 4 g of freeze-dried sample + 40 mL methanol (vortex); dryness → 5 mL methanol; online derivatization with BSTFA + 1% TMCS at room T for 60 min	– HP-5-MS column (30 m, 0.25 mm i.d. and 0.25 μm film thickness) – 1 μL of injection volume (splitless) – Carrier gas: He 1 mL/min – Injector temperature: 250 °C	27 analytes (organic acids, phytohormones, vitamins, flavonoids and phenolic acids)	(Hurtado-Fernandez *et al.*, 2014)

Table 3.2 (Continued)

Analytical technique	Extraction protocol	Conditions of analysis	Number of analytes (classes)	Reference
		– T gradient: 140 °C for 5 min, 3 °C min^{-1} until 220 °C (1 min), 10 °C min^{-1} till 241 °C (1 min), at 3 °C min^{-1} till 310 °C (kept 5 min) – FID detector: 300 °C, 40 mL/min H_2, 370 mL/min air – APCI-MS operated in positive mode (+2000 V, +4000 nA); 50–1000 m/z; transfer line and source temperature: 300°C		
CE-UV	Solid-liquid extraction: 4 g of freeze-dried sample + 40 mL methanol (vortex); dryness → 1 mL methanol/water (50:50, v/v)	– Uncoated fused-silica capillary 50 μm i.d., 50 cm effective length – Running buffer: 40 mM sodium tetraborate at pH 9.4, 30 kV, 25 °C – 10 s hydrodynamic injection (0.5 psi) – UV detection at 254 nm	21 phenolic acids (8 hydroxycinnamic acids + 13 hydroxybenzoic acids)	(Hurtado-Fernandez *et al.*, 2013)

CE-ESI-IT MS		– Bare fused-silica capillary 50 µm i.d., 85 cm total length – Running buffer: 40 mM ammonium acetate at pH 9.5, 30 kV, 25 °C – 5 s hydrodynamic injections (0.5 psi) – MS in negative ion mode (+3200 V); 50–800 m/z in full scan and MRM	10 compounds (perseitol, quinic, chlorogenic, *trans*-cinnamic, pantothenic and abscisic acids, epicatechin, catechin, ferulic and *p*-coumaric acids)	(Contreras-Gutierrez *et al.*, 2013)
		Propolis		
LC-UV	Ultrasound assisted extraction: 0.1 g of frozen sample + 3 mL ethanol 95% (v/v) (1 h ultrasound, 360 W, 25 KHz); 10 min centrifugation at 3000 rpm; supernatant dilution (1:10, v/v) with ethanol; filtration (0.45 µm)	– Kromasil C18 column (4.6 × 250 mm, 5 µm particle size) – Water with 0.1 % formic acid (phase A) and methanol (phase B); gradient: 0 min, 75% B; 22 min, 30% B; 45 min, 45% B; 58 min, 30% B; 75 min, 20% B; 80 min, 5% B; 95 min, 75% B (kept isocratic for 10 min) – 2 µL of injection volume – 1 mL/min flow rate – UV detection at 310 nm	7 phenolic acids and derivatives (chlorogenic, isochlorogenic A, B and C and caffeic acids; caffeic acid phenethyl ester; and artepillin C)	(Sun *et al.*, 2019)

Table 3.2 (Continued)

Analytical technique	Extraction protocol	Conditions of analysis	Number of analytes (classes)	Reference
LC-ESI-QTOF MS	Ultrasound-assisted extraction: 0.1 g of frozen sample + 10 mL methanol (30 min ultrasound, 360 W, 25 KHz); filtration (0.45 µm)	– C18 Agilent SB column (2.1 × 100 mm, 1.8 µm particle size) – Water with 0.1 % formic acid (phase A) and methanol (phase B); gradient: 0 min, 25% B; 11 min, 45% B; 22 min, 55% B; 29 min, 70% B; 40 min, 95% B (kept isocratic for 4 min) – 2 µL of injection volume – 0.3 mL/min flow rate – MS in positive and negative ion mode (3500 V); 100–1700 m/z MS mode, 50–1000 m/z MS/MS mode		
LC-ESI-QqQ MS/MS	1) Solid-liquid extraction: 8 g of frozen sample + 100 mL ethanol 80% (v/v) (30 min in a water bath at 70 °C); dryness → 100 mL of methanol	– Acquity BEH C18 column (100 × 2.1 mm, 1.7 µm particle size) – Water with 0.1% formic acid (phase A) and acetonitrile (phase B); linear gradient: 3–100% B in 10 min	12 polyphenols (6 phenolic acids + 6 flavonoids)	(Xavier *et al.*, 2017)

	2) Fractionation by liquid–liquid extraction: (a) methanolic extract/hexane (1:1, v/v); (b) [methanolic extract + 20 mL water]/chloroform (1:1, v/v) ; vacuum concentration of the three fractions (hexane, chloroform and hydromethanol)	– 2 μL of injection volume – 0.2 mL/min flow rate – MS in negative ion mode (+4500 V); MRM mode		
GC-EI-Q MS	Solid-liquid extraction: 200 g of frozen and ground sample + 1000 mL ethanol/water (80:20, v/v) (24 h stirring at room T); derivatization with BSTFA at 70 °C for 20 min; silylated products extracted with 2 mL 2,2,4-trimethylpentane	– HP-5MS column (30 m, 0.25 mm i.d. and 0.25 μm film thickness) – 1 μL of injection volume (1:20 split ratio) – Carrier gas: He 0.7 mL/min – Injector temperature: 280 °C – T gradient: 100 °C initial T, 5 °C min^{-1} until 280 °C (kept 8 min) – EI (70 eV) MS in positive mode; 50–800 Da in full scan and SIM; transfer line temperature: 280 °C	21 polyphenols (9 phenolic acids + 2 simple phenols + 1 stilbene + 9 flavonoids)	(Molnár *et al.*, 2017)
CZE-UV	Solid-liquid extraction: 10 g of frozen and ground sample + 100 mL 0.1 M carbonate buffer (pH=9)	– Bare fused-silica capillary 50 μm i.d., 72 cm total length	20 polyphenols (1 stilbene + 11 flavonoids + 8 phenolic acids)	(Gatea *et al.*, 2015)

Table 3.2 (Continued)

Analytical technique	Extraction protocol	Conditions of analysis	Number of analytes (classes)	Reference
	(15 min stirring at 70 °C + 10 days maceration at room T); filtration (Whatman no. 1)	– Running buffer: 45 mM sodium tetraborate with 0.9 mM sodium dodecyl sulfate at pH 9.35, 30 kV, 30 °C – 12 s hydrodynamic injections (35 mbar) – UV detection (200 to 360 nm), quantification at 280 nm		
		Wine		
LC-UV	1) QuEChERS extraction: 5 mL of sample + water with 0.1% formic acid + 2.5 mL acetonitrile in a 15 mL PTFE centrifuge tube (hand-shaken for 30 s); phase separation with 1.5 g of NaCl and 4 g of $MgSO_4$ (10 min centrifugation at 3000 rpm) 2) SPE clean-up: 1 mL aliquot in an SPE tube containing 150 mg $CaCl_2$, 50 mg PSA and 50 mg C18; dryness → LC initial mobile phase	– Symmetry C18 column (4.6 × 250 mm, 5 μm particle size) – Water with 0.1% formic acid (phase A) and methanol (phase B); gradient: 0 min, 20% B; 2 min, 30% B; 8 min, 32% B; 14 min, 80% B; 22 min, 20% B (kept isocratic for 10 min) – 10 μL of injection volume – 1 mL/min flow rate – UV detection at 280, 320 and 370 nm	10 polyphenols (1 stilbene + 3 flavonoids + 6 phenolic acids)	(Fontana and Bottini, 2014)

LC-ESI-QqQ/Q Orbitrap MS	Direct analysis of filtered samples (0.2 μm, polypropylene)	– Kinetex core–shell C18 column (100 mm × 2.1 mm, 2.6 μm particle size) – Water with 0.1% acetic acid (phase A) and acetonitrile with 0.1% acetic acid (phase B); gradient: 0 min, 5% B (kept for 3 min); 13 min, 15% B; 28 min, 35% B; 33 min, 50% B; 99% B (kept for 5 min); 5% B (kept for 8 min) – 5 μL of injection volume – 600 μL/min flow rate – MS in positive (−3500 V) and negative ion mode (+2500 V); QqQ working in SRM mode; Q Orbitrap 100–600 m/z MS and MS/MS mode	50 polyphenols (7 phenolic acids + 1 stilbene + 42 flavonoids)	(Cavaliere *et al.*, 2019)
LC-ESI-QqQ MS	a) Direct analysis of filtered samples (0.2 μm, polytetrafluoroethylene) b) Fractions obtained through Sephadex LH-20 gel chromatography	– Acquity UPLC BEH phenyl column (100 mm × 2.1 mm, 1.7 μm particle size) – Water with 0.1% formic acid (phase A) and acetonitrile (phase B); gradient: 0 min, 0.1% B (kept for 0.5 min); 5 min, 30% B; 8 min, 45% B; 90% B (wash for 3.5 min); 0.1% B (stabilization)	≈50 individual monomeric anthocyanins and anthocyanin adducts + 2D fingerprints of proanthocyanidin derivatives	(Laitila *et al.*, 2019)

Table 3.2 (Continued)

Analytical technique	Extraction protocol	Conditions of analysis	Number of analytes (classes)	Reference
		– 5 µL of injection volume – 0.5 mL/min flow rate – MS in positive and negative ion mode (3400 V); MRM mode		
GC-IT MS/MS	SPE extraction: 1 mL sample* + 1 mL water (pH=2 with HCl 6 M) in a HLB Supelco SPE cartridge (3 mL, 60 mg); elution with 2 × 2 = mL of tetrahydrofurane; dryness; derivatization with 100 µL MSTFA + 50 µL pyridine at 80 °C for 1 h; dilution with toluene *Wine before and after alkaline hydrolysis with 9 mL NaOH 2 M for 2 h at room T (to break bound forms)	– VF-5MS column (30 m, 0.25 mm i.d. and 0.25 µm film thickness) – 1 µL of injection volume (1:10 split ratio) – Carrier gas: He 1 mL/min – Injector temperature: 250 °C – T gradient: 40 °C for 1 min, 10 °C min^{-1} until 320 °C (kept 3 min) – EI (70 eV) MS in positive mode; 50–640 m/z in full scan and SIM; transfer line temperature: 170 °C	Free and bound forms of 5 phenolic acids (including *cis* and *trans* forms of some of them)	(Ivanović *et al.*, 2016)
CE-UV	Liquid-liquid extraction: 2 mL sample + 4 mL ethyl ether; hydrolysis (300 µL HCl) for 30 min in ultrasound bath; dryness → 1 mL of ethanol/water (60:40, v/v)	– Fused-silica capillary 50 µm i.d., 45 cm effective length – Running buffer: 40 mM sodium tetraborate at pH 8.9, 26 kV, 25 °C – 5 s hydrodynamic injections (10 psi) – UV detection at 215, 280 and 320 nm	8 polyphenols (3 phenolic acids + 5 flavonoids)	(Şanli *et al.*, 2016)

→, reconstitution; BSTFA, N,O-bis(trimethylsilyl)trifluoroacetamide; MRM, multiple reaction monitoring; PSA, primary-secondary amine; SIM, selective ion monitoring; TMS, trimethylchlorosilane.

recently, APCI has also been used to determine phenolic components from foods (Hurtado-Fernandez *et al.*, 2015b). The broad range of application (in terms of polarity and size of analytes) and the robustness of APCI as interface (in particular, coupled to high-resolution MS analyzers) demonstrate that GC-APCI is definitively a necessary and valuable addition to the "polyphenols toolbox." Several examples are included in Table 3.2.

Capillary electrophoresis is an alternative separation technique to LC and GC, especially suitable for the determination of low to medium molecular weight polar and charged compounds in food matrices (Acunha *et al.*, 2016; Alvarez *et al.*, 2018; Hurtado-Fernandez *et al.*, 2010; Ramautar *et al.*, 2015). In some cases, the resultant CE separations can be faster and more efficient than the corresponding LC and GC ones; however, CE normally leads to worse sensitivity and reproducibility if compared with the other techniques.

Capillary zone electrophoresis (CZE) and micellar electrokinetic chromatography (MEKC) are the "classic" CE modes and the most applied to the analysis of phytochemicals in plant materials (Jáč *et al.*, 2006). Different detection systems can be coupled to CE; they can be classified into three main groups: detection based on optical techniques; electrochemical techniques; and a miscellaneous group, where MS and radiochemical techniques can be included. UV-Vis absorption is clearly the detection system used to a larger extent (Arries *et al.*, 2017; Marti *et al.*, 2017; Spisso *et al.*, 2018), although nowadays, CE coupled to MS is very popular and a bit more affordable (Acunha *et al.*, 2016; Alvarez *et al.*, 2018; Hurtado-Fernandez *et al.*, 2010).

In order to obtain a good separation in CE, it is necessary to optimize several parameters, such as buffer nature, pH and concentration, type and dimensions of the capillary, additives (kind and concentration), temperature, voltage and injection mode, etc. The influence of every parameter on the separation has to be evaluated by the analyst and will depend on the selected CE methodology, the class of phenolic molecule/s under study, and the matrix being analyzed (Hurtado-Fernandez *et al.*, 2010). Table 3.2 includes several examples that illustrate the use of CE to determine polyphenols from diverse matrices.

3.3.3 Other Analytical Strategies

Nuclear magnetic resonance (NMR) spectroscopy is becoming more popular in the field of food analysis. Its main limitation is equipment cost, but its advantages are very obvious: structural elucidation power, very suitable approach for untargeted screening, minimal sample preparation required, simplicity of the measurement procedures, and instrumental stability (Ignat *et al.*, 2011; Le Gall and Colquhoun, 2003).

For structural elucidation of any new polyphenol isolated from the food matrix, NMR spectroscopy has become a powerful and routinely used analytical tool (Diaz-de-Cerio *et al.*, 2018; Ye *et al.*, 2015). Moreover, it also finds many other applications in food science and technology, such as quality control and metabolomics studies (Le Gall and Colquhoun, 2003; Villa-Ruano *et al.*, 2019).

Near infrared (NIR) spectroscopy is another fast, powerful, accurate, and nondestructive analytical tool which can be used for analysis of foods containing polyphenols (McGoverin *et al.*, 2010). For instance, NIR has been used to develop an industry-friendly quantitative method for determining total polyphenol content in Chinese dates by combining NIR and mid-infrared (MIR) spectroscopy data (Arslan *et al.*, 2019). Another recent example described the potential of NIR to build different spectroscopic calibration models able to predict total antioxidant capacity, total phenolic substances, total flavonoids and condensed tannins content in a sample-set composed of 147 sorghum grain samples (Alfieri *et al.*, 2019).

Interesting approaches based on direct MS analysis have been proposed trying to overcome problems derived from the use of LC (time consuming and poor resolution as polymer chain length and structural diversity increase). Fulcrand and co-workers published a very complete review discussing the potential of direct MS approaches to characterize polyphenol composition of complex matrices (Fulcrand *et al.*, 2008). Direct flow injection ESI MS analysis can be used, for instance, to establish polyphenol fingerprints of food extracts (Cooper and Marshall, 2001). Moreover, matrix-assisted laser desorption/ionization (MALDI)-MS is emerging as a potentially suitable method for assessing cocoa content and quality through the evaluation of phenolic analytes (Bonatto and Silva, 2015).

Microfluidic methods have also been developed to determine polyphenols in apples and pears (Crevillen *et al.*, 2009), to quantify polyphenol content in white wine employing the FC reaction method (Sandoval-Ventura *et al.*, 2017) or to evaluate the concentration levels of polyphenolic antioxidants in green tea and red wine (Nikovaev *et al.*, 2015).

3.4 Some Considerations Regarding the Determination of Polyphenols in Biological Samples

To explore and determine the mechanisms of action of food polyphenols and their role in disease prevention, it is crucial to understand the factors that constrain their release from the foodstuff, their extent of absorption, and their fate in the organism. In this regard, it is essential to investigate polyphenol bioavailability, which can be defined as: "that fraction of an oral dose, either parent compound or active metabolite, from a particular preparation that reaches the systemic circulation" (Stahl *et al.*, 2002). D'Archivio *et al.* (2010) simplified that definition a little, explaining that it simply means how much of the ingested amount of polyphenols is able to exert its beneficial effects in the target tissues (D'Archivio *et al.*, 2010).

At this point, it is important to notice that the most abundant phenolic compounds in our diet are not necessarily those that have the best bioavailability profile, either because they have a lower intrinsic activity or because they are poorly absorbed from the intestine, highly metabolized, or rapidly eliminated. In addition, the metabolites that are found in blood and target organs, resulting from digestive or hepatic activity, may differ from the native compounds in terms of biological activity (Manach *et al.*, 2004, 2009). In other words, to properly address the bioavailability of food and plant phenolic compounds, we should review *in vitro* and *in vivo* (both animal and human) studies on the absorption, transport, metabolism, and excretion of food polyphenols. That is beyond the scope of this chapter, which in this section is just trying to briefly present and discuss the analytical approaches that are useful for the study of food and plant polyphenols in clinical samples (body fluids and organ tissues). We will address the sample collection, preservation and extraction of the analytes of interest, as well as the determination of those compounds by applying different analytical tools. It is recommended to study previously written thorough contributions (Jenab *et al.*, 2009; Manach *et al.*, 2005; Spencer *et al.*, 2008; Vacek *et al.*, 2010).

As stated above, the study of the metabolization of phenolic compounds in the human body is mandatory. Ingested polyphenols are absorbed in the digestive tract (primarily in the small and large intestine) where they undergo chemical transformations. The deglycosylation by intestinal enzymes is pivotal in polyphenol absorption; lactase-phlorizin hydrolase and cytosolic β-glucosidase are key enzymes participating in the hydrolysis processes. Phenolic substances that were not absorbed in the small intestine can be further metabolized by bacteria in the large intestine, via deglycosylation, reduction, and hydrolysis reactions. When absorption of polyphenols occurs in the digestive tract, they are enzymatically metabolized in individual tissues; substances not absorbed are logically eliminated. Polyphenols are conjugated to glucuronide, sulfate, and methyl groups in the gut mucosa and inner tissues. The liver is the primary organ of metabolic transformation of phenolic substances enabling their biological inactivation prior to their excretion from the body. The polyphenols reaching the colon are extensively metabolized by the microflora into a wide array of low molecular weight phenolic acids. Blood or plasma samples can be very useful to study individual metabolites (and their formation), whilst urine or feces will obviously contain the final metabolization products. To obtain a deeper insight into polyphenol absorption and metabolism phenomena, read the following studies (Chen *et al.*, 2018; Scalbert *et al.*, 2002; Vacek *et al.*, 2010).

Plasma and urine are probably the most studied biological samples; there are few applications (if the total number is compared) evaluating the presence of polyphenols and their metabolites in organ tissues. Indeed, obtaining plasma or urine (or feces) samples is much easier; however, it is necessary to bear in mind the great variability between the samples taken (even though they are coming from the same individual) depending on a great number of factors.

As far as sample preparation is concerned, the extraction protocols are specific for each kind of biological sample and differ depending on the final goal of the analysis. Plasma samples are usually combined with organic solvents (acetonitrile, methanol, etc.), acids or salts to achieve protein precipitation and to break protein–metabolite binding. Other strategies such as ultrafiltration and SPE with special cartridges are also useful to remove proteins (Vuckovic, 2012). Tissues are typically washed (usually with phosphate-buffered saline), quenched in liquid nitrogen, ground and homogenized, and extracted with hydroorganic mixtures of solvents or using SPE (Vuckovic, 2012). In the case of urine, direct injection into the separation system after centrifugation, filtration, and dilution is usually applied. When a preconcentration step is required, SPE (with C18 stationary phases) or LLE (usually with ethyl acetate or ethyl ether) are recommended (Fernández-Peralbo and Luque de Castro, 2012). Regarding feces, the addition of antimicrobial agents (e.g., NaN_3) to the homogenized samples followed by solid-liquid extraction with adequate solvents is commonplace (Deda *et al.*, 2015).

Two main approaches can be used when polyphenol determination is intended (Lu *et al.*, 2015; Scalbert *et al.*, 2002; Vacek *et al.*, 2010). The first pursues the determination of the total polyphenol content in homogenates and, therefore, requires the hydrolysis (either enzymatic or chemical) of conjugate forms. It is very common to use an enzyme mixture of β-D-glucuronidase and sulfatase. The sample without conjugates will be further treated in an acidified solvent to release the polyphenols bound to proteins. This preparation containing aglycones will be centrifuged and filtered and afterwards analysed. The second approach, which is complementary, tries to evaluate both free polyphenols and their metabolites (conjugates). In this case, hydrolysis and elimination of proteins are not required stages of the sample treatment. The resulting extract will be more complex than the previous one; thus, the analysis of the second preparation will probably need to use more sensitive methodologies. The two described approaches are presented in Figure 3.3.

At the end of the process, polyphenols coming from any of the described preparations can be further purified by means of a

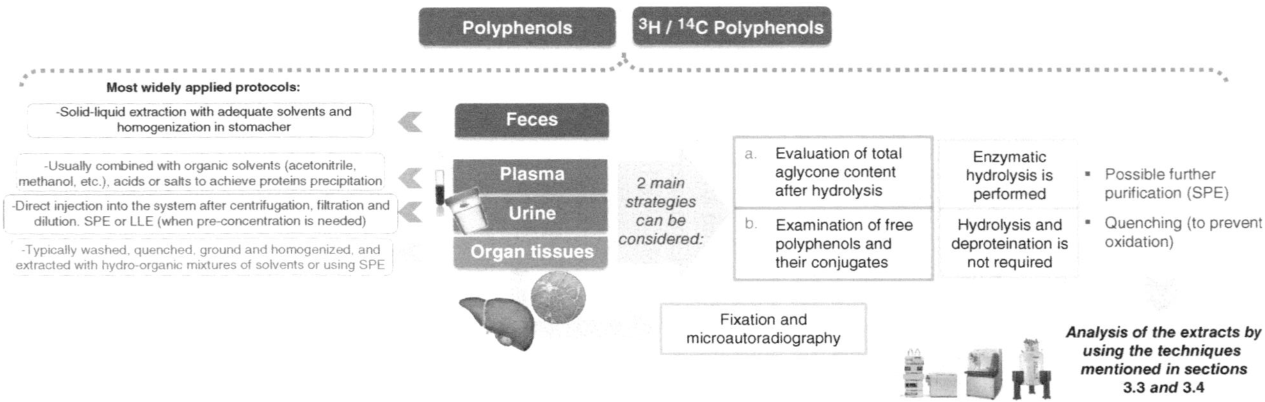

Figure 3.3 Experimental schema for the analysis of phenolic compounds from foods and beverages in biological samples.

variety of protocols, SPE being the most recurrent (with sorbents based on RP or ion-exchange interactions). The use of reducing agents (such as ascorbic acid) to prevent oxidation is also a common practice, so-called quenching. Additional details about purification, quenching or other aspects concerning sample conservation and stability evaluation can be found in the following reports (Álvarez-Sánchez *et al.*, 2010; Dietmair *et al.*, 2010; Saude and Sykes, 2007; Volmer *et al.*, 2011).

The analytical techniques that we have enumerated here can be applied to the analysis of biological extracts. Indeed, these techniques have been applied in the study of absorption, metabolism, and excretion of diverse families of phenolic substances. Several noteworthy examples have been included in Table 3.3.

When we focus on methods aiming to monitor the tissue uptake of polyphenols, most of the applications try to assess the distribution of phenolic compounds in various organs in rodent models after ingestion as a part of their diet. For instance, polyphenols radioactively labeled with ^{3}H or ^{14}C have been used in *in vivo* studies and their distribution directly investigated in fixed tissue by means of microautoradiography. In other projects, the phenolic substances are autoradiographically quantified in the extracts obtained from homogenized tissue samples. When LC (with diverse detectors) is selected as the analytical platform, individual radioactive metabolites might be identified in appointed fractions (Vacek *et al.*, 2010).

3.5 Conclusions and Future Directions

The many reports on the determination of polyphenols in food and biological samples provide a very reliable indication of the relevance of this family of substances. To determine these analytes in food or biological samples is quite challenging, bearing in mind the great variety of compounds that can be present in a single sample and their heterogeneity in terms of structure, polarity, and relative abundance.

In this chapter, we have described the current situation in this field, paying attention to the different steps of the analytical procedure. LLE and SPE remain the most popular strategies for sample treatment, but they co-exist with other innovative

Table 3.3 Some illustrative examples of polyphenol determination in biological samples

Matrix (purpose)	Analytical technique	Sample preparation	Conditions of analysis	Analytes	Reference
Urine (metabolism)	LC-ESI-QTOF MS	– Samples frozen at −80 °C – Thawed samples centrifuged (5 min at 12 000 g); 50 μL of supernatant diluted with 50 μL water (vortex mixed)	– Luna RP 18 column (2 × 50 mm, 5 μm particle size, 100 Å pore size) – Water with 0.1% formic acid (phase A) and acetonitrile with 0.1% formic acid (phase B); gradient (1): 0 min, 1% B; 4 min, 20% B; 6 min, 95% B (kept for 1.5 min); 8 min, 1% B (kept for 4 min) – 15 μL of injection volume – 600 μL/min flow rate – MS in positive ion mode; 70–700 m/z in full scan	Fingerprint of the products of grape skin polyphenol metabolism	(Khymenets *et al.*, 2015)
Feces (metabolism)	LC-ESI-QTOF MS	– Fresh feces + sodium phosphate buffer (homogenized in a stomacher bag for 30 s); slurry filtration (nylon mesh)	– Kinetex F5 column (2.1 × 100 mm, 1.7 μm particle size, 100 Å pore size) – Water with 0.1% formic acid (phase A) and methanol (phase B); gradient (1): 0 min, 20% B (kept 3 min); 6 min, 50% B (kept for 3 min); 15 min, 100% B; 15–20 min, 20% B	6 stilbenoids and their metabolic transformation products	(Jarosova *et al.*, 2019)

		– Liquid-liquid extraction: 400 μL supernatant (centrifuged fermented fecal slurry) diluted with 2 mL water + 2.5 mL ethyl acetate (three extraction steps); dryness → 1 mL methanol with 0.1% formic acid	– 5 μL of injection volume – 0.2 mL/min flow rate – MS in negative ion mode		
Plasma (metabolism)	a) LC-ESI-TOF MS for unknown metabolite characterization b) LC-ESI-QqQ MS/MS for metabolite quantification	– Samples frozen at −80 °C – SPE extraction: 200 μL sample + 300 μL acidified saline (9 g of NaCl in 1 L of water with 0.1% formic acid) (vortex); application in an Oasis HLB column SPE cartridge (30 mg); washing with (1) 1 mL 1.5 M formic acid and (2) 1 mL water/methanol (95:5, v/v); elution with 2 mL methanol with 0.1% formic acid	– XBridge BEH shield RP-C18 XP column (2.1 × 100 mm, 2.5 μm particle size) – Water with 0.1% formic acid (phase A) and acetonitrile with 0.1% formic acid (phase B); gradient (a): 0 min, 5% B; 12 min, 15% B (kept for 3 min); 24 min, 40% B; 30 min, 50% B; 31 min, 5% B (kept for 5 min)/gradient (b): 0 min, 10% B; 9 min, 35% B; 10 min, 65% B; 12 min, 10% B (kept for 3 min) – 10 μL of injection volume – 0.26 mL/min flow rate – MS in negative ion mode (+3500 V); TOF 100–1000 m/z; QqQ working in SRM mode	7 flavan-3-ols derivatives (methylated and glucuronidated catechin and epicatechin)	(Blount *et al.*, 2015)

Table 3.3 (Continued)

Matrix (purpose)	Analytical technique	Sample preparation	Conditions of analysis	Analytes	Reference
	$^{1}H/^{13}C$ NMR for unknown metabolite characterization	SPE extract evaporated and dissolved in CD_3OD (5 mm Shigemi tube)	Bruker Avance DRX 500 MHz spectrometer; 2D NMR experiments using HSQC, HMBC and NOESY		
Urine (metabolism)	^{1}H NMR	– Samples frozen at −80 °C – Thawed samples centrifuged (5 min at 13 200 rpm); 600 μL supernatant + 120 μL potassium phosphate buffer 1.5 M in D_2O (pH 7.0) with 0.1% 3-(trimethylsilyl) propionate-2,2,3,3-d_4 and sodium azide (2 mM); 720 μL transferred to a 5 mm NMR tube	Varian Inova 500 MHz NMR spectrometer; NOESY experiments: 298 K internal temperature; 128 transients collected into 32k data points; 14 ppm spectral width; 2 s acquisition time; 100 ms mixing time; 5 s relaxation delay.	Untargeted study of catechins and their conjugated metabolites	(Madrid-Gambin *et al.*, 2019)
Urine (metabolism)	GC-EI-Q MS	– Samples acidified with 5% ascorbic acid, centrifuged (10 min, 2000 g, 4 °C) and stored at −80 °C	– HP-5MS column (30 m, 0.25 mm i.d. and 0.25 μm film thickness) – 1 μL of injection volume (splitless)	Metabolic profile (114 metabolites: 17 amino acids + 47 organic acids + 7 other compounds + 43 unknown).	(Teixeira *et al.*, 2019)

		– 150 µL sample + 20 µL NaOH 5 M + 167 µL methanol + 34 µL pyridine (mixed); derivatization with 2 × 20 µL methyl chloroformate	– Carrier gas: He 0.71 mL/min – Temperature: 290 °C (injector), 250 °C (source) – T gradient: 45 °C for 2 min, 9 °C min^{-1} until 180 °C (kept 5 min), 40 °C min^{-1} until 220 °C (kept 5 min), 40 °C min^{-1} until 240 °C (kept 11.5 min), 40 °C min^{-1} until 280 °C (kept 10 min) – EI (70 eV) MS in positive mode; 38–5500 m/z in full scan	Metabolic profile (114 metabolites: 17 amino acids + 47 organic acids + 7 other compounds + 43 unknown).	(Teixeira *et al.*, 2019)
Tissues (biodistribution of radiolabeled compounds)	1) Gamma-counter measurements 2) Microauto-radiography	1) Injection of 200 µL of ^{125}I-(-)-epigallocatechin gallate via tail vein; mice sacrifice; collection of organs 2) Organs fix in 4% paraformaldehyde, paraffin embedding; tissue sections (10 µm) coated with nuclear emulsion (48 h in a darkroom at −20 °C); slides were developed with D-19 developer; tissues were stained with hematoxylin-eosin	1) Radioactivity detected in organ tissues by using a gamma-counter 2) Silver grains were observed by microscopic examination	^{125}I-(-)-epigallocatechin gallate	(Diao *et al.*, 2014)

HMBC, heteronuclear multiple bond correlation spectroscopy; HSQC, heteronuclear single quantum coherence spectroscopy; NOESY, nuclear Overhauser effect spectroscopy; SRM: selective reaction monitoring mode.

approaches. LC, CE, and GC, mainly coupled to MS, are among the most widely used techniques for both structural characterization and quantification of polyphenols, although other detection systems (e.g., fluorescence detector and PDA for CE or LC, and FID for GC) are still commonplace. As far as MS is concerned, the use of diverse ionization sources as well as different MS analyzers (of low and high resolution) produces complementary information that can be very useful for the comprehensive characterization of a sample.

In the past, the development of targeted methodologies aimed at the reliable quantification of particular polyphenols was predominantly pursued. In contrast, the trend nowadays is to develop innovative and powerful methods to determine a great variety of polyphenols (or their metabolites) within a single run (nontargeted approaches) with the aim of characterizing, classifying or authenticating the samples under study.

Acknowledgments

L. O.-G. wishes to thank the University of Granada for a postdoctoral research grant “Ayuda puente” from the Special Research Programme.

References

Acunha, T., Ibañez, C., Garcia-Cañas, V., Simo, C., and Cifuentes, A. (2016) Recent advances in the application of capillary electromigration methods for food analysis and Foodomics. *Electrophoresis*, **37**, 111–41.

Aires, A. (2017) Phenolics in foods: extraction, analysis and measurements. In M. Soto-Hernández, M. Palma Tenango and R. García-Mateos (Eds.), *Phenolic Compounds – Natural Sources, Importance and Applications*. London: InTechOpen.

Ajila, C.M., Brar, S.K., Verma, M., Tyagi, R.D., Godbout, S., and Valéro, J.R. (2011) Extraction and analysis of polyphenols: recent trends. *Critical Reviews in Biotechnology*, **31**, 227–49.

Alfieri, M., Cabassi, G., Habyarimana, E., Quaranta, F., Balconi, C., and Redaelli, R. (2019) Discrimination and prediction of polyphenolic compounds and total antioxidant capacity in sorghum grains. *Journal of Near Infrared Spectroscopy*, **27**, 46–53.

Alvarez, G., Montero, L., Llorens, L., Castro-Puyana, M., and Cifuentes, A. (2018) Recent advances in the application of capillary electromigration methods for food analysis and Foodomics. *Electrophoresis*, **39**, 136–59.

Álvarez-Sánchez, B., Priego-Capote, F., and Luque de Castro, M.D. (2010) Metabolomics analysis II. Preparation of biological samples prior to detection. *Trends in Analytical Chemistry*, **29**, 120–7.

Ameer, K., Shahbaz, H.M., and Kwon, J. (2017) Green extraction methods for polyphenols from plant matrices and their byproducts: a review. *Comprehensive Reviews in Food Science and Food Safety*, **16**, 295–315.

Andrés-Lacueva, C., Medina-Remon, A., Llorach, R., *et al.* (2009) Phenolic compounds: chemistry and occurrence in fruits and vegetables. In L.A. de la Rosa, E. Alvarez-Parrilla, and G.A. González-Aguilar (Eds.), *Fruit and Vegetable Phytochemicals: Chemistry, Nutritional Value, and Stability*. Hoboken: Wiley-Blackwell.

Antolovich, M., Prenzler, P., Robards, K., and Ryan, D. (2000) Sample preparation in the determination of phenolic compoundsin fruits. *Analyst*, **125**, 989–1009.

Association of Official Analytical Chemists (AOAC). (1980) *Official Methods of Analysis*, 12th edition. Washington, DC: AOAC.

Arries, W.J., Tredoux, A.G.J., de Beer, D., Joubert, E., and de Villiers, A. (2017) Evaluation of capillary electrophoresis for the analysis of rooibos and honeybush tea phenolics. *Electrophoresis*, **38**, 897–905.

Arslan, M., Zou, X.B., Tahir, H.E., Zareef, M., Hu, X.T., and Rakha, A. (2019) Total polyphenol quantitation using integrated NIR and MIR spectroscopy: a case study of Chinese dates (*Ziziphus jujuba*). *Phytochemical Analysis*, **30**, 357–63.

Ayres, D.C. and Loike J.D. (1990) *Lignans: Chemical, Biological and Clinical Properties*. New York: Cambridge University Press.

Azmir, J., Zaidul, Rahman, M.M., *et al.* (2013) Techniques for extraction of bioactive compounds from plant materials: a review. *Journal of Food Engineering*, **117**, 426–36.

Beale, R.N. and Roe, E.M.F. (1953) Ultra-violet absorption spectra of trans- and cis-stilbenes stilbenes and their derivatives. Part I. trans- and cis-stilbenes. *Journal of the Chemical Society (Resumed)*, 2755–63.

Belwal, T., Ezzat, S.M., Rastrelli, L., *et al.* (2018) A critical analysis of extraction techniques used for botanicals: trends, priorities, industrial uses and optimization strategies. *Trends in Analytical Chemistry*, **100**, 82–102.

Blount, J.W., Redan, B.W., Ferruzzi, M.G., *et al.* (2015) Synthesis and quantitative analysis of plasma-targeted metabolites of catechin and epicatechin. *Journal of Agricultural and Food Chemistry*, **63**, 2233–40.

Bonatto, C.C. and Silva, L.P. (2015) Cocoa content influences chocolate molecular profile investigated by MALDI-TOF mass spectrometry. *Journal of the Science of Food and Agriculture*, **95**, 1753–6.

Bravo, L. (1998) Polyphenols: chemistry, dietary sources, metabolism, and nutritional significance. *Nutrition Reviews*, **56**, 317–33.

Cabrera-Banegil, M., Schaide, T., Manzano, R., Delgado-Adamez, J., Duran-Meras, I., and Martin-Vertedor, D. (2017) Optimization and validation of a rapid liquid chromatography method for determination of the main polyphenolic compounds in table olives and in olive paste. *Food Chemistry*, **233**, 164–73.

Cacciola, F., Dugo, P., and Mondello, L. (2017) Multidimensional liquid chromatography in food analysis. *Trends in Analytical Chemistry*, **96**, 116–23.

Cavaliere, C., Capriotti, A.L., La Barbera, G., Montone, C.M., Piovesana, S., and Laganà, A. (2018) Liquid chromatographic strategies for separation of bioactive compounds in food matrices. *Molecules*, **23**, 3091.

Cavaliere, C., Antonelli, M., Capriotti, A.L., *et al.* (2019) A triple quadrupole and a hybrid quadrupole Orbitrap mass spectrometer in comparison for polyphenol quantitation. *Journal of Agricultural and Food Chemistry*, **67**, 4885–6.

Chen, L., Cao, H., and Xiao, J. (2018) Polyphenols: absorption, bioavailability, and metabolomics. In C.M. Galanakis (Ed.), *Polyphenols: Properties, Recovery, and Applications*. Cambridge: Woodhead Publishing.

Contreras-Gutierrez, P.K., Hurtado-Fernandez, E., Gomez-Romero, M., Hormaza, J.I., Carrasco-Pancorbo, A., and Fernandez-Gutierrez, A. (2013) Determination of changes in the metabolic profile of avocado fruits (*Persea americana*) by two CE-MS approaches (targeted and non-targeted). *Electrophoresis*, **34**, 2928–42.

Cooper, H.J. and Marshall, A.G. (2001) Electrospray ionization Fourier transform mass spectrometric analysis of wine. *Journal of Agricultural and Food Chemistry*, **49**, 5710–18.

Cortés-Herrera, C., Artavia, G., Leiva, A., and Granados-Chinchilla, F. (2018) Liquid chromatography analysis of common nutritional components, in feed and food. *Foods*, **8**, 1.

Cory, H., Passarelli, S., Szeto, J., Tamez, M., and Mattei, J. (2018) The role of polyphenols in human health and food systems: a mini-review. *Frontiers in Nutrition*, **5**, 87.

Crevillen, A.G., Pumera, M., Gonzalez, M.C., and Escarpa, A. (2009) Towards lab-on-a-chip approaches in real analytical domains based on microfluidic chips/electrochemical multi-walled carbon nanotube platforms. *Lab Chip*, **9**, 346–53.

Crozier, A., Jaganath, I.B., and Clifford, M.N. (2009) Dietary phenolics: chemistry, bioavailability and effects on health. *Natural Product Reports*, **26**, 1001–43.

D'Archivio, M., Filesi, C., Vari, R., Scazzocchio, B., and Masella, R. (2010) Bioavailability of the polyphenols: status and controversies. *International Journal of Molecular Sciences*, **11**, 1321–42.

Da Silva, R.P.F.F., Rocha-Santos. T.A.P., and Duarte, A.C. (2016) Supercritical fluid extraction of bioactive compounds. *Trends in Analytical Chemistry*, **76**, 40–51.

Deda, O., Gika, H.G., Wilson, I.D., and Theodoridis, G.A. (2015) An overview of fecal sample preparation for global metabolic profiling. *Journal of Pharmaceutical and Biomedical Analysis*, **113**, 137–50.

Deslaurien, I. (2000) Recovery, separation and characterization of phenolic compounds and flavonoids from maple products. Master's thesis, McGill University, Montreal, Canada.

Diao, Y., Zhao, W., Li, Y., *et al.* (2014) Radiolabeling of EGCG with ^{125}I and its biodistribution in mice. *Journal of Radioanalytical and Nuclear Chemistry*, **301**, 167–73.

Diaz-de-Cerio, E., Aguilera-Saez, L.M., Gomez-Caravaca, A.M., *et al.* (2018) Characterization of bioactive compounds of *Annona cherimola* L. leaves using a combined approach based on HPLC-ESI-TOF-MS and NMR. *Analytical and Bioanalytical Chemistry*, **410**, 3607–19.

Dietmair, S., Timmins, N.E., Gray, P.P., Nielsen, L.K., and Krömer, J.O. (2010) Towards quantitative metabolomics of mammalian cells: development of a metabolite extraction protocol. *Analytical Biochemistry*, **404**, 155–64.

Euterpio, M.A., Pagano, I., Piccinelli, A.L., Rastrelli, L., and Crescenzi, C. (2013) Development and validation of a method for the determination of (E)-resveratrol and related phenolic compounds in beverages using molecularly imprinted solid phase extraction. *Journal of Agriculture and Food Chemistry*, **61**, 1640–5.

Fernández-Peralbo, M.A. and Luque de Castro, M.D. (2012) Preparation of urine samples prior to targeted or untargeted metabolomics mass-spectrometry analysis. *Trends in Analytical Chemistry*, **41**, 75–85.

Fontana, A.R. and Bottini, R. (2014) High-throughput method based on quick, easy, cheap, effective, rugged and safe followed by liquid chromatography-multi-wavelength detection for the quantification of multiclass polyphenols in wines. *Journal of Chromatography A*, **1342**, 44–53.

Fulcrand, H., Mané, C., Preys, S., *et al.* (2008) Direct mass spectrometry approaches to characterize polyphenol composition of complex samples. *Phytochemistry*, **69**, 3131–8.

Gabaston, J., Leborgne, C., Valls, J., *et al.* (2018) Subcritical water extraction of stilbenes from grapevine by-products: a new green chemistry approach. *Industrial Crops and Products*, **126**, 272–9.

Galarce-Bustos, O., Novoa, L., Pavon-Perez, J., Henriquez-Aedo, K., and Aranda, M. (2019) Chemometric optimization of QuEChERS extraction method for polyphenol determination in beers by liquid chromatography with ultraviolet detection. *Food Analytical Methods*, **12**, 448–57.

Garcia-Salas, P., Morales-Soto, A., Segura-Carretero, A., and Fernandez-Gutiérrez, A. (2010) Phenolic-compound-extraction systems for fruit and vegetable samples. *Molecules*, **15**, 8813–26.

Gatea, F., Teodor, E.D., Matei, A.O., Badea, G.I., and Radu, G.L. (2015) Capillary electrophoresis method for 20 polyphenols

separation in propolis and plant extracts. *Food Analytical Methods*, **8**, 1197–206.
Giacometti, J., Kovačević, D.B., Putnik, P., *et al.* (2018) Extraction of bioactive compounds and essential oils from mediterranean herbs by conventional and green innovative techniques: a review. *Food Research International*, **113**, 245–62.
Giusti, M.M. and Wrolstad, R.E. (2003) Acylated anthocyanins from edible sources and their applications in food systems. *Biochemical Engineering Journal*, **14**, 217–25.
Gomes, C., Sadoyan, G., Dias, R.C.S., and Costa, M.R.P.F.N. (2017) Development of molecularly imprinted polymers to target polyphenols present in plant extracts. *Processes*, **4**, 72.
Harborne, J. B. (1984) *Phytochemical Methods*, 2nd edition. London: Chapman and Hall.
Hurtado-Fernandez, E., Gomez-Romero, M., Carrasco-Pancorbo, A., and Fernandez-Gutierrez, A. (2010) Application and potential of capillary electroseparation methods to determine antioxidant phenolic compounds from plant food material. *Journal of Pharmaceutical and Biomedical Analysis*, **53**, 1130–60.
Hurtado-Fernandez, E., Contreras-Gutierrez, P.K., Cuadros-Rodriguez, L., Carrasco-Pancorbo, A., and Fernandez-Gutierrez, A. (2013) Merging a sensitive capillary electrophoresis-ultraviolet detection method with chemometric exploratory data analysis for the determination of phenolic acids and subsequent characterization of avocado fruit. *Food Chemistry*, **141**, 3492–503.
Hurtado-Fernandez, E., Pacchiarotta, T., Mayboroda, O.A., Fernandez-Gutierrez, A., and Carrasco-Pancorbo, A. (2014) Quantitative characterization of important metabolites of avocado fruit by gas chromatography coupled to different detectors (APCI-TOF MS and FID). *Food Research International*, **62**, 801–11.
Hurtado-Fernandez, E., Bajoub, A., Morales, J.C., Fernandez-Gutierrez, A., and Carrasco-Pancorbo, A. (2015a) Exploratory analysis of avocado extracts by GC-MS: new insights into the avocado fruit ripening process. *Analytical Methods*, **7**, 7318–26.
Hurtado-Fernandez, E., Pacchiarotta, T., Mayboroda, O.A., Fernandez-Gutierrez, A., and Carrasco-Pancorbo, A. (2015b)

Metabolomic analysis of avocado fruits by GC-APCI-TOF MS: effects of ripening degrees and fruit varieties. *Analytical and Bioanalytical Chemistry*, **407**, 547–55.

Hurtado-Fernandez, E., Gonzalez-Fernandez, J.J., Hormaza, J.I., Bajoub, A., Fernandez-Gutierrez, A., and Carrasco-Pancorbo, A. (2016) Targeted LC-MS approach to study the evolution over the harvesting season of six important metabolites in fruits from different avocado cultivars. *Food Analytical Methods*, **9**, 3479–91.

Ignat, I., Volf, I., and Popa, V.I. (2011) A critical review of methods for characterisation of polyphenolic compounds in fruits and vegetables. *Food Chemistry*, **126**, 1821–35.

Ivanović, M., Razboršek, M.I., and Kolar, M. (2016) Simultaneous GC-MS determination of free and bound phenolic acids in slovenian red wines and chemometric characterization. *Acta Chimica Slovenica*, **63**, 661–9.

Jáč, P., Polášek, M., and Pospíšilová, M. (2006) Recent trends in the determination of polyphenols by electromigration methods. *Journal of Pharmaceutical and Biomedical Analysis*, **40**, 805–14.

Jarosova, V., Vesely, O., Marsik, P., *et al.* (2019) Metabolism of stilbenoids by human faecal microbiota. *Molecules*, **24**, 1155.

Jenab, M., Slimani, N., Bictash, M., Ferrari, P., and Bingham, S.A. (2009) Biomarkers in nutritional epidemiology: applications, needs and new horizons. *Human Genetics*, **125**, 507–25.

Jesus, M.S., Genisheva, Z., Romani, A., Pereira, R.N., Teixeira, J.A., and Domingues, L. (2019) Bioactive compounds recovery optimization from vine pruning residues using conventional heating and microwave-assisted extraction methods. *Industrial Crops and Products*, **132**, 99–110.

Kaufmann, A. (2014) Combining UHPLC and high-resolution MS: a viable approach for the analysis of complex samples? *Trends in Analytical Chemistry*, **63**, 113–28.

Kelly, N.P., Kelly, A.L., and O'Mahony, J.A. (2019) Strategies for enrichment and purification of polyphenols from fruit-based materials. *Trends in Food Science & Technology*, **83**, 248–58.

Khoddami, A., Wilkes, M.A., and Roberts, T.H. (2013) Techniques for analysis of plant phenolic compounds. *Molecules*, **18**, 2328–75.

Khymenets, O., Andres-Lacueva, C., Urpi-Sarda, M., *et al.* (2015) Metabolic fingerprint after acute and under sustained consumption of a functional beverage based on grape skin extract in healthy human subjects. *Food & Function*, **6**, 1288–98.

Kyle, J.A.M. and Duthie, G.G. (2005) Flavonoids in foods. In O.M. Andersen and K.R. Markham (Eds.), *Flavonoids. Chemistry, Biochemistry and Applications*. Boca Raton: CRC Press.

La Barbera, G., Capriotti, A.L., Cavaliere, C., *et al.* (2017) Liquid chromatography-high resolution mass spectrometry for the analysis of phytochemicals in vegetal-derived food and beverages. *Food Research International*, **100**, 28–52.

Laitila, J.E., Suvanto, J., and Salminen, J.-P. (2019) Liquid chromatography–tandem mass spectrometry reveals detailed chromatographic fingerprints of anthocyanins and anthocyanin adducts in red wine. *Food Chemistry*, **294**, 138–51.

Le Gall, G. and Colquhoun, I.J. (2003) NMR spectroscopy in food authentication. In M. Lees (Ed.), *Food Science and Technology*. Cambridge: Woodhead Publishing.

Li, Y.Y., Li, L.X., Cui, Y., Zhang, S.T., and Sun, B.S. (2017) Separation and purification of polyphenols from red wine extracts using high speed counter current chromatography. *Journal of Chromatography B*, **1054**, 105–13.

Ligor, M., Ratiu, I.A., Kiełbasa, A., Al-Suod, H., and Buszewski, B. (2018) Extraction approaches used for the determination of biologically active compounds (cyclitols, polyphenols and saponins) isolated from plant material. *Electrophoresis*, **39**, 1860–74.

Lu, D., Feng, C., Wang, D., *et al.* (2015) Analysis of twenty phenolic compounds in human urine: hydrochloric acid hydrolysis, solid-phase extraction based on K_2CO_3-treated silica, and gas chromatography tandem mass spectrometry. *Analytical and Bioanalytical Chemistry*, **407**, 4131–41.

Lucci, P., Saurina, J., and Núñez, O. (2017) Trends in LC-MS and LC-HRMS analysis and characterization of polyphenols in food. *Trends in Analytical Chemistry*, **88**, 1–24.

Madrid-Gambin, F., Garcia-Aloy, M., Vázquez-Fresno, R., *et al.* (2019) Metabolic signature of a functional high-catechin tea after acute and sustained consumption in healthy volunteers through ^{1}H NMR based Metabolomics analysis of urine. *Journal of Agricultural and Food Chemistry*, **67**, 3118–24.

Manach, C., Scalbert, A., Morand, C., Rémésy, C., and Jiménez, L. (2004) Polyphenols: food sources and bioavailability. *American Journal of Clinical Nutrition*, **79**, 727–47.

Manach, C., Williamson, G., Morand, C., Scalbert, A., and Remesy, C. (2005) Bioavailability and bioefficacy of polyphenols in humans. I. Review of 97 bioavailability studies. *American Journal of Clinical Nutrition*, **81**, 230S–242S.

Manach, C., Hubert, J., Llorach, R., and Scalbert, A. (2009) The complex links between dietary phytochemicals and human health deciphered by Metabolomics. *Molecular Nutrition & Food Research*, **53**, 1303–15.

Manzoor, M.F., Zeng, X.A., Rahaman, A., *et al.* (2019) Combined impact of pulsed electric field and ultrasound on bioactive compounds and FT-IR analysis of almond extract. *Journal of Food Science and Technology*, **56**, 2355–64.

Marrubini, G., Appelblad, P., Maietta, M., and Papetti, A. (2018) Hydrophilic interaction chromatography in food matrices analysis: an updated review. *Food Chemistry*, **257**, 53–66.

Marti, R., Valcarcel, M., Herrero-Martinez, J.M., Cebolla-Cornejo, J., and Rosello, S. (2017) Simultaneous determination of main phenolic acids and flavonoids in tomato by micellar electrokinetic capillary electrophoresis. *Food Chemistry*, **221**, 439–46.

McGoverin, C.M., Weeranantanaphan, J., Downey, G., and Manley, M. (2010) The application of near infrared spectroscopy to the measurement of bioactive compounds in food commodities. *Journal of Near Infrared Spectroscopy*, **18**, 87–111.

Molnár, S., Mikuska, K., Patonay, K., *et al.* (2017) Comparative studies on polyphenolic profile and antimicrobial activity of propolis samples selected from distinctive geographical areas of Hungary. *Food Science and Technology International*, **23**, 349–57.

Moreira, S.A., Alexandre, E.M.C., Pintado, M., and Saraiva, J.A. (2019) Effect of emergent non-thermal extraction technologies on bioactive individual compounds profile from different plant materials. *Food Research International*, **115**, 177–90.

Motilva, M.J., Serra, A., and Macià, A. (2013) Analysis of food polyphenols by ultra high-performance liquid chromatography coupled to mass spectrometry: an overview. *Journal of Chromatography A*, **1292**, 66–82.

Naczk, M. and Shahidi, F. (2004) Extraction and analysis of phenolics in food. *Journal of Chromatography A*, **1054**, 95–111.

Naczk, M. and Shahidi, F. (2006) Phenolics in cereals, fruits and vegetables: occurrence, extraction and analysis. *Journal of Pharmaceutical and Biomedical Analysis*, **41**, 1523–42.

Nikovaev, A.V., Kartsova, L.A., and Filimonov, V.V. (2015) A microfluidic chip for the determination of polyphenolic antioxidants. *Journal of Analytical Chemistry*, **70**, 983–8.

Nolvachai, Y. and Marriott, P.J. (2013) GC for flavonoids analysis: past, current, and prospective trends. *Journal of Separation Science*, **36**, 20–36.

Oroian, M. and Escriche, I. (2015) Antioxidants: characterization, natural sources, extraction and analysis. *Food Research International*, **74**, 10–36.

Panja, P. (2017) Green extraction methods of food polyphenols from vegetable materials. *Current Opinion in Food Science*, **17**, 173–82.

Pereira, D.T.V., Tarone, A.G., Cazarin, C.B.B., Barbero, G.F., and Martinez, J. (2019) Pressurized liquid extraction of bioactive compounds from grape marc. *Journal of Food Engineering*, **240**, 105–13.

Ramautar, R., Somsen, G.W., and de Jong, G.J. (2015) CE-MS for metabolomics: developments and applications in the period 2012–2014. *Electrophoresis*, **36**, 212–24.

Robards, K. and Antolovich, M. (1997) Analytical chemistry of fruit bioflavonoids. *A review. Analyst*, **122**, 11R–34R.

Rodrigues, I.D., Gomes, S.M.C., Fernandes, I.P.G., and Oliveira-Brett, A.M. (2019) Phenolic composition and total antioxidant capacity by electrochemical, spectrophotometric and HPLC-EC evaluation in Portuguese red and white wines. *Electroanalysis*, **31**, 936–45.

Sadtler, S.P. (1889) The debt of medical and sanitary science to synthetic chemistry. *Journal of the Franklin Institute*, **127**, 203–19.

Sánchez-Rangel, J.C., Benavides, J., Heredia, J.B., Cisneros-Zevallos, L., and Jacobo-Velázquez, D.A. (2013) The Folin–Ciocalteu assay revisited: improvement of its specificity for total phenolic content determination. *Analytical Methods*, **5**, 5990–9.

Sandoval-Ventura, O., Olguín-Contreras, L.F., and Cañizares-Macías, M.P. (2017) Total polyphenols content in white wines on a microfluidic flow injection analyzer with embedded optical fibers. *Food Chemistry*, **221**, 1062–8.

Şanli, S., Şanli, N., Ozkan, S.A., and Lunte, C. (2016) Development and validation of a green capillary electrophoretic method for determination of polyphenolic compounds in red wine samples. *Chromatographia*, **79**, 1351–8.

Santana, A.L. and Macedo, G.A. (2019) Effects of hydroalcoholic and enzyme-assisted extraction processes on the recovery of catechins and methylxanthines from crude and waste seeds of guarana (*Paullinia cupana*). *Food Chemistry*, **281**, 222–30.

Saude, E.J. and Sykes, B.D. (2007) Urine stability for metabolomic studies: effects of preparation and storage. *Metabolomics*, **3**, 19–27.

Saurina, J. and Sentellas, S. (2015) Determination of phenolic compounds in food matrices: applications to characterization and authentication. In O. Núñez, H. Gallart-Ayala, C.P.B., Martins, and P. Lucci (Eds.), *Fast Liquid Chromatography-Mass Spectrometry Methods in Food and Environmental Analysis*. London: Imperial College Press.

Scalbert, A., Morand, C., Manach, C., and Rémésy, C. (2002) Absorption and metabolism of polyphenols in the gut and impact on health. *Biomedicine & Pharmacotherapy*, **56**, 276–82.

Shahidi, F. and Naczk, M. (2004) *Phenolics in Food and Nutraceuticals: Sources, Applications and Health Effects*. Boca Raton: CRC Press.

Singleton, V.L. and Rossi, J.A. (1965) Colorimetry of total phenolics with phosphomolybdic-phosphotungstic acid reagents. *American Journal of Enology and Viticulture*, **16**, 144–58.

Spencer, J.P.E., Abd El Mohsen, M.M., Minihane, A.M., and Mathers, J.C. (2008) Biomarkers of the intake of dietary polyphenols: strengths, limitations and application in nutrition research. *British Journal of Nutrition*, **99**, 12–22.

Spisso, A., Gomez, F.J.V., and Silva, M.F. (2018) Determination of ellagic acid by capillary electrophoresis in Argentinian wines. *Electrophoresis*, **39**, 1621–7.

Stahl, W., van den Berg, H., Arthur, J., *et al.* (2002) Bioavailability and metabolism. *Molecular Aspects of Medicine*, **23**, 39–100.

Sun, S., Liu, M., He, J., Li, K., Zhang, X., and Yin, G. (2019) Identification and determination of seven phenolic acids in Brazilian green propolis by UPLC-ESI-QTOF-MS and HPLC. *Molecules*, **24**, 1791.

Teixeira, L.L., Dörr, F., Dias, C.T.S., *et al.* (2019) Human urine metabolomic signature after ingestion of polyphenol-rich juice of purple grumixama (*Eugenia brasiliensis Lam.*). *Food Research International*, **120**, 544–52.

Tura, D. and Robards, K. (2002) Simple handling strategies for the determination of biophenols in food and plants. *Journal of Chromatography A*, **975**, 71–93.

Vacek, J., Ulrichova, J., Klejdus, B., and Simanek, V. (2010) Analytical methods and strategies in the study of plant polyphenolics in clinical samples. *Analytical Methods*, **2**, 604–13.

Villa-Ruano, N., Ramirez-Meraz, M., Mendez-Aguilar, R., *et al.* (2019) ^{1}H NMR-based metabolomics profiling of ten new races from *Capsicum annuum* cv. serrano produced in Mexico. *Food Research International*, **119**, 785–92.

Volmer, M., Northoff, S., Scholz, S., Thüte, T., Büntemeyer, H., and Noll, T. (2011) Fast filtration for metabolome sampling of suspended animal cells. *Biotechnology Letters*, **33**, 495–502.

Vuckovic, D. (2012) Current trends and challenges in sample preparation for global metabolomics using liquid chromatography–mass spectrometry. *Analytical and Bionalytical Chemistry*, **403**, 1523–48.

Xavier, J.A., Barros Valentim, I., Camatari, F.O.S., *et al.* (2017) Polyphenol profile by UHPLC-MS/MS, anti-glycation, antioxidant and cytotoxic activities of several samples of propolis from the northeastern semi-arid region of Brazil. *Pharmaceutical Biology*, **55**, 1884–93.

Xu, C.C., Wang, B., Pu, Y.Q., Tao, J.S., and Zhang, T. (2017) Advances in extraction and analysis of phenolic compounds from plant materials. *Chinese Journal of Natural Medicines*, **15**, 0721–0731.

Ye, Q., Wang, H., and Xu, J. (2015) Application of NMR spectroscopy for the characterization of dietary polyphenols. In Atta-ur-Rahman and M.I. Chaudhary (Eds.), *Applications of NMR Spectroscopy*, volume 3. Sarja: Bentham Science Publishers.

Zhang, G., Hu, M., He, L., Fu, P., Wang, L., and Zhou, J. (2013) Optimization of microwave assisted enzymatic extraction of polyphenols from waste peanut shells and evaluation of its antioxidant and antibacterial activities in vitro. *Food and Bioproducts Processing*, **91**, 158–68.

4

Hydroxycinnamates

Iziar A. Ludwig, Laura Rubió, Alba Macià, and Maria P. Romero*

Departament de Tecnologia dels Aliments-Àrea Nutrició, Universitat de Lleida, Lleida, Spain

4.1 Introduction

Hydroxycinnamates (HCs) are ubiquitous in the plant kingdom and thus are present in all plant food groups. The major hydroxycinnamic acids are *p*-coumaric, caffeic, ferulic, and sinapic acids, caffeic and ferulic being the most important ones. These hydroxycinnamic acids are present in dietary foods in their free form but mainly esterified with organic acids (quinic or tartaric acid), polysaccharides (arabinoside or galactoside residues), and lipids (sterols), through their carboxylic group.

The main HC esters are the chlorogenic acids (CGAs), which include different groups of compounds and related isomers formed by esterification of one molecule of quinic acid and 1–3 molecules of hydroxycinnamic acids. The main groups of CGAs include caffeoylquinic acids (CQAs), dicaffeoylquinic acids (diCQAs), feruloyquinic acids (FQAs), *p*-coumaroylquinic acids (*p*-CoQAs), and mixed diesters of caffeoyl-feruloyl-quinic acids (Clifford, 2000; Clifford *et al.*, 2003). These esters are present in various fruits (apples, peaches, apricots, blueberries, and pears), and vegetables (artichoke, tomato, eggplant, and broccoli), and in beverages such as coffee and mate. Other minor HC esters include (i) caftaric acid (CTA) and feftaric

*Corresponding author.

Dietary Polyphenols: Metabolism and Health Effects, First Edition.
Edited by Francisco A. Tomás-Barberán, Antonio González-Sarrías, and Rocío García-Villalba.

acid (FTA), formed by esterification of one molecule of tartaric acid and one molecule of caffeic or ferulic acid, respectively (Clifford, 2000; Monagas *et al.*, 2005), present in wine, (ii) rosmarinic acid, which is formed by caffeic acid esterified with 3-(3,4-dihydroxyphenyl)lactic acid and is found in many herbs and spices such as thyme, rosemary and basil (Clifford, 2000), and (iii) γ-oryzanol, a mixture of ferulic acid esters of phytosterols, present in rice bran (Zhimin and Godberg, 1999). Hydroxycinnamates are also present in high amounts in cereals where they are covalently linked to polysaccharides (Acosta-Estrada *et al.*, 2014; van Hung, 2016). These latter together with coffee HCs (CGA) represent the major dietary sources of hydroxycinnamates.

Several epidemiological studies have linked hydroxycinnamate-rich food intake to a reduction in chronic health diseases such as cancer, cardiovascular diseases, and type 2 diabetes (Aune *et al.*, 2013; Bøhn *et al.*, 2012; Ludwig *et al.*, 2014; Nkondjock, 2009; Tang *et al.*, 2015). However, the exact action of these phenolic compounds has not yet been elucidated, principally because mechanistic studies have been focused on the action of hydroxycinnamates present in food without taking into account the complex metabolism they undergo once ingested.

In this chapter, the metabolism of the HCs in the upper and lower gastrointestinal tract is discussed and metabolic pathways are proposed. Bioavailability in different food matrixes and at different doses are described together with the main metabolites present in urine and plasma. Further, critically interpreted *in vivo* and *in vitro* studies that have taken into account metabolism of hydroxycinnamates and, thus, the actual bioactive compounds are discussed.

4.2 Metabolism of Hydroxycinnamates and Metabolic Pathways

After ingestion, hydroxycinnamates are subjected to different reactions which can take place in the upper gastrointestinal tract (stomach and small intestines) and/or in the large intestine. These reactions profoundly modify the structure of the phenolic compounds present in food, leading to the generation of many different compounds (metabolites) which are the actual compounds that reach the peripheral circulation and tissues. The chemical modifications that take place may improve the absorption and bioavailability of the phenolic compounds (Karakaya, 2004; Scalbert and Williamson, 2000).

The fate of the HCs – if they are absorbed in the small or large intestine – is highly influenced by the chemical structure (free or esterified forms) in the food (Scalbert *et al.*, 2012). Briefly, once these phenolics are ingested, principally free hydroxycinnamic acids undergo an intensive phase II metabolism in the intestinal epithelium (small intestine) and the liver (hepatic cells), and form mainly sulfated, glucuronidated, and/or methylated metabolites through the reaction of the enzymes sulfotransferase (SULF), uridine-5′-diphosphate glucuronosyl-transferases (UGT), and catechol-O-methyltransferase (COMT), respectively. Then, the generated metabolites are transported through blood bound mainly to albumin and distributed to different tissues and organs. Nevertheless, an important fraction of the hydroxycinnamates (principally esterified forms) are not absorbed in the upper gastrointestinal tract (GIT) and reach the colon (large intestine), where they are extensively metabolized by microbiota enzymes, resulting in a further modification of their chemical structure. The action of colonic bacteria involves hydrolysis of the ester bond and the transformation of the hydroxycinnamic acids into smaller phenolic acids. The resulting microbial catabolites may then be excreted in feces, undergo phase II metabolism locally and/or be absorbed and reach the liver, where they would also be subject to phase II metabolism before reentering the systemic blood circulation, and finally being excreted in urine.

As caffeic and ferulic acids are the major hydroxycinnamic acids in dietary foods, the metabolism described in this chapter is focused principally on these two phenolic acids and their principal esters. Figure 4.1 shows a schematic representation of the metabolic pathways and the reactions that take place by mammalian or microbial enzymes, from the ingestion of the phenolics to the generated metabolites in the small intestine or their passage to the large intestine. Table 4.1 summarizes

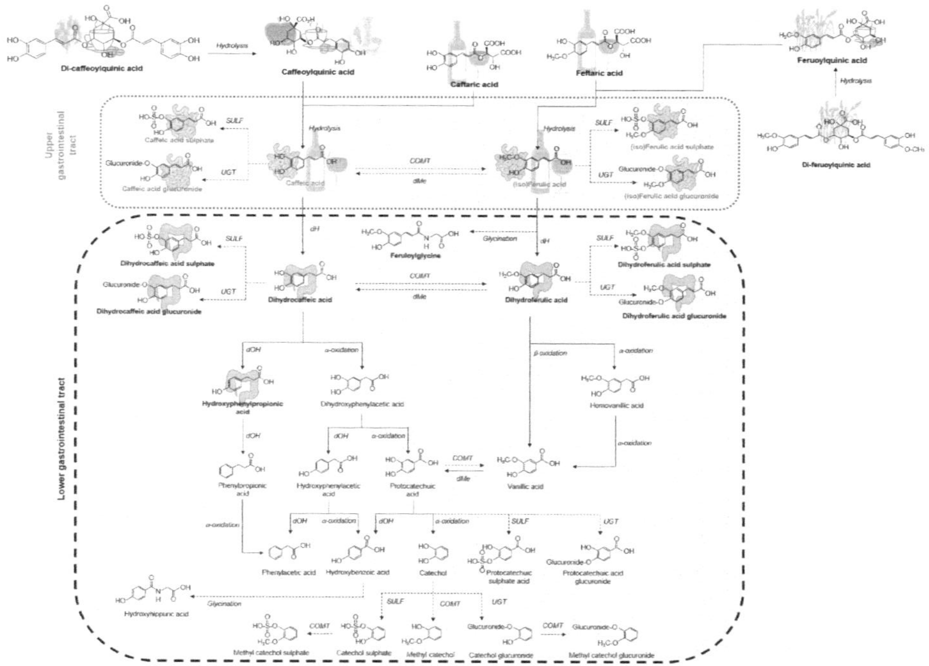

Figure 4.1 Metabolic pathways of major hydroxycinnamates in the diet (CQAs, diCQAs, FQAs, diFQAs, CTA and FTA). Dotted arrows represent transformations mediated by mammalian enzymes. Compound names in bold indicate major metabolites. COMT, catechol-O-methyltransferase-mediated methylation; dH, dehydrogenation; dMe, demethylation; dOH, dehydroxylation; SULF, sulfuryl-O-transferase-mediated sulfation; UGT, UDP glucuronosyl-transferase-mediated glucuronidation.

Table 4.1 Brief summary of the metabolism of hydroxycinnamic acids according to chemical structure

Dietary hydroxy-cinnamic acids	Dietary sources	Examples	Absorption sites	Metabolism (main enzymatic reactions)	Main metabolites
Monomers	Coffee Wine Cereals	Caffeic acid Ferulic acid	Small intestine Colon	Glucuronidation Methylation Sulfation Glycination Dehydrogenation Dehydroxylation Demethylation Decarboxylation	Caffeic acid sulfate Caffeic acid glucuronide Ferulic acid sulfate Ferulic acid glucuronide Feruloylglycine Dihydrocaffeic acid (sulfate) Dihydroferulic acid (sulfate) Hydroxyphenylpropionic acid (sulfate) Hydroxyhippuric acid
Dimers	Cereals Thyme	Diferulic acid Rosmarinic acid	Small intestine Colon	Hydrolysis Glucuronidation Methylation Sulfation Dehydrogenation Dehydroxylation Demethylation Decarboxylation	Caffeic acid sulfate Caffeic acid glucuronide Ferulic acid sulfate Methyl rosmarinic acid Dihydroxyphenylpropionic acid (sulfate)

(*continued*)

Table 4.1 (Continued)

Dietary hydroxy-cinnamic acids	Dietary sources	Examples	Absorption sites	Metabolism (main enzymatic reactions)	Main metabolites
Soluble esters	Coffee Apple Tomato Mate Artichoke Wine Rice	Caffeoylquinic acid Dicaffeoylquinic acid Feruoylquinic acid Diferuoylquinic acid Caftaric acid Feftaric acid γ-Oryzanol Feruoyl-saccharides	Small intestine Colon	Hydrolysis Glucuronidation Methylation Sulfation Dehydrogenation Dehydroxylation Demethylation Decarboxylation	The same metabolites as monomers
Bound insoluble forms (polymers)	Cereals	Feruoyl-polysaccharides	Colon	Hydrolysis Glucuronidation Sulfation Dehydrogenation Dehydroxylation Demethylation Decarboxylation	The same metabolites as monomers

metabolism of hydroxycinnamates according to their chemical structure and the main metabolites detected in urine and plasma after HC-rich food ingestion.

4.2.1 Absorption in the Upper Gastrointestinal Tract

After HC ingestion, the esterified forms are, at least partly, subjected to the action of lactase-phlorizin hydrolase (LPH) and mammalian esterase enzymes, respectively, in order to release their free forms. The free forms, and also a small amount of esterified forms, are absorbed in the upper gastrointestinal tract (small intestine) and extensively conjugated by phase II enzymes into methylated, sulfated, and glucuronidated metabolites (the latter to a lesser extent). The reported phenolic metabolites are caffeic acid sulfate, caffeic acid glucuronide, ferulic acid sulfate, and ferulic acid glucuronide. Caffeic acid can also be methylated producing (iso)ferulic acid, which in turn can generate the sulfated and glucuronidated forms (see Figure 4.1). Different authors have reported the identification of these plasma metabolites at early absorption times, from one to three hours, after the ingestion of coffee (Stalmach *et al.*, 2009), wine (Motilva *et al.*, 2016; Nardini *et al.*, 2009), apple (Yuste *et al.*, 2019), and tomato (Martinez-Huélamo *et al.*, 2015), indicating that HC absorption starts in the upper GIT.

As mentioned above, esterified forms, such as caffeoylquinic acid and feruoylquinic acid, are only partially absorbed in the upper gastrointestinal tract (Calinoiu and Vodnar, 2018; Zhao and Moghadasian, 2010). Different studies have reported this fact after the ingestion of coffee (Redeuil *et al.*, 2011; Renouf *et al.*, 2010; Stalmach *et al.*, 2009, 2010, 2014), and also in apples (Yuste *et al.*, 2019), artichoke (Wittemer *et al.*, 2005), tomato (Martinez-Huélamo *et al.*, 2015), and mate (de Oliveira *et al.*, 2017). This partial absorption was explained by the lack of esterase enzymes in human tissues (intestinal mucosa, liver) and biological fluids (plasma, gastric juice, duodenal fluid) able to completely hydrolyze these phenolic compounds, and also by the relatively slow hydrolysis (da Encarnacao *et al.*, 2015).

In the same way, when ferulic acid was esterified with mono-, di, and some oligosaccharides (Zhao *et al.*, 2003), or with sterols (γ-oryzanol) (Perez Ternero *et al.*, 2017), which are the common forms in cereals, only a part of the ferulic acid was released by the esterase enzymes.

4.2.2 Absorption in the Lower Gastrointestinal Tract

As HCs are present in dietary foods mostly bound to organic acids, polysaccharides and lipids through ester links, and due to the lack of esterases able to completely cleave these bonds in the small intestine, most of these compounds are not absorbed in the upper GIT but reach the colon, and are extensively metabolized by colonic microbiota in the large intestine (Stalmach *et al.*, 2009, 2010). The microbiota present in the colon has great catabolic activity and the main reactions that take place are hydrolysis (by cleavage of the ester bond), dehydrogenation (reduction of carbon-carbon double bond), decarboxylation (α-oxidation and β-oxidation), demethylation, and dehydroxylation. Subsequently, the resulting microbial catabolites may undergo phase II metabolism locally and/or be absorbed and reach the liver, where they are subject to enzymatic metabolism before reentering the systemic blood circulation and finally being excreted in urine.

In this regard, once the HCs reach the colon, the bacterial esterase enzymes cleave the ester bond yielding hydroxycinnamic acids in their free forms. Released ferulic acid may be absorbed and glycinated in the liver to form ferulouylglycine. However, most free phenolic acids are readily converted by the microbiota to the dihydroforms by reduction of the side-chain double bond by reductase activity (dehydrogenation), and the main metabolites generated are dihydrocaffeic acid and dihydroferulic acid. These metabolites can be absorbed through the colonic epithelium and conjugated by phase II enzymes, resulting in dihydroferulic acid and dihydrocaffeic acid sulfates and glucuronides (see Figure 4.1). These dihydrometabolites and feruloylglycine have been determined after the acute intake of coffee (Lang *et al.*, 2013; Stalmach *et al.*, 2009, 2010, 2014; Redeuil *et al.*, 2011; Renouf et al., 2010), apple (Yuste *et al.*, 2019), artichoke leaf extract (Wittemer *et al.*, 2005), tomato (Martinez-Huélamo *et al.*, 2015), and mate (de Oliveira *et al.*, 2017), resulting from (di)caffeoylquinic and (di)feruloylquinic acids present in the foods; after the intake of rice bran enzymatic extract (Perez Ternero *et al.*, 2017) resulting from γ-oryzanol; and bread (Bresciani *et al.* 2016) resulting from feruloyl-polysaccharides.

Further, the nonabsorbed dihydrometabolites may be further microbially metabolized to phenylpropionic acid, phenylacetic acid, and catechol derivatives. According to the results presented by different authors (Aura, 2008; Calinoiu and Vodnar, 2018; Clifford, 2000; Del Rio *et al.*, 2010; Karakaya, 2004; Mosele *et al.*, 2015; Scalbert and Williamson, 2000; Scalbert *et al.*, 2012; Zhao and Moghadasian, 2010), Figure 4.1 shows that dihydrocaffeic acid may be methylated, dehydroxylated, and decarboxylated (α-oxidation), and generate dihydroferulic acid, 3-hydroxyphenylpropionic acid, and dihydroxyphenylacetic acid, respectively. On the other hand, dihydroferulic acid may be demethylated, and generate dihydrocaffeic acid. (Dihydro)caffeic acid and (dihydro)ferulic acid, which is its methylated form, are interconvertible and share metabolites.

The 3-hydroxyphenylpropionic acid can be further metabolized and generate phenylpropionic acid by dehydroxylation, and then phenylacetic acid by α-oxidation. Dihydroxyphenylacetic acid can also be dehydroxylated to generate hydroxyphenylacetic acid and further decarboxylated (α-oxidation) to generate hydroxybenzoic acid, which can generate hydroxyhippuric acid by conjugation to glycine in the liver.

On the other hand, dihydroxyphenylacetic acid can also be decarboxylated and generate protocatechuic (α-oxidation) acid and catechol derivatives (α-oxidation). Dihydroxyphenylacetic acid has been reported to be a microbial metabolite of diferulic acid (Braune *et al.*, 2009; Koistinen *et al.*, 2017), which is present at high concentration levels in cereals. Additionally, methyl catechol, which was described as a metabolite of dihydroxyphenylacetic acid (generated by β-oxidation or two subsequent α-oxidations followed by a methylation) (Koistinen *et al.*, 2017), may also be linked to diferulic acid metabolism (see Table 4.1).

Additionally, protocatechuic acid can also be methylated and generate vanillic acid, which could also be formed by β-oxidation of dihydroferulic acid (Zhao and Moghadasian, 2008) (see Figure 4.1).

However, although all these catabolites have been identified in urine and/or plasma after HC consumption, as a result of the great catabolic activity of the microbiota, in terms of quantity, dihydrocaffeic acid, dihydroferulic acid, and hydroxyphenylpropionic acid have been reported to be the major microbial metabolites after the ingestion of HC-rich food sources such as coffee (Bresciani *et al.*, 2016; Gonthier *et al.*, 2003; Redeuil *et al.*, 2011; Renouf *et al.*, 2010; Stalmach *et al.*, 2009, 2010, 2014), cereals (Koistinen *et al.*, 2017; Mateo-Anson *et al.*, 2011), apple (Yuste *et al.*, 2019), artichoke (Wittemer *et al.*, 2005), mate (de Oliveira *et al.*, 2017), thyme extract (Mosele *et al.*, 2014), and rice bran enzymatic extract (Perez Ternero *et al.*, 2017). These catabolites, which are known to be better absorbed than their parent hydroxycinnamic esters, reach the liver via the portal vein, where further phase II metabolism takes place (see Table 4.1).

4.3 Bioaccessibility and Bioavailability of Hydroxycinnamates: Influence of Food Matrix, Processing, Dose, and Interindividual Differences

While bioaccessibility is defined as the amount of a bioactive compound potentially absorbable from the gut lumen and considers principally the transformations occurring during gastrointestinal passage of the food, bioavailability generally includes gastrointestinal digestion, absorption, metabolism, and tissue distribution (Holst and Williamson, 2008).

Bioaccessibility and bioavailability of a compound can be affected by many factors, some related to the food (e.g., chemical structure of the compound, food matrix, and dose), while others depend on the individual (e.g., gastric emptying, intestinal transit time, composition of the microbiota).

In the following sections, we will first discuss the reported bioavailability of the typical HCs in fruits, vegetables, and beverages, represented by simple esters with organic acids, the most abundant group being the chlorogenic acids (CGA), hydroxycinnamic esters with quinic acid. The bioavailability of hydroxycinnamic acids in cereals-based products, principally ferulic acid bound to polysaccharides, will be discussed separately because of their tight association to the food matrix and the consequently distinctive bioavailability.

4.3.1 Bioavailability of Hydroxycinnamates in Fruits, Vegetables, and Beverages

In vitro methods that simulate oral, gastric, and intestinal phases have been performed in various chlorogenic acid-rich foods to determine their bioaccessibility. Monente *et al.* (2015) analyzed the effects of *in vitro* digestion on the stability of chlorogenic acids from Robusta and Arabica coffee, including caffeoylquinic acids, feruloylquinic acids, and their respective lactones produced during the coffee-roasting process. Overall, 92% of the initial amount of chlorogenic acids was recovered after the last step of the digestion process, showing that chlorogenic acids in coffee are stable and thus potentially bioavailable in the stomach or small intestine. However, stability was not the same for all compounds. While caffeoylquinic and feruloylquinic acids were hardly degraded by the simulated gastrointestinal juices, only 17–19% and 45–59% of the lactones were recovered in Arabica and Robusta coffee, respectively. Lower recoveries rates of caffeoylquinic acids of 55–67% have been reported for blanched artichoke heads (d'Antouno *et al.*, 2015) and fried or griddled artichoke stems (Júaniz *et al.*, 2017), both rich sources of diCQAs. It is of note that in the latter study, raw artichoke stems were also analyzed and showed a much lower recovery rate of just 2%.

Although heat treatments are generally associated with a degradation of phenolic compounds in foods, results obtained so far point to an enhanced bioaccessibility of chlorogenic acids in the upper GIT after frying, griddling or blanching. A similar approximately 56–63% recovery rate of chlorogenic acids after *in vitro* digestion has been reported in apple juice, a rich source of 5-caffeoylquinic and 4-*p*-coumaroylquinic acids (Kahle *et al.*, 2011). Authors also observed an increase of minor coumaroylquinic isomers (5- and 3-*p*-CoQA) and caffeoylquinic isomers (3- and 4-CQA) due to interesterification reactions during the duodenal phase of the *in vitro* digestion, probably caused by the pH conditions during this phase (Deshpande *et al.*, 2014).

Although *in vitro* methods are a quick and inexpensive way to estimate bioaccessibility in the upper GIT, they cannot completely simulate *in vivo* conditions. More realistic results may be obtained from feeding studies with ileostomists and the analysis of ileostomy bag contents. Results so obtained provide information on substances that are absorbed in the small intestine, those that are metabolized before reaching the colon, and those that reach the colon unaltered.

In this sense, changes in the CQA amounts in the upper GIT have been investigated by feeding ileostomists coffee, cloudy apple juice, and an apple smoothie and analyzing the amounts excreted in ileal fluid over an eight-hour period (Erk *et al.*, 2014). With coffee and apple smoothie, there was an overall recovery of 76% and 77% of intake, respectively, while for cloudy apple juice it was only 26%. Further, in coffee, the ratios of the individual CQA isomers were not affected by interesterification reactions during passage through the proximal GIT. In contrast, whereas the two apple drinks contained only 5-CQA and 4-CQA isomers, the ileal fluid also contained 3-CQA and 1-CQA. The matrix in which the CQAs were ingested, therefore, appears to have had an impact on their fate as they pass through the upper GIT, with interesterification of CQAs occurring with the apple products but not coffee. Erk *et al.* (2014) suggest that the CQA profile of the coffee was closer to the interesterification equilibrium reported by Trugo and Macrae (1984) and so did not alter substantially during gastric transport.

However, more remarkable was the three-fold lower recovery of CQAs in ileal fluid after cloudy apple juice consumption compared with the apple smoothie. Both drinks showed a similar CQA content and differed only in that cloudy apple juice was pressed and unfiltered whereas the smoothie comprised 60% cloudy juice and 40% apple purée which contains a much higher proportion of cell wall constituents. Although neither plasma nor urine was analyzed in this study to confirm the point, it appears unlikely that it is a consequence of enhanced absorption in the small intestine of the cloudy apple juice CQAs.

In another study on ileostomists conducted by Stalmach *et al.* (2010), volunteers consumed coffee containing 385 μmol chlorogenic acids. In keeping with the previously mentioned results, similar overall recoveries of 71% were found in ileal fluids. In addition to the ileal fluid, urine samples were collected over 24 hours and bioavailability of chlorogenic acids was calculated as the percentage of the renal excretion of metabolites after ingestion. Renal excretion reported corresponded to 8.1% of the ingested chlorogenic acids, the main compounds being ferulic acid sulfate, caffeic acid sulfate, and isoferulic acid glucuronide, in keeping with the pathways presented in Figure 4.1 for upper GIT absorption and metabolism.

In a similar study conducted by Erk *et al.* (2012), volunteers drank coffee containing three different doses of chlorogenic acids, 1053 μmol (low), 2218 μmol (medium), and 4525 μmol (high), to further study possible dose dependency in upper GIT metabolism. Between 69% (high) and 77% (low) of the ingested chlorogenic acids were excreted via the ileal fluid with no dose-dependent significant differences observed. Renal excretion reported was 8% (high), 12% (medium), and 15% (low), pointing to a lower absorption rate at high doses. However, these differences were not statistically significant, mainly due to the great interindividual differences observed. More specifically, one volunteer showed a much longer GIT transit time and renal excretion was twice as high as for the other subjects at each trial (low, medium, and high dose). The authors concluded that the relation between the GIT transit time and renal excretion observed for this volunteer gives a hint that decelerated passage of CGA in the gastrointestinal tract leads to higher absorption.

In vitro and ileostomist studies discussed so far point to matrix-dependent degradation/transformation of the hydroxycinnamate esters during passage through the upper GIT and to a limited absorption of 8–15% with around two-thirds of the ingested hydroxycinnamates passing to the lower GIT. Studies on bioavailability of hydroxycinnamic esters in healthy volunteers which take into account microbiota-derived metabolites and colonic absorption have been performed principally with coffee. This may be explained, first, because coffee is the principal dietary source of hydroxycinnamates, and second, because microbiota-derived metabolites of hydroxycinnamates are common to many other (poly)phenol subclasses, making it impossible to distinguish between those originating from the parent HCs and other (poly)phenols present in food. Thus, accurate results on bioavailability, calculated as percentage of excreted metabolites to ingested amounts, can only be obtained from food sources that contain HC as the principal source of phenolic compounds, such as in the case of coffee.

Bioavailability was investigated in an acute feeding study in which human subjects drank coffee containing 412 μmol of chlorogenic acids and plasma and urine were collected 0–24 hours after ingestion (Stalmach *et al.*, 2009). A total of 17 hydroxycinnamate derivatives, mainly sulfate and glucuronide conjugates, were identified and quantified in the urine, corresponding to 29% of the total intake. Major components excreted 0–24 hours after coffee intake were dihydroferulic acid, its sulfate, dihydrocaffeic acid sulfate, ferulic acid sulfate, dihydroferulic acid glucuronide, and feruloylglycine. Almost 50% of the metabolites detected were excreted 5–8 hours post ingestion. Plasma analysis lead to the identification of 12 metabolites with low plasma C_{max} values peaking within one hour of intake, as a result of small intestinal absorption, detected for unmetabolized 5-CQA, three FQAs, two sulfated CQALs, and, at higher concentrations, caffeic acid sulfate (92 nmol/L) and ferulic acid sulfate (76 nmol/L). However, highest C_{max} values were observed for dihydroferulic acid (385 nmol/L), dihydroferulic acid sulfate (145 nmol/L), and dihydrocaffeic acid sulfate (325 nmol/L) peaking 4.7–4.8 hours post ingestion, indicative of large intestinal absorption and in keeping with the pathways illustrated in Figure 4.1. It is of note that variations in the urinary excretion and plasma concentrations of these latter metabolites were markedly higher than those described for small intestinal absorption. Results described by Stalmach *et al.* (2009), together with the one-half to one-third lower bioavailability reported in ileostomists (Erk *et al.*, 2012; Stalmach *et al.*, 2010), suggest that quantitatively much more absorption of chlorogenic acids and their metabolites occurs in the colon than in the small intestine.

Dose–response effects on bioavailability were investigated after low (412 μmol), medium (635 μmol), and high (795 μmol) amounts of chlorogenic acids were administered through coffee beverages to healthy volunteers (Stalmach *et al.*, 2014). Urine and plasma were analyzed 0–24 hours post ingestion. In urine, median amounts of metabolites excreted after 24 hours accounted for 24% (low), 25% (medium), and 15% (high), showing a trend towards reduced bioavailability of chlorogenic acids associated with the highest dose ingested. This trend was, however, not significant, mainly due to the high interindividual differences observed. However, when divided into metabolites absorbed in the small intestine (mainly caffeic acid sulfate, ferulic acid sulfate and isoferulic acid glucuronide) and large intestine (dihydroferulic acid, dihydrocaffeic acid, dihydroferulic acid sulfate and dihydrocaffeic acid sulfate), at the highest dose there was a significant trend towards lower absorption in the small intestine that was not reflected to the same extent with the colonic metabolites. In addition, some volunteers might be classified as "high excretors" and others as "low excretors," which further maintained this condition at all three doses ingested.

Much of this person-to-person variation probably reflects variation in the composition and biochemical competence of the gut microbiota. In this regard, several *in vitro* colonic fermentation studies, which aimed to explore the microbially mediated breakdown of HC esters in the colon, reported great interindividual differences when coffee (Ludwig *et al.*, 2013), artichoke (Juániz *et al.*, 2017), caffeoylquinic acid (Gonthier *et al.*, 2006; Tomás-Barberán *et al.*, 2014) and caftaric acid (Gonthier *et al.*, 2006) were incubated with human fecal samples. Results thus obtained showed that, overall, caffeic and/or ferulic acid appeared as first catabolites at early timepoints as transient peaks with rapid turnover to their dihydroforms (dihydrocaffeic and/or dihydroferulic acids), which represented major end metabolites together with 3-(3-hydroxy)phenylpropionic acid. Along with these compounds, several minor catabolites were detected including phenylpropionic, phenylacetic, benzoic, and catechol derivatives, in keeping with the pathways presented in Figure 4.1. However, all studies reported high interindividual differences with significant differences in catabolite amounts and profiles corroborating the gut-mediated person-to-person variation observed in *in vivo* bioavailability studies.

In summary, it can be concluded that *in vitro* digestion and *in vivo* ileostomist studies have shown that matrix (apple juice vs apple smoothie), processing (raw vs heat treated), and dose (low vs high) may impact on HC ester bioaccessibility in the upper GIT. However, bioavailability studies that take into account lower GIT absorption show that probably the most important factor influencing HC ester bioavailability is the microbially mediated breakdown of these compounds in the colon. The interindividual variation observed in the bioavailability of dietary (poly)phenols in general is an area of increasing interest because understanding of the variation in the capacity to absorb and metabolize these compounds is thought to be a key factor in obtaining better knowledge of the beneficial effects of plant bioactive compounds against diseases (Manach *et al.*, 2017).

4.3.2 Bioavailability in Cereal-Based Products

Cereals are a rich source of hydroxycinnamic acids (ferulic acid, caffeic acid, and sinapic acid) with ferulic acid (FA) representing 70–90% of the total HCs (Holtekjølen *et al.*, 2006). They are found principally covalently bound to oligosaccharides and polysaccharides in the bran of the grain, with a minor contribution of soluble free or conjugated compounds (Li *et al.*, 2008). Although the hydroxycinnamic–fiber linkage in cereal-based products is thought to mostly prevent the absorption of these compounds in the upper GIT (Manach *et al.*, 2004), some authors have suggested that hydroxycinnamic acids may be partially liberated from the food matrix and absorbed in the small intestine.

Among the wide range of cereal based-products, bread is one of the most consumed (Angelino *et al.*, 2017) and three studies performed *in vitro* digestion with white bread to determine the FA bioaccessibility in the upper GIT as the percentage recovered in the dialyzate after the duodenal step in relation to the original sample. Results showed in general low but contradictory percentages of recovery. While in the first study conducted by Mateo Anson *et al.* (2009a), no FA was detected in the dialyzate, a second study conducted by the same group reported a low 4.9% bioavailability of FA (Mateo Anson *et al.*, 2009b). The third study on FA bioaccessibility in white bread found an approximately two-fold higher 10.2% bioavailability of FA after *in vitro* digestion (Hemery *et al.*, 2010). Both latter studies compared the bioaccessibility of hydroxycinnamic acids in white bread in comparison to wholemeal bread. The wholemeal breads had a 9–12-fold higher initial FA content compared to the white bread due to the presence of the outer layers of the kernel, which include the bran and thus the main fraction of bound HC. Although the wholemeal breads delivered a much higher amount of hydroxycinnamic acids, interestingly, the bioaccessibility was lower when compared to white bread. Mateo Anson *et al.* (2009b) found 1.1% bioavailability for wholemeal bread vs 4.9% for white bread and Hemery *et al.* (2010) found 2.9% vs 10.2%.

This discrepancy might be the consequence of upper GIT bioavailability of only soluble free or conjugated compounds, as proposed by Rondini *et al.* (2004). If only free and conjugated HC are readily available for upper GIT absorption, these results support the consideration that free compounds are the major contributors to the bioaccessibility of HCs in the upper GIT, while HCs bound to undigested cell wall polysaccharides reach the colon unaltered. In this regard, several preprocessing techniques applied to wheat bran have been proposed to enhance cereal HC bioaccessibility in the upper GIT, including germination, fermentation, and enzymatic treatments (Angelino *et al.*, 2017). Bioaccessibility of HCs was investigated by Mateo Anson *et al.* (2009b) who compared *in vitro* digestion of three differently treated wholemeal breads with the same initial content of ferulic, caffeic, and sinapic acid. One was a wholemeal bread with native bran, one contained bran that had been fermented, and the third contained bran that was fermented and enzymatically treated with xylanase, β-glucanase, α-amylase, cellulase, and ferulic acid esterase. Compared to native wheat bran, fermented and fermented and enzymatically treated bran showed respectively two- and five-fold higher bioaccessibility of hydroxycinnamic acids in the dialyzate after *in vitro* digestion. The authors concluded that the bioprocessing techniques applied (fermentation and enzymatic treatment) led to the hydrolysis of wheat fiber polymers and a structural breakdown of bran cell walls, facilitating HC bioaccessibility in the upper GIT.

In vivo, however, the unabsorbed portion of HCs will pass to the large intestine where the enzymatic cocktail excreted by microbiota is able to hydrolyze the nonextractable bound hydroxycinnamic acids that human digestive enzymes cannot. Bioavailability of hydroxycinnamic acids in bread has been investigated to evaluate the effect of enrichment (aleurone-enriched bread with high content of bound HCs vs wholemeal bread) (Brescani *et al.*, 2016) and bioprocessing (bread containing fermented and enzymatically treated bran vs native bran) (Mateo Anson *et al.*, 2011).

In the first study, healthy volunteers consumed, on three separate days, a wholemeal bread (87 mg FA), and a commercial wheat bread enriched with 6% (w/w) aleurone fraction at two different servings of 94 and 190 g, containing respectively 43 mg and 87 mg total FA (Brescani *et al.* 2016). Plasma (0–24h) and urine (0–48h) were analyzed for FA metabolites. A total of nine metabolites related to HCs were identified and quantified in the urine samples, with high concentrations found of ferulic acid metabolites (ferulic acid sulfate, dihydroferulic acid sulfate, dihydroferulic acid glucuronide, and feruloylglycine). By observing the slopes of the 48-hour cumulative urine excretion curves, only ferulic acid sulfate showed a steep slope within the 0–3-hour time period, suggestive of early absorption in the small intestine, while dihydroferulic sulfate and glucuronide along with feruloylglycine, caffeic acid sulfate, phenylpropionic, dihydroxyphenylacetic, vanillic and hydroxybenzoic acid sulfates exhibited an increased rate of appearance within 3–48 hours after bread consumption, indicative of microbiota catabolism and slower release of less accessible fiber-linked phenolics. Bioavailability for the single portion aleurone-enriched commercial bred (43 mg FA) was 8%, evaluated by focusing specifically on its direct metabolites, such as ferulic acid sulfate, dihydroferulic acid sulfate, and dihydroferulic acid glucuronide. Intriguingly, no significant difference was found in urinary FA bioavailability between the double portion of aleurone-enriched and wholegrain breads (~5% and ~4%, respectively), both delivering 87 mg FA. The authors commented that the higher bioavailability derived from the lower ferulic consumption in the single compared to double portion of aleurone-enriched bread may be due to a reduction in the capacity to metabolize and absorb HCs as intake increases. Five metabolites were identified and quantified in plasma but only two could be strictly linked with FA metabolism: ferulic acid sulfate and dihydroferulic acid sulfate. The first peaked at one hour post consumption, related to upper GIT absorption, with mean C_{max} of 84 nmol/L for the control bread (87 mg FA), 56 nmol/L for the single portion aleurone-enriched bread (43 mg FA), and 77 nmol/L for the double portion aleurone-enriched bread (87 mg FA). Dihydroferulic acid sulfate showed a maximum peak at seven hours, in accordance with its colonic formation and lower C_{max} of 9 nmol/L for the control bread (87 mg FA), 10 nmol/L for the single portion aleurone-enriched bread (43 mg FA), and 12 nmol/L for the double portion aleurone-enriched bread (87 mg FA). Pharmacokinetic profiles were, however, not significantly different among the three tested bread products for either of the two plasma metabolites. Although interindividual differences were not discussed by the authors, high standard error of the means obtained for each metabolite, in urine and plasma, implies person-to-person variability in the metabolization of HC from cereals too, as has been discussed for HC esters in fruits, vegetables, and beverages.

Mateo Anson *et al.* (2011) conducted a similar study comparing wholemeal bread with native bran and wholemeal bread with bran that was fermented and enzymatically treated. Conjugated metabolites were not studied but urine samples were enzymatically treated with a mixture of β-glucuronidase and sulfatase to cleave the glucuronide and sulfate moieties of phase II metabolism and detect only the aglycones, to which bioavailability was accounted for. Results obtained showed that bread with bioprocessed bran showed a 10% FA bioavailability compared to 4% found in control bread with native bran, indicating that these techniques may enhance bioavailability of hydroxycinnamates in cereal-based products.

It can be concluded that HCs present in cereal-based products show low upper GIT absorption due to their covalent bond to cell wall polysaccharides. Only free and soluble conjugated HC may be absorbed in the small intestine while the vast majority of HCs pass to the large intestine where they are slowly hydrolyzed and liberated HC are further degraded by the gut microbiota to lower molecular weight compounds, which then undergo phase II metabolism. Overall bioavailability seems, however, to be lower for HC in cereal-based products than in HC esters present in fruits, vegetables, and beverages. Strategies to improve their bioavailability include germination, fermentation, and enzymatic treatment of wholemeal flour.

4.4 Biological Activity of Hydroxycinnamates and Their Derivatives

Estimating HC intake is the first step towards documenting the protective effects of these compounds against the risk of chronic diseases. In the European Prospective Investigation into Cancer and Nutrition (EPIC) study, Zamora-Ros *et al.* (2016) estimated HC intake for 36 037 subjects aged 35–74 years recruited in 10 European countries. An increasing south–north gradient of consumption was found and coffee was the main food source of hydroxycinnamates and accounted for 55.3–80.7% of the total HC intake. Ferulic acid, to a lesser extent, also contributes in great part to the HCs intake, which is mainly found complexed with polysaccharides in high-fiber foods, and consequently, other important dietary sources are wheat bran, whole-grain flour products, and cereals (Kroon *et al.*, 1997). Other minor food sources were fruits, vegetables, nuts, and seeds.

The effective *in vivo* potential of HCs depends not only on the quantity of the compound ingested but also the amount absorbed and/or metabolized, and consequently the plasma and/or tissue concentrations achieved after HC consumption. So, to have a biological effect, the HCs and/or their *in vivo* metabolites must reach an organ or tissue in sufficient concentration that is maintained for an adequate period of time (Ludwig *et al.*, 2014). We present in this section carefully designed and critically interpreted *in vivo* and *in vitro* studies that have taken these aspects into account and can provide valuable insights into the potential biological activity of HCs.

Focusing on the principal dietary source of HCs, which is coffee, the main components are CGAs, among which the caffeoylquinic acids (CQAs), especially 5-CQA, dominate along with lower amounts of feruloylquinic and dicaffeoylquinic acids. To test their biological activities, one must take into account that substantial amounts of the ingested CGAs are not absorbed and pass from the small to the large intestine and are likely to be confined to the gastrointestinal tract (Stalmach *et al.*, 2010). Effects elsewhere in the body are probably mediated by the metabolites generated in the small intestine wall and/or systemically after absorption, as well as colonic catabolites of CQAs absorbed in the lower GIT. So, regularly consumed foods rich in HCs are particularly intriguing for their potential beneficial effects in the GIT, where they can reach concentrations of hundreds of micromoles per liter (Clifford *et al.*, 2017; Halliwell *et al.*, 2005; Vitaglione *et al.*, 2012).

In this regard, it has been suggested that CGAs could act as chemopreventive agents in the GIT by modulating the expression of genes encoding enzymes involved in phase II metabolism, which form part of the endogenous antioxidant defenses, playing a critical role in conversion of reactive electrophiles and xenobiotics into less toxic products (Feng *et al.*, 2005). As recently reviewed, many *in vitro* studies indicate that CGAs might diminish the growth of human colorectal cancer cell lines, mainly by cell phase arrest and apoptosis boost (Bułdak *et al.*, 2018). Moreover, the anticarcinogenic effect of dihydrocaffeic and dihydroferulic acids against proliferation of Caco-2 and SW480 cell lines has been reported, which is of paramount importance since these compounds are the main metabolites detectable at the colonic level after the consumption of hydroxycinnamate-rich foods (Martini *et al.*, 2019). This study suggests that these metabolites could be the real active molecules involved in the reported biological effects of hydroxycinnamic acid-rich foods, especially those effects related to gastrointestinal pathologies such as colon cancer.

Concerning the HCs bound to cell wall materials, such as ferulic acid present mainly in whole grains, they also can survive gastrointestinal digestion to reach the colon intact, as previously discussed, where they might provide an antioxidant environment and exert a major anticarcinogenic action in the colon (Pérez-Jiménez and Saura-Calixto, 2005). Thus, bound antioxidant HCs might act along the whole length of the digestive tract by trapping oxidative compounds and soluble radicals, which are responsible for many common pathologies (Fardet, 2010). *In vivo*, ferulic acid also significantly blocked free radicals, therefore preventing the oxidative stress correlated with alcohol and polyunsaturated fatty acid (PUFA)-induced toxicity in rat model (Rukkumani *et al.*, 2004).

Considering that intestinal epithelial cells are exposed to high concentrations of HCs, these compounds may also be implicated in the prevention of intestinal inflammation. A recent study demonstrated that dietary CGAs and caffeic acid inhibited both

TNF-α- and H_2O_2-induced IL-8 production *in vitro* in Caco-2 cells, and attenuated DSS-induced colitis in mice (Shin *et al.*, 2015). The same authors have recently elucidated these antiinflammatory mechanisms, reporting that chlorogenic acids and caffeic acid inhibited activation of the PKD-IKK-NF-κB-IL-8 signaling pathway by scavenging intracellular ROS generated by oxidative stress (Shin *et al.*, 2017). This study also demonstrated that the antiinflammatory effects of CGA and caffeic acid were dependent on their catechol groups. Verzelloni *et al.* (2011) also measured the protective effect of a series of colonic catabolites, at physiological high nmol/L to low mmol/L concentrations, against oxidative stress induced in cultured human neuroblastoma SK-N-MC cells. A mixture of three CQA catabolites, dihydrocaffeic acid, dihydroferulic acid, and feruloylglycine, each at 0.5 mmol/L, induced significant protection.

Ferulic acid also exhibited effects on inflammatory messengers in several *in vitro* and *in vivo* models of intestinal inflammation. In an *in vitro* study, the significant antiinflammatory effect of the HCs present in wholemeal flour was reported in human colon cells (HT-29) by measuring the levels of IL-8 and TGF-1, resulting in an important intestinal health benefit (Laddomada *et al.*, 2015). *In vivo*, a clinical study based on a randomized controlled group of obese subjects underlined a similar antiinflammatory potential of HCs from wholewheat grain meals (Vitaglione *et al.*, 2015). According to this study, the wholewheat-based food resulted in an increased content of fecal ferulic and dihydroferulic acid.

Given that most phenolic acids diminish oxidative stress and inflammation, factors involved in the pathogenesis of type 2 diabetes (T2D), several human studies reported an inverse correlation between the intake of wheat bran (bran-enriched products; cereal fibers) and the risk of T2D with a 20–40% lower risk (Cho *et al.*, 2013; de Munter *et al.*, 2007; Montonen *et al.*, 2003). It is also suggested that HCs, specifically CGAs in coffee, could reduce the glucose absorption due to inhibition of glucose release by preventing hepatic glucose-6-phosphatase activity and inhibition of glucose absorption in the small intestine by preventing glucose-6-phosphate translocase 1 (Johnston *et al.*, 2003, 2005).

In terms of cardiovascular health, several studies suggest that HCs could improve several risk factors (Onakpoya *et al.*, 2015). CGAs may act as antithrombotic agents by inhibiting platelet activity; 5-CQA and caffeic acid showed antiplatelet activity with mice platelets *in vitro* at a concentration of 50 nM and *in vivo* after oral administration to mice at 400 mg per 30 g body weight by suppressing the expression of P-selectin, a biomarker for platelet activation (Park, 2009). For humans, while this intake is achievable after drinking 2–3 cups of a CQA-rich coffee, it will result in only transient, low nM concentrations of CQAs and caffeic acid derivatives in the bloodstream (Stalmach *et al.*, 2009) since CQAs are poorly absorbed in their intact form in the upper GIT and extensively degraded in the colon by the microbiota, as discussed previously.

Inhibition *in vitro* of platelet activity has been also reported for the CQA colonic catabolites dihydroferulic acid, dihydrocaffeic acid, and 3-(3-O-hydroxyphenyl)propionic acid at a concentration of 10 mM (Rechner and Kroner, 2005). Ferulic acid could also play an important role in antiinflammation by inhibiting cyclooxygenase (COX-2), prostaglandin E2 (PGE2), and TNF-α, as evidenced by improved vascular endothelial function both *in vivo* and *in vitro* (Ghosh *et al.*, 2017; Srinivasan *et al.*, 2007). Ferulic acid, together with caffeic acid and coumaric acid, could also reduce the level of HMG-CoA reductase, the key enzyme of cholesterol synthesis (Bhaskaragoud *et al.*, 2016; Ghosh *et al.*, 2017; Yeh *et al.*, 2009).

In accordance with these results, a recent study demonstrated that ferulic acid supplementation (1000 mg/day) can improve lipid profiles, oxidative stress, and inflammation in hyperlipidemic subjects (Bumrungpert *et al.*, 2018). A few *in vitro* mechanistic studies focused on the potential effects on cardiovascular health have been conducted with sulfated and glucuronidated colonic catabolites, which are the main circulating metabolites. Van Rymenant *et al.* (2017) compared the effects of ferulic acid and ferulic acid sulfate on vasorelaxation *ex vivo* using mouse saphenous artery, femoral artery, and aorta. While ferulic acid was inactive, its ferulic acid sulfate caused a concentration-dependent relaxation in all three tissues that were significant at sub-mmol/L concentrations. Amin *et al.* (2015) studied the antiinflammatory effects of ferulic acid and protocatechuic acid (3,4-dihydroxybenzoic acid) and its glucuronide and sulfate conjugates in human umbilical vein endothelial cells (HUVECs) stimulated with either oxidized LDL or a cluster of differentiation 40 ligand. Protocatechuic acid and its phase II metabolites were effective in modulating the production of key inflammatory mediators IL-6 and vascular cell adhesion molecule-1 (VCAM-1) at dietary relevant concentrations as low as 100 nmol/L, with maximum reduction observed for the sulfate conjugates.

In summary, it can be said that HCs are the most common dietary phenolic compounds, coffee and cereal-based products being the main sources. Once ingested, they undergo extensive metabolism starting in the upper GIT where mostly free hydroxycinnamic acids are absorbed and conjugated with sulfur, glucuronide or methyl groups before entering the bloodstream. However, most HCs pass unabsorbed to the colon where the catabolic activity of the microbiota mediates hydrolysis of the HC esters and further breakdown to lower molecular phenolic acids, which are more easily absorbed than the parent compounds and undergo further phase II metabolism. Bioavailability depends on many factors such as chemical structure, food matrix, processing technologies, and dose. Nonetheless, the most influential factor seems to be the composition and biochemical competence of the gut microbiota which reflects great interindividual differences. *In vivo* human and animal studies have connected consumption of HC-rich foods with a lower risk of suffering several chronic diseases such as type 2 diabetes, cardiovascular disease, and cancer. The beneficial health effects of HCs may be exerted by the parent compounds present in food but most probably by their degradation products generated as they pass through the gastrointestinal tract, as well as by their metabolites produced after absorption and phase II metabolism.

It is unfortunate that there are still comparatively few studies of relevant *in vitro* catabolites and metabolites at concentrations close to those known to occur in humans consuming real-world diets. Nevertheless, the results summarized here suggest that at least some gut flora catabolites and their phase II metabolites may indeed have antioxidant, chemopreventive, antiinflammatory, and antiplatelet effects at concentrations which could be related to the health benefits associated with a HC-rich diet.

References

Acosta-Estrada, B.A., Gutiérrez-Uribe, J.A., and Serna-Saldívar, S.O. (2014) Bound phenolics in foods, a review. *Food Chemistry*, **152,** 46–55.

Amin, H.P., Czank, C., Raheem, S., *et al.* (2015) Anthocyanins and their physiologically relevant metabolites alter the expression of IL-6 and VCAM-1 in CD40L and oxidized LDL challenged vascular endothelial cells. *Molecular Nutrition and Food Research*, **59**, 1095–1106.

Angelino, D., Cossu, M., Marti, A., *et al.* (2017) Bioaccessibility and bioavailability of phenolic compounds 1 in bread: a review. *Food and Function*, **8**, 2368–93.

Aune, D., Norat, T., Romundstad, P., and Vatten, L.J. (2013) Whole grain and refined grain consumption and the risk of type 2 diabetes: a systematic review and dose-response meta-analysis of cohort studies. *European Journal of Epidemiology*, **28**, 845–58.

Aura, A.M. (2008) Microbial metabolism of dietary phenolic compounds in the colon. *Phytochemistry Reviews*, **7**, 407–29.

Bhaskaragoud, G., Rajath, S., Mahendra, V., Kumar, G.S., Krishna, A.G., and Kumar, G.S. (2016) Hypolipidemic mechanism of oryzanol components-ferulic acid and phytosterols. *Biochemical and Biophysical Research Communications*, **476**, 82–9.

Bøhn, S.K., Ward, N.C., Hodgson, J.M., and Croft, C.D. (2012) Effects of tea and coffee on cardiovascular disease risk. *Food and Function*, **3**, 575–91.

Braune, A., Bunzel, M., Yonekura, R., and Blaut, M. (2009) Conversion of dehydrodiferulic acids by human intestinal microbiota. *Journal of Agricultural and Food Chemistry*, **57**, 3356–62.

Bresciani, L., Scazzina, F., Leonardi, R., *et al.* (2016) Bioavailability and metabolism of phenolic compounds from wholegrain wheat and aleurone-rich wheat bread. *Molecular Nutrition and Food Research*, **60**, 2343–54.

Bułdak, R.J., Hejmo, T., Osowski, M., Bułdak, L. Kukla, M., Polaniak, R., Birkner, E. (2018) The impact of coffee and its selected bioactive compounds on the development and progression of colorectal cancer in vivo and in vitro. *Molecules*, **23**, 3309.

Bumrungpert, A., Lilitchan, S., Tuntipopipat, S., Tirawanchai, N., and Komindr, S. (2018) Ferulic acid supplementation improves lipid profiles, oxidative stress, and inflammatory status in hyperlipidemic subjects: a randomized, double-blind, placebo-controlled clinical trial. *Nutrients*, **10**, art. no. 713.

Calinoiu, L.F. and Vodnar, D.C. (2018) Whole grains and phenolic acids: a review on bioactivity, functionality, health benefits and bioavailability. *Nutrients*, **10**, 1615.

Cho, S.S., Qi, L., Fahey, G.C., and Klurfeld, D.M. (2013) Consumption of cereal fiber, mixtures of whole grains and bran, and whole grains and risk reduction in type 2 diabetes, obesity, and cardiovascular disease. *American Journal of Clinical Nutrition*, **98**, 594–619.

Clifford, M.N. (2000) Chlorogenic acids and other cinnamates - nature, occurrence, dietary burden, absorption and metabolism. *Journal of the Science of Food and Agriculture*, **80**, 1033–43.

Clifford, M.N., Johnston, K.L., Knight, S., and Kuhnert, N. (2003) Hierarchical schene for LC-MSn identification of chlorogenic acids. *Journal of Agricultural and Food Chemistry*, **51**, 2900–11.

Clifford, M.N., Jaganath, I.B., Ludwig, I.A., and Crozier, A. (2017) Chlorogenic acids and the acyl-quinic acids: discovery, biosynthesis, bioavailability and bioactivity. Natural Products Reports, **34**, 1391–1421.

Da Encarnacao, J.A., Farrell, T.L., Ryder, A., Kraut, N.U., and Williamson, G. (2015) In vitro enzymatic hydrolysis of chlorogenic acids in coffee. *Molecular Nutrition and Food Research*, **59**, 231–9.

D'Antouno, I., Garbetta, A., Linsalata, V., Minerva, F., and Cardinale, A. (2015) Polyphenols from artichoke heads (Cynara cardunculus (L.) subsp. scolymus Hayek): in vitro bio-accessibility, intestinal uptake and bioavailability. *Food and Function*, **6**, 1268–77.

Del Rio, D., Stalmach, A., Calani, L., and Crozier, A. (2010) Bioavailability of coffee chlorogenic acids and green tea flavan-3-ols. *Nutrients*, **2**, 820–33.

De Munter, J.S.L., Hu, F.B., Spiegelman, D., Franz, M., and van Dam, R.M. (2007) Whole grain, bran, and germ intake and risk

of type 2 diabetes: a prospective cohort study and systematic review. *PLoS Medicine*, **4**, e261.
De Oliveira, D.M., Sampaio, G.R., Pinto, C.B., Catharino, R.R., and Bastos, D.H. (2017) Bioavailability of chlorogenic acids in rats after acute ingestion of maté tea (*Ilex paraguariensis*) or 5-caffeoylquinic acid. *European Journal of Nutrition*, **56**, 2541–56.
Deshpande, S., Jaiswal, R., Matei, M.F., and Kuhnert, N. (2014) Investigation of acyl migration in mono- and dicaffeoylquinic acids under aqueous basic, aqueous acidic, and dry roasting conditions. *Journal of Agricultural and Food Chemistry*, **62**, 9160–70.
Erk, T., Williamson, G., Renouf, M., *et al.* (2012) Dose-dependent absorption of chlorogenic acids in the small intestine assessed by coffee consumption in ileostomists. *Molecular Nutrition and Food Research*, **56**, 1488–1500.
Erk, T., Renouf, M., Williamson, G., Melcher, R., Steiling, H., and Richling, E. (2014) Absorption and isomerization of caffeoylquinic acids from different foods using ileostomist volunteers. *European Journal of Nutrition*, **53**, 159–166.
Fardet, A. (2010) New hypotheses for the health-protective mechanisms of whole-grain cereals: what is beyond fibre? *Nutrition Research Reviews*, **23**, 65–134.
Feng, R., Lu, Y., Bowman, L.L., Qian, Y., Castranova, V., and Ding, M. (2005) Inhibition of activator Protein-1, NF-kB, and MAPKs and induction of phase 2 detoxifing enzyme activity by chlorogenic acid. *Journal of Biological Chemistry*, **280**, 27888–95.
Ghosh, S., Basak, P., Dutta, S., Chowdhury, S., and Sil, P.C. (2017) New insights into the ameliorative effects of ferulic acid in pathophysiological conditions. *Food and Chemical Toxicology*, **103**, 41–55.
Gonthier, M.P., Verny, M.A., Besson, C., Rémésy, C., and Scalbert, A. (2003) Chlorogenic acid bioavailability largely depends on its metabolism by the gut microflora in rats. *Journal of Nutrition*, **133**, 1853–9.
Gonthier, M.P., Remesy, C., Scalbert, A., *et al.* (2006) Microbial metabolism of caffeic acid and its esters chlorogenic and caftaric acids by human faecal microbiota *in vitro*. *Biomedicine and Pharmacotherapy*, **60**, 536–40.

Halliwell, B., Rafter, J., and Jenner, A. (2005) Health promotion by flavonoids, tocopherols, tocotrienols, and other phenols: direct or indirect effects? Antioxidant or not? *American Journal of Clinical Nutrition*, **81**, 268S–276S.

Hemery, Y.M., Anson, N.M., Havenaar, R., Maenen., G.R., Noort, M.W., and Rouau, X. (2010) Dry fractionation of wheat bran increases the bioaccessibility of phenolic acids in breads made from processed bran fractions. *Food Research International*, **43**, 1429–38.

Holst, B. and Williamson, G. (2008) Nutrients and phytochemicals: from bioavailability to bioefficacy beyond antioxidants. *Current Opinion in Biotechnology*, **19**, 73–82.

Holtekjølen, A.K., Kinitz, C., and Knutsen, S.H. (2006) Flavanol and bound phenolic acid contents in different barley varieties. *Journal of Agricultural and Food Chemistry*, **56**, 2253–60.

Johnston, K.L., Clifford, M.N., and Morgan, L.M. (2003) Coffee acutely modifies gastrointestinal hormone secretion and glucose tolerance in humans: glycemic effects of chlorogenic acid and caffeine. *American Journal of Clinical Nutrition*, **78**, 728–33.

Johnston, K.L., Sharp, P., Clifford, M.N., and Morgan, M. (2005) Dietary polyphenols decrease glucose uptake by human intestinal Caco-2 cells. *FEBS Letters*, **579**, 1653–7.

Júaniz, I., Ludwig, I.A., Bresciani, L., *et al.* (2017) Bioaccessibility of (poly)phenolic compounds of raw and cooked cardoon (Cynara cardunculus L.) after simulated gastrointestinal digestion and fermentation by human colonic microbiota. *Journal of Functional Foods*, **32**, 195–207.

Kahle, K., Kempf, M., Schreier, P., *et al.* (2011) Intestinal transit and systemic metabolism of apple polyphenols. *European Journal of Nutrition*, **50**, 507–22.

Karakaya, S. (2004) Bioavailability of phenolic compounds. *Critical Reviews in Food Science and Nutrition*, **44**, 453–64.

Koistinen, V.M., Nordlund, E., Katina, K., *et al.* (2017) Effect of bioprocessing on the in vitro colonic microbial metabolism of phenolic acids from rye bran fortified breads. *Journal of Agricultural and Food Chemistry*, **65**, 1854–64.

Kroon, P.A., Faulds, C.B., Ryden, P, Robertson, J.A., and Williamson, G. (1997) Release of covalently bound ferulic acid from fiber in the human colon. *Journal of Agricultural and Food Chemistry*, **45**, 661–7.

Laddomada, B., Durante, M., Minervini, F., *et al.* (2015) Phytochemical composition and anti-inflammatory activity of extracts from the whole-meal flour of Italian durum wheat cultivars. *International Journal of Molecular Sciences*, **16**, 3512–27.

Lang, R., Dieminger, N., Beusch, A., *et al.* (2013) Bioappearance and pharmacokinetics of bioactives upon coffee consumption. *Analytical and Bioanalytical Chemistry*, **405**, 8487–503.

Li, L., Shewry, P.R., and Ward, J.L. (2008) Phenolic acids in wheat varieties in the HEALTHGRAIN Diversity Screen. *Journal of Agricultural and Food Chemistry*, **56**, 9732–9.

Ludwig, I.A., de Peña, M.P., Cid, C., and Crozier, A. (2013) Catabolism of coffee chlorogenic acids by human colonic microbiota. *BioFactors*, **39**, 623–32.

Ludwig I.A., Clifford M.N., Lean M.E.J., Ashihara H., and Crozier A. (2014) Coffee: biochemistry and potential impact on health. *Food and Function*, **5**, 1695–717.

Manach, C., Scalbert, A., Morand, C., Rémésy, C., and Jimenéz, L. (2004) Polyphenols: food sources and bioavailability. *American Journal of Clinical Nutrition*, **78**, 727–47.

Manach, C., Milenkovic, D., Van de Wiele, T., *et al.* (2017) Addressing the inter-individual variation in response to consumption of plant food bioactives: towards a better understanding of their role in healthy aging and cardiometabolic risk reduction. *Molecular Nutrition and Food Research*, **61**, 1600557.

Mateo Anson, N., Selinheimo, E., Havenaar., R., *et al.* (2009a) Bioprocessing of wheat bran improves in vitro bioaccessibility and colonic metabolism of phenolic compounds. *Journal of Agricultural and Food Chemistry*, **57**, 6148–55.

Mateo Anson, N., van den Berg, R., Havenaar, R., Bast, A., and Haenen, G.R.M. (2009b) Bioavailability of ferulic acid is determined by its bioaccessibility. *Journal of Cereal Sciences*, **49**, 296–300.

Mateo Anson, N., Aura, A.M., Selinheimo, E., *et al.* (2011) Bioprocessing of wheat bran in whole wheat bread increases the bioavailability of phenolic acids in men and exerts anti-inflammatory effects ex vivo. *Journal of Nutrition*, **141**, 137–43.

Martínez-Huélamo, M., Tulipani, S., Estruch, R., *et al.* (2015) The tomato sauce making process affects the bioaccessibility and bioavailability of tomato phenolics: a pharmacokinetic study. *Food Chemistry*, **173**, 864–72.

Martini, S., Conte, A., and Tagliazucchi, D. (2019) Antiproliferative activity and cell metabolism of hydroxycinnamic acids in human colon adenocarcinoma cell lines. *Journal of Agricultural and Food Chemistry*, **67**, 3919–31.

Monagas, M., Bartolomé, B., and Gómez-Cordovés, C. (2005) Updated knowledge about the presence of phenolic compounds in wine. *Critical Reviews in Food Science and Nutrition*, **45**, 84–118.

Monente, C., Ludwig, I. A., Stalmach, A., de Peña, M.P., Cid, C., and Crozier, A. (2015) In vitro studies on the stability in the proximal gastrointestinal tract and bioaccessibility in Caco-2 cells of chlorogenic acids from spent coffee grounds. *International Journal of Food Sciences and Nutrition*, **66**, 657–64.

Montonen, J., Knekt, P., Järvinen, R., Aromaa, A., and Reunanen, A. (2003) Whole-grain and fiber intake and the incidence of type 2 diabetes. *American Journal of Clinical Nutrition*, **77**, 622–9.

Mosele, J.I., Macià, A., and Motilva. M.J. (2015) Metabolic and microbial modulation of the large intestine ecosystem by non-absorbed diet phenolic compounds: a review. *Molecules*, **20**, 17429–68.

Mosele, J.I., Martín-Peláez, S., Macià, A., *et al.* (2014) Study of the catabolism of thyme phenols combining in vitro fermentation and human intervention. *Journal of Agricultural and Food Chemistry*, **62**, 10954–61.

Motilva, M.J., Macià, A., Romero, M.P., Rubió. L., Mercader, M., and González-Ferrero, C. (2016) Human bioavailability and metabolism of phenolic compounds from red wine enriched with free or nano-encapsulated phenolic extract. *Journal of Functional Foods*, **25**, 80–93.

Nardini, M., Forte, M., Vrhovsek, U., Mattivi, F., Viola, R., and Scaccini, C. (2009) White wine phenolics are absorbed and extensively metabolized in humans. *Journal of Agricultural and Food Chemistry*, **57**, 2711–18.

Nkondjock, A. (2009) Coffee consumption and the risk of cancer: an overview. *Cancer Letters*, **277**, 121–5.

Onakpoya, I., Spencer, E., Thompson, M., and Heneghan, C. (2015) The effect of chlorogenic acid on blood pressure: a systematic review and meta-analysis of randomized clinical trials. *Journal of Human Hypertension*, **29**, 77.

Park, B. (2009) 5-Caffeoylquinic acid and caffeic acid orally administered suppress P-selectin expression on mouse platelets. *Journal of Nutrition and Biochemistry*, **20**, 800–5.

Pérez-Jiménez, J. and Saura-Calixto, F. (2005) Literature data may underestimate the actual antioxidant capacity of cereals. *Journal of Agricultural and Food Chemistry*, **53**, 5036–40.

Perez-Ternero, C., Macià, A., Alvarez de Sotomayor, M., Parrado, J., Motilva, M.J., and Herrera, D. (2017) Bioavailability of the ferulic acid-derived phenolic compounds of a rice bran enzymatic extract and their activity against superoxide production. *Food and Function*, **8**, 2165–74.

Rechner, A.R. and Kroner, C. (2005) Anthocyanins and colonicmetabolites of dietary polyphenols inhibit platelet function. *Thrombosis Research*, **116**, 327–34.

Redeuil, K., Smarrito-Menozzi, C., Guy, P., *et al.* (2011) Identification of novel circulating coffee metabolites in human plasma by liquid chromatography-mass spectrometry. *Journal of Chromatography A*, **1218**, 4678–88.

Renouf, M., Guy, P.A., Marmet, C., *et al.* (2010) Measurement of caffeic and ferulic acid equivalents in plasma after coffee consumption: small intestine and colon are key sites for coffee metabolism. *Molecular Nutrition and Food Research*, **54**, 760–6.

Rondini, L., Peyrat-Maillard, M.N., Marsset, A., *et al.* (2004) Bound ferulic acid from bran is more bioavailable than the free compound in rat. *Journal of Agricultural and Food Chemistry*, **52**, 4338–43.

Rukkumani, R., Aruna, K., Varma, P.S., and Menon, V.P. (2004) Influence of ferulic acid on circulatory prooxidant-antioxidant status during alcohol and PUFA induced toxicity. *Journal of Physiology and Pharmacology*, **55**, 551–61.

Scalbert, A. and Williamson, G. (2000) Dietary intake and bioavailability of polyphenols. *Journal of Nutrition*, **130**, 2073S–2085S.

Scalbert, A., Morand, C., Mannach, C., and Rémésy, C. (2012) Absorption and metabolism of polyphenols in the gut and impact on heath. *Biomedicine and Pharmacotherapy*, **56**, 276–82.

Shin, H.S., Satsu, H., Bae, M.-J., *et al.* (2015) Anti-inflammatory effect of chlorogenic acid on the IL-8 production in Caco-2 cells and the dextran sulphate sodium-induced colitis symptoms in C57BL/6 mice. *Food Chemistry*, **168**, 167–75.

Shin H.S., Satsu H., Bae M.-J., Totsuka M., and Shimizu M. (2017) Catechol groups enable reactive oxygen species scavenging-mediated suppression of PKD-NFkappaB-IL-8 signaling pathway by chlorogenic and caffeic acids in human intestinal cells. *Nutrients*, **9**, art. no. 165.

Srinivasan, M., Sudheer, A.R., and Menon, V.P. (2007) Ferulic acid: therapeutic potential through its antioxidant property. *Journal of Clinical Biochemistry and Nutrition*, **40**, 92–100.

Stalmach, A., Mullen, W., Barron, D., *et al.* (2009) Metabolite profiling of hydroxycinnamate derivatives in plasma and urine after the ingestion of coffee by humans: identification of biomarkers of coffee consumption. *Drug Metabolism and Disposition*, **37**, 1749–58.

Stalmach, A., Steiling, H., Williamson, G., and Crozier, A. (2010) Bioavailability of chlorogenic acids following acute ingestion of coffee by humans with an ileostomy. *Archives of Biochemistry and Biophysics*, **501**, 98–105.

Stalmach, A., Williamson, G., and Crozier, A. (2014) Impact of dose on the bioavailability of coffee chlorogenic acids in humans. *Food and Function*, **5**, 1727–37.

Tang, G., Wang, D., Long, J., Yang, F., and Si, L. (2015) Meta-analysis of the association between whole grain intake and coronary heart disease risk. *American Journal of Cardiology*, **115**, 625–9.

Tomás-Barberán, F., García-Villalba, R., Quartieri, A., *et al.* (2014) In vitro transformation of chlorogenic acid by human gut microbiota. *Molecular Nutrition and Food Research*, **58**, 1122–31.

Trugo, L.C. and Macrae, R. (1984) A study of the effect of roasting on the chlorogenic acid composition of coffee using HPLC. *Food Chemistry*, **15**, 219–27.

Van Hung, P. (2016) Phenolic compounds of cereals and their antioxidant capacity. *Critical Reviews in Food Science and Nutrition*, **56**, 25–35.

Van Rymenant, E., Van Camp, J., Boydens, C., *et al.* (2017) Ferulic acid-4-O-sulfate rather than ferulic acid relaxes arteries and lowers blood pressure in mice. *Journal of Nutritional Biochemistry*, **44**, 44–51.

Verzelloni, E., Pellacani, C., Tagliazucchi, D., *et al.* (2011) Antiglycative and neuroprotective activity of colon-derived polyphenols catabolites. *Molecular Nutrition and Food Research*, **55**, S35–S43.

Vitaglione, P., Fogliano, V., and Pellegrini, N. (2012) Coffee, colon function and colorectal cancer. *Food and Function*, **3**, 916–22.

Vitaglione, P., Mennella, I., Ferracane, R., *et al.* (2015) Whole-grain wheat consumption reduces inflammation in a randomized controlled trial on overweight and obese subjects with unhealthy dietary and lifestyle behaviors: role of polyphenols bound to cereal dietary fiber. *American Journal of Clinical Nutrition*, **101**, 251–61.

Wittemer, S.M., Ploch, M., Windeck, T., *et al.* (2005) Bioavailability and pharmacokinetics of caffeoylquinic acids and flavonoids after oral administration of Artichoke leaf extracts in humans. *Phytomedicine*, **12**, 28–38.

Yeh, Y.H., Lee, Y.T., Hsieh, H.S., and Hwang, D.F. (2009) Dietary caffeic acid, ferulic acid and coumaric acid supplements on cholesterol metabolism and antioxidant activity in rats. *Journal of Food and Drug Analysis*, **17**, 123–32.

Yuste, S., Ludwig, I., Rubió. L., *et al.* (2019) In vivo biotransformation of (poly)phenols and anthocyanins of red-fleshed apple and identification of intake biomarkers. *Journal of Functional Foods*, **55**, 146–55.

Zamora-Ros R., Knaze V., Rothwell J.A., *et al.* (2016) Dietary polyphenol intake in europe: the European prospective investigation into cancer and nutrition (EPIC) study. *European Journal of Nutrition*, **55**, 1359–75.

Zhao, Z. and Moghadasian, M.H. (2008) Chemistry, natural sources, dietary intake and pharmacokinetic properties of ferulic acid: a review. *Food Chemistry*, **109**, 691–702.

Zhao, Z. and Moghadasian. M.H. (2010) Bioavailability of hydroxycinnamates: a brief review of an in vivo and in vitro studies. *Phytochemistry Reviews*, **9**, 133–45.

Zhao, Z., Egashira, Y., and Sanada, H. (2003) Digestion and absorption of ferulic acid sugar ester in rat gastrointestinal tract. *Journal of Agricultural and Food Chemistry*, **51**, 5534–9.

Zhimin, X. and Godberg, S. (1999) Purification and identification of components of γ-oryzanol in rice bran oil. *Journal of Agricultural and Food Chemistry*, **47**, 2724–8.

5

Flavonols and Flavones

Cláudia Nunes dos Santos[1], Regina Menezes[1], Diogo Carregosa[1], Katerina Valentova[2], Alexandre Foito[3], Gordon McDougall[3], and Derek Stewart[3,4]*

[1] *CEDOC, NOVA Medical School, Faculdade de Ciências Médicas, Universidade NOVA de Lisboa, Lisboa, Portugal*
[2] *Laboratory of Biotransformation, Institute of Microbiology of the Czech Academy of Sciences, Prague, Czech Republic*
[3] *The James Hutton Institute, Invergowrie, Dundee Scotland, UK*
[4] *School of Engineering and Physical Sciences, Heriot Watt University, Edinburgh Scotland, UK*

5.1 Introduction

Flavone and flavonol glycosides are one of the largest classes of flavonoid with over 2000 known structures. Both groups share a 2-phenyl-chrome-4-one backbone (Figure 5.1), composed of a three-ring skeleton structure (C6-C3-C6) with two benzene rings (A- and B- ring) and an oxygen-containing ring with a C2-C3 double bond (C-ring) (Cheng *et al.*, 2014).

Flavonols are 3-hydroxyflavones (3-hydroxy-2-phenyl-chromen-4-one), and the most studied flavonols are quercetin, kaempferol, and myricetin. They all have a hydroxyl group at C-5, C-7, and C-4′ but they differ in the hydroxylation pattern of the ring B (Table 5.1). Onions, kale, lettuce, tomatoes, apples, grapes, and berries are rich sources of flavonols. Apart from fruits and vegetables, tea and red wine are also common sources of flavonols (see Table 5.1) (Panche *et al.*, 2016).

*Corresponding author.

Dietary Polyphenols: Metabolism and Health Effects, First Edition.
Edited by Francisco A. Tomás-Barberán, Antonio González-Sarrías, and Rocío García-Villalba.

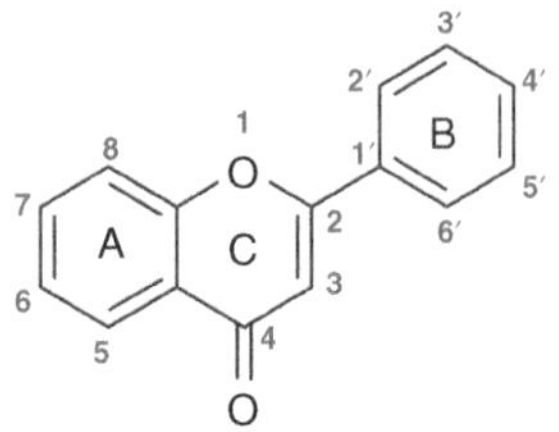

Figure 5.1 Structure of the 2-phenyl-chrome-4-one backbone common to all flavones and flavonols.

Quercetin is probably the most studied flavonoid. Its structure is characterized by the ring B in the form of catechol. Quercetin was first isolated in 1857 from oak (*Quercus*) but it is found in a variety of fruits and vegetables (see Table 5.1). The aglycone quercetin is poorly soluble in water and is mainly found in the form of glycosides (Dueñas *et al.*, 2012; Purchartová *et al.*, 2015). Kaempferol is found in various edible plants (see Table 5.1). Compared to quercetin, kaempferol has one fewer OH group in the B-ring, having a phenol rather than a catechol structure. Kaempferol is relatively common in the plant kingdom, mostly as glycosides such as astragalin (kaempferol-3-*O*-glucoside) (Calderón-Montaño *et al.*, 2011). Myricetin is characterized by a B-ring with a pyrogallol structure (Park *et al.*, 2016). Thus, compared to quercetin, it has one additional hydroxyl group and is sometimes called hydroxyquercetin. Myricetin was first isolated in the late eighteenth century in India from the bark of *Myrica nagi* Thunb. (Myricaceae) as light yellow-colored crystals and isolation was primarily motivated by an interest in its use as a dye (Semwal *et al.*, 2016). As quercetin, myricetin is ubiquitous in food plants (see Table 5.1), occurring in both free and glycosylated forms.

Flavones, distinct from the flavonols due to the lack of hydroxylation at C3 (see Figure 5.1), are widely present in leaves, flowers, and fruits mainly as glucosides. Celery, parsley, red peppers, chamomile, mint, and *Ginkgo biloba* are among the major sources of flavones (Panche *et al.*, 2016). The flavones luteolin and apigenin have both hydroxyl groups at C-5, C-7 and C-4′ (see Table 5.1), but luteolin also has a hydroxyl group at C-3′. Luteolin and apigenin are important dietary flavones, which are naturally present in various fruits and vegetables, and due to their structural similarity they share similar metabolic fates. Dietary flavones generally occur in their glycosylated

Table 5.1 Hydroxylation pattern of kaempferol, quercetin, myricetin, apigenin, and luteolin with reference to Figure 5.1 structural notation

Class	Compound	Molecular formula	3	5	6	7	8	3′	4′	5′	Food sources	References
Flavonols	Kaempferol	$C_{15}H_{10}O_6$	-OH	-OH	-H	-OH	-H	-H	-OH	-H	Apples, onion, tea, cabbage, broccoli, endive, kale, beans, tomato, strawberries, leek and grapes, plants used in traditional medicine (e.g., *Equisetum* spp., *Ginkgo biloba*, *Moringa oleifera*, *Sophora japonica*, *Tilia* spp. or propolis)	(Calderón-Montaño *et al.*, 2011; Imran *et al.*, 2019)
	Quercetin	$C_{15}H_{10}O_7$	-OH	-OH	-H	-OH	-H	-OH	-OH	-H	Apples, citrus fruits, blueberries, cranberries, leafy vegetables, onions and tubers, herbs and spices, but also in tea, cocoa, and wine	(Harwood *et al.*, 2007; d'Andrea, 2015)
	Myricetin	$C_{15}H_{10}O_8$	-OH	-OH	-H	-OH	-H	-OH	-OH	-OH	Blackcurrant, grapes, cranberry, blueberry, beans, red wine, rose petals, oranges, sea buckthorn, chia seeds, carob, pistachio, grape seeds, cruciferous vegetables, peppers, garlic, black tea, and herbs	(Devi *et al.*, 2015; Park *et al.*, 2016; Semwal *et al.*, 2016)

Table 5.1 (Continued)

Class	Compound	Molecular formula	3	5	6	7	8	3′	4′	5′	Food sources	References
Flavones	Apigenin	$C_{15}H_{10}O_5$	-H	-OH	-H	-OH	-H	-H	-OH	-H	Artichoke, broccoli, celery, onion, maize, pepper, rice and parsley	(Ali *et al.*, 2017; Hostetler *et al.*, 2017)
	Luteolin	$C_{15}H_{10}O_6$	-H	-OH	-H	-OH	-H	-OH	-OH	-H	Artichoke, carrots, cucumber, pepper, olives, and turnips. aromatic plants (parsley and thyme)	(Bhagwat *et al.*, 2014; Nabavi *et al.*, 2015)

forms as glycosylation increases their solubility in the aqueous cellular environment (Amić *et al.*, 2007; Pandurangan and Esa, 2014). Most occur in their *O*-glycosylated forms, but cereals such as millet and wheat contain *C*-glycosides of flavones (Hostetler *et al.*, 2017).

Luteolin and its derivatives are present in various edible plants (see Table 5.1) and can also be often found in leaves, rinds, barks, clover blossom, and ragweed pollen (Calderón-Montaño *et al.*, 2011). Apigenin and its derivatives are found in various edible and aromatic plants (see Table 5.1).

5.2 Uptake and Metabolism of Flavonols and Flavones

5.2.1 Flavonols or 3-Hydroxyflavones (Quercetin, Kaempferol, Myricetin)

The mean intake of flavonols in adults in Europe is estimated to be 23 mg/day, but it can be increased to grams/day using various food supplements containing, for example, quercetin or its glycosides (Harwood *et al.*, 2007; Valentová *et al.*, 2014; Vogiatzoglou *et al.*, 2015). Thanks to the wide scientific interest in quercetin, its uptake and metabolism in humans and other mammals are quite well understood (Figure 5.2). After consumption, quercetin glycosides are first deglycosylated mainly in the small intestine. Rhamnose-containing glycosides such as rutin and quercitrin cannot be cleaved by human enzymes and are deglycosylated by the action of colonic bacterial enzymes (Slámová *et al.*, 2018). Therefore, they can only be absorbed in the colon, and this later absorption leads to longer presence of the aglycone in body fluids (Almeida *et al.*, 2018). Furthermore, not all intestinal bacteria have this ability and, in some people, such strains seem to be almost absent, contributing to large interindividual variability in the serum bioavailability of quercetin (Almeida *et al.*, 2018). The absorbed aglycone is then predominantly subject to phase II metabolism, i.e., sulfation, glucuronidation, and *O*-methylation (Dueñas *et al.*, 2012), in the small intestine and liver. Quercetin can also be degraded by colonic microbiota via C-ring

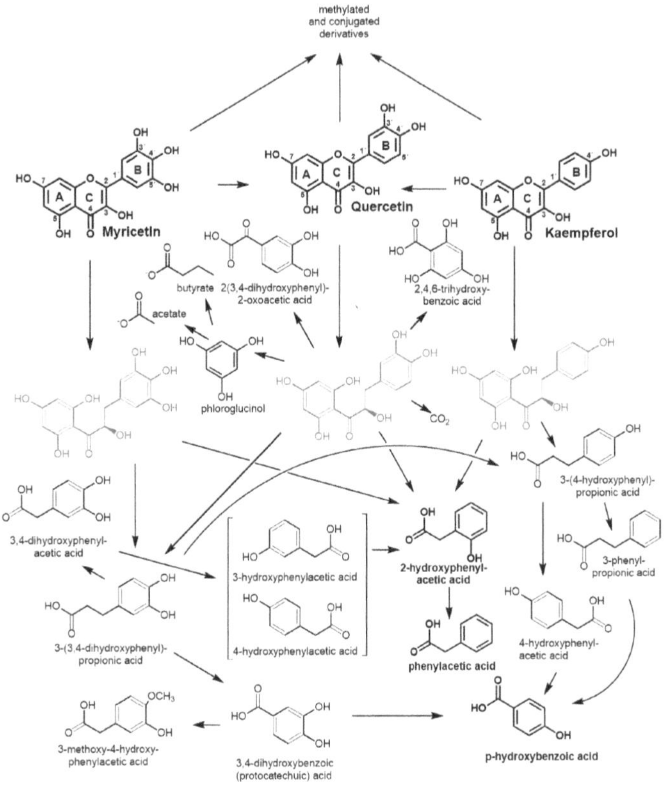

Figure 5.2 Plausible metabolic fate of quercetin, kaempferol, and myricetin including bacterial C-ring fission.

fission, forming other low molecular weight phenolics, e.g., 3-hydroxybenzoic, 3,4-dihydroxybenzoic, homovanilic, phenylacetic and 3,4-dihydroxyphenylacetic acids (3,4-DHPAA). Many of these have been found in the plasma and/or urine of human volunteers after consumption of quercetin in the form of tablets or quercetin-rich food (Almeida *et al.*, 2018; Serra *et al.*, 2012).

Like other flavonoids, kaempferol is mainly ingested in the form of glycosides and deglycosylated in the small intestine. Kaempferol can then be absorbed in the small intestine by passive diffusion due to its lipophilicity, but probably also by facilitated diffusion or active transport. Both in the small intestine and the liver, kaempferol is metabolized to glucuronides and sulfoconjugates (Vollmer *et al.*, 2018). Interestingly, kaempferol can be transformed to quercetin, probably by CYP1A1, in rat liver microsomes (Nielsen *et al.*, 1998). Kaempferol and kaempferol glycosides are extensively metabolized by colon microbiota, which both hydrolyze the glycosides to aglycones and break the C-ring to form simple phenolic compounds such as 4-hydroxyphenyl acetic acid, phloroglucinol, and 4-methyl phenol (Vollmer *et al.*, 2018). These phenolic compounds, but also conjugated kaempferol, kaempferol aglycone, and even some kaempferol glycosides can reach systemic circulation and tissues and are then excreted in the urine. The percentage of kaempferol excreted in urine is approximately 2% of the total amount ingested; in plasma, kaempferol was detected at nanomolar concentrations (Calderón-Montaño *et al.*, 2011).

The metabolic fate of myricetin has been studied to a more limited extent. More is known about the impact of myricetin on the pharmacokinetics of various drugs (Park *et al.*, 2016). The few pharmacokinetic studies available focused only on the level of the parent compound in human plasma (Park *et al.*, 2016), showing its low (<10%) bioavailability (Yu *et al.*, 2014). *In vitro* fermentation models using rat or human colonic microflora showed myricetin to be transformed into 2-(4-hydroxyphenyl)acetic, 2-(3,5-dihydroxyphenyl)acetic, phenylacetic, and 3-(3,5-dihydroxyphenyl)propionic acids (Selma *et al.*, 2009; Serra *et al.*, 2012). Therefore, it seems to follow the typical pathway of flavonoid degradation, with the formation of the trihydroxylated chalcone as a precursor of the detected phenolic acids (Serra *et al.*, 2012). In another study, myricitrin (myricetin rhamnoside) was deglycosylated and dehydroxylated by human microbiota forming, among others, quercetin (Du *et al.*, 2014). In line with the proposed degradation pathway above, 3,5-dihydroxyphenylacetic and 3,4,5-trihydroxyphenylacetic acids together with methylated and glucuronidated metabolites were found in rat plasma, urine, and feces after consumption of a single dose of myricetin at 300 mg/kg body weight by oral gavage (Lin *et al.*, 2012).

5.2.2 Flavones (Luteolin, Apigenin)

The mean intake of flavones in adults in Europe is estimated to be 4 mg/day (Vogiatzoglou *et al.*, 2015). The dietary mean intake of luteolin has been estimated in different populations (Cao *et al.*, 2010; Otaki *et al.*, 2009; Somerset and Johannot, 2008; Zamora-Ros *et al.*, 2010) and varied significantly, ranging from a mean intake of 0.08 mg/day (Somerset and Johannot, 2008) up to 8.08 mg/day (Cao *et al.*, 2010). A similar trend was noted for apigenin (Cao *et al.*, 2010; Otaki *et al.*, 2009; Somerset and Johannot, 2008; Zamora-Ros *et al.*, 2016) with dietary intakes ranging from a mean intake of 0.25 mg/day (Zamora-Ros *et al.*, 2010) up to 4.23 mg/day (Cao *et al.*, 2010). There were clear differences between the foods contributing to the total flavone intake in populations from different geographical areas. In a Spanish cohort, the intake of red wine, oranges, other fruits, lettuce, and peppers provided the majority of dietary flavone sources (Zamora-Ros *et al.*, 2010). A Chinese study reported celery and eggplant as main contributors (Cao *et al.*, 2010), while a study in a Japanese population indicated that Malabar spinach, green pepper, grapefruit, parsley, and celery were the main contributors (Otaki *et al.*, 2009).

Flavones are typically detected in plasma or urine as glucuronide or sulfate conjugates rather than the original flavone-*O*-glucosides (Chalet *et al.*, 2018). As with many flavonoid glycosides, dietary flavone glycosides must first be hydrolyzed to release the aglycone before they can be absorbed (Németh *et al.*, 2003). There is evidence indicating that apigenin and other flavones could be absorbed in the stomach of rodent models although the absorption was limited (Pforte *et al.*, 2000). Additionally, an *ex vivo* study indicated that flavonoid glucosides could be absorbed through the human stomach and subsequently hydrolyzed by a stomach-specific *β*-glucosidase (Pforte *et al.*, 2000).

However, with its greater surface area, the small intestine is much more effective at absorbing compounds. Indeed, there is widespread evidence indicating that the small intestine is a major site of hydrolysis and metabolism of flavone glucosides. The small intestinal cells produce lactase-phlorizin hydrolase, which is a membrane-bound enzyme that exhibits broad substrate specificity for flavonoid-O-β-glucosides and is capable of cleaving flavone glucosides (Németh *et al.*, 2003). While these cells also produce a cytosolic β-glucosidase which can hydrolyze flavone glucosides, these do not play a role in flavone deglycosylation as there is no known mechanism of transport of flavone glycosides into the enterocytes (Hostetler *et al.*, 2017).

Flavone aglycones can be subject to conjugation reactions in the lining of the small intestine by enterocytes (Chalet *et al.*, 2018) but can also cross to the basolateral side of the small intestine and be transported to the liver where they can be further conjugated by phase II metabolic processes (Gradolatto, 2004). Different isoforms of UDP-glucuronosyltransferases (UGT) show distinct regioselectivity for transfer of glucuronides to flavones/flavonols/flavanones: positions 3′>7>3>5 were favored for UGT1A1; 7>3>5 for UGT1A3; 3>7>6 for UGT1A9; and 3′>4′ for UGT1A10 (Wu *et al.*, 2011). Interestingly, *in vitro* studies indicate that luteolin was preferably glucuronidated at the 3′ and 4′ positions in the enterocytes and at the 7 position in hepatocytes (Hostetler *et al.*, 2017). Catechol-*O*-methyl transferase catalyzes the transfer of a methyl group from *S*-adenosyl-L-methionine to polyphenols containing a catechol moiety predominantly at 3′ but also at 4′ (Williamson and Manach, 2005). Consequently, luteolin but not apigenin was reported to be methylated in human liver cells (Kanazawa *et al.*, 2006) and rats (H. Chen *et al.*, 2012a). Sulfotransferase 1A3 was found to predominantly sulfate flavones at position 7 (Wu *et al.*, 2011), and the general order of sulfation stereospecificity in flavones/flavonols was at positions 7>3′>4′>6>8 (Teles *et al.*, 2018).

The conjugated flavones have different physicochemical properties compared to the aglycone molecules, which often leads to a facilitated excretion out of the enterocytes. As a result, once metabolized by intestinal cells, the glucuronide and sulfate conjugates can be transported to both the intestinal lumen and the bloodstream (Hostetler *et al.*, 2017).

The major metabolic conjugation pathways for luteolin are methylation and glucuronidation, which result in the production of diosmetin, chrysoeriol, luteolin 7-*O*-glucuronide (Shi *et al.*, 2018), and luteolin 3′-*O*-glucuronide (Z. Chen *et al.*, 2012a). Luteolin 3′-*O*-glucuronide was reported to be abundant

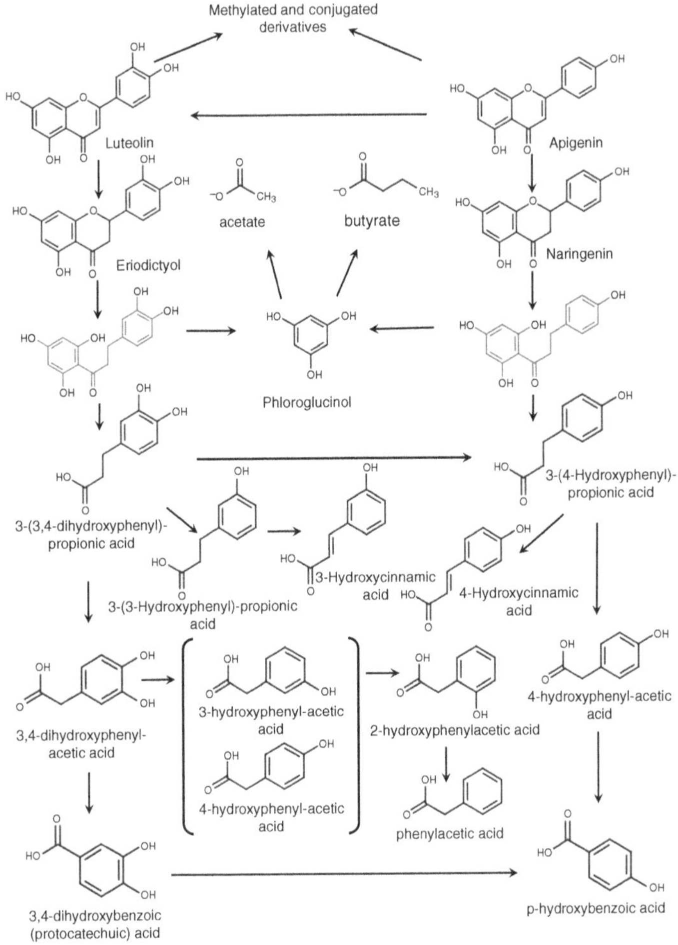

Figure 5.3 Plausible metabolic fate of apigenin and luteolin including bacterial ring fission.

in most of the tissues and plasma, whereas metabolites with 7-*O*-glucuronidation were excreted in the bile (Shi *et al.*, 2018). Sulfation may also occur, and recent evidence has suggested that luteolin 3′-*O*-sulfate was the predominant plasma metabolite in humans as opposed to luteolin glucuronide conjugates in rats, indicating that animal species differ in their pathways of luteolin metabolism (Hayasaka *et al.*, 2018).

Apigenin can be glucuronidated and sulfated in intestinal cells (Hu *et al.*, 2016) or in liver cells where glucuronidation occurs preferentially over sulfation (Cai *et al.*, 2007). Interestingly, apigenin does not experience methylation, but there is evidence of hepatic cytochrome P450-mediated hydroxylation of apigenin into luteolin in rats (Gradolatto, 2004) and human liver microsomes (Nielsen *et al.*, 1998). While this can explain the presence of luteolin-related metabolites after apigenin exposure, phase II metabolism is still considered the main pathway for apigenin metabolization (Lu *et al.*, 2011).

Flavone aglycones and conjugates that are resistant to the action of the brush border lactase-phlorizin hydrolase reach the colon and can be further metabolized by the microbiota as described below and indicated in Figure 5.3.

5.3 Microbiota Formation of Low Molecular Weight Phenolic, Common Colonic Metabolites

As described previously, methylation, sulfation, and glucuronidation are the major metabolic pathways of flavonoids in the liver, while in the intestine, flavonoids can be transformed by the intestinal bacteria into a wide range of low molecular weight phenolic metabolites.

5.3.1 Flavonols (Quercetin, Kaempferol, Myricetin)

In humans, quercetin metabolites with an intact flavonoid structure have been the most studied, and our understanding of quercetin degradation products formed by human gut microbiota after C-ring fission and about metabolites excreted in feces is still incomplete. However, significant increases in urinary concentrations of 4-ethylphenol, benzoic acid, and 4-ethylbenzoic acid were noted in healthy men after oral consumption of 200 mg of pure quercetin (Loke *et al.*, 2009). After supplementation with quercetin-3-*O*-rutinoside, mainly phenylacetic acids, namely 3-hydroxyphenylacetic acid (36% of the dose ingested), homovanillic acid (8%) and 3,4-dihydroxyphenyl acetic acid (5%), were excreted into the urine of healthy humans with an intact colon. The absence of conventional colonic microbiota, as in ileostomy subjects, abolished the formation of the majority of the phenolic acid metabolites, indicating the importance of bacterial biotransformation (Olthof *et al.*, 2003). Some microbial metabolites of quercetin were identified when rutin was given in pure form or tomato juice, including phenylacetic and hydroxyhippuric acid derivatives (Jaganath *et al.*, 2006; Olthof *et al.*, 2003).

Studies on other mammals fill in the picture on the involvement of microbiota action and indicating other metabolites. In 1956, Booth *et al.* provided evidence that quercetin could be metabolized into phenolic compounds in the urine of rabbits after oral ingestion of rutin or its aglycone quercetin, reviewed in Feng *et al.* (2018). Three of the metabolites were identified to be 3,4-dihydroxyphenylacetic acid (3,4-DHPAA), 3-hydroxyphenylacetic acid (3-HPAA), and homovanillic acid. Afterwards (Griffiths and Barrow, 1972), the authors concluded that these metabolites could not be detected in germ-free rats, indicating that intestinal bacteria were responsible for their formation, reviewed in Feng *et al.* (2018).

One of the first organisms reported as being capable of converting rutin to phloroglucinol, 3,4-dihydroxybenzaldehyde, 3,4-DHPAA, and CO_2 was Butyrivibrio sp. C3, isolated from the bovine rumen in the 1960s (Feng *et al.*, 2018). Certain strains of *Pediococcus* spp., *Streptococcus* spp., *Lactobacillus* spp., *Bifidobacterium* spp., *Eubacterium* spp., *Enterococcus* spp., *Clostridium* spp. and *Bacteroides* spp. are described to transform quercetin to various phenolic acids (e.g., 3-hydroxybenzoic, 3,4-dihydroxybenzoic, and 3,4-dihydroxyphenylacetic) (Cermak *et al.*, 2006). Quercetin was also metabolized by porcine hindgut contents *in vitro* (Cermak *et al.*, 2006). After *in vitro* colonic fermentation of quercetin with rat feces for 48 hours, the main product was protocatechuic acid with lower amounts of homovanillic, phenylacetic, and *p*-hydroxybenzoic acids (Serra *et al.*, 2012). Degradation of quercetin by the rat gut microbiota therefore probably involves C-ring fission, formation of 3-(3,4-dihydroxyphenyl)propionic acid (3,4-DHPPA), and subsequent transformation to 3,4-DHPAA. Further transformation leads to protocatechuic acid and then to

4-hydroxybenzoic acid. 3,4-DHPAA can also be dehydroxylated to 3- or 4-hydroxyphenylacetic and phenylacetic acids (see Figure 5.2) (Serra *et al.*, 2012). These compounds can be further degraded into various simpler products and finally to carbon dioxide (Walle, 2004; Walle *et al.*, 2001).

To date, only a limited number of bacteria have been identified with the capacity to metabolize flavonoids and most studies aiming to isolate and identify bacteria were performed *in vitro*. The production of numerous colonic polyphenol metabolites is affected by considerable interindividual differences. This variability extends to the production of different metabolites, different levels of intermediates and end-products as well as variable time courses (Feng *et al.*, 2018). In a recent revision on quercetin bioavailability in humans, the coefficient of variation that affects the C_{max} was calculated for all metabolites and a pattern was observed depending on whether the site of absorption was the small intestine or the colon (including the action of the microbiota). Compounds, which underwent microbial metabolism in the colon exhibit a greater interindividual variation than compounds absorbed in the small intestine, presumably due to variation in personal microbiota (Almeida *et al.*, 2018).

Importantly, this study also highlighted that an individual who would be considered a low responder as judged by evaluating levels of glucuronides, glucosides, and methylated metabolites of quercetin may actually be revealed as a faster metabolizer when assessing the concentration of low molecular weight phenolic acids and therefore a higher responder (Almeida *et al.*, 2018). This highlights that a precise understanding of interindividual variability of quercetin bioavailability requires measurement of all metabolic routes, including the gut microbiota.

5.3.2 Flavones (Luteolin, Apigenin)

The intestinal bacterial degradation of flavones starts with the ring fission of the C-ring as it is shown for luteolin in Figure 5.3. During this process, the reduction of the double bond in the 2,3-position can happen in a similar way to flavonols. The steps that follow differ due to the 3-hydroxyl group in flavonol molecules. The degradation metabolites of flavones are mainly 3-(hydroxyphenyl)propionic acids like 3-(3,4-dihydroxyphenyl)propionic acid, also known as dihydrocaffeic acid, in the case of luteolin or 3-(4-hydroxyphenyl)propionic acids in the case of apigenin (see Figure 5.3). These molecules can then undergo oxidation and be converted to low molecular weight metabolites such as 2-(hydroxy)phenylacetic acid and benzoic acids like protocatechuic acid (see Figure 5.3). Moreover, these molecules can be the subject of further metabolism leading to the presence of sulfate and glycine conjugates, reviewed in Feng *et al.* (2018).

On the other hand, the reduction of the double bond in the 2,3-position without ring fission is expected. Luteolin reduction of this double bond can lead to the formation of eriodictyol. In any case, this molecule can also suffer ring fission, which leads to the formation of a chalcone subsequently reduced to dihydrochalcone. Also, luteolin can be directly degraded into 3,4-DHPPA and other phenolic compounds.

Hanske *et al.* (2009) evaluated the bioavailability and excretion of apigenin by intragastrically applying apigenin-7-*O*-glucoside in either germ-free rats or rats with human-like intestinal microbiota (Hanske *et al.*, 2009). The total excretion of apigenin-7-*O*-glucoside was similar in both types of rats. However, in germ-free rats, apigenin-7-*O*-glucoside was mainly excreted with feces, while in human microbiota-associated rats, it was predominantly excreted with urine (Hanske *et al.*, 2009). Rats with human intestinal microbiota excreted more of the following metabolites in their urine: free and conjugated naringenin, free phloretin, conjugated luteolin, free 3,4-dihydroxyphenylpropionic acid, free and conjugated 3-(4-hydroxyphenyl)propionic acid, and 4-hydroxycinnamic acid, and free 3-hydroxyphenylpropionic acid (Hanske *et al.*, 2009).

In vitro experiments with *Eubacterium ramulus*, an obligate anaerobe found in the human colon, showed that eriodictyol and 3,4-DHPPA were metabolites of luteolin (Braune *et al.*, 2001). Studies with *Clostridium orbiscindens*, which is found in numbers similar to those for *E. ramulus* in the intestinal tract, also showed that eriodictyol was a product of luteolin metabolism and that naringenin was a product of apigenin. Phloretin, 3-(4-dihydroxyphenyl)propionic acid, and phloroglucinol were end-products of both apigenin and luteolin degradation (Schoefer *et al.*, 2003).

5.4 Health Effects of Flavonol and Flavone Metabolites

The flavonols and flavones are widely reported to exert health-promoting activities against various pathologies, including inflammation, diabetes, cancer, heart disease, and viral infections. In this section, we will focus on the beneficial effects in chronic processes associated with neurodegenerative and cardiometabolic disorders as well as cancers.

5.4.1 Flavonols or 3-Hydroxyflavones

Many studies have addressed the potential biological effects of flavonols; however, the potential effects of the three most abundant flavonols, quercetin, kaempferol, and myricetin, have been explored in greater detail. The biological activity of quercetin has been evaluated in several studies showing a wide array of effects both *in vitro* and *in vivo* (d'Andrea, 2015; Formica and Regelson, 1995; Kerimi and Williamson, 2018; Russo *et al.*, 2012). Quercetin contributes to the protection of body tissues against oxidative stress and the prevention of diseases such as cancer, cardiovascular and neurodegenerative diseases or inflammation (Purchartová *et al.*, 2015; Raffa *et al.*, 2017; Russo *et al.*, 2012). For many years, quercetin had been considered to be carcinogenic due to positive Ames test, but it was approved in 2010 by the FDA as "Generally recognized as safe" (GRAS Notice No. GRN 000341). Kaempferol displays protective actions against cancer, inflammation, diabetes, obesity, cardiovascular diseases, oxidative stress, asthma, and microbial infection (Imran *et al.*, 2019). The presence of an additional hydroxyl group in the B-ring of the myricetin structure may account for its higher biological activity compared to quercetin and other flavonols (Park *et al.*, 2016). Myricetin exhibits antioxidant, antiinflammatory, antiviral, anticarcinogenic, antihypertensive, antimicrobial, antithrombotic, antidiabetic, cytoprotective, antiobesity, antihyperlipidemic and antiaging actions and has been shown to slow down Parkinson's disease (Abdel-Raouf *et al.*, 2011; Devi *et al.*, 2015; Raffa *et al.*, 2017; Salvamani *et al.*, 2014; Sarian *et al.*, 2017; Wang *et al.*, 2018).

Inflammation

Regarding the potential of flavonols to modulate inflammation, these compounds have shown the ability to inhibit the PI3K pathway, ERK-1/2, AKT, p38 and IκB degradation, thereby upregulating cyclooxygenase (COX), lipoxygenase, IL-1β, TNF-α, and nitric oxide (NO) production in macrophages (Endale *et al.*, 2013; Kim *et al.*, 1998, 2013; Ko, 2012). The antiinflammatory potential of quercetin and kaempferol was supported by studies using microglial cells, which revealed their potential to downregulate IL-1β, TNF-α, TLR4, NF-kB, p38 MAPK, JNK, and AKT (Bureau *et al.*, 2008; Park *et al.*, 2011). Animal studies, mainly in rodents, have demonstrated that quercetin could modulate the production of cytokines (Li *et al.*, 2016). Moreover, quercetin presented a neuroprotective effect on experimental allergic encephalomyelitis by blocking IL-12 signaling through the JAK-STAT pathway in T lymphocytes (Muthian and Bright, 2004). Myricetin displayed antiinflammatory activity by inhibiting the production of lipopolysaccharide (LPS)-induced prostaglandins in rat peritoneal macrophages (Takano-Ishikawa *et al.*, 2006).

Neurodegenerative Diseases

Myricetin was described as having neuroprotective activities against neurotoxins in animal and cellular models. In a cellular model of Parkinson's disease, myricetin was found to exert a neuroprotective effect against 1-methyl-4-phenylpyridinium (MPP^+)-induced damage in MES23.5 cells, by reducing cell loss and nuclear condensation. It also suppressed the production of intracellular ROS, restored mitochondria function, and decreased apoptosis induced by MPP^+. Furthermore, it decreased the phosphorylation of MAPK kinase-4 and JNK, the signaling cascades induced by MPP^+ (Zhang *et al.*, 2011). On the other hand, the toxicity of another neurotoxin, 6-hydroxydopamine, was found to be reduced by myricetin, by decreasing the dopamine content in the substantia nigra striatum-induced system of rats (Ma *et al.*, 2007). Myricetin also prevented 6-hydroxydopamine-induced decrease of tyrosine hydroxylase-positive neurons and tyrosine hydroxylase mRNA expression in the same region of rat brain.

Quercetin has been reported to protect neuronal cells from the toxicity of amyloid β (Aβ) peptide. In a mouse model of Alzheimer's disease, administration of kaempferol reduced Aβ-induced impaired performance in a behavioral test (Ansari *et al.*, 2009). Furthermore, in a model of Huntington's disease, administration of kaempferol was shown to afford efficient protection against neurodegeneration in Wistar rats, attenuating motor deficit and delaying mortality (Lagoa *et al.*, 2009).

Cardiovascular Diseases

Cardiovascular diseases are another important area where flavonols have revealed a potential effect. In elderly humans, a diet containing quercetin suppressed NO production and other cardiovascular risk factors. A double-blind, randomized clinical trial with daily supplementation with 500 mg of quercetin significantly reduced systolic blood pressure and serum concentrations of TNF-α and IL-6 in humans (Zahedi *et al.*, 2013). In fact, out of seven clinical studies using quercetin, five reported an antihypertensive effect (Ożarowski *et al.*, 2018).

Pretreatment with kaempferol significantly improved the recovery of left ventricular developed pressure and its maximum up/down rate, as well as increasing the levels of intracellular antioxidant defenses (SOD, P-GSK-3β and GSH/GSSG ratio). Furthermore, the pretreatment reduced myocardial infarct size and decreased the apoptosis induced by ischemia/reperfusion injury *via* GSK-3β inhibition. Kaempferol inhibited SK1/MAPK signaling pathways induced in hearts of aortic-banding mouse models. In *in vitro* experiments, kaempferol also inhibited the activity of the ASK1/JNK1/2/p38 signaling pathway and the enlargement of cardiomyocytes and protected the mouse heart and cardiomyocytes from pathological oxidative stress (Feng *et al.*, 2017). It also prevented and reversed cardiac remodeling induced by angiotensin II. In this model, kaempferol exerted no basal effects but attenuated cardiac fibrosis, hypertrophy, and dysfunction induced by angiotensin II (Liu *et al.*, 2017).

Myricetin also presents cardioprotective effects. Oral administration of 100 and 300 mg/kg doses of myricetin to Wistar rats resulted in a reduction in heart rate and the levels of cardiac marker enzymes (lactate dehydrogenase, creatine kinase, aspartate aminotransferase, SOD, and CAT), as well as changes in vascular reactivity and electrocardiographic patterns caused by isoproterenol (a potent β-adrenergic agonist with peripheral vasodilator, bronchodilator, and cardiac stimulating properties) (Tiwari *et al.*, 2009).

Cancer

Quercetin is known to impair proliferation of carcinoma cell lines such as human breast cancer cell lines (SK-Br3, MDA-MB-453, and MDA-MB-231) (Abdal Dayem *et al.*, 2016). Quercetin can block the cell cycle at the G_2/M or G_1/S transition (Hashemzaei *et al.*, 2017). An important effect of quercetin is to regulate cell cycle by modulating several molecular targets, including p21, cyclin B, p27, cyclin-dependent kinases, and topoisomerase II. In human lung cancer cells, quercetin glucuronides induced cell cycle arrest at the G_2/M phase by increasing the expression of specific proteins such as cyclin B, Cdc25c-ser-216-p, and Wee1 (Yang, 2005). In HepG2 human hepatoma cells, quercetin blocked cell cycle progression at the G_1 phase and exerted this effect through the increase of p21 and p27 and p53 (Mu *et al.*, 2007). Quercetin is also a potent enhancer of TNF-related apoptosis-inducing ligand (TRAIL)-induced apoptosis, a phenomenon that specifically occurs in prostate cancer cells (Jung *et al.*, 2010). It was observed that this increase was through the induction of the expression of death receptor (DR)-5. TRAIL and its receptors have been used as targets of several anticancer therapeutics. In lung tumor induced by benzo(a)pyrene, administration of quercetin before the initiation stage of carcinogenesis reduced tumor burden in mice, which showed increased activity of antioxidant enzymes and a decrease in the levels of lipid peroxides (Houyel *et al.*, 1988; Kamaraj *et al.*, 2007).

Kaempferol impairs cancer angiogenesis both *in vitro* and *in vivo* through inhibiting VEGF secretion in human ovarian cancer cell lines, in a mechanism that involves the ERK-NFκB-cMyc-p21 pathway (Luo *et al.*, 2009). Kaempferol has also been shown to reduce cyclin-dependent kinases levels, to cause a surge in p53 levels in breast cancer MDA-MB-453 cells. Moreover, phosphorylation of p53 at serine 15, a post-translation modification associated with apoptosis of cancer cells, was induced by kaempferol in the same cells (Choi and Ahn, 2008). In the highly invasive breast cancer cell line MDA-MB-231, kaempferol has shown great promise in disrupting cancer metastasis, apparently by inhibiting matrix metalloproteinase-3 (MMP-3) protein activity (Qin *et al.*, 2016).

Extensive research into the anticancer activities of myricetin has indicated that the compound is cytotoxic towards several human cancer cell lines, including hepatic, skin, pancreatic, and colon cancer cells (Hackman *et al.*, 1973). Myricetin protected against skin cancer by strongly inhibiting the process of tumor promotion through restriction of MEK, JAK1, Akt, and MKK4 kinase activity (Kang *et al.*, 2011).

Summary

In conclusion, the potential effects of flavonols have been widely studied in inflammation, neurodegenerative diseases, cardiovascular diseases, and cancer. These studies have revealed the pleiotropic ability of quercetin, kaempferol, and myricetin in the modulation of several regulatory pathways (Figure 5.4). However, most of these studies have been conducted in cellular models with concentrations many times higher than those found in circulation. Rodent and other animal models have been crucial for understanding the potential physiological effects of flavonols, although they can give different outcomes depending on the concentrations of these compounds. In the future, the use of lower, more physiological concentrations of these compounds as well as examining physiologically relevant metabolites could help to validate the effects of flavonols in a physiological context.

5.4.2 Flavones (Luteolin, Apigenin)

Numerous *in vitro* and *in vivo* studies have demonstrated that apigenin, luteolin, and their derivatives are associated with antiinflammatory, neuroprotective, anticancer, and cardioprotective effects (Ali *et al.*, 2017; Hu *et al.*, 2016; Imran *et al.*, 2019; Jiang *et al.*, 2016; Luo *et al.*, 2017; Nabavi *et al.*, 2015; Salehi *et al.*, 2019). A simplified view of the main regulatory pathways modulated by flavones is shared with flavonols and depicted in Figure 5.4.

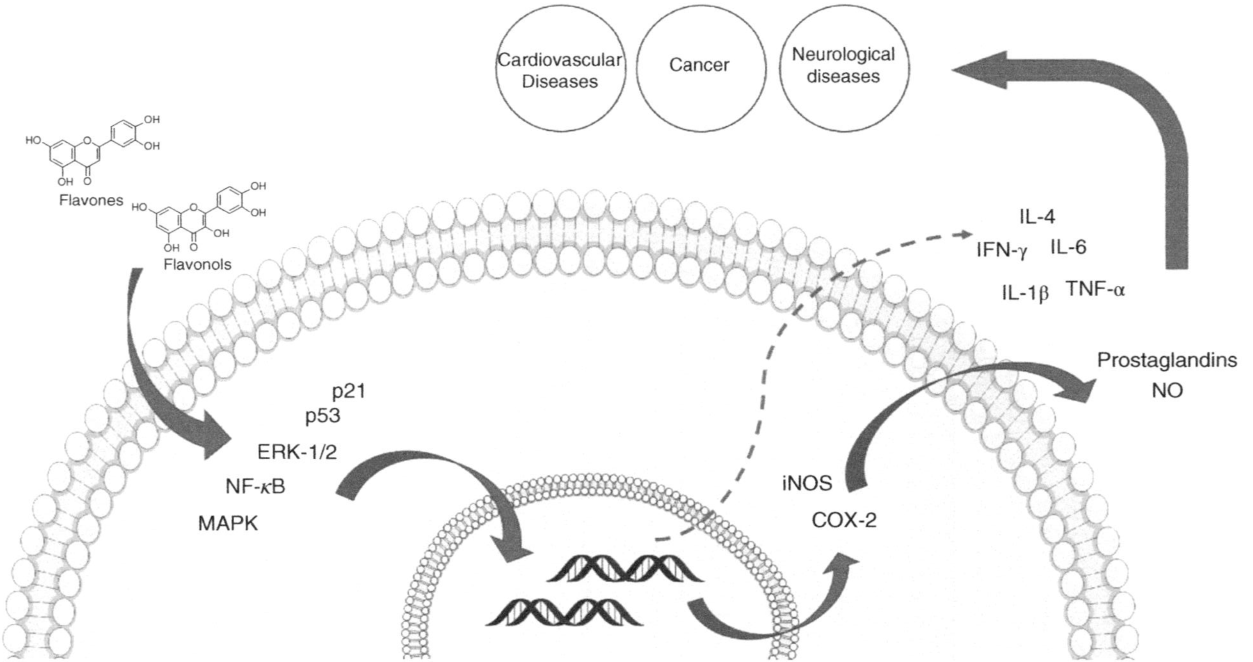

Figure 5.4 Pleiotropic action of flavonols (quercetin, myricetin, and kaempferol) and flavones (apigenin and luteolin) in the modulation of central regulatory pathways (e.g., NF-κB, MAPK, and ERK) which in turn modulate the expression of enzymes such as COX-2 and iNOS associated with the release of proinflammatory cytokines.

Generally, most *in vitro* cell studies addressing bioactivity of flavones, and other polyphenols, evaluate the protective activity of aglycone precursors at supraphysiological serum concentrations (i.e., 10–1000 μM) (Vinson, 2019). Thus, investigating the mechanisms underlying flavone metabolites bioactivity is crucial for understanding the impact of these compounds on health.

As noted before, the metabolism of apigenin generates luteolin and naringenin, thus the biological effects attributed to these compounds are potentially associated with the intake of apigenin-rich foods (Ali *et al.*, 2017; Jiang *et al.*, 2016; Pandurangan and Esa, 2014; Zaidun *et al.*, 2018; Zeng *et al.*, 2018; Zhao *et al.*, 2019). Like luteolin aglycone, the phase II luteolin metabolites are associated with antiinflammatory activities using *in vitro* cell assays, rat models, and humans (Hostetler *et al.*, 2017; Li *et al.*, 2012). However, luteolin glucuronides are reported to reduce the expression of proinflammatory genes in cell models to a lesser extent than luteolin aglycone (Li *et al.*, 2012). Indeed, this fits with the notion that luteolin glucuronides are deconjugated at inflammatory sites (Shimoi *et al.*, 2000). In addition, the inhibitory effects of luteolin-sulfate, the main luteolin metabolite found in human plasma after oral administration of luteolin aglycone, were even lower than that of luteolin-glucuronide. This may be partially explained by the fact that intracellular amounts of luteolin metabolites are very low in cells treated with luteolin-sulfate, compared with luteolin glucuronide-treated cells, suggesting that luteolin-sulfate is not readily taken up by macrophage cells (Hayasaka *et al.*, 2018).

To overcome experimental limitations associated with the diversity of polyphenol metabolites, Dellafiora *et al.* (2014) developed an *in silico* approach for the systematic analysis of these compounds by first identifying promising candidates for further extensive biological characterization. In a pilot study, the authors modeled the effect of phase II-flavonoid metabolites, including those of luteolin, on the poisoning of human topoisomerases, which are recognized chemopreventive targets (Pommier, 2013). It was shown that glucuronidation potentially enhances luteolin effects, whereas sulfation generally decreased bioactivity towards topoisomerase poisoning (Dellafiora *et al.*, 2014). The application of this predictive model to other phenolic structures may potentially assist the study of several metabolic modifications and their biological effects.

5.4.3 Flavonols, Flavones and Their Low Molecular Weight Colonic Metabolites in Health

Flavonols and flavones can be extensively catabolized by microbiota, as described in detail in section 5.3, resulting in an elaborate number of low molecular weight metabolites. These metabolites add a crucially important piece to the evidence on health effects of flavonoids. Moreover, they are generally present at higher physiological serum concentrations. They share many of the effects already observed for their parent compounds: antiinflammatory, anticarcinogenic, and cardiovascular modulators.

A common metabolite from ring fission of flavonols and flavones is 3,4-DHPPA or dihydrocaffeic acid. 3,4-DHPPA showed antiinflammatory activity, which was associated with the inhibition of TNF-α, IL-1β, and IL-6 secretion in LPS-stimulated peripheral blood mononuclear cells from healthy human volunteers (Monagas *et al.*, 2009). 3,4-DHPPA has also been reported to induce antioxidant defences in HepG2 cells (Baeza *et al.*, 2016). Regarding cardiovascular diseases 3,4-DHPPA and the methoxylated metabolite dihydroferulic acid are more efficient inhibitors of *in vitro* platelet activation than their phenolic parent compounds (Baeza *et al.*, 2017).

The contribution of these low molecular weight metabolites of flavonols and flavones in neurological diseases is also of interest due to their ability to balance neuroinflammation (Carregosa *et al.*, 2019). In microglia cells, ferulic acid reduced NO, IL-6, and IL-1β through NF-κB and MAPK pathways (Kim *et al.*, 2015). Furthermore, 3-(4-hydroxyphenyl)propionic acid reduced MAPK and ERK activation mediated by 3-morpholinosydnonimine, a neuronal damage inductor, in SH-SY5Y neuronal cells (Esteban-Fernández *et al.*, 2017).

As an important luteolin metabolite, the biological activities of eriodictyol have been associated with the alleviation of LPS-oxidative stress and synaptic dysfunctions in BV-2 microglial cells and mouse brain through mechanisms that involve MAPKs, NF-κB mediated by ROS, Sirt1, and

Nrf2/Keap1 signaling pathways (He *et al.*, 2019). It was also shown to modulate AKT/FOXO1 signaling, thereby inhibiting the survival and inflammatory responses and promoting apoptosis in rheumatoid arthritis fibroblast-like synoviocytes (Liu and Yan, 2019). Other biological effects of eriodictyol were related to the alleviation of adiposity, hepatic steatosis, insulin resistance, and inflammation in obese mice, protection of cardiomyocytes against hypoxia/reoxygenation-induced injury by improving mitochondrial dysfunction, inhibition of high glucose-induced oxidative stress and inflammation in retinal ganglia cells, and other IL-1β-induced bioactivities (He *et al.*, 2019; Lv *et al.*, 2019; Xie *et al.* 2018).

The biological effect of phloretin, a natural phenolic compound present in apples and also a common metabolite of apigenin and luteolin, was recently reviewed by Mariadoss *et al.* (2019). Among the wide variety of effects attributed to phloretin, its antioxidative, antiinflammatory, antimicrobial, antiallergic, anticarcinogenic, antithrombotic, and hepatoprotective activities stand out (Choi, 2019; Mariadoss *et al.*, 2019; Yang, 2005; Ying *et al.*, 2018). Also, phloretin bioactivity has been linked to the activation of apoptotic-associated gene expression and signal transduction (Mariadoss *et al.*, 2019). The modulatory role of phloretin in Alzheimer's disease conditions was evaluated in a rat model exposed to Aβ25-35 where it improved spatial memory formation, alleviated oxidative stress thereby reducing TNF-α-mediated neuroinflammation, and decreased amyloid β accumulation in rat brains (Ghumatkar *et al.*, 2019).

5.5 Conclusions and Future Perspectives

It is very clear that flavanols and flavones have an important place in our diet. Indeed, the common sources of these compounds are species that humans have coevolved with and eaten as part of our hunter-gatherer past. That being said, we now have a food system that has evolved at a significant pace and, with the ready availability of cheap, highly processed food containing minimal nutrition and health-beneficial components, we have seen an explosion in most of the pathologies identified earlier including dietary tract cancers, cardio- and other vascular-related diseases, and type 2 diabetes. However, the body of evidence presented from the cell to full human-scale studies paints an almost universally positive picture for the inclusion of these compounds as part of our diet.

There are still milestones to achieve regarding the knowledge of flavonols and flavones health effects. The full mechanisms of action remain unclear, and a labeled compound approach may help to facilitate the identification of compound metabolic fates, and health and disease state, enzyme signaling, and cascades. Linked to this is the need for studies to use compound levels that are more representative of what a body or tissue may be presented with *via* dietary means. For example, acute interventions with 500 mg shots of quercetin are equivalent to 7.2 kg of onions. The relevance here must be questioned unless the plan is for the compound to be used as a supplement or as the basis of new pharmaceutical compounds.

Acknowledgments

This work was supported by Fundação para a Ciência e Tecnologia (FCT) grant UID/Multi/04462/2013, I&D 2015-2020 iNOVA4Health – Programme in Translational Medicine and PTDC/BIA-MOL/31104/2017 (RM) and by the Scottish Government Rural and Environment Science and Analytical Services Division (AF and DS). KV acknowledges the Czech Ministry of Education, Youth and Sport, project no. LTC19039.

References

Abdal Dayem, A., Choi, H.Y., Yang, G.M., *et al.* (2016) The anti-cancer effect of polyphenols against breast cancer and cancer stem cells: molecular mechanisms. *Nutrients*, **8**, 581.

Abdel-Raouf, N., Ibraheem, I.B., Abdel-Tawab, S., and Naser, Y.A. (2011) Antimicrobial and antihyperlipidemic activities of isolated quercetin from anabaena aequalis. *Journal of Phycology*, **47**, 955–62.

Ali, F., Rahul, Naz, F., Jyoti, S., and Siddique, Y.H. (2017) Health functionality of apigenin: a review. *International Journal of Food Properties*, **20**, 1197–238.

Almeida, A.F., Borge, G.I.A., Piskula, M., *et al.* (2018) Bioavailability of quercetin in humans with a focus on interindividual variation. *Comprehensive Reviews in Food Science and Food Safety*, **17**, 714–31.

Amić, D., Davidović-Amić, D., Beslo, D., *et al.* (2007) SAR and QSAR of the antioxidant activity of flavonoids. *Current Medicinal Chemistry*, **14**, 827–45.

Ansari, M.A., Abdul, H.M., Joshi, G., Opii, W.O., and Butterfield, D.A. (2009) Protective effect of quercetin in primary neurons against Aβ(1-42): relevance to Alzheimer's disease. *Journal of Nutritional Biochemistry*, **20**, 269–25.

Baeza, G., Sariá, G., Mateos, R., and Bravo, L. (2016) Dihydrocaffeic acid, a major microbial metabolite of chlorogenic acids, shows similar protective effect than a yerba mate phenolic extract against oxidative stress in HepG2 cells. *Food Research International*, **87**, 25–33.

Baeza, G., Bachmair, E.M., Wood, S., *et al.* (2017) The colonic metabolites dihydrocaffeic acid and dihydroferulic acid are more effective inhibitors of in vitro platelet activation than their phenolic precursors. *Food & Function*, **8**, 1333–42.

Bhagwat, S., Haytowitz, D.B., and Holden, J.M. (2014) USDA Database for the Flavonoid Content of Selected Foods. Nutrient Data Laboratory, Beltsville Human Nutrition Research Center, ARS, USDA, Release 3. Available at: www.ars.usda.gov/northeast-area/beltsville-md-bhnrc/beltsville-human-nutrition-research-center/methods-and-application-of-food-composition-laboratory/mafcl-site-pages/flavonoids/.

Booth, A.N., Deeds, F., Jones, F., and Murray, C. (1956) The metabolic fate of rutin and quercetin in the animal body. *Journal of Biological Chemistry*, **223**, 251–7.

Braune, A., Gütschow, M., Engst, W., and Blaut, M. (2001) Degradation of quercetin and luteolin by eubacterium ramulus. *Applied and Environmental Microbiology*, **67**, 5558–67.

Bureau, G., Longpré, F., and Martinoli, M.-G. (2008) Resveratrol and quercetin, two natural polyphenols, reduce apoptotic neuronal cell death induced by neuroinflammation. *Journal of Neuroscience Research*, **86**, 403–10.

Cai, H., Boocock, D.J., Steward, W.P., and Gescher, A.J. (2007) Tissue distribution in mice and metabolism in murine and human liver of apigenin and tricin, flavones with putative cancer chemopreventive properties. *Cancer Chemotherapy and Pharmacology*, **60**, 257–66.

Calderón-Montaño, J.M., Burgos-Morón, E., Pérez-Guerrero, C., and López-Lázaro, M. (2011) A review on the dietary flavonoid kaempferol. *Mini Reviews in Medicinal Chemistry*, **11**, 298–344.

Cao, J., Zhang, Y., Chen, W., and Zhao, X. (2010) The relationship between fasting plasma concentrations of selected flavonoids and their ordinary dietary intake. *British Journal of Nutrition*, **103**, 249–55.

Carregosa, D., Carecho, R., Figueira, I., and N Santos, C. (2019) Low molecular weight metabolites from polyphenols as effectors for attenuating neuroinflammation. *Journal of Agricultural and Food Chemistry*, **68**, 1790–1807.

Cermak, R. and Breves, G.M. (2006) In vitro degradation of the flavonol quercetin and of quercetin glycosides in the porcine hindgut. *Archives of Animal Nutrition*, **60**, 180–9.

Chalet, C., Rubbens, J., Tack, J., Duchateau, G.S., and Augustijns, P. (2018) Intestinal disposition of quercetin and its phase-II metabolites after oral administration in healthy volunteers. *Journal of Pharmacy and Pharmacology*, **70**, 1002–8.

Chen, H., Yao, K., Nadas, J., *et al.* (2012a) Prediction of Molecular Targets of Cancer Preventing Flavonoid Compounds Using Computational Methods. PLoS ONE. N.J. Haslam (Ed.). Available at: https://journals.plos.org/plosone/article?id=10.1371/journal.pone.0038261

Chen, Z., Kong, S., Song, F., Li, L., and Jiang, H. (2012b) Pharmacokinetic study of luteolin, apigenin, chrysoeriol and diosmetin after oral administration of Flos Chrysanthemi extract in rats. *Fitoterapia*, **83**, 1616–22.

Cheng, A.X., Han, X.J., Wu, Y.F., and Lou, H.X. (2014) The function and catalysis of 2-oxoglutarate-dependent oxygenases involved in plant flavonoid biosynthesis. *International Journal of Molecular Sciences*, **15**, 1080–95.

Choi, B.Y. (2019) Biochemical basis of anti-cancer-effects of phloretin - a natural dihydrochalcone. *Molecules*, **24**, 278.

Choi, E.J. and Ahn, W.S. (2008) Kaempferol induced the apoptosis via cell cycle arrest in human breast cancer MDA-MB-453 cells. *Nutrition Research and Practice*, **2**, 322.

D'Andrea, G. (2015) Quercetin: a flavonol with multifaceted therapeutic applications? *Fitoterapia*, **106**, 256–71.

Dellafiora, L., Mena, P., Del Rio, D., and Cozzini, P. (2014) Modeling the effect of phase ii conjugations on topoisomerase i poisoning: pilot study with luteolin and quercetin. *Journal of Agricultural and Food Chemistry*, **62**, 5881–6.

Devi, K.P., Rajavel, T., Habtemariam, S., Nabavi, S.F., and Nabavi, S.M. (2015) Molecular mechanisms underlying anticancer effects of myricetin. *Life Sciences*, **142**, 19–25.

Du, L., Zhao, M., Xu, J., *et al.* (2014) Identification of the metabolites of myricitrin produced by human intestinal bacteria in vitro using ultra-performance liquid chromatography/quadrupole time-of-flight mass spectrometry. *Expert Opinion on Drug Metabolism & Toxicology*, **10**(7), 921–31.

Dueñas, M., Gonzalez-Manzano, S., Surco-Laos, F., *et al.* (2012) Characterization of sulfated quercetin and epicatechin metabolites. *Journal of Agricultural and Food Chemistry*, **60**, 3592–8.

Endale, M., Park, S.C., Kim, S. (2013) Quercetin disrupts tyrosine-phosphorylated phosphatidylinositol 3-kinase and myeloid differentiation factor-88 association, and inhibits MAPK/AP-1 and IKK/NF-κB-induced inflammatory mediators production in RAW 264.7 cells. *Immunobiology*, **218**, 1452–67.

Esteban-Fernández, A., Rendeiro, C., Spencer, J.P., *et al.* (2017) Neuroprotective effects of selected microbial-derived phenolic metabolites and aroma compounds from wine in human sh-sy5y neuroblastoma cells and their putative mechanisms of action. *Frontiers in Nutrition*, **4**, 3.

Feng, H., Cao, J., Zhang, G., and Wang, Y. (2017) Kaempferol attenuates cardiac hypertrophy via regulation of ask1/mapk signaling pathway and oxidative stress. *Planta Medica*, **83**, 837–45.

Feng, X., Li, Y., Brobbey Oppong, M., and Qiu, F. (2018) Insights into the intestinal bacterial metabolism of flavonoids and the bioactivities of their microbe-derived ring cleavage metabolites. *Drug Metabolism Reviews*, **50**, 343–56.

Formica, J.V. and Regelson, W. (1995) Review of the biology of quercetin and related bioflavonoids. *Food and Chemical Toxicology*, **33**, 1061–80.

Ghumatkar, P.J., Patil, S.P., Peshattiwar, V., *et al.* (2019) The modulatory role of phloretin in Aβ25-35 induced sporadic Alzheimer's disease in rat model. *Naunyn-Schmiedeberg's Archives of Pharmacology*, **392**, 327–39.

Gradolatto, A. (2004) Metabolism of apigenin by rat liver phase I and phase II enzymes and by isolated perfused rat liver. *Drug Metabolism and Disposition*, **32**, 58–65.

Griffiths, L.A. and Barrow, A. (1972) Metabolism of flavonoid compounds in germ-free rats. *Biochemical Journal*, **130**, 1161–2.

Hackman, R., Pentikainen, P., Neuvonen, P.J., and Vapaatalo, H. (1973) Inhibition of the apomorphine gnawing compulsion by amantadine. *Experientia*, **29**, 1524–5.

Hanske, L., Loh, G., Sczesny, S., Blaut, M., and Braune, A. (2009) The bioavailability of apigenin-7-glucoside is influenced by human intestinal microbiota in rats. *Journal of Nutrition*, **139**, 1095–1102.

Harwood, M., Danielewska-Nikiel, B., Borzelleca, J.F., *et al.* (2007) A critical review of the data related to the safety of quercetin and lack of evidence of in vivo toxicity, including lack of genotoxic/carcinogenic properties. *Food and Chemical Toxicology*, **45**, 2179–205.

Hashemzaei, M., Far, A.D., Yari, A., *et al.* (2017) Anticancer and apoptosis-inducing effects of quercetin in vitro and in vivo. *Oncology Reports*, **38**, 819–28.

Hayasaka, N., Shimizu, N., Komoda, T., *et al.* (2018) Absorption and metabolism of luteolin in rats and humans in relation to in vitro anti-inflammatory effects. *Journal of Agricultural and Food Chemistry*, **66**, 11320–9.

He, P., Yan, S., Wen, X., *et al.* (2019) Eriodictyol alleviates lipopolysaccharide-triggered oxidative stress and synaptic dysfunctions in BV-2 microglial cells and mouse brain. *Journal of Cellular Biochemistry*, **120**, 14756–70.

Hostetler, G.L., Ralston, R.A., and Schwartz, S.J. (2017) Flavones: food sources, bioavailability, metabolism, and bioactivity. *Advances in Nutrition*, **8**, 423–35.

Houyel, L., Petit, J., Langlois, J., *et al.* (1988) [Surgery for incomplete atrioventricular canal. Mid-term follow-up apropos of 128 patients]. *Archives des Maladies du Coeur et des Vaisseaux*, **81**, 501–5.

Hu, W., Wang, X., Wu, L., *et al.* (2016) Apigenin-7-O-*β*- d -glucuronide inhibits LPS-induced inflammation through the inactivation of AP-1 and MAPK signaling pathways in RAW 264.7 macrophages and protects mice against endotoxin shock. *Food & Function*, **7**, 1002–13.

Imran, M., Rauf, A., Shah, Z.A., *et al.* (2019) Chemo-preventive and therapeutic effect of the dietary flavonoid kaempferol: a comprehensive review. *Phytotherapy Research*, **33**, 263–75.

Jaganath, I.B., Mullen, W., Edwards, C.A., Crozier, A. (2006) The relative contribution of the small and large intestine to the absorption and metabolism of rutin in man. *Free Radical Research*, **40**, 1035–46.

Jiang, N., Doseff, A., and Grotewold, E. (2016) Flavones: from biosynthesis to health benefits. *Plants*, **5**, 27.

Jung, Y.-H., Heo, J., Lee, Y.J., *et al.* (2010) Quercetin enhances TRAIL-induced apoptosis in prostate cancer cells via increased protein stability of death receptor 5. *Life Sciences*, **86**, 351–7.

Kamaraj, S., Vinodhkumar, R., Anandakumar, P., *et al.* (2007) The effects of quercetin on antioxidant status and tumor markers in the lung and serum of mice treated with benzo(a)pyrene. *Biological & Pharmaceutical Bulletin*, **30**, 2268–73.

Kanazawa, K., Uehara, M., Yanagitani, H., Hashimoto, T. (2006) Bioavailable flavonoids to suppress the formation of 8-OHdG in HepG2 cells. *Archives of Biochemistry and Biophysics*, **455**, 197–203.

Kang, N.J., Jung, S.K., Lee, K.W., and Lee, H.J. (2011) Myricetin is a potent chemopreventive phytochemical in skin carcinogenesis. *Annals of the New York Academy of Sciences*, **1229**, 124–32.

Kerimi, A. and Williamson, G. (2018) Differential impact of flavonoids on redox modulation, bioenergetics, and cell signaling in normal and tumor cells: a comprehensive review. *Antioxidants & Redox Signaling*, **29**, 1633–59.

Kim, B.-W., Koppula, S., Park, S.Y., *et al.* (2015) Attenuation of neuroinflammatory responses and behavioral deficits by Ligusticum officinale (Makino) Kitag in stimulated microglia and MPTP-induced mouse model of Parkinson's disease. *Journal of Ethnopharmacology*, **164**, 388–97.

Kim, H.H., Kim, D.H., Kim, M.H., *et al.* (2013) Flavonoid constituents in the leaves of Myrica rubra sieb. et zucc. with anti-inflammatory activity. *Archives of Pharmacal Research*, **36**, 1533–40.

Kim, H.P., Mani, I., Iversen, L., and Ziboh, V.A. (1998) Effects of naturally-occurring flavonoids and biflavonoids on epidermal cyclooxygenase and lipoxygenase from guinea-pigs. *Prostaglandins, Leukotrienes, and Essential Fatty Acids*, **58**, 17–24.

Ko, S.-Y. (2012) Myricetin suppresses LPS-induced MMP expression in human gingival fibroblasts and inhibits osteoclastogenesis by downregulating NFATc1 in RANKL-induced RAW 264.7 cells. *Archives of Oral Biology*, **57**, 1623–32.

Lagoa, R., Lopez-Sanchez, C., Samhan-Arias, A.K., *et al.* (2009) Kaempferol protects against rat striatal degeneration induced by 3-nitropropionic acid. *Journal of Neurochemistry*, **111**, 473–87.

Li, Y.-C., Yeh, C.H., Yang, M.L., and Kuan, Y.H. (2012) Luteolin suppresses inflammatory mediator expression by blocking the akt/nf κ b pathway in acute lung injury induced by lipopolysaccharide in mice. *Evidence-Based Complementary and Alternative Medicine*, **2012**, 1–8.

Li, Y., Yao, J., Han, C., *et al.* (2016) Quercetin, Inflammation and Immunity. *Nutrients*, **8**, 167.

Lin, Y., Wu, B., Li, Z., *et al.* (2012) Metabolite Identification of Myricetin in Rats Using HPLC Coupled with ESI-MS. *Chromatographia*, **75**, 655–60.

Liu, Y., Gao, L., Guo, S., *et al.* (2017) Kaempferol alleviates angiotensin ii-induced cardiac dysfunction and interstitial fibrosis in mice. *Cellular Physiology and Biochemistry*, **43**, 2253–63.

Liu, Y. and Yan, X. (2019) Eriodictyol inhibits survival and inflammatory responses and promotes apoptosis in rheumatoid arthritis fibroblast-like synoviocytes through AKT/FOXO1 signaling. *Journal of Cellular Biochemistry*, **120**, 14628-35.

Loke, W.M., Jenner, A.M., Proudfoot, J.M., *et al.* (2009) A metabolite profiling approach to identify biomarkers of flavonoid intake in humans. *Journal of Nutrition*, **139**, 2309–14.

Lu, X., Sun, D., Chen, Z., *et al.* (2011) Evaluation of hepatic clearance and drug-drug interactions of luteolin and apigenin by using primary cultured rat hepatocytes. *Die Pharmazie*, **66**, 600–5.

Luo, H., Rankin, G.O., Liu, L., *et al*. (2009) Kaempferol inhibits angiogenesis and VEGF expression through both HIF dependent and independent pathways in human ovarian cancer cells. *Nutrition and Cancer*, **61**, 554–63.

Luo, Y., Shang, P., and Li, D. (2017) Luteolin: a flavonoid that has multiple cardio-protective effects and its molecular mechanisms. *Frontiers in Pharmacology*, **8**, 692.

Lv, P., Yu, J., Xu, X., Lu, T., and Xu, F. (2019) Eriodictyol inhibits high glucose-induced oxidative stress and inflammation in retinal ganglial cells. *Journal of Cellular Biochemistry*, **120**, 5644–51.

Ma, Z.-G., Wang, J., Jiang, H., Liu, T.W., and Xie, J.X. (2007) Myricetin reduces 6-hydroxydopamine-induced dopamine neuron degeneration in rats. *NeuroReport*, **18**, 1181–5.

Mariadoss, A.V.A., Vinyagam, R., Rajamanickam, V., *et al*. (2019) Pharmacological aspects and potential use of phloretin: a systemic review. *Mini-Reviews in Medicinal Chemistry*, **19**, 1060–7.

Monagas, M., Khan, N., Andres-Lacueva, C., *et al*. (2009) Effect of cocoa powder on the modulation of inflammatory biomarkers in patients at high risk of cardiovascular disease. *American Journal of Clinical Nutrition*, **90**, 1144–50.

Mu, C., Jia, P., Yan, Z., *et al*. (2007) Quercetin induces cell-cycle G1 arrest through elevating Cdk inhibitors p21 and p27 in human hepatoma cell line (HepG2). *Methods and Findings in Experimental and Clinical Pharmacology*, **29**, 179.

Muthian, G. and Bright, J.J. (2004) Quercetin, a flavonoid phytoestrogen, ameliorates experimental allergic encephalomyelitis by blocking il-12 signaling through jak-stat pathway in t lymphocyte. *Journal of Clinical Immunology*, **24**, 542–52.

Nabavi, S.F., Russo, G.L., Daglia, G.M., and Nabavi, S.M. (2015) Role of quercetin as an alternative for obesity treatment: you are what you eat! *Food Chemistry*, **179**, 305–10.

Németh, K., Plumb, G.W., Berrin, J.G., *et al*. (2003) Deglycosylation by small intestinal epithelial cell beta-glucosidases is a critical step in the absorption and metabolism of dietary flavonoid glycosides in humans. *European Journal of Nutrition*, **42**, 29–42.

Nielsen, S.E., Breinholt, V., Justesen, U., *et al.* (1998) In vitro biotransformation of flavonoids by rat liver microsomes. *Xenobiotica*, **28**, 389–401.

Olthof, M.R., Hollman, P.C., Buijsman, M.N., *et al.* (2003) Chlorogenic acid, quercetin-3-rutinoside and black tea phenols are extensively metabolized in humans. *Journal of Nutrition*, **133**, 1806–14.

Otaki, N., Kimira, M., Katsumata, S., *et al.* (2009) Distribution and major sources of flavonoid intakes in the middle-aged Japanese women. *Journal of Clinical Biochemistry and Nutrition*, **44**, 231–8.

Ożarowski, M., Mikolajczak, P.L., Kujawski, R., *et al.* (2018) Pharmacological effect of quercetin in hypertension and its potential application in pregnancy-induced hypertension: review of *in vitro*, *in vivo*, and clinical studies. *Evidence-Based Complementary and Alternative Medicine*, **2018**, 1–19.

Panche, A.N., Diwan, A.D., and Chandra, S.R. (2016) Flavonoids: an overview. *Journal of Nutritional Science*, **5**, e47.

Pandurangan, A.K. and Esa, N.M. (2014) Luteolin, a bioflavonoid inhibits colorectal cancer through modulation of multiple signaling pathways: a review. *Asian Pacific Journal of Cancer Prevention*, **15**, 5501–8.

Park, K.-S., Chong, Y., and Kim, M.K. (2016) Myricetin: biological activity related to human health. *Applied Biological Chemistry*, **59**, 259–69.

Park, S., Sapkota, K., Kim, S., *et al.* (2011) Kaempferol acts through mitogen-activated protein kinases and protein kinase B/AKT to elicit protection in a model of neuroinflammation in BV2 microglial cells. *British Journal of Pharmacology*, **164**, 1008–25.

Pforte, H., Naser, T., Jacobasch, G., and Buhr, H.J. (2000) Absorption and modification of rutin in the human stomach. In G.R. Fenwick and I.T. Johnson (Eds.), *Dietary Anticarcinogens and Antimutagens: Chemical and Biological Aspects*. London: Elsevier.

Pommier, Y. (2013) Drugging topoisomerases: lessons and challenges. *ACS Chemical Biology*, **8**, 82–95.

Purchartová, K., Valentová, K., Pelantová, H., et al. (2015) Prokaryotic and eukaryotic aryl sulfotransferases: sulfation of quercetin and its derivatives. *ChemCatChem*, **7**, 3152–62.

Qin, Y., Cui, W., Yang, X., and Tong, B. (2016) Kaempferol inhibits the growth and metastasis of cholangiocarcinoma in vitro and in vivo. *Acta Biochimica et Biophysica Sinica*, **48**, 238–45.

Raffa, D., Maggio, B., Raimondi, M.V., *et al.* (2017) Recent discoveries of anticancer flavonoids. *European Journal of Medicinal Chemistry*, **142**, 213–28.

Russo, M., Spagnuolo, C., Tedesco, I., *et al.* (2012) The flavonoid quercetin in disease prevention and therapy: facts and fancies. *Biochemical Pharmacology*, **83**, 6–15.

Salehi, B., Venditti, A., Sharifi-Rad, M., *et al.* (2019) The therapeutic potential of Apigenin. *International Journal of Molecular Sciences*, **20**, 1305.

Salvamani, S., Gunasekaran, B., Shaharuddin, N.A., *et al.* (2014) Antiartherosclerotic effects of plant flavonoids, *BioMed Research International*, **2014**, 1–11.

Sarian, M.N., Ahmed, Q.U., Mat So'ad, S.Z., *et al.* (2017) Antioxidant and antidiabetic effects of flavonoids: a structure-activity relationship based study. *BioMed Research International*, **2017**, 1–14.

Schoefer, L., Mohan, R., Schwiertz, A., Braune, A., and Blaut, M. (2003) Anaerobic degradation of flavonoids by clostridium orbiscindens. *Applied and Environmental Microbiology*, **69**, 5849–54.

Selma, M.V., Espín, J.C., and Tomás-Barberán, F.A. (2009) Interaction between phenolics and gut microbiota: role in human health. *Journal of Agricultural and Food Chemistry*, **57**, 6485–501.

Semwal, D.K., Semwal, R.B., Combrinck, S., and Viljoen, A. (2016) Myricetin: a dietary molecule with diverse biological activities. *Nutrients*, **8**, 90.

Serra, A., Macià, A., Romero, M.-P., *et al.* (2012) Metabolic pathways of the colonic metabolism of flavonoids (flavonols, flavones and flavanones) and phenolic acids. *Food Chemistry*, **130**, 383–93.

Shi, Y., Pang, X., Wang, J., and Liu, G. (2018) NanoTRAIL-oncology: a strategic approach in cancer research and therapy. *Advanced Healthcare Materials*, **7**, e1800053.

Shimoi, K., Saka, N., Kaji, K., *et al.* (2000) Metabolic fate of luteolin and its functional activity at focal site. *BioFactors*, **12**, 181–6.

Slámová, K., Kapešová, J., and Valentová, K. (2018) "Sweet flavonoids": glycosidase-catalyzed modifications. *International Journal of Molecular Sciences*, **19**(7), 2126.

Somerset, S.M. and Johannot, L. (2008) Dietary flavonoid sources in Australian adults.. *Nutrition and Cancer*, **60**, 442–9.

Takano-Ishikawa, Y., Goto, M., and Yamaki, K. (2006) Structure-activity relations of inhibitory effects of various flavonoids on lipopolysaccharide-induced prostaglandin E2 production in rat peritoneal macrophages: comparison between subclasses of flavonoids. *Phytomedicine*, **13**, 310–17.

Teles, Y., Souza, M., and Souza, M. (2018) Sulphated flavonoids: biosynthesis, structures, and biological activities. *Molecules*, **23**, 480.

Tiwari, R., Mohan, M., Kasture, S., *et al.* (2009) Cardioprotective potential of myricetin in isoproterenol-induced myocardial infarction in wistar rats. *Phytotherapy Research*, **23**, 1361–6.

Valentová, K., Vrba, J., Bancirová, M., *et al.* (2014) Isoquercitrin: pharmacology, toxicology, and metabolism. *Food and Chemical Toxicology*, **68**, 267–82.

Vinson, J.A. (2019) Intracellular polyphenols: how little we know. *Journal of Agricultural and Food Chemistry*, **67**, 3865–70.

Vogiatzoglou, A., Mulligan, A.A., Lentjes, M.A.H., *et al.* (2015) Flavonoid intake in european adults (18 to 64 years). *PLoS ONE*, **10**, e0128132.

Vollmer, M., Esders, S., Farquharson, F.M., *et al.* (2018) Mutual interaction of phenolic compounds and microbiota: metabolism of complex phenolic apigenin-c- and kaempferol-o-derivatives by human fecal samples. *Journal of Agricultural and Food Chemistry*, **66**, 485–97.

Walle, T. (2004) Absorption and metabolism of flavonoids. *Free Radical Biology and Medicine*, **36**, 829–37.

Walle, T., Walle, U.K., and Halushka, P.V. (2001) Carbon dioxide is the major metabolite of quercetin in humans. *Journal of Nutrition*, **131**, 2648–52.

Wang, S., Yao, J., Zhou, B., *et al.* (2018) Bacteriostatic effect of quercetin as an antibiotic alternative in vivo and its antibacterial mechanism in vitro. *Journal of Food Protection*, **81**, 68–78.

Williamson, G. and Manach, C. (2005) Bioavailability and bioefficacy of polyphenols in humans. II. Review of 93 intervention studies. *American Journal of Clinical Nutrition*, **81**, 243S–255S.

Wu, B., Kulkarni, K., Basu, S., *et al.* (2011) First-pass metabolism via udp-glucuronosyltransferase: a barrier to oral bioavailability of phenolics. *Journal of Pharmaceutical Sciences,* **100**, 3655–81.

Xie, Y., Ji, R., and Han, M. (2018) Eriodictyol protects H9c2 cardiomyocytes against the injury induced by hypoxia/reoxygenation by improving the dysfunction of mitochondria. *Experimental and Therapeutic Medicine,* **17**, 551–7.

Yang, J.-H. (2005) Inhibition of lung cancer cell growth by quercetin glucuronides via G2/M arrest and induction of apoptosis. *Drug Metabolism and Disposition,* **34**, 296–304.

Ying, Y., Jin, J., Ye, L., *et al.* (2018) Phloretin prevents diabetic cardiomyopathy by dissociating keap1/nrf2 complex and inhibiting oxidative stress. *Frontiers in Endocrinology,* **9**, 774.

Yu, W., Wang, Y., Zhou, D.-X., *et al.* (2014) Equol is neuroprotective during focal cerebral ischemia and reperfusion that involves p-src and gp91phox. *Current Neurovascular Research,* **11**, 367–77.

Zahedi, M., Ghiasvand, R., Feizi, A., *et al.* (2013) Does quercetin improve cardiovascular risk factors and inflammatory biomarkers in women with type 2 diabetes: a double-blind randomized controlled clinical trial. *International Journal of Preventive Medicine,* **4**, 777–85.

Zaidun, N.H., Thent, Z.C., and Latiff, A.A. (2018) Combating oxidative stress disorders with citrus flavonoid: naringenin. *Life Sciences,* **208**, 111–22.

Zamora-Ros, R., Andres-Lacueva, C., Lamuela-Raventós, R.M., *et al.* (2010) Estimation of dietary sources and flavonoid intake in a Spanish adult population (EPIC-Spain). *Journal of the American Dietetic Association,* **110**, 390–8.

Zamora-Ros, R., Achaintre, D., Rothwell, J.A., *et al.* (2016) Urinary excretions of 34 dietary polyphenols and their associations with lifestyle factors in the EPIC cohort study. *Scientific Reports,* **6**, 26905.

Zeng, W., Jin, L., Zhang, F., *et al.* (2018) Naringenin as a potential immunomodulator in therapeutics. *Pharmacological Research,* **135**, 122–6.

Zhang, K., Ma, Z., Wang, J., *et al.* (2011) Myricetin attenuated MPP+-induced cytotoxicity by anti-oxidation and inhibition of

MKK4 and JNK activation in MES23.5 cells. *Neuropharmacology*, **61**, 329–35.
Zhao, L., Yuan, X., Wang, J., *et al.* (2019) A review on flavones targeting serine/threonine protein kinases for potential anticancer drugs. *Bioorganic & Medicinal Chemistry*, **27**, 677–85.

6

Isoflavones

Cara L. Frankenfeld

Department of Global and Community Health, George Mason University, Fairfax, VA, USA

6.1 Uptake and Metabolism of Isoflavones

Isoflavones are polyphenolic compounds that structurally resemble mammalian estrogens (Figure 6.1). Isoflavones can be found in glycoside (bound with sugar molecule) or aglycone forms. The most common forms found in the human diet include daidzin and genistin (glycoside) or daidzein and genistein (aglycone), and these are found in high amounts in soybeans and soy products, and to a lesser extent in other members of the Leguminosae family (Horn-Ross *et al.*, 2000; Liggins *et al.*, 2000a,b, 2002; Messina, 1999; Reinli and Block, 1996; Wakai *et al.*, 1999). Biochanin A and formononetin are commonly found in clover and are parent compounds to genistein and daidzein respectively (Heinonen *et al.*, 2004; Nair *et al.*, 1991), which is why isoflavones are also observed in animal milk (Kasparovska *et al.*, 2016; Njastad *et al.*, 2014). See Chapter 1 for more information about dietary sources of isoflavones.

6.1.1 Gut Microbial Metabolism

Isoflavones that are present as glycosides are bound to a sugar molecule. Enzymes present in enterocytes and gut microbial metabolism are responsible for converting glycosides to aglycones (Chuankhayan *et al.*, 2007; Day *et al.*, 1998;

Dietary Polyphenols: Metabolism and Health Effects, First Edition.
Edited by Francisco A. Tomás-Barberán, Antonio González-Sarrías, and Rocío García-Villalba.

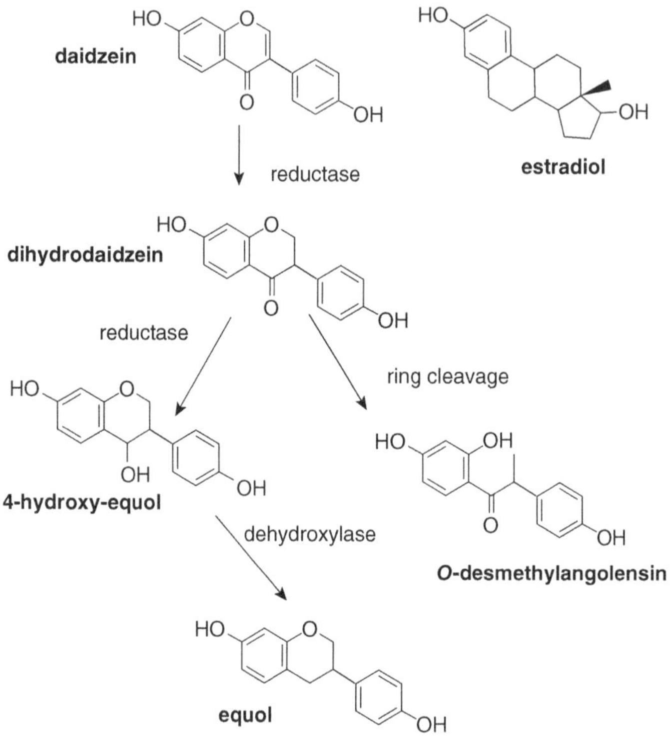

Figure 6.1 Chemical structures of daidzein and gut microbiota metabolites equol and *O*-desmethylangolensin in a simplified illustration of metabolism. 17β-estradiol is presented to illustrate structural similarity of the compounds. Source: Adapted from Heinonen *et al.* (1999) and reproduced from Frankenfeld *et al.* (2014).

Franke *et al.*, 2014; Kaya *et al.*, 2008; Rowland *et al.*, 2003). Common isoflavone glycosides include daidzin, genistin, and glycitin. The aglycones are referred to as daidzein, genistein, and glycitein, respectively. Gut microbial metabolism is also responsible for biotransformation to other compounds, such as equol and *O*-desmethylangolensin (Atkinson *et al.*, 2005; Heinonen *et al.*, 1999; Rafii *et al.*, 2003). Isoflavones and metabolites are then absorbed, circulated in the body, and then excreted in urine. Information about the relationship between the gut microbiome and flavonoids is provided in Chapter 13.

6.1.2 Pharmacokinetic Studies

There are several pharmacokinetic studies of isoflavones in rodent and other animal models. However, animal models are challenged for evaluating isoflavone exposure in humans due to differences in gut microbial metabolism (Atkinson *et al.*, 2005), and discussion about the pharmacokinetics of isoflavones will focus primarily on human studies, with some supporting information from other model systems.

In order to be absorbed from the gastrointestinal tract into the body, the sugar moiety needs to be removed. There are different ways and places in which this occurs, including by lactase-phlorizin hydrolase which is located in the enterocyte brush border (Day *et al.*, 2000; Nielsen and Williamson, 2007). As a result, isoflavones have a biphasic appearance in urine and blood after consumption (Franke *et al.*, 2014; Setchell *et al.*, 2002a; Vergne *et al.*, 2008; Zubik and Meydani, 2003). The initial peak appearance in urine and blood occurs approximately 1–2 hours after isoflavone consumption and is followed by a second, higher peak approximately 4–6 hours after consumption. It is believed that mucosal glucosidases may be responsible for the removal of the sugar moiety to allow some isoflavone absorption to occur in the small intestine, which corresponds to the first peak. β-Glucosidase activity by gut microorganisms is believed to contribute to the second peak in urine and blood. The contributors to these patterns were confirmed in participants who consumed both soy and oral antibiotics (Franke *et al.*, 2004).

Dose and frequency of isoflavone consumption do not appear to influence bioavailability (Yang *et al.*, 2005). However, there are other factors that affect the bioavailability of isoflavones in humans. Glycoside and aglycone isoflavone forms are both found in foods, and for equivalent doses, the half-life of the compounds is not different (Nielsen and Williamson, 2007). However, isoflavones consumed in fermented products (contain primarily aglycones) than in unfermented soy products (contains primary glycosides) have higher bioavailability, as assessed by percent of isoflavone dose excreted in urine (Hutchins *et al.*, 1995; Izumi *et al.*, 2000; Richelle *et al.*, 2002; Zubik and Meydani, 2003). In some studies, liquid forms of isoflavones, such as soymilk, appear to have higher bioavailability when compared with texturized vegetable protein or tempeh (Faughnan *et al.*, 2004). However, other studies do not observe differences in bioavailability associated with food source (Tew *et al.*, 1996; Xu *et al.*, 2000). There may be differences associated with food versus supplement, and some evidence suggests that bioavailability may be higher in supplements (Anupongsanugool *et al.*, 2005; Vergne *et al.*, 2008), but this is not well established, and may be specific to isoflavone (Anupongsanugool *et al.*, 2005), or supplement or extract preparation (Rodriguez-Morato *et al.*, 2015; Setchell *et al.*, 2001). Apart from differences in bioavailability in the very young, which may be due to immature gut microbiomes, there do not appear to be strong differences in bioavailability associated with inherent personal characteristics of age or sex (Nielsen and Williamson, 2007).

Isoflavones are distributed throughout the body via the circulation (Franke *et al.*, 2014). Isoflavones that enter the bloodstream are transported to the liver, and in the liver, methylation, glycine conjugation, sulfation, and β-glucuronidation can occur (Barnes *et al.*, 2011). Using tissue biopsies, it is has been observed that isoflavones can enter various tissues in the human body, such as breast tissue (Bolca *et al.*, 2010; Huser *et al.*, 2018; Maubach *et al.*, 2004). For many tissues, it is only possible to study distribution using animal models. Based on animal models, it is expected that isoflavones can be widely distributed throughout the body and can interact with cells in many tissues, including brain, bone, liver, kidney, spleen, heart, muscle, thyroid, pituitary gland, and reproductive organs (Urpi-Sarda *et al.*, 2008; Yasuda *et al.*, 2005).

Tissue distribution is unequal, with higher concentrations in rats observed in liver and kidney than spleen and reproductive organs (Yasuda *et al.*, 2005), and notably higher kidney concentrations than other tissues in ewes (Urpi-Sarda *et al.*, 2008). Isoflavones are excreted in urine and feces (Franke *et al.*, 2014; Rowland *et al.*, 2003; Wiseman *et al.*, 2004; Xu *et al.*, 1995), and it has been observed that isoflavones have a relatively short half-life in blood and overnight urine concentrations are believed to provide a better marker of soy consumption (Franke *et al.*, 2014). Urinary recovery of isoflavones varies by isoflavone and source. As an illustration, in a study of Korean women 45.8% of daidzein was recovered in urine after fermented soybean consumption, but 33.8% of daidzein was recovered in urine after soymilk consumption (Chang and Choue, 2013). However, genistein recovery in urine was 23.4% from fermented soybeans and 22.1% from soymilk consumption.

6.2 Biological Mechanisms of Isoflavones

Isoflavones are considered phytoestrogens because of their ability to bind to estrogen receptors and elicit responses. The majority of our knowledge about the biological mechanisms of isoflavones is hormonal. Other well-studied mechanisms include antioxidant and antiinflammatory activities . Isoflavones may exert other effects, such as involvement in cellular processes (e.g., growth, replication, proliferation) (Akiyama *et al.*, 1987; Hua *et al.*, 2018; Mahmoud *et al.*, 2014; Markovits *et al.*, 1989) and exhibit antimicrobial properties (Cannalire *et al.*, 2017; das Neves *et al.*, 2016; Hanski *et al.*, 2014; Hong *et al.*, 2006; Lauwaet *et al.*, 2010; Park *et al.*, 2014), which are not yet as well characterized. This section will focus on hormonal, antioxidant, and antiinflammatory activities.

6.2.1 Hormonal

There is an extensive body of research that demonstrates the hormonal actions of isoflavones. Estrogen receptors (ERs) are considered to be promiscuous, in that they are capable of binding to a variety of structurally diverse physiological steroids (Baker and Lathe, 2018). This is important because it is through this action that estrogens exert influence on or regulate physiological processes in bone, brain, cardiovascular system, and reproductive organs (Arimoto *et al.*, 2013; Gao and Dahlman-Wright, 2011; Hamilton *et al.*, 2017; Leitman *et al.*, 2010; Prins and Korach, 2008; Sugiyama *et al.*, 2010; Zuloaga *et al.*, 2012). Isoflavones can bind to ERs and elicit responses, albeit at lower levels than estradiol (Kostelac *et al.*, 2003), the primary endogenous estrogen in humans. There are two main isoforms of ERs: ER-α and ER-β (Kuiper *et al.*, 1997; Minutolo *et al.*, 2011; Paterni *et al.*, 2013), and isoflavones have a preference for the ER-β isoform (Leclercq and Jacquot, 2014).

However, they bind and activate both receptors. The half-maximal effective concentration (EC_{50}) of 17β-estradiol for ER-α was observed to be 0.03 μM, whereas genistein and daidzein are less effective at 15 μM and >300 μM respectively (Kostelac *et al.*, 2003). For ER-β, the EC_{50} of 17β-estradiol was observed to be 0.01 μM, with genistein and daidzein being less effective at 0.03 μM and 0.35 μM, respectively (Kostelac *et al.*, 2003). These observations indicate the ability for isoflavones to bind to ERs, but less effectively than estradiol.

Given these observations, a consideration for hormonal actions is that in environments with high circulating estrogens (e.g., premenopausal women), isoflavones may exert antiestrogenic effects by competing for binding to ERs, and in environments with low circulating estrogens (e.g., post-menopausal women or men), isoflavones may exert weakly estrogenic effects.

6.2.2 Antioxidant

Isoflavones exert antioxidant activities in *in vitro* models. In a study comparing six isoflavones (genistein, daidzein, glycitein, formononetin, biochanin A, and prunetin), antioxidant capacity was evaluated with *in vitro* assays assessing LDL oxidation and oxygen radical absorbance capacity (ORAC) (Rufer and Kulling, 2006). In these analyses, all of the isoflavones tested exhibited some level of antioxidant capacity. Genistein, daidzein, and equol exhibited the highest antioxidant capacity and were higher than the control of quercetin. Other *in vitro* studies also support antioxidant activities of isoflavones (Liang *et al.*, 2019; Liu *et al.*, 2005; Park *et al.*, 2011; Romani *et al.*, 2010). Isoflavones have also been observed to protect cancerous and noncancerous cells from free radical-induced DNA damage (Foti *et al.*, 2005; Khan *et al.*, 2008; Leung *et al.*, 2009; Pugalendhi *et al.*, 2009). However, the evidence is mixed, with some studies reporting favorable impacts (Djuric *et al.*, 2001; Wiseman *et al.*, 2000) or no impact (Hanwell *et al.*, 2009; Heneman *et al.*, 2007; Hutchins *et al.*, 2005). There is a need for further human studies for different oxidative stress conditions to evaluate whether and under what circumstances the antioxidant mechanisms observed *in vitro* translate to human health.

6.2.3 Antiinflammatory

Inflammation can result when the body is exposed to harmful stimuli and is involved in some disease development or progression, including heart disease, obesity, and cancer (Calabro and Yeh, 2008; Kaplan and Frishman, 2001; Mathieu *et al.*, 2010; Norwitz *et al.*, 1999; O'Connor *et al.*, 2010; Rubio-Jurado *et al.*, 2018; Sudhir, 2006; Sun *et al.*, 2007; Tahara and Arisawa, 2012; Thompson *et al.*, 2015; Tracy, 2003; Willerson and Ridker, 2004).

There are several mechanisms by which isoflavones exert antiinflammatory effects. *In vitro* studies have evaluated human chondrocytes and observed that pretreatment with genistein reduces LPS-stimulated inflammatory responses (Hooshmand *et al.*, 2007). Daidzein has also been observed to reduce TNF-α-induced inflammatory responses in mouse epithelial cells (Li *et al.*, 2014) and LPS-induced inflammatory responses in mouse monocyte THP-1 cells (Tanaka *et al.*, 2014). Animal models also support the antiinflammatory potential of genistein and daidzein (Duan *et al.*, 2003; Ganai *et al.*, 2015; Kao *et al.*, 2007; Lim *et al.*, 2013; Paradkar *et al.*, 2004). These observations suggest that there is mechanistic potential for isoflavones to protect against inflammation. However, studies of inflammatory markers in humans have been mixed (Dong *et al.*, 2011; Fanti *et al.*, 2006; Ferguson *et al.*, 2014; Mangano *et al.*, 2013; Song *et al.*, 2016; Wu *et al.*, 2012). Similar to the strong observations in mechanistic studies for antioxidant potential, there is a need for further human studies of different inflammatory conditions to evaluate whether and under what conditions the antiinflammatory mechanisms observed *in vitro* translate to human health.

6.3 Physiological and Health Effects of Isoflavones

It is estimated that higher isoflavone intakes are associated with lower all-cause mortality, when compared to lower isoflavone intakes in observational studies (Nachvak *et al.*, 2019). When considering evidence across studies through meta-analysis, this observation was statistically significant for isoflavone intake but not for soy intake, although the magnitude of association for soy and isoflavone intakes was similar, with a 10% lower risk of death in individuals in the highest category of isoflavone or soy intake compared to the lowest category. These observations suggest that isoflavones influence mortality that may be independent of soy intake, but it is not possible to separate these effects from these observations.

Hormonal, oxidative, or inflammatory mechanisms underlie several health conditions or diseases, and isoflavones have the biological potential to influence a variety of health conditions through these mechanisms. The focus of this section will be on the most commonly studied conditions and diseases for which there is a deeper body of evidence, which includes bone health, cancer, reproductive hormones, cardiovascular disease (CVD), diabetes, obesity, cardiometabolic risk biomarkers, menopausal symptoms, and neurological outcomes.

6.3.1 Bone

Reproductive-related hormones, particularly estrogens and androgens, are involved in the growth, differentiation, and function of bone (Compston, 2001). Some of the proposed mechanisms include inhibition of bone resorption and increased production of 1-25-dihydroxy vitamin D, growth hormone, and insulin-like growth factor 1 (Pacifici, 1998). Given this known association with reproductive hormones and bone, it is biologically plausible that isoflavones would exert influence on bone.

Isoflavone intake in randomized controlled trials has been associated with increases in lumbar spine bone mineral density (BMD) (Lambert *et al.*, 2017; Liu *et al.*, 2009; Ma *et al.*, 2008; Ricci *et al.*, 2010; Taku *et al.*, 2010b). Metaanalyses of the studies suggest that interventions with isoflavone supplements or soy protein are associated with a 0.01–0.04 g/cm^2 increase in BMD, with similar associations for isoflavone supplements, soy protein, isolated isoflavones, and soy extracts (Lambert *et al.*, 2017; Liu *et al.*, 2009; Ma *et al.*, 2008; Ricci *et al.*, 2010a). However, soy products containing isoflavones were positively but not significantly associated with BMD (Ricci *et al.*, 2010a), suggesting that isoflavones may be exerting an influence in the absence of other components of soy, or that higher doses that can be administered with supplements or extracts may be needed. For hip or femoral neck BMD, most studies observe a positive or null association between isoflavone intake and BMD in controlled studies (Lambert *et al.*, 2017; Liu *et al.*, 2009; Taku *et al.*, 2010b). The type of isoflavone intake – extract, supplement, or food – does not appear to differentiate between positive and null associations.

Cortical bone is predominant in the appendicular skeleton (limbs and supporting pectoral and pelvic girdle), and a greater proportion of trabecular bone is found in the axial skeleton (head, trunk, and spine) and trabecular bone is more metabolically responsive than cortical bone (Cummings *et al.*, 1985). These differences observed across spine and hip BMD may be informative of mechanisms of action and may indicate that more metabolically active bone sites (e.g., spine) are more likely to be responsive to isoflavone intake than less metabolically active sites (e.g., hip).

There are several biomarkers of bone that can be informative about mechanisms by which bone is responsive to isoflavone intake. Bone alkaline phosphatase (BAP) reflects the activity of bone-forming cells and serum BAP is considered a reliable and sensitive indicator of bone metabolism (Kress *et al.*, 1999). Deoxypyridinoline (DPD) also provides a measure of systemic bone resorption, and urinary DPD is considered a specific index of bone resorption by osteoclasts (Robins *et al.*, 1994; Rubinacci *et al.*, 1999). Higher concentrations of BAP or DPD are indicators of a higher rate of bone resorption relative to formation. Across studies of isoflavone intake, either through supplements or isoflavone-containing food, most do not support that isoflavone intake is associated with BAP in controlled trials (Taku *et al.*, 2010a; Wei *et al.*, 2012). However, one metaanalysis that considered soy products or soy isoflavones estimated a significantly lower concentration of 1.2 nmol/mmol of BAP associated with intake in controlled trials (Ma *et al.*, 2008), whereas urinary DPD is favorably responsive to isoflavone intake, through supplements or isoflavone-containing food, with response ranges from lowering DPD from 0.02 to 18.% (Ma *et al.*, 2008; Taku *et al.*, 2010a; Wei *et al.*, 2012).

Serum osteocalcin is another biomarker for bone that is produced by osteoblasts and is believed to provide a marker of the mineralization process in bone formation (Farrugia *et al.*, 1989; Lian and Gundberg, 1988). Fewer controlled trials have evaluated serum osteocalcin than the other bone biomarkers, and isoflavone intake appears to be either favorably or not related to serum osteocalcin (Taku *et al.*, 2010a). While there is not overwhelming evidence to suggest that isoflavone intake is favorably associated with bone metabolism biomarkers, these studies support a lack of unfavorable impacts.

6.3.2 Cancer

Isoflavone intake is inversely associated with all-cancer mortality across observational studies (Nachvak *et al.*, 2019). When studies are statistically combined, it is estimated that there is approximately a 20% lower all-cancer mortality risk associated with the highest category of isoflavone intake compared with the lowest. These observations suggest a broad impact on cancer. It is not entirely possible to rule out that these observations may be the result of uncontrolled confounding, in that individuals who consume sources of isoflavones are different from individuals who do not consume, and there may be other characteristics of isoflavone consumers that influence their cancer risk. These studies did control for some of those factors and the summary estimate was based on several studies, suggesting that isoflavones are a relevant exposure for reduced cancer mortality.

It is essential to consider that these observations for all-cancer mortality are being driven by more common cancers, such as breast and prostate cancers, and impacts on other cancers may be more limited. However, the overall body of evidence suggests that there is a lower cancer mortality risk associated with isoflavone intake, which contributes additional evidence towards the beneficial effects of isoflavone intake. The most studied cancers include breast and endometrial among women, prostate among men, and colorectal in both sexes, and the upcoming paragraphs will specifically look at the evidence associated with those cancer subtypes.

Breast and Endometrial Cancers

Breast and endometrial cancers are considered hormonally responsive (Ariazi *et al.*, 2002; Boyd *et al.*, 2006; Brinton and Felix, 2014; Busch *et al.*, 2017; Westley and May, 2006), and given this aspect, it is biologically plausible that development or progression could be influenced by isoflavone intake. Across observational studies of breast cancer incidence in women looking at isoflavone intake from foods and/or supplements or as measured by serum biomarkers of intake (daidzein or genistein concentrations), isoflavone exposure has been inversely or not associated with breast cancer incidence (Chen *et al.*, 2014; Rienks *et al.*, 2017; Xie *et al.*, 2013; Zhao *et al.*, 2019). There is some evidence to support that there is a stronger protective factor in Asian women than Western women, with a summary estimate of 30% lower risk in Asian populations compared to a null association among Western populations (Xie *et al.*, 2013). Evidence is mixed about whether there is a differential effect for premenopausal and postmenopausal breast cancer (Chen *et al.*, 2014; Dong and Qin, 2011; Xie *et al.*, 2013). There is not a clear pattern of differential associations across the source of isoflavones (Chen *et al.*, 2014; Dong and Qin, 2011; Rienks *et al.*, 2017; Xie *et al.*, 2013; Zhao *et al.*, 2019), suggesting that isoflavones may be relevant regardless of source. There is one metaanalysis of four studies that indicates that isoflavones may be protective against recurrence, with a summary estimate of a 16% lower risk of recurrence for women in the highest versus lowest category of exposure (Dong and Qin, 2011).

Higher breast density is a risk factor for breast cancer (Boyd *et al.*, 2006), and controlled trials have evaluated whether isoflavones from supplements or diet are associated with changes in breast density (Hooper *et al.*, 2010). In a meta-analysis of eight studies that evaluated isoflavone intake, there was no association in change in percent density in response to isoflavone interventions (Hooper *et al.*, 2010), suggesting that the effects on breast cancer may work through other mechanisms besides structural change to breast architecture.

With regard to another hormonally responsive cancer in women, evidence across 13 observational studies supports that isoflavone intake is associated with a lower risk of endometrial cancer incidence (Zhong *et al.*, 2018), with no differences observed in association across isoflavone source or Asian versus Western populations. Evidence does not support that isoflavones influence endometrial thickness (Liu *et al.*, 2016), suggesting that they may be influencing endometrial cancer through mechanisms other than structural changes to the endometrium.

Prostate Cancer

In addition to possible hormone-related risk factors (Gann, 2002), one of the possible, albeit not definitively established, risk factors for prostate cancer is prostatitis (inflammation of the prostate) (Nelson *et al.*, 2004). Given the hormonal and antiinflammatory mechanisms exerted by isoflavones, there is biological plausibility that prostate cancer development or progression could be influenced by isoflavone intake. Observational studies of prostate cancer incidence have considered Western (primarily European) and Asian (primarily Japanese) populations, and evaluated circulating concentrations of genistein and daidzein (Perez-Cornago *et al.*, 2018; Rienks *et al.*, 2017). Daidzein, but not genistein, levels were associated with a lower risk of prostate cancer when comparing the highest with the lowest category (Rienks *et al.*, 2017). In a pooled analysis of six studies of European men and two studies of Japanese men, there appeared to be more of a beneficial effect in Japanese men for circulating daidzein and genistein concentrations, but neither population group had statistically significantly lower risk (Perez-Cornago *et al.*, 2018).

Interestingly, across two studies evaluating prostate cancer diagnosis in high-risk men and isoflavone intake from supplements, extracts, or foods, there was a 50% reduction in risk associated with isoflavone intake (van Die *et al.*, 2014), suggesting that there may be benefit to higher-risk individuals. However, given that this observation was based on only two studies, this is an area for further work before definitive statements about high-risk populations can be made.

Prostate-specific antigen (PSA) is a protein produced by normal and malignant cells of the prostate gland, and elevated blood PSA levels have been seen in men with prostate cancer (Platz *et al.*, 2004), and may be used to track disease progression in individuals diagnosed with prostate cancer (Kirby, 2016).

PSA levels have also been used for prostate cancer screening, but this is more controversial (Ahlering *et al.*, 2019; Carter, 2018; Tikkinen *et al.*, 2018). PSA levels provide an indicator of prostate gland activity, and across seven controlled trials evaluating isoflavone intake from supplements, extracts, or soy foods, the evidence did not support that isoflavone intake is associated with PSA levels (van Die *et al.*, 2014). These observations suggest that any impacts on prostate cancer may be working through mechanisms not tied to prostate gland activity.

Gastrointestinal and Colorectal Cancers

Inflammation is a potential mechanism by which gastrointestinal and colorectal cancers are believed to initiate or progress (Dite, 2010; Grivennikov, 2013; Kraus and Arber, 2009; Marusawa and Jenkins, 2014; Rizzo *et al.*, 2011; Vazzana *et al.*, 2012; Ward and Mrsny, 2009), and given the antiinflammatory mechanisms exerted by isoflavones (Duan *et al.*, 2003; Ganai *et al.*, 2015; Hooshmand *et al.*, 2007; Kao *et al.*, 2007; Li *et al.*, 2014; Lim *et al.*, 2013; Paradkar *et al.*, 2004; Tanaka *et al.*, 2014), it is biologically plausible that isoflavones could influence gastrointestinal or colorectal cancer incidence or mortality. Evidence from observational studies support that isoflavone intake is associated with lower risk for colorectal and gastrointestinal cancers (Tse and Eslick, 2016; Yu *et al.*, 2016) when comparing highest and lowest categories of intake. However, when evaluating just gastric cancers, observational evidence did not support that isoflavone intake was associated with gastric cancers across all studies, and there may be some differential risk associated with population (You *et al.*, 2018). Specifically, there may lower risk for gastric cancers associated with isoflavone intake in Asian populations. These observations for incidence align with studies of mortality. In a metaanalysis of four colorectal cancer studies and six gastric cancer studies, there were significantly lower summary estimates of risk for both types of cancers associated with the highest versus the lowest category of isoflavone intake (Nachvak *et al.*, 2019).

There are a limited number of studies that have evaluated colorectal or gastrointestinal cancer mechanisms associated with isoflavones. In one controlled trial, soy protein containing isoflavones was evaluated for 12 months in adults recently diagnosed with adenomatous polyps, and, contrary to expectations, the intervention group experienced more significant cell proliferation in the sigmoid colon, along with no differences in the height of proliferating cells (Adams *et al.*, 2005). However, more work is needed to identify possible mechanisms by which isoflavone intake may reduce risk of colorectal and gastrointestinal cancers.

6.3.3 Reproductive Hormones

Reproductive hormones are involved in numerous physiological processes, and high or low levels have been considered for their impacts on various diseases (Cornish *et al.*, 2019; de Jong *et al.*, 2013; Hsu *et al.*, 2015; Joffe and Bromberger, 2016; Krishnamachari *et al.*, 2014; Matthews, 1989; Michaud *et al.*, 2010; Qing *et al.*, 2017; Sutton-Tyrrell *et al.*, 2005). Isoflavones are considered phytoestrogens because they are structurally similar to mammalian estrogens, and they can bind to ER and elicit a response (Baker and Lathe, 2018; Kostelac *et al.*, 2003). Altogether, as noted below, there appears to be a minimal impact of isoflavone intake on reproductive hormones in women or men, but the evidence at this stage may be insufficient to draw conclusions.

In women, there has been one metaanalysis that evaluated reproductive hormones in pre- and postmenopausal women (Hooper *et al.*, 2009). The researchers considered studies that evaluated isoflavone intake in a variety of forms, including supplements and diet, and had a variety of comparison groups, including placebo or usual diet. Premenopausal and postmenopausal women have considerably different endogenous reproductive hormone concentrations, and it is essential to consider them separately, as isoflavone could theoretically induce different effects depending on the endogenous environment. In premenopausal women, there were nonsignificantly lower concentrations of estrone (E1), estradiol (E2), and sex hormone binding globulin (SHBG) in response to isoflavone intervention. In postmenopausal women, there was a nonsignificantly lower concentration of E1 in response to isoflavone intervention, nonsignificantly higher concentration of E2, and no difference in SHBG. However, even though these were combined estimates from several studies, there was a notable lack of precision in the estimates for both analyses.

Studies of reproductive hormones in men have been summarized in two metaanalyses (Hamilton-Reeves *et al.*, 2010; van Die *et al.*, 2014). In the broader metaanalysis that included isoflavones from extract, supplements, or food, there were no associations between isoflavone intake and dihydrotestosterone, estradiol, free testosterone, or SHBG (van Die *et al.*, 2014). When considering just isolated soy protein and soy foods for testosterone and SHBG in a different metaanalysis, similar nonsignificant associations were observed (Hamilton-Reeves *et al.*, 2010). However, the direction of association was towards positive associations, in that there appeared to be increases in SHBG and testosterone in response to isoflavone intervention, but there was a lack of precision in the summary estimate, particularly for SHBG. These observations suggest that isoflavone intake does not alter reproductive hormone concentrations in adult males, but there may still be an insufficient amount of evidence.

6.3.4 Cardiovascular Disease, Blood Triglycerides and Cholesterol, and Inflammatory Markers

Cardiovascular disease is a major public health concern worldwide. The results of observational studies of soy or isoflavone intake and cardiovascular disease mortality are, on average, null (Nachvak *et al.*, 2019). The results of one metaanalysis of six studies estimated that soy isoflavone intake was associated with a 2% lower risk of CVD mortality in individuals with the highest versus the lowest category of consumption. Similarly, the summary estimate for CVD mortality in relation to soy intake was a nonsignificant 9% lower risk. A challenge with CVD, like other diseases, is that it is a heterogeneous category overall, and it is possible that isoflavone intake is associated with particular CVD outcomes but not all CVD outcomes. Evidence from controlled trials suggests that particular CVD-related endpoints are associated with isoflavone intake.

Commonly studied CVD-related endpoints include blood pressure and endothelial dysfunction. High blood pressure (hypertension) is considered an adverse cardiovascular condition, but is also a risk factor for other cardiovascular outcomes, including stroke, coronary artery disease, heart failure, atrial fibrillation, and peripheral vascular disease (Kannel *et al.*, 1996; Lawes *et al.*, 2008; Lewington *et al.*, 2002; Rapsomaniki *et al.*, 2014; Shen *et al.*, 2013; Staessen *et al.*, 2000). In controlled trials of isoflavone intake (supplement or isoflavone-containing soy protein), isoflavone supplements were associated with lower systolic, but not diastolic, blood pressure (Taku *et al.*, 2010c). When analyses of hypertensive and normotensive individuals were considered separately, hypertensive individuals had improved (lower) systolic and diastolic blood pressure in response to isoflavone intake from supplements or soy protein, but normotensive individuals had a null response (Liu *et al.*, 2012).

Flow-mediated dilation (FMD) is a standard measure of endothelial dysfunction (Harris *et al.*, 2010), and is considered a marker of nitric oxide bioavailability (Thijssen *et al.*, 2011). Lower values of FMD are associated with increased risk for hypertension and CVD (Rossi *et al.*, 2004; Shechter *et al.*, 2009). Controlled trials of isoflavone intake suggest that intake is associated with improved FMD (Beavers *et al.*, 2012; Li *et al.*, 2010) and that the effect may be specific to isolated isoflavones but not isoflavone-containing soy protein (Beavers *et al.*, 2012). These observations for blood pressure and FMD suggest that isoflavones may be beneficial through physiological impacts associated with blood flow.

High blood triglycerides are a risk factor for cardiovascular diseases (Austin, 1998; Hokanson and Austin, 1996). There have been several controlled trials that have evaluated soy protein (Zhan and Ho, 2005) or isoflavone supplements (Simental-Mendia *et al.*, 2018; Yeung and Yu, 2003). Across trials of soy protein, a significant decrease in blood triglycerides was observed (Zhan and Ho, 2005), with a summary estimate of trials lowering blood triglyceride by 0.2 mmol/L. However, no significant associations were observed for studies of isoflavone supplements (Simental-Mendia *et al.*, 2018; Yeung and Yu, 2003).

High total or LDL cholesterol and low HDL cholesterol in blood are also risk factors for cardiovascular diseases (Blaha, n.d.; Kamstrup, 2017; Tada *et al.*, 2019). There have been numerous studies of controlled trials of isoflavone supplements, isolated isoflavones, or soy protein and cholesterol levels that have been summarized in various metaanalyses (Qin *et al.*, 2013; Simental-Mendia *et al.*, 2018; Taku *et al.*, 2007, 2008; Weggemans and Trautwein, 2003; Yeung and Yu, 2003; Zhan and Ho, 2005; Zhuo *et al.*, 2004). For total cholesterol, soy protein has not been associated with changes in controlled trials (Taku *et al.*, 2007; Zhan and Ho, 2005), and observations for isolated isoflavones or supplements have been mixed, with one summary estimate demonstrating a significantly lowered total cholesterol (Simental-Mendia *et al.*, 2018), but others estimating a null association (Taku *et al.*, 2008; Yeung and Yu, 2003). There was some evidence that effects may be stronger for males or individuals with higher baseline blood triglycerides and with isoflavone doses higher than 80 mg/day or shorter duration trials (Zhan and Ho, 2005). Results are similarly mixed towards null or inverse for LDL cholesterol (Simental-Mendia *et al.*, 2018; Taku *et al.*, 2007, 2008; Weggemans and Trautwein, 2003; Yeung and Yu, 2003; Zhuo *et al.*, 2004). There is limited evidence available, but there appears to be no difference in cholesterol effect across normocholesterolemic and hypercholesterolemic groups in response to soy protein isolate interventions (Zhuo *et al.*, 2004). For HDL cholesterol, controlled trials of soy protein have observed beneficial responses with increased HDL cholesterol (Weggemans and Trautwein, 2003; Zhan and Ho, 2005) and a summary null association was observed for studies of isoflavone supplements (Simental-Mendia *et al.*, 2018; Yeung and Yu, 2003).

Altogether, these observations suggest that isoflavones have a null association or are beneficially associated with cholesterol levels. The source of isoflavones may be important, and perhaps there is an independent influence of soy protein, or particular subgroups of individuals may have differential response.

There are some biomarkers, including homocysteine and C-reactive protein (CRP), that are considered to be risk factors or risk markers for cardiometabolic diseases (Grundy *et al.*, 2004; Tracy, 2003). CRP is produced by the liver in response to inflammation (Pepys and Hirschfield, 2003). Dong *et al.* (2011) evaluated six controlled trials of isoflavone extracts and

CRP concentrations and observed no significant association. Homocysteine is an amino acid, and high levels may have a role in artery damage (McCully, 2015). Song *et al.* (2016) evaluated 16 controlled trials of isoflavone or soy protein intake, which had a variety of control groups, and homocysteine concentrations (Song *et al.*, 2016). There was a small, nonsignificant decrease in homocysteine concentrations in response to the interventions, and there was no suggestion of particular subgroups that might be more responsive.

6.3.5 Diabetes, Insulin Resistance, and Blood Glucose and Insulin

Type 2 diabetes mellitus (T2DM) is an adverse health outcome common to developed populations (Bhattarai, 2009; Zimmet *et al.*, 2016), and is characterized by impaired glucose metabolism and insulin resistance. The mechanisms by which isoflavones may be associated with T2DM or associated metabolic dysregulation are less straightforward than those associated with conditions related to reproductive hormones. However, there is some evidence to suggest that isoflavones may target cell signaling pathways involved in T2DM pathogenesis and they are proposed to act therapeutically in ameliorating T2DM, but whether isoflavones can act preventively is unclear (Duru *et al.*, 2018).

Observational studies have examined isoflavone intake and serum biomarkers of isoflavone intake (daidzein and genistein concentrations) in relation to T2DM risk (Ding *et al.*, 2016; Rienks *et al.*, 2017). Although there were small numbers of studies considered in a pooled analysis of isoflavone (n=3 studies) and metaanalysis of serum isoflavone concentrations (n=3 studies), results support that isoflavone exposure is associated with a reduced risk of T2DM. In the pooled analysis, highest vs lowest category of isoflavone intake was associated with a statistically significantly 11% reduction in risk, with similar magnitudes of association for daidzein and genistein, when they were considered separately (Ding *et al.*, 2016). In the metaanalysis of serum isoflavone concentrations, highest vs lowest categories of daidzein and genistein concentrations were significantly associated with approximately 20% reduction in T2DM risk (Rienks *et al.*, 2017). These observations suggest that isoflavone intake is associated with T2DM incidence, but do not provide information about mechanisms and, given that they are based on a limited number of observational studies contributing to these pooled and metaanalysis studies, the overall evidence is insufficient for definitive conclusions about isoflavones and T2DM.

There have been three metaanalyses that have summarized studies of insulin and fasting blood glucose (FBG) in response to soy isoflavone supplements (Fang *et al.*, 2016; Ricci *et al.*, 2010b; Zhang *et al.*, 2013). Across the studies, there was supportive evidence for significantly decreased blood insulin concentrations in response to soy isoflavone supplement interventions, ranging from decreases of 0.43 uIU/mL to 1.37 uIU/mL. For FBG, two of the summary estimates were nonsignificantly decreased (Ricci *et al.*, 2010b; Zhang *et al.*, 2013), and the other observed a small, significant decrease of −0.22 mmol/L across the studies (Fang *et al.*, 2016). High insulin is believed to lead to insulin resistance, and these observations suggest a potential mechanistic role by which isoflavones may act therapeutically for T2DM.

However, studies of insulin resistance are not supportive of this mechanism. Homeostatic model assessment of insulin resistance (HOMA-IR) is a calculation that incorporates insulin and glucose concentrations to evaluate the presence and extent of insulin resistance (Wallace *et al.*, 2004). Higher HOMA-IR values correspond to higher insulin resistance. Two of these metaanalyses also summarized studies that evaluated insulin resistance (Fang *et al.*, 2016; Ricci *et al.*, 2010b), specifically for HOMA-IR. These studies provided summary estimates across 12 and four controlled trials of soy isoflavone supplements, and, for both evaluations, supplementation was nonsignificantly inversely associated with HOMA-IR.

While there may be a role for isoflavone intake in T2DM prevention or management, the evidence base is still relatively small, and the mechanisms are not fully elucidated.

6.3.6 Obesity

Obesity is an adverse health outcome common to developed populations, but obesity prevalence is growing worldwide across all populations (Kelly *et al.*, 2008). Obesity is also a risk factor for other diseases, including T2DM and CVD (Cercato and Fonseca, 2019; Igel *et al.*, 2018; Kotsis *et al.*, 2018; Ladhani *et al.*, 2017; Niswender, 2010; Said *et al.*, 2016). There is a diversity of ways to classify obesity, including body weight, measures of weight to height ratio such as body mass index (BMI), waist circumference, and fat mass. Across nine controlled trials of isoflavone supplements, evidence suggested that intervention was associated with reductions in body weight (Zhang *et al.*, 2013). However, controlled trials of isoflavone-containing soy protein were not associated with BMI, fat mass, or waist circumference (Akhlaghi *et al.*, 2017).

6.3.7 Menopausal Symptoms

Numerous studies have evaluated isoflavone intake and climacteric symptoms in peri- or postmenopausal women under controlled conditions (Bolaños *et al.*, 2010; Howes *et al.*, 2006; Taku *et al.*, 2012). The results of these studies are supportive that, in a variety of forms, isoflavones are beneficial in reducing the frequency or severity of symptoms. Studies have included isoflavone concentrates, extracts, and supplements, and observed an overall small reduction (n=0.2–0.5 fewer hot flashes) in frequency (Bolaños *et al.*, 2010; Howes *et al.*, 2006). It is not possible to evaluate whether there is a clear benefit of one type of isoflavone over another. Other work, using a modeling approach, predicts that isoflavone intake is similarly effective to estradiol, but takes longer to achieve the same effects (Li *et al.*, 2015).

6.3.8 Neurological Outcomes

Some controlled trials in older adults have considered the relationships between cognitive function and isoflavone supplements or dietary-based interventions with isoflavones, and these are summarized in two metaanalyses (Cheng *et al.*, 2015; Cui *et al.*, 2019). The underlying mechanism of interest is that isoflavones may have similar effects on the brain to estrogens, which are believed to have a positive impact on cognitive function (Lee *et al.*, 2005). Cognitive function can be considered in general domains depending on the tool being used, and there are often summary scores for total cognitive function and subscores for different components. Across the controlled trials summarized, isoflavones had a positive effect on total score and memory (Cheng *et al.*, 2015; Cui *et al.*, 2019), and those effects did not differ in subgroups on region, dose, sex or menopausal status, or duration of intervention (Cui *et al.*, 2019). However, there were null associations with global cognition, executive function, psychomotor speed, attention, language, and visuospatial reasoning subdomains (Cui *et al.*, 2019), suggesting that the effects of isoflavones on cognitive function may be limited to overall memory.

6.4 Physiological and Health Effects of Isoflavone Metabolites and Metabotypes

Equol and *O*-desmethylangolensin (ODMA) are metabolites of the isoflavone daidzein (Figure 6.1). For equol and ODMA, individuals either do or do not produce the compounds based on the gut microbial community (Atkinson *et al.*, 2005). These are referred to as metabotypes or phenotypes. Various bacteria have been identified in *in vitro* studies to metabolize daidzein to equol or ODMA (Frankenfeld, 2011a; Rafii *et al.*, 2007). Being an equol or ODMA nonproducer indicates a lack of particular bacteria, consortium of bacteria, or other gut microbial environment factor that influences the capability to produce the metabolite (producers vs nonproducers). It has been observed that individual producer vs nonproducer metabotype can shift over time (Franke *et al.*, 2012; Frankenfeld *et al.*, 2005), but there are no clearly established personal or dietary factors that predict metabotype or metabotype change.

The gut microbial environment can act as an effect modifier of flavonoid intake and health outcomes (see Chapter 13 for more information).

Concerning isoflavones specifically and using ODMA as an example, Table 6.1 illustrates the exposure patterns that arise for ODMA with diet–microbiota interactions. The same exposure pattern applies to equol, except that equol can be independently obtained as a supplement in Japan (Ishiwata *et al.*, 2009). In daidzein consumers, all individuals are exposed to daidzein,

Table 6.1 Daidzein and *O*-desmethylangolensin (ODMA) exposure patterns based on intake of daidzein and ODMA producer metabotype. The same exposure pattern applies to daidzein metabolism to equol. Source: Adapted from Frankenfeld (2011a).

		Daidzein intake	
		Daidzein consumers	**Daidzein nonconsumers**
Meabotype	**ODMA producer**	Daidzein, ODMA, and ODMA-producing microbial community	ODMA-producing gut microbial community
	ODMA nonproducer	Daidzein	None

and depending on the gut microbial environment, some individuals will be exposed to the metabolite. With regard to equol-producer and ODMA-producer metabotypes, soy is one of the richest sources of daidzein and soy consumption is generally high in some parts of Asia, such as Japan and China (Frankenfeld *et al.*, 2004; Lee *et al.*, 2003; Liu *et al.*, 2004; Nishio *et al.*, 2007; Oba *et al.*, 2007; Yamasaki *et al.*, 2015); thus, chronic high exposure to the metabolites (in producers) is likely, and health effects are likely due to a combination of daidzein, metabolite, and gut microbial environment. In contrast, soy consumption is generally low in the US population (Frankenfeld, 2011b; Maskarinec *et al.*, 1998), so the group most likely represented by adults in the US is daidzein nonconsumers. Thus, chronic high exposure to the metabolites will not have occurred, and long-term health effects associated with being a producer or nonproducer are more likely to be due to differences in the gut microbial environment.

In terms of the importance of the gut microbial environment for dietary isoflavone intake, the underlying metabotype may be relevant as an effect modifier, and while ODMA has low physiological activity in the body (Frankenfeld, 2011a), equol has been observed to have stronger binding affinity to some hormone receptors than the parent compound daidzein under some circumstances (Kostelac *et al.*, 2003; Munoz *et al.*, 2009; Setchell and Clerici, 2010). Thus, there is some belief that equol producers vs nonproducers may have different health situations associated with isoflavone intake (Lampe, 2009; Magee, 2011; Setchell *et al.*, 2002b). However, the evidence for this is currently mixed and many observations were based on *post hoc* analyses of studies that were not necessarily designed to specifically evaluate the equol producer metabotype (Mayo *et al.*, 2019).

6.5 Summary of Isoflavone Intake and Health

There is limited evidence for isoflavone intake, in some situations, for particular populations that may have a different baseline risk or metabolism, such as developing children or people with preexisting health conditions. However, results of controlled trials and observational studies in most healthy adults suggest that isoflavone intake may be beneficially associated with several health conditions or with reduced risk of disease. Overall, isoflavone exposure was associated favorably or not at all, which suggests a beneficial, or at least nonharmful, effect of isoflavones in the general adult population. Potential mechanisms of action include impact on inflammatory processes, cardiometabolic biomarkers, and circulating endogenous hormones, but more work is needed to elucidate pathways of effects in humans. The source or form of isoflavones may differentially influence health, and mechanisms for these observations are not fully understood. Evidence supports that gut microbial metabolism of isoflavones and associated metabotypes may also be significant contributors to overall health, but this is an area requiring more extensive human studies to elucidate relationships and mechanisms.

References

Adams, K.F., Lampe, P.D., Newton, K.M., *et al.* (2005) Soy protein containing isoflavones does not decrease colorectal epithelial cell proliferation in a randomized controlled trial. *American Journal of Clinical Nutrition*, **82**, 620–6.

Ahlering, T., Huynh, L.M., Kaler, K.S., *et al.* (2019) Unintended consequences of decreased PSA-based prostate cancer screening. *World Journal of Urology*, **37**, 489–96.

Akhlaghi, M., Zare, M., and Nouripour, F. (2017) Effect of soy and soy isoflavones on obesity-related anthropometric measures: a systematic review and meta-analysis of randomized controlled clinical trials. *Advances in Nutrition*, **8**, 705–17.

Akiyama, T., Ishida, J., Nakagawa, S., *et al.* (1987) Genistein, a specific inhibitor of tyrosine-specific protein kinases. *Journal of Biological Chemistry*, **262**, 5592–5.

Anupongsanugool, E., Teekachunhatean, S., Rojanasthien, N., Pongsatha, S., and Sangdee, C. (2005) Pharmacokinetics of isoflavones, daidzein and genistein, after ingestion of soy beverage compared with soy extract capsules in postmenopausal Thai women. *BMC Clinical Pharmacology*, **5**, 2.

Ariazi, E.A., Clark, G.M., and Mertz, J.E. (2002) Estrogen-related receptor alpha and estrogen-related receptor gamma associate with unfavorable and favorable biomarkers, respectively, in human breast cancer. *Cancer Research*, **62**, 6510–18.

Arimoto, J.M., Wong, A., Rozovsky, I., Lin, S.W., Morgan, T.E., and Finch, C.E. (2013) Age increase of estrogen receptor-alpha (ERalpha) in cortical astrocytes impairs neurotrophic support in male and female rats. *Endocrinology*, **154**, 2101–13.

Atkinson, C., Frankenfeld, C.L., and Lampe, J.W. (2005) Gut bacterial metabolism of the soy isoflavone daidzein: exploring the relevance to human health. *Experimental Biology and Medicine (Maywood).* **230**, 155–70.

Austin, M.A. (1998) Plasma triglyceride as a risk factor for cardiovascular disease. *Canadian Journal of Cardiology*, **14 Suppl B**, 14B–17B.

Baker, M.E. and Lathe, R. (2018) The promiscuous estrogen receptor: evolution of physiological estrogens and response to phytochemicals and endocrine disruptors. *Journal of Steriod Biochemistry and Molecular Biology*, **184**, 29–37.

Barnes, S., Prasain, J., d'Alessandro, T., *et al.* (2011) The metabolism and analysis of isoflavones and other dietary polyphenols in foods and biological systems. *Food & Function*, **2**, 235–44.

Beavers, D.P., Beavers, K.M., Miller, M., Stamey, J., and Messina, M.J. (2012) Exposure to isoflavone-containing soy products and endothelial function: a Bayesian meta-analysis of randomized controlled trials. *Nutrition, Metabolism, and Cardiovascular Diseases*, **22**, 182–91.

Bhattarai, M.D. (2009) Three patterns of rising type 2 diabetes prevalence in the world: need to widen the concept of prevention in individuals into control in the community. *Journal of the Nepal Medical Association*, **48**, 173–9.

Bláha, V. (2019) Lipoprotein(a) - the cardiovascular risk factor: significance and therapeutic possibilities. *Vnitrni Lekarstvi*, **64**, 1160–8.

Bolaños, R., Del Castillo, A., Francia, J., Bolanos, R., del Castillo, A., and Francia, J. (2010) Soy isoflavones versus placebo in the treatment of climacteric vasomotor symptoms: systematic review and meta-analysis. *Menopause*, **17**, 660–6.

Bolca, S., Urpi-Sarda, M., Blondeel, P., *et al.* (2010) Disposition of soy isoflavones in normal human breast tissue. *American Journal of Clinical Nutrition*, **91**, 976–84.

Boyd, N.F., Martin, L.J., Yaffe, M.J., and Minkin, S. (2006) Mammographic density: a hormonally responsive risk factor for breast cancer. *Journal of the British Menopause Society*, **12**, 186–93.

Brinton, L.A. and Felix, A.S. (2014) Menopausal hormone therapy and risk of endometrial cancer. *Journal of Steriod Biochemistry and Molecular Biology*, **142**, 83–9.

Busch, E.L., Crous-Bou, M., Prescott, J., *et al.* (2017) Endometrial cancer risk factors, hormone receptors, and mortality prediction. *Cancer Epidemiology, Biomarkers & Prevention*, **26**, 727–35.

Calabro, P. and Yeh, E.T.H. (2008) Intra-abdominal adiposity, inflammation, and cardiovascular risk: new insight into global cardiometabolic risk. *Current Hypertension Reports*, **10**, 32–8.

Cannalire, R., Machado, D., Felicetti, T., *et al.* (2017) Natural isoflavone biochanin A as a template for the design of new and potent 3-phenylquinolone efflux inhibitors against Mycobacterium avium. *European Journal of Medical Chemistry*, **140**, 321–30.

carter, h.b. (2018) prostate-specific antigen (psa) screening for prostate cancer: revisiting the evidence. *JAMA*, **319**, 1866–8.

Cercato, C. and Fonseca, F.A. (2019) Cardiovascular risk and obesity. *Diabetology & Metabolic Syndrome*, **11**, 74.

Chang, Y. and Choue, R. (2013) Plasma pharmacokinetics and urinary excretion of isoflavones after ingestion of soy products with different aglycone/glucoside ratios in South Korean women. *Nutrition Research and Practice*, 7, 393–9.

Chen, M., Rao, Y., Zheng, Y., *et al.* (2014) Association between soy isoflavone intake and breast cancer risk for pre- and post-menopausal women: a meta-analysis of epidemiological studies. *PLoS One*, **9**, e89288.

Cheng, P.-F., Chen, J.-J., Zhou, X.-Y., *et al.* (2015) Do soy isoflavones improve cognitive function in postmenopausal women? A meta-analysis. *Menopause*, **22**, 198–206.

Chuankhayan, P., Rimlumduan, T., Svasti, J., and Cairns, J.R.K. (2007) Hydrolysis of soybean isoflavonoid glycosides by Dalbergia beta-glucosidases. *Journal of Agricultural and Food Chemistry*, **55**, 2407–12.

Compston, J.E. (2001) Sex steroids and bone. *Physiological Reviews*, **81**, 419–47.

Cornish, R., Staff, A.C., Boyd, A., *et al.* (2019) Maternal reproductive hormones and angiogenic factors in pregnancy and subsequent breast cancer risk. *Cancer Causes & Control*, **30**, 63–74.

Cui, C., Birru, R.L., Snitz, B.E., *et al.* (2019) Effects of soy isoflavones on cognitive function: a systematic review and meta-analysis of randomized controlled trials. *Nutrition Reviews*, **78**, 134–44.

Cummings, S.R., Kelsey, J.L., Nevitt, M.C., and O'Dowd, K.J. (1985) Epidemiology of osteoporosis and osteoporotic fractures. *Epidemiologic Reviews*, 7, 178–208.

das Neves, M.V.M., da Silva, T.M.S., Lima, E. de O., da Cunha, E.V.L., and Oliveira, E. de J. (2016) Isoflavone formononetin from red propolis acts as a fungicide against Candida sp. *Brazilian Journal of Microbiology*, **47**, 159–66.

Day, A.J., DuPont, M.S., Ridley, S., *et al.* (1998) Deglycosylation of flavonoid and isoflavonoid glycosides by human small intestine and liver beta-glucosidase activity. *FEBS Letters*, **436**, 71–5.

Day, A.J., Canada, F.J., Diaz, J.C., *et al.* (2000) Dietary flavonoid and isoflavone glycosides are hydrolysed by the lactase site of lactase phlorizin hydrolase. *FEBS Letters*, **468**, 166–70.

de Jong, S., Huisman, M., Sutedja, N., *et al.* (2013) Endogenous female reproductive hormones and the risk of amyotrophic lateral sclerosis. *Journal of Neurology*, **260**, 507–12.

Ding, M., Pan, A., Manson, J.E., *et al.* (2016) Consumption of soy foods and isoflavones and risk of type 2 diabetes: a pooled analysis of three US cohorts. *European Journal of Clinical Nutrition*, **70**, 1381–7.

Dite, P. (2010) Chronic inflammation is an important risk factor for the development of gastrointestinal cancer. *Preface. Digestive Diseases*, **28**, 573.

Djuric, Z., Chen, G., Doerge, D.R., Heilbrun, L.K., and Kucuk, O. (2001) Effect of soy isoflavone supplementation on markers of oxidative stress in men and women. *Cancer Letters*, **172**, 1–6.

Dong, J.-Y. and Qin, L.-Q. (2011) Soy isoflavones consumption and risk of breast cancer incidence or recurrence: a meta-analysis of prospective studies. *Breast Cancer Research and Treatment*, **125**, 315–23.

Dong, J.-Y., Wang, P., He, K., and Qin, L.-Q. (2011) Effect of soy isoflavones on circulating C-reactive protein in postmenopausal women: meta-analysis of randomized controlled trials. *Menopause*, **18**, 1256–62.

Duan, W., Kuo, I.C., Selvarajan, S., Chua, K.Y., Bay, B.H., and Wong, W.S.F. (2003) Antiinflammatory effects of genistein, a tyrosine kinase inhibitor, on a guinea pig model of asthma. *American Journal of Respiratory and Critical Care Medicine*, **167**, 185–92.

Duru, K.C., Kovaleva, E.G., Danilova, I.G., van der Bijl, P., and Belousova, A. V (2018) The potential beneficial role of isoflavones in type 2 diabetes mellitus. *Nutrition Research*, **59**, 1–15.

Fang, K., Dong, H., Wang, D., Gong, J., Huang, W., and Lu, F. (2016) Soy isoflavones and glucose metabolism in menopausal women: a systematic review and meta-analysis of randomized controlled trials. *Molecular Nutrition & Food Research*, **60**, 1602–14.

Fanti, P., Asmis, R., Stephenson, T.J., Sawaya, B.P., and Franke, A.A. (2006) Positive effect of dietary soy in ESRD patients with systemic inflammation – correlation between blood levels of the soy isoflavones and the acute-phase reactants. *Nephrology, Dialysis, Transplantation*, **21**, 2239–46.

Farrugia, W., Fortune, C.L., Heath, J., Caple, I.W., and Wark, J.D. (1989) Osteocalcin as an index of osteoblast function during and after ovine pregnancy. *Endocrinology*, **125**, 1705–10.

Faughnan, M.S., Hawdon, A., Ah-Singh, E., Brown, J., Millward, D.J., and Cassidy, A. (2004) Urinary isoflavone kinetics: the effect of age, gender, food matrix and chemical composition. *British Journal of Nutrition*, **91**, 567–74.

Ferguson, J.F., Ryan, M.F., Gibney, E.R., Brennan, L., Roche, H.M., and Reilly, M.P. (2014) Dietary isoflavone intake is associated with evoked responses to inflammatory cardiometabolic stimuli and improved glucose homeostasis in healthy volunteers. *Nutrition, Metabolism, and Cardiovascular Diseases*, **24**, 996–1003.

Foti, P., Erba, D., Riso, P., Spadafranca, A., Criscuoli, F., and Testolin, G. (2005) Comparison between daidzein and genistein antioxidant activity in primary and cancer lymphocytes. *Archives of Biochemistry and Biophysics*, **433**, 421–7.

Franke, A.A., Custer, L.J., and Hundahl, S.A. (2004) Determinants for urinary and plasma isoflavones in humans after soy intake. *Nutrition and Cancer*, **50**, 141–54.

Franke, A.A., Lai, J.F., Pagano, I., Morimoto, Y., and Maskarinec, G. (2012) Equol production changes over time in pre-menopausal women. *British Journal of Nutrition*, **107**, 1201–6.

Franke, A.A., Lai, J.F., and Halm, B.M. (2014) Absorption, distribution, metabolism, and excretion of isoflavonoids after soy intake. *Archives of Biochemistry and Biophysics*, **559**, 24–8.

Frankenfeld, C.L (2011a) O-Desmethylangolensin: the importance of equol's lesser known cousin to human health. *Advances in Nutrition*, **2**, 317–24.

Frankenfeld, C.L. (2011b) Dairy consumption is a significant correlate of urinary equol concentration in a representative sample of US adults. *American Journal of Clinical Nutrition*, **93**, 1109–16.

Frankenfeld, C.L., Lampe, J.W., Shannon, J., *et al.* (2004) Frequency of soy food consumption and serum isoflavone concentrations among Chinese women in Shanghai. *Public Health Nutrition*, **7**, 765–72.

Frankenfeld, C.L., Atkinson, C., Thomas, W.K., *et al.* (2005) High concordance of daidzein-metabolizing phenotypes in individuals measured 1 to 3 years apart. *British Journal of Nutrition*, **94**, 873–6.

Frankenfeld, C.L., Atkinson, C., Wähälä, K., *et al.* (2014) Obesity prevalence in relation to gut microbial environments capable of producing equol or O-desmethylangolensin from the isoflavone daidzein. *European Journal of Clinical Nutrition*, **68**, 526–30.

Ganai, A.A., Khan, A.A., Malik, Z.A., and Farooqi, H. (2015) Genistein modulates the expression of NF-kappaB and MAPK (p-38 and ERK1/2), thereby attenuating d-Galactosamine induced fulminant hepatic failure in Wistar rats. *Toxicology and Applied Pharmacology*, **283**, 139–46.

Gann, P.H. (2002) Risk factors for prostate cancer. *Reviews in Urology*, **4** Suppl 5, S3–S10.

Gao, H. and Dahlman-Wright, K. (2011) The gene regulatory networks controlled by estrogens. *Molecular and Cellular Endocrinology*, **334**, 83–90.

Grivennikov, S.I. (2013) Inflammation and colorectal cancer: colitis-associated neoplasia. *Semin. Immunopathology*, **35**, 229–44.

Grundy, S.M., Brewer Jr., H.B., Cleeman, J.I., Smith Jr., S.C., and Lenfant, C. (2004) Definition of metabolic syndrome: report of the National Heart, Lung, and Blood Institute/American Heart Association conference on scientific issues related to definition. *Circulation*, **109**, 433–8.

Hamilton, K.J., Hewitt, S.C., Arao, Y., and Korach, K.S. (2017) Estrogen hormone biology. *Current Topics in Developmental Biology*, **125**, 109–46.

Hamilton-Reeves, J.M., Vazquez, G., Duval, S.J., Phipps, W.R., Kurzer, M.S., and Messina, M.J. (2010) Clinical studies show no effects of soy protein or isoflavones on reproductive hormones in men: results of a meta-analysis. *Fertility and Sterility*, **94**, 997–1007.

Hanski, L., Genina, N., Uvell, H., *et al.* (2014) Inhibitory activity of the isoflavone biochanin A on intracellular bacteria of genus Chlamydia and initial development of a buccal formulation. *PLoS One*, **9**, e115115.

Hanwell, H.E.C., Kay, C.D., Lampe, J.W., Holub, B.J., and Duncan, A.M. (2009) Acute fish oil and soy isoflavone supplementation increase postprandial serum (n-3) polyunsaturated fatty acids and isoflavones but do not affect triacylglycerols or biomarkers of oxidative stress in overweight and obese hypertriglyceridemic men. *Journal of Nutrition*, **139**, 1128–34.

Harris, R.A., Nishiyama, S.K., Wray, D.W., and Richardson, R.S. (2010) Ultrasound assessment of flow-mediated dilation. *Hypertension*, **55**, 1075–85.

Heinonen, S., Wähälä, K., and Adlercreutz, H. (1999) Identification of isoflavone metabolites dihydrodaidzein, dihydrogenistein, 6'-OH-O-dma, and cis-4-OH-equol in human urine by gas chromatography-mass spectroscopy using authentic reference compounds. *Analytical Biochemistry*, **274**, 211–19.

Heinonen, S.-M., Wahala, K., and Adlercreutz, H. (2004) Identification of urinary metabolites of the red clover isoflavones formononetin and biochanin A in human subjects. *Journal of Agricultural and Food Chemistry*, **52**, 6802–9.

Heneman, K.M., Chang, H.C., Prior, R.L., and Steinberg, F.M. (2007) Soy protein with and without isoflavones fails to substantially increase postprandial antioxidant capacity. *Journal of Nutritional Biochemistry*, **18**, 46–53.

Hokanson, J.E. and Austin, M.A. (1996) Plasma triglyceride level is a risk factor for cardiovascular disease independent of high-density lipoprotein cholesterol level: a meta-analysis of population-based prospective studies. *Journal of Cardiovascular Risk*, **3**, 213–19.

Hong, H., Landauer, M.R., Foriska, M.A., and Ledney, G.D. (2006) Antibacterial activity of the soy isoflavone genistein. *Journal of Basic Microbiology*, **46**, 329–35.

Hooper, L., Ryder, J.J., Kurzer, M.S., *et al.* (2009) Effects of soy protein and isoflavones on circulating hormone concentrations in pre- and post-menopausal women: a systematic review and meta-analysis. *Human Reproduction (Oxford, England)*, **Update 15**, 423–40.

Hooper, L., Madhavan, G., Tice, J.A., Leinster, S.J., and Cassidy, A. (2010) Effects of isoflavones on breast density in pre- and post-menopausal women: a systematic review and meta-analysis of randomized controlled trials. *Human Reproduction*, **Update 16**, 745–60.

Hooshmand, S., Soung, D.Y., Lucas, E.A., Madihally, S. V, Levenson, C.W., Arjmandi, B.H. (2007) Genistein reduces the production of proinflammatory molecules in human chondrocytes. *Journal of Nutritional Biochemistry*, **18**, 609–14.

Horn-Ross, P.L., Lee, M., John, E.M., and Koo, J. (2000) Sources of phytoestrogen exposure among non-Asian women in California, USA. *Cancer Causes & Control*, **11**, 299–302.

Howes, L.G., Howes, J.B., and Knight, D.C. (2006) Isoflavone therapy for menopausal flushes: a systematic review and meta-analysis. *Maturitas*, **55**, 203–11.

Hsu, B., Cumming, R.G., Seibel, M.J., *et al.* (2015) Reproductive hormones and longitudinal change in bone mineral density and incident fracture risk in older men: the Concord Health and Aging in Men Project. *Journal of Bone and Mineral Research*, **30**, 1701–8.

Hua, F., Li, C.-H., Chen, X.-G., and Liu, X.-P. (2018) Daidzein exerts anticancer activity towards SKOV3 human ovarian cancer cells by inducing apoptosis and cell cycle arrest, and inhibiting the Raf/MEK/ERK cascade. *International Journal of Molecular Medicine*, **41**, 3485–92.

Huser, S., Guth, S., Joost, H.G., *et al.* (2018) Effects of isoflavones on breast tissue and the thyroid hormone system in humans: a comprehensive safety evaluation. *Archives of Toxicology*, **92**, 2703–48.

Hutchins, A.M., Slavin, J.L., and Lampe, J.W. (1995) Urinary isoflavonoid phytoestrogen and lignan excretion after consumption of fermented and unfermented soy products. *Journal of the American Dietetic Association*, **95**, 545–51.

Hutchins, A.M., McIver, I.E., and Johnston, C.S. (2005) Soy isoflavone and ascorbic acid supplementation alone or in combination minimally affect plasma lipid peroxides in healthy postmenopausal women. *Journal of the American Dietetic Association*, **105**, 1134–7.

Igel, L.I., Saunders, K.H., and Fins, J.J. (2018) Why weight? An analytic review of obesity management, diabetes prevention, and cardiovascular risk reduction. *Current Atherosclerosis Reports*, **20**, 39.

Ishiwata, N., Melby, M.K., Mizuno, S., and Watanabe, S. (2009) New equol supplement for relieving menopausal symptoms: randomized, placebo-controlled trial of Japanese women. *Menopause*, **16**, 141–8.

Izumi, T., Piskula, M.K., Osawa, S., *et al.* (2000) Soy isoflavone aglycones are absorbed faster and in higher amounts than their glucosides in humans. *Journal of Nutrition*, **130**, 1695–9.

Joffe, H. and Bromberger, J.T. (2016) Shifting paradigms about hormonal risk factors for postmenopausal depression: age at menopause as an indicator of cumulative lifetime exposure to female reproductive hormones. *JAMA Psychiatry*, **73**, 111–2.

Kamstrup, P.R. (2017) Lipoprotein(a): the common, likely causal, yet elusive risk factor for cardiovascular disease. *Journal of Lipid Research*, **58**, 1731–2.

Kannel, W.B., Wolf, P.A., Verter, J., and McNamara, P.M. (1996) Epidemiologic assessment of the role of blood pressure in stroke: the Framingham Study. 1970. *JAMA*, **276**, 1269–78.

Kao, T.H., Wu, W.M., Hung, C.F., Wu, W.B., and Chen, B.H. (2007) Anti-inflammatory effects of isoflavone powder produced from soybean cake. *Journal of Agricultural and Food Chemistry*, **55**, 11068–79.

Kaplan, R.C. and Frishman, W.H. (2001) Systemic inflammation as a cardiovascular disease risk factor and as a potential target for drug therapy. *Heart Disease*, **3**, 326–32.

Kasparovska, J., Krizova, L., Lochman, J., Dadakova, K., and Kasparovsky, T. (2016) Soybean-derived isoflavone determination in rumen fluid and milk by LC-MS-(TOF). *Journal of Chromatographic Science*, **54**, 997–1003.

Kaya, M., Ito, J., Kotaka, A., *et al.* (2008) Isoflavone aglycones production from isoflavone glycosides by display of beta-glucosidase from Aspergillus oryzae on yeast cell surface. *Applied Microbiology and Biotechnology*, **79**, 51–60.

Kelly, T., Yang, W., Chen, C.-S., Reynolds, K., and He, J. (2008) Global burden of obesity in 2005 and projections to 2030. *International Journal of Obesity (London)*, **32**, 1431–7.

Khan, N., Afaq, F., and Mukhtar, H. (2008) Cancer chemoprevention through dietary antioxidants: progress and promise. *Antioxidants & Redox Signaling*, **10**, 475–510.

Kirby, R. (2016) The role of PSA in detection and management of prostate cancer. *The Practitioner*, **260**, 17-21.

Kostelac, D., Rechkemmer, G., and Briviba, K. (2003) Phytoestrogens modulate binding response of estrogen receptors alpha and beta to the estrogen response element. *Journal of Agricultural and Food Chemistry*, **51**, 7632–5.

Kotsis, V., Jordan, J., Micic, D., *et al.* (2018) Obesity and cardiovascular risk: a call for action from the European Society of Hypertension Working Group of Obesity, Diabetes and the

High-Risk Patient and European Association for the Study of Obesity: part A: mechanisms of obesity induced hypertension. *Journal of Hypertension*, **36**, 1427–40.

Kraus, S. and Arber, N. (2009) Inflammation and colorectal cancer. *Current Opinion in Pharmacology*, **9**, 405–10.

Kress, B.C., Mizrahi, I.A., Armour, K.W., Marcus, R., Emkey, R.D., and Santora, A.C. 2nd, (1999) Use of bone alkaline phosphatase to monitor alendronate therapy in individual postmenopausal osteoporotic women. *Clinical Chemistry*, **45**, 1009–17.

Krishnamachari, B., Il'yasova, D., Scheurer, M.E., Bondy, M.L., Wrensch, M., and Davis, F.G. (2014) A pooled multisite analysis of the effects of female reproductive hormones on glioma risk. *Cancer Causes & Control*, **25**, 1007–13.

Kuiper, G.G., Carlsson, B., Grandien, K., *et al.* (1997) Comparison of the ligand binding specificity and transcript tissue distribution of estrogen receptors alpha and beta. *Endocrinology*, **138**, 863–70.

Ladhani, M., Craig, J.C., Irving, M., Clayton, P.A., and Wong, G. (2017) Obesity and the risk of cardiovascular and all-cause mortality in chronic kidney disease: a systematic review and meta-analysis. *Nephrology, Dialysis, Transplantation*, **32**, 439–49.

Lambert, M.N.T., Hu, L.M., and Jeppesen, P.B. (2017) A systematic review and meta-analysis of the effects of isoflavone formulations against estrogen-deficient bone resorption in peri- and postmenopausal women. *American Journal of Clinical Nutrition*, **106**, 801–11.

Lampe, J.W. (2009) Is equol the key to the efficacy of soy foods? *American Journal of Clinical Nutrition*, **89**, 1664S-1667S.

Lauwaet, T., Andersen, Y., van de Ven, L., Eckmann, L., and Gillin, F.D. (2010) Rapid detachment of Giardia lamblia trophozoites as a mechanism of antimicrobial action of the isoflavone formononetin. *Journal of Antimicrobial Chemotherapy*, **65**, 531–4.

Lawes, C.M.M., Vander Hoorn, S., and Rodgers, A. (2008) Global burden of blood-pressure-related disease *Lancet*, **371**, 1513–18.

Leclercq, G. and Jacquot, Y. (2014) Interactions of isoflavones and other plant derived estrogens with estrogen receptors for prevention and treatment of breast cancer-considerations concerning related efficacy and safety. *Journal of Steriod Biochemistry and Molecular Biology*, **139**, 237–44.

Lee, M.M., Gomez, S.L., Chang, J.S., Wey, M., Wang, R.T., and Hsing, A.W. (2003) Soy and isoflavone consumption in relation to prostate cancer risk in China. *Cancer Epidemiology, Biomarkers & Prevention*, **12**, 665–8.

Lee, Y.-B., Lee, H.J., and Sohn, H.S. (2005) Soy isoflavones and cognitive function. *Journal of Nutritional Biochemistry*, **16**, 641–9.

Leitman, D.C., Paruthiyil, S., Vivar, O.I., *et al.* (2010) Regulation of specific target genes and biological responses by estrogen receptor subtype agonists. *Current Opinion in Pharmacology*, **10**, 629–36.

Leung, H.Y., Yung, L.H., Poon, C.H., Shi, G., Lu, A.-L., and Leung, L.K. (2009) Genistein protects against polycyclic aromatic hydrocarbon-induced oxidative DNA damage in non-cancerous breast cells MCF-10A. *British Journal of Nutrition*, **101**, 257–62.

Lewington, S., Clarke, R., Qizilbash, N., Peto, R., and Collins, R. (2002) Age-specific relevance of usual blood pressure to vascular mortality: a meta-analysis of individual data for one million adults in 61 prospective studies. *Lancet*, **360**, 1903–13.

Li, H., Pan, L., Ke, Y., *et al.* (2014) Daidzein suppresses pro-inflammatory chemokine Cxcl2 transcription in TNF-alpha-stimulated murine lung epithelial cells via depressing PARP-1 activity. *Acta Pharmacologica Sinica*, **35**, 496–503.

Li, L., Lv, Y., Xu, L., and Zheng, Q. (2015) Quantitative efficacy of soy isoflavones on menopausal hot flashes. *British Journal of Clinical Pharmacology*, **79**, 593–604.

Li, S.-H., Liu, X.-X., Bai, Y.-Y., *et al.* (2010) Effect of oral isoflavone supplementation on vascular endothelial function in postmenopausal women: a meta-analysis of randomized placebo-controlled trials. *American Journal of Clinical Nutrition*, **91**, 480–6.

Lian, J.B. and Gundberg, C.M. (1988) Osteocalcin. Biochemical considerations and clinical applications. *Clinical Orthopaedics and Related Research*, **226**, 267–91.

Liang, F., Cao, W., Huang, Y., *et al.* (2019) Isoflavone biochanin A, a novel nuclear factor erythroid 2-related factor 2 (Nrf2)-antioxidant response element activator, protects against oxidative damage in HepG2 cells. *BioFactors*, **45**, 563–74.

Liggins, J., Bluck, L.J., Runswick, S., Atkinson, C., Coward, W.A., and Bingham, S.A. (2000a) Daidzein and genistein contents of vegetables. *British Journal of Nutrition*, **84**, 717–25.

Liggins, J., Bluck, L.J., Runswick, S., Atkinson, C., Coward, W.A., and Bingham, S.A. (2000b) Daidzein and genistein content of fruits and nuts. *Journal of Nutritional Biochemistry*, **11**, 326–31.

Liggins, J., Mulligan, A., Runswick, S., and Bingham, S.A. (2002) Daidzein and genistein content of cereals. *European Journal of Clinical Nutrition*, **56**, 961–6.

Lim, D.W., Lee, C., Kim, I.-H., and Kim, Y.T. (2013) Anti-inflammatory effects of total isoflavones from Pueraria lobata on cerebral ischemia in rats. *Molecules*, **18**, 10404–12.

Liu, J., Chang, S.K.C., and Wiesenborn, D. (2005) Antioxidant properties of soybean isoflavone extract and tofu in vitro and in vivo. *Journal of Agricultural and Food Chemistry*, **53**, 2333–40.

Liu, J., Ho, S.C., Su, Y., Chen, W., Zhang, C., and Chen, Y. (2009) Effect of long-term intervention of soy isoflavones on bone mineral density in women: a meta-analysis of randomized controlled trials. *Bone*, **44**, 948–53.

Liu, J., Yuan, F., Gao, J., *et al*. (2016) Oral isoflavone supplementation on endometrial thickness: a meta-analysis of randomized placebo-controlled trials. *Oncotarget*, **7**, 17369–79.

Liu, X.X., Li, S.H., Chen, J.Z., *et al*. (2012) Effect of soy isoflavones on blood pressure: a meta-analysis of randomized controlled trials. *Nutrition, Metabolism, and Cardiovascular Diseases*, **22**, 463–70.

Liu, Z., Li, W., Sun, J., *et al*. (2004) Intake of soy foods and soy isoflavones by rural adult women in China. *Asia Pacic Journal of Clinical Nutrition*, **13**, 204–9.

Ma, D-F., Qin, L.-Q., Wang, P.-Y., and Katoh, R. (2008a) Soy isoflavone intake increases bone mineral density in the spine of menopausal women: meta-analysis of randomized controlled trials. *Clinical Nutrition*, **27**, 57–64.

Ma, D-F., Qin, L.-Q., Wang, P.-Y., and Katoh, R. (2008b) Soy isoflavone intake inhibits bone resorption and stimulates bone formation in menopausal women: meta-analysis of randomized controlled trials. *European Journal of Clinical Nutrition*, **62**, 155–61.

Magee, P.J. (2011) Is equol production beneficial to health? *Proceedings of the Nutrition Society*, **70**, 10–18.

Mahmoud, A.M., Yang, W., and Bosland, M.C. (2014) Soy isoflavones and prostate cancer: a review of molecular mechanisms. *Journal of Steroid Biochemistry and Molecular Biology*, **140**, 116–32.

Mangano, K.M., Hutchins-Wiese, H.L., Kenny, A.M., *et al.* (2013) Soy proteins and isoflavones reduce interleukin-6 but not serum lipids in older women: a randomized controlled trial. *Nutrition Research*, **33**, 1026–33.

Markovits, J., Linassier, C., Fosse, P., *et al.* (1989) Inhibitory effects of the tyrosine kinase inhibitor genistein on mammalian DNA topoisomerase II. *Cancer Research*, **49**, 5111–17.

Marusawa, H., and Jenkins, B.J. (2014) Inflammation and gastrointestinal cancer: an overview. *Cancer Letters*, **345**, 153–6.

Maskarinec, G., Singh, S., Meng, L., and Franke, A.A. (1998) Dietary soy intake and urinary isoflavone excretion among women from a multiethnic population. *Cancer Epidemiology, Biomarkers & Prevention*, **7**, 613–9.

Mathieu, P., Lemieux, I., and Despres, J.-P. (2010) Obesity, inflammation, and cardiovascular risk. *Clinical Pharmacology and Therapeutics*, **87**, 407–16.

Matthews, K.A. (1989) Interactive effects of behavior and reproductive hormones on sex differences in risk for coronary heart disease. *Health Psychology*, **8**, 373–87.

Maubach, J., Depypere, H.T., Goeman, J., *et al.* (2004) Distribution of soy-derived phytoestrogens in human breast tissue and biological fluids. *Obstetrics and Gynecology*, **103**, 892–8.

Mayo, B., Vazquez, L., and Florez, A.B. (2019) Equol: a bacterial metabolite from the daidzein isoflavone and its presumed beneficial health effects. *Nutrients*, **11**, e2231.

McCully, K.S. (2015) Homocysteine and the pathogenesis of atherosclerosis. *Expert Review of Clinical Pharmacology*, **8**, 211–19.

Messina, M.J. (1999) Legumes and soybeans: overview of their nutritional profiles and health effects. *American Journal of Clinical Nutrition*, **70**, 439S–450S.

Michaud, D.S., Gallo, V., Schlehofer, B., *et al.* (2010) Reproductive factors and exogenous hormone use in relation to risk of glioma and meningioma in a large European cohort study. *Cancer Epidemiology, Biomarkers & Prevention*, **19**, 2562–9.

Minutolo, F., Macchia, M., Katzenellenbogen, B.S., and Katzenellenbogen, J.A. (2011) Estrogen receptor beta ligands: recent advances and biomedical applications. *Medicinal Research Reviews*, **31**, 364–442.

Munoz, Y., Garrido, A., and Valladares, L. (2009) Equol is more active than soy isoflavone itself to compete for binding to thromboxane A(2) receptor in human platelets. *Thrombosis Research*, **123**, 740–4.

Nachvak, S.M., Moradi, S., Anjom-Shoae, J., *et al.* (2019) Soy, soy isoflavones, and protein intake in relation to mortality from all causes, cancers, and cardiovascular diseases: a systematic review and dose-response meta-analysis of prospective cohort studies. *Journal of the Academy of Nutrition and Dietetics*, **119**, 1483–500.

Nair, M.G., Safir, G.R., and Siqueira, J.O. (1991) Isolation and identification of vesicular-arbuscular mycorrhiza-stimulatory compounds from clover (trifolium repens) roots. *Applied and Environmental Microbiology*, **57**, 434–9.

Nelson, W.G., de Marzo, A.M., DeWeese, T.L., and Isaacs, W.B. (2004) The role of inflammation in the pathogenesis of prostate cancer. *Journal of Urology*, **172**, S6–S11.

Nielsen, I.L.F. and Williamson, G. (2007) Review of the factors affecting bioavailability of soy isoflavones in humans. *Nutrition and Cancer*, **57**, 1–10.

Nishio, K., Niwa, Y., Toyoshima, H., *et al.* (2007) Consumption of soy foods and the risk of breast cancer: findings from the Japan Collaborative Cohort (JACC) Study. *Cancer Causes & Control*, **18**, 801–8.

Niswender, K. (2010) Diabetes and obesity: therapeutic targeting and risk reduction – a complex interplay. *Diabetes, Obesity & Metabolism*, **12**, 267–87.

Njastad, K.M., Adler, S.A., Hansen-Moller, J., Thuen, E., Gustavsson, A.-M., and Steinshamn, H. (2014) Gastrointestinal metabolism of phytoestrogens in lactating dairy cows fed silages with different botanical composition. *Journal of Dairy Science*, **97**, 7735–50.

Norwitz, E.R., Jeong, K.H., and Chin, W.W. (1999) Molecular mechanisms of gonadotropin-releasing hormone receptor gene regulation. *Journal of the Society for Gynecologic Investigation*, **6**, 169–78.

Oba, S., Nagata, C., Shimizu, N., *et al.* (2007) Soy product consumption and the risk of colon cancer: a prospective study in Takayama, Japan. *Nutrition and Cancer*, **57**, 151–7.

O'Connor, P.M., Lapointe, T.K., Beck, P.L., and Buret, A.G. (2010) Mechanisms by which inflammation may increase intestinal cancer risk in inflammatory bowel disease. *Inflammatory Bowel Diseases*, **16**, 1411–20.

Pacifici, R. (1998) Cytokines, estrogen, and postmenopausal osteoporosis – the second decade. *Endocrinology*, **139**, 2659–61.

Paradkar, P.N., Blum, P.S., Berhow, M.A., Baumann, H., and Kuo, S.-M. (2004) Dietary isoflavones suppress endotoxin-induced inflammatory reaction in liver and intestine. *Cancer Letters*, **215**, 21–8.

Park, J.-S., Jung, J.-S., Jeong, Y.-H., *et al.* (2011) Antioxidant mechanism of isoflavone metabolites in hydrogen peroxide-stimulated rat primary astrocytes: critical role of hemeoxygenase-1 and NQO1 expression. *Journal of Neurochemistry*, **119**, 909–19.

Park, K., Kim, Y.-I., Shin, K.-O., *et al.* (2014) The dietary ingredient, genistein, stimulates cathelicidin antimicrobial peptide expression through a novel S1P-dependent mechanism. *Journal of Nutritional Biochemistry*, **25**, 734–40.

Paterni, I., Bertini, S., Granchi, C., Macchia, M., and Minutolo, F. (2013) Estrogen receptor ligands: a patent review update. *Expert Opinion on Therapeutic Patents*, **23**, 1247–71.

Pepys, M.B. and Hirschfield, G.M. (2003) C-reactive protein: a critical update. *Journal of Clinical Investigation*, **111**, 1805–12.

Perez-Cornago, A., Appleby, P.N., Boeing, H., *et al.* (2018) Circulating isoflavone and lignan concentrations and prostate cancer risk: a meta-analysis of individual participant data from seven prospective studies including 2,828 cases and 5,593 controls. *International Journal of Cancer*, **143**, 2677–86.

Platz, E.A., de Marzo, A.M., and Giovannucci, E. (2004) Prostate cancer association studies: pitfalls and solutions to cancer misclassification in the PSA era. *Journal of Cellular Biochemistry*, **91**, 553–71.

Prins, G.S. and Korach, K.S. (2008) The role of estrogens and estrogen receptors in normal prostate growth and disease. *Steroids*, **73**, 233–44.

Pugalendhi, P., Manoharan, S., Panjamurthy, K., Balakrishnan, S., and Nirmal, M.R. (2009) Antigenotoxic effect of genistein against 7,12-dimethylbenz[a]anthracene induced genotoxicity in bone marrow cells of female Wistar rats. *Pharmacological Reports*, **61**, 296–303.

Qin, Y., Niu, K., Zeng, Y., *et al.* (2013) Isoflavones for hypercholesterolaemia in adults. *Cochrane Database of Systematic Reviews*, **6**, CD009518.

Qing, X.R., Wan, C.C., Shang, X.J., *et al.* (2017) Relative contributions of testosterone deficiency and metabolism syndrome at the risk of reduced quality of life: A cross-sectional study among Chinese mid-aged and elderly men. *Andrologia*, **49**, e12736.

Rafii, F., Davis, C., Park, M., Heinze, T.M., and Beger, R.D. (2003) Variations in metabolism of the soy isoflavonoid daidzein by human intestinal microfloras from different individuals. *Archives of Microbiology*, **180**, 11–16.

Rafii, F., Jackson, L.D., Ross, I., *et al.* (2007) Metabolism of daidzein by fecal bacteria in rats. *Comparative Medicine*, **57**, 282–6.

Rapsomaniki, E., Timmis, A., George, J., *et al.* (2014) Blood pressure and incidence of twelve cardiovascular diseases: lifetime risks, healthy life-years lost, and age-specific associations in 1.25 million people. *Lancet*, **383**, 1899–911.

Reinli, K. and Block, G. (1996) Phytoestrogen content of foods – a compendium of literature values. *Nutrition and Cancer*, **26**, 123–48.

Ricci, E, Cipriani, S., Chiaffarino, F., Malvezzi, M., and Parazzini, F. (2010a) Soy isoflavones and bone mineral density in perimenopausal and postmenopausal Western women: a systematic review and meta-analysis of randomized controlled trials. *Journal of Women's Health and Law*, **19**, 1609–17.

Ricci, E., Cipriani, S., Chiaffarino, F., Malvezzi, M., and Parazzini, F. (2010b) Effects of soy isoflavones and genistein on glucose metabolism in perimenopausal and postmenopausal non-Asian women: a meta-analysis of randomized controlled trials. *Menopause*, **17**, 1080–6.

Richelle, M., Pridmore-Merten, S., Bodenstab, S., Enslen, M., and Offord, E.A. (2002) Hydrolysis of isoflavone glycosides to aglycones by beta-glycosidase does not alter plasma and urine isoflavone pharmacokinetics in postmenopausal women. *Journal of Nutrition*, **132**, 2587–92.

Rienks, J., Barbaresko, J., and Nothlings, U. (2017) Association of isoflavone biomarkers with risk of chronic disease and mortality: a systematic review and meta-analysis of observational studies. *Nutrition Reviews*, **75**, 616–41.

Rizzo, A., Pallone, F., Monteleone, G., and Fantini, M.C. (2011) Intestinal inflammation and colorectal cancer: a double-edged sword? *World Journal of Gastroenterology*, **17**, 3092–100.

Robins, S.P., Woitge, H., Hesley, R., Ju, J., Seyedin, S., and Seibel, M.J. (1994) Direct, enzyme-linked immunoassay for urinary deoxypyridinoline as a specific marker for measuring bone resorption. *Journal of Bone and Mineral Research*, **9**, 1643–9.

Rodriguez-Morato, J., Farre, M., Perez-Mana, C., *et al.* (2015) Pharmacokinetic comparison of soy isoflavone extracts in human plasma. *Journal of Agricultural and Food Chemistry*, **63**, 6946–53.

Romani, A., Vignolini, P., Tanini, A., Pampaloni, B., and Heimler, D. (2010) HPLC/DAD/MS and antioxidant activity of isoflavone-based food supplements. *Natural Product Communications*, **5**, 1775–80.

Rossi, R., Chiurlia, E., Nuzzo, A., Cioni, E., Origliani, G., and Modena, M.G. (2004) Flow-mediated vasodilation and the risk of developing hypertension in healthy postmenopausal women. *Journal of the American College of Cardiology*, **44**, 1636–40.

Rowland, I., Faughnan, M., Hoey, L., Wähälä, K., Williamson, G., and Cassidy, A. (2003) Bioavailability of phyto-oestrogens. *British Journal of Nutrition*, **89** Suppl 1, S45–S58.

Rubinacci, A., Melzi, R., Zampino, M., Soldarini, A., and Villa, I. (1999) Total and free deoxypyridinoline after acute osteoclast activity inhibition. *Clinical Chemistry*, **45**, 1510–16.

Rubio-Jurado, B., Balderas-Pena, L.-M.-A., Garcia-Luna, E.E., *et al.* (2018) Obesity, thrombotic risk, and inflammation in cancer. *Advances in Clinical Chemistry*, **85**, 71–89.

Rufer, C.E. and Kulling, S.E. (2006) Antioxidant activity of isoflavones and their major metabolites using different in vitro assays. *Journal of Agricultural and Food Chemistry*, **54**, 2926–31.

Said, S., Mukherjee, D., and Whayne, T.F. (2016) Interrelationships with metabolic syndrome, obesity and cardiovascular risk. *Current Vasculare Pharmacology*, **14**, 415–25.

Setchell, K.D. and Clerici, C. (2010) Equol: pharmacokinetics and biological actions. *Journal of Nutrition*, **140**, 1363S–8S.

Setchell, K.D., Brown, N.M., Desai, P., *et al*. (2001) Bioavailability of pure isoflavones in healthy humans and analysis of commercial soy isoflavone supplements. *Journal of Nutrition*, **131**, 1362s–75s.

Setchell, K.D.R., Brown, N.M., Zimmer-Nechemias, L., *et al*. (2002a) Evidence for lack of absorption of soy isoflavone glycosides in humans, supporting the crucial role of intestinal metabolism for bioavailability. *American Journal of Clinical Nutrition*, **76**, 447–53.

Setchell, K.D., Brown, N.M., and Lydeking-Olsen, E. (2002b) The clinical importance of the metabolite equol-a clue to the effectiveness of soy and its isoflavones. *Journal of Nutrition*, **132**, 3577–84.

Shechter, M., Issachar, A., Marai, I., *et al*. (2009) Long-term association of brachial artery flow-mediated vasodilation and cardiovascular events in middle-aged subjects with no apparent heart disease. *International Journal of Cardiology*, **134**, 52–8.

Shen, L., Ma, H., Xiang, M.-X., and Wang, J.-A. (2013) Meta-analysis of cohort studies of baseline prehypertension and risk of coronary heart disease. *American Journal of Cardiology*, **112**, 266–71.

Simental-Mendia, L.E., Gotto, A.M.J., Atkin, S.L., Banach, M., Pirro, M., and Sahebkar, A. (2018) Effect of soy isoflavone supplementation on plasma lipoprotein(a) concentrations: a meta-analysis. *Journal of Clinical Lipidology*, **12**, 16–24.

Song, X., Zeng, R., Ni, L., and Liu, C. (2016) The effect of soy or isoflavones on homocysteine levels: a meta-analysis of randomised controlled trials. *Journal of Human Nutrition and Dietetics*, **29**, 797–804.

Staessen, J.A., Gasowski, J., Wang, J.G., *et al*. (2000) Risks of untreated and treated isolated systolic hypertension in the elderly: meta-analysis of outcome trials. *Lancet*, **355**, 865–72.

Sudhir, K. (2006) Lipoprotein-associated phospholipase A2, vascular inflammation and cardiovascular risk prediction. *Vascular Health and Risk Management*, **2**, 153–6.

Sugiyama, N., Barros, R.P.A., Warner, M., and Gustafsson, J.-A. (2010) ERbeta: recent understanding of estrogen signaling. *Trends in Endocrinology and Metabolism*, **21**, 545–52.

Sun, J., Turner, A., Xu, J., Gronberg, H., and Isaacs, W. (2007) Genetic variability in inflammation pathways and prostate cancer risk. *Urologic Oncology*, **25**, 250–9.

Sutton-Tyrrell, K., Wildman, R.P., Matthews, K.A., *et al.* (2005) Sex-hormone-binding globulin and the free androgen index are related to cardiovascular risk factors in multiethnic premenopausal and perimenopausal women enrolled in the Study of Women Across the Nation (SWAN). *Circulation*, **111**, 1242–9.

Tada, H., Takamura, M., Kawashiri, M.-A. (2019) Lipoprotein(a) as an old and new causal risk factor of atherosclerotic cardiovascular disease. *Journal of Atherosclerosis and Thrombosis*, **26**, 583–91.

Tahara, T. and Arisawa, T. (2012) Potential usefulness of DNA methylation as a risk marker for digestive cancer associated with inflammation. *Expert Review of Molecular Diagnostics*, **12**, 489–97.

Taku, K., Umegaki, K., Sato, Y., Taki, Y., Endoh, K., Watanabe, S. (2007) Soy isoflavones lower serum total and LDL cholesterol in humans: a meta-analysis of 11 randomized controlled trials. *American Journal of Clinical Nutrition*, **85**, 1148–56.

Taku, K., Umegaki, K., Ishimi, Y., and Watanabe, S. (2008) Effects of extracted soy isoflavones alone on blood total and LDL cholesterol: meta-analysis of randomized controlled trials. *Therapeutics and Clinical Risk Management*, **4**, 1097–103.

Taku, K, Melby, M.K., Kurzer, M.S., Mizuno, S., Watanabe, S., and Ishimi, Y. (2010a) Effects of soy isoflavone supplements on bone turnover markers in menopausal women: systematic review and meta-analysis of randomized controlled trials. *Bone*, **47**, 413–23.

Taku, K, Melby, M.K., Takebayashi, J., *et al.* (2010b) Effect of soy isoflavone extract supplements on bone mineral density in menopausal women: meta-analysis of randomized controlled trials. *Asia Pacific Journal of Clinical Nutrition*, **19**, 33–42.

Taku, K., Lin, N., Cai, D., *et al.* (2010c) Effects of soy isoflavone extract supplements on blood pressure in adult humans: systematic review and meta-analysis of randomized placebo-controlled trials. *Journal of Hypertension*, **28**, 1971–82.

Taku, K., Melby, M.K., Kronenberg, F., Kurzer, M.S., and Messina, M. (2012) Extracted or synthesized soybean isoflavones reduce menopausal hot flash frequency and severity: systematic review and meta-analysis of randomized controlled trials. *Menopause*, **19**, 776–90.

Tanaka, T., Tahara-Hanaoka, S., Nabekura, T., *et al*. (2014) PPARbeta/delta activation of CD300a controls intestinal immunity. *Scientific Reports*, **4**, 5412.

Tew, B.Y., Xu, X., Wang, H.J., Murphy, P.A., and Hendrich, S. (1996) A diet high in wheat fiber decreases the bioavailability of soybean isoflavones in a single meal fed to women. *Journal of Nutrition*, **126**, 871–7.

Thijssen, D.H.J., Black, M.A., Pyke, K.E., *et al*. (2011) Assessment of flow-mediated dilation in humans: a methodological and physiological guideline. *American Journal of Physiology. Heart and Circulatory Physiology*, **300**, H2–H12.

Thompson, P.A., Khatami, M., Baglole, C.J., *et al*. (2015) Environmental immune disruptors, inflammation and cancer risk. *Carcinogenesis*, **36** Suppl 1, S232–S53.

Tikkinen, K.A.O., Dahm, P., Lytvyn, L., *et al*. (2018) Prostate cancer screening with prostate-specific antigen (PSA) test: a clinical practice guideline. *BMJ*, **362**, k3581.

Tracy, R.P. (2003) Inflammation, the metabolic syndrome and cardiovascular risk. *International Journal of Clinical Practice. Supplement*, **134**, 10–17.

Tse, G. and Eslick, G.D. (2016) Soy and isoflavone consumption and risk of gastrointestinal cancer: a systematic review and meta-analysis. *European Journal of Nutrition*, **55**, 63–73.

Urpi-Sarda, M., Morand, C., Besson, C., *et al*. (2008) Tissue distribution of isoflavones in ewes after consumption of red clover silage. *Archives of Biochemistry and Biophysics*, **476**, 205–10.

van Die, M.D., Bone, K.M., Williams, S.G., and Pirotta, M.V. (2014) Soy and soy isoflavones in prostate cancer: a systematic review and meta-analysis of randomized controlled trials. *BJU International*, **113**, E119–E30.

Vazzana, N., Riondino, S., Toto, V., *et al*. (2012) Obesity-driven inflammation and colorectal cancer. *Current Medicinal Chemistry*, **19**, 5837–53.

Vergne, S., Bennetau-Pelissero, C., Lamothe, V., *et al*. (2008) Higher bioavailability of isoflavones after a single ingestion of a soya-based supplement than a soya-based food in young healthy males. *British Journal of Nutrition*, **99**, 333–44.

Wakai, K., Egami, I., Kato, K., et al. (1999) Dietary intake and sources of isoflavones among Japanese. *Nutrition and Cancer*, **33**, 139–45.

Wallace, T.M., Levy, J.C., and Matthews, D.R. (2004) Use and abuse of HOMA modeling. *Diabetes Care*, **27**, 1487–95.

Ward, S.G. and Mrsny, R. (2009) New insights into mechanisms of gastrointestinal inflammation and cancer. *Current Opinion in Pharmacology*, **9**, 677–9.

Weggemans, R.M. and Trautwein, E.A. (2003) Relation between soy-associated isoflavones and LDL and HDL cholesterol concentrations in humans: a meta-analysis. *European Journal of Clinical Nutrition*, **57**, 940–6.

Wei, P., Liu, M., Chen, Y., and Chen, D.-C. (2012) Systematic review of soy isoflavone supplements on osteoporosis in women. *Asian Pacific Journal of Tropical Medicine*, **5**, 243–8.

Westley, B.R. and May, F.E.B. (2006) Identification of steroid hormone-regulated genes in breast cancer. *Methods in Molecular Medicine*, **120**, 363–88.

Willerson, J.T. and Ridker, P.M. (2004) Inflammation as a cardiovascular risk factor. *Circulation*, **109**, II-2–II-10.

Wiseman, H., O'Reilly, J.D., Adlercreutz, H., *et al.* (2000) Isoflavone phytoestrogens consumed in soy decrease F(2)-isoprostane concentrations and increase resistance of low-density lipoprotein to oxidation in humans. *American Journal of Clinical Nutrition*, **72**, 395–400.

Wiseman, H., Casey, K., Bowey, E.A., *et al.* (2004) Influence of 10 wk of soy consumption on plasma concentrations and excretion of isoflavonoids and on gut microflora metabolism in healthy adults. *American Journal of Clinical Nutrition*, **80**, 692–9.

Wu, S.H., Shu, X.O., Chow, W.-H., *et al.* (2012) Soy food intake and circulating levels of inflammatory markers in Chinese women. *Journal of the Academy of Nutrition and Dietetics*, **112**, 996–1004.

Xie, Q., Chen, M.-L., Qin, Y., *et al.* (2013) Isoflavone consumption and risk of breast cancer: a dose-response meta-analysis of observational studies. *Asia Pacific Journal of Clinical Nutrition*, **22**, 118–27.

Xu, X., Harris, K.S., Wang, H.J., Murphy, P.A., and Hendrich, S. (1995) Bioavailability of soybean isoflavones depends upon gut microflora in women. *Journal of Nutrition*, **125**, 2307–15.

Xu, X., Wang, H.J., Murphy, P.A., and Hendrich, S. (2000) Neither background diet nor type of soy food affects short-term isoflavone bioavailability in women. *Journal of Nutrition*, **130**, 798–801.

Yamasaki, K., Kayaba, K., and Ishikawa, S. (2015) Soy and soy products intake, all-cause mortality, and cause-specific mortality in Japan: the Jichi Medical School Cohort Study. *Asia Pacific Journal of Public Health*, **27**, 531–41.

Yang, Z., Kulkarni, K., Zhu, W., *et al.* (2005) Higher bioavailability of isoflavones after a single ingestion of a soya-based supplement than a soya-based food in young healthy males. *British Journal of Nutrition*, **5**, 333–44.

Yasuda, S., Wu, P.-S., Okabe, M., Tachibana, H., and Yamada, K. (2005) Tissue-specific distribution of genistein, daidzein and bisphenol a in male sprague-dawley rats after intragastric administration. *Food Science and Technology Research*, **11**, 187–93.

Yeung, J. and Yu, T. (2003) Effects of isoflavones (soy phyto-estrogens) on serum lipids: a meta-analysis of randomized controlled trials. *Nutrition Journal*, **2**, 15.

You, J., Sun, Y., Bo, Y., *et al.* (2018) The association between dietary isoflavones intake and gastric cancer risk: a meta-analysis of epidemiological studies. *BMC Public Health*, **18**, 510.

Yu, Y., Jing, X., Li, H., Zhao, X., and Wang, D. (2016) Soy isoflavone consumption and colorectal cancer risk: a systematic review and meta-analysis. *Scientific Report*, **6**, 25939.

Zhan, S. and Ho, S.C. (2005) Meta-analysis of the effects of soy protein containing isoflavones on the lipid profile. *American Journal of Clinical Nutrition*, **81**, 397–408.

Zhang, Y.-B., Chen, W.-H., Guo, J.-J., *et al.* (2013) Soy isoflavone supplementation could reduce body weight and improve glucose metabolism in non-Asian postmenopausal women – a meta-analysis. *Nutrition*, **29**, 8–14.

Zhao, T.-T., Jin, F., Li, J.-G., *et al.* (2019) Dietary isoflavones or isoflavone-rich food intake and breast cancer risk: a meta-analysis of prospective cohort studies. *Clinical Nutrition*, **38**, 136–45.

Zhong, X.-S., Ge, J., Chen, S.-W., Xiong, Y.-Q., Ma, S.-J., and Chen, Q. (2018) Association between dietary isoflavones in soy and legumes and endometrial cancer: a systematic review and meta-analysis. *Journal of Academic Nutrition and Dietetics*, **118**, 637–51.

Zhuo, X.-G., Melby, M.K., and Watanabe, S. (2004) Soy isoflavone intake lowers serum LDL cholesterol: a meta-analysis of 8 randomized controlled trials in humans. *Journal of Nutrition*, **134**, 2395–400.

Zimmet, P., Alberti, K.G., Magliano, D.J., and Bennett, P.H. (2016) Diabetes mellitus statistics on prevalence and mortality: facts and fallacies. *Nature Reviews. Endocrinology*, **12**, 616–22.

Zubik, L. and Meydani, M. (2003) Bioavailability of soybean isoflavones from aglycone and glucoside forms in American women. *American Journal of Clinical Nutrition*, **77**, 1459–65.

Zuloaga, K.L., O'Connor, D.T., Handa, R.J., and Gonzales, R.J. (2012) Estrogen receptor beta dependent attenuation of cytokine-induced cyclooxygenase-2 by androgens in human brain vascular smooth muscle cells and rat mesenteric arteries. *Steroids*, **77**, 835–44.

7

Dietary Anthocyanins

Iva Fernandes[1], *Hélder Oliveira*[1], *Cláudia Marques*[2,3], *Ana Faria*[2,3,4], *Conceição Calhau*[2,3], *Nuno Mateus*[1], and *Victor de Freitas*[1*]

[1] *REQUIMTE/LAQV, Department of Chemistry and Biochemistry, Faculty of Sciences, University of Porto, Porto, Portugal*
[2] *Nutrição e Metabolismo, NOVA Medical School, Faculdade de Ciências Médicas, Universidade Nova de Lisboa, Lisboa, Portugal*
[3] *CINTESIS – Center for Research in Health Technologies and Information Systems, Porto, Portugal*
[4] *Comprehensive Health Research Centre, Universidade NOVA de Lisboa, Lisboa, Portugal*

7.1 Absorption and Metabolism of Anthocyanins

Anthocyanins are present in many plants, fruits and vegetables, being a daily presence in several diet habits around the world (Figure 7.1). The absorption of these compounds cannot be assessed from a purely nutritional perspective. To understand their absorption and metabolism, a merge of different science fields is necessary.

These compounds have a structural pH dependence that may change their overall biological behavior (Figure 7.2) (Oliveira *et al.*, 2006). Therefore, an in-depth physical and chemical knowledge is essential to understand their absorption at the gastrointestinal level. Besides that, other factors such as the interaction with other phenols and proteins can affect their absorption (discussed later).

*Corresponding author.

Dietary Polyphenols: Metabolism and Health Effects, First Edition.
Edited by Francisco A. Tomás-Barberán, Antonio González-Sarrías, and Rocío García-Villalba.

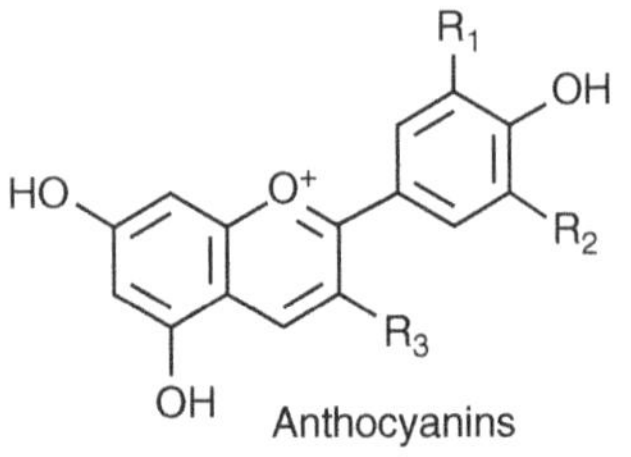

Anthocyanins	R_1	R_2
Pg3glc	H	H
Pn3glc	OCH_3	H
Cy3glc	OH	H
Mv3glc	OCH_3	OCH_3
Pt3glc	OCH_3	OH
Dp3glc	OH	OH

Figure 7.1 Representation of the general structure of anthocyanins (flavylium form). R_3 = sugar moiety. Pg3glc: pelargonidin-3-glucoside; Dp3glc: delphinidin-3-glucoside; Cy3glc: cyanidin-3-glucoside, Pn3glc: peonidin-3-glucoside, Mv3glc: malvidin-3-glucoside; Pt3glc: petunidin-3-glucoside.

These molecules have a 2-phenylbenzopyrilium core structure (flavylium cation) with different patterns of hydroxylation/methoxylation. Anthocyanins are glucoside derivatives of anthocyanidins linked through OH groups to different sugars depending on the natural source. Due to their characteristic pyrilium cation, they can adapt to different pH conditions by changing their structure in a series of reversible equilibrium reactions, giving rise to secondary forms (see Figure 7.2) (Pina *et al.*, 2015).

At pH 1, anthocyanins are found in their most stable form, the flavylium cation with a positively charged pyranic ring. When the pH rises, a proton transfer reaction occurs, and a quinoidal base appears. This reaction forms a dynamic system that happens extremely fast. In moderately acidic solutions (pH >2) the hydration (slower than proton transfer) of the flavylium cation occurs through a nucleophilic attack of water molecules at position C2, originating the hemiacetal neutral form that becomes the predominant species for the slightly acidic pH ranges. However, other reactions occur such as the opening of the B ring on the hemiacetal neutral base to yield the chalcone. At pH values higher than 7, the anthocyanins suffer a deprotonation reaction to produce the respective anionic base (Oliveira

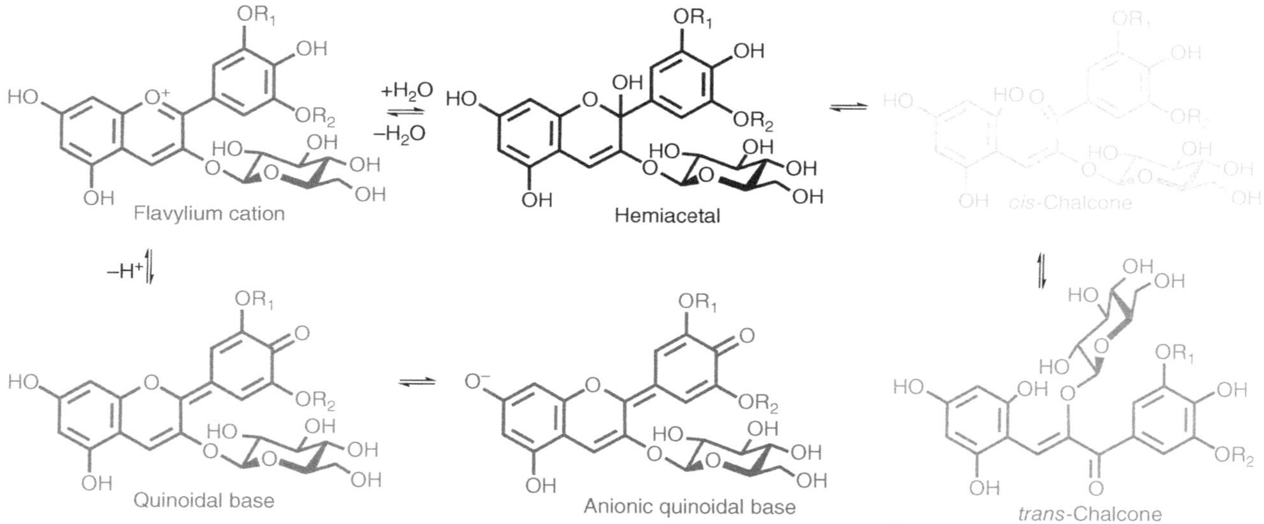

Figure 7.2 Anthocyanin equilibrium form at different pH.

et al., 2019a). From a biological perspective, this is extremely important since the pH network of the gastrointestinal tract is very wide, ranging from acidic to neutral conditions. However, in the literature, few studies take this feature into consideration (Fernandes *et al.*, 2012a).

7.1.1 Oral Cavity Absorption

The fate of anthocyanins after oral administration follows a very particular pattern. The food digestion process starts in the oral cavity, where salivary enzymes begin to digest food at an optimal pH of 5.6–7.9. At this pH, anthocyanins are mainly present as hemiacetal and chalcone forms. The differences found in the absorption of anthocyanins by oral tissues suggest that the non-charged structures may be more easily absorbed by passive diffusion (Mallery *et al.*, 2007).

The oral tissues are then exposed to the maximum amount of these compounds, which will, theoretically, facilitate their absorption. The oral bioavailability of anthocyanins is reported to be affected by several internal factors at the oral cavity, but their chemical structure and dietary factors may also influence their overall stability and absorption. The main focus of the studies related to this theme has been the influence of the oral environment on the stability of anthocyanins.

Several components present in saliva, such as proteins, enzymes, metal ions or salts, may affect the stability of anthocyanins, in a positive or a negative way. Anthocyanins have a higher degree of degradation in Na^{+}/K^{+} phosphate-buffered solutions than in water for the same pH (Woodward *et al.*, 2009). In one study, anthocyanins from blueberries, chokeberries, red grapes, and strawberries were incubated in saliva collected from 14 healthy human subjects. All the anthocyanins identified were partially degraded (Kamonpantana *et al.*, 2012). The degradation was attributed to enzymatic action and, to a large extent, oral microbiota. Also, the binding of anthocyanins to salivary proteins has been considered a minor factor, and in fact, interactions of different types of anthocyanins with this kind of proteins are reported, causing mainly astringency and bitterness sensations (Ferrer-Gallego *et al.*, 2015; Garcia-Estévez *et al.*,

2017; Vieira *et al.*, 2018). The formation of these complexes will affect the stability and absorption efficiency of anthocyanins.

A recent study found that both 3-glucoside anthocyanins from red wine and 3-sophoroside-5-glucoside anthocyanins from purple-fleshed sweet potato, after incubation with human saliva isolated from healthy volunteers (Oliveira *et al.*, 2019b), showed no significant degradation, which differs from a previous report (Kamonpantana *et al.*, 2012). This may be explained by the different structures of anthocyanins present in the food sources tested. It was shown that the degree of glycosylation improves the stability of anthocyanins due to the formation of an intramolecular H-bonding network within the anthocyanin moiety (He and Giusti, 2010). Kamonpatana *et al.* (2012) demonstrated a significant increase in the stability of di- and trisaccharide conjugates of anthocyanidins compared to the monosaccharide conjugates. This may explain the previous results concerning 3-sophoroside-5-glucoside anthocyanins. Also, it was shown that other dietary components, such as starch and glucose, enhanced the stability of anthocyanins in both oral and gastrointestinal cavities (Oliveira *et al.*, 2019b). This is of high importance since anthocyanins usually are ingested as part of complex food matrices and not as isolated compounds.

On the other hand, the interaction of anthocyanins with the oral enzymatic machinery may reduce their stability. Oral glucosidases may cleave the bond between the aglycone and glycosyl moieties. β-Glucosidases have been proposed as being responsible for the cleavage of anthocyanins to anthocyanidins, although the studies were not performed at the oral cavity (Kay *et al.*, 2009). The influence of the aglycone structure was also shown to be very important for anthocyanin stability. The incubation of different anthocyanins with saliva revealed that the stability of these compounds decreased with the degree of hydroxylation of the B ring (Kamonpantana *et al.*, 2012). Indeed, glycosides of delphinidin and petunidin were more susceptible to degradation than those of cyanidin, pelargonidin, peonidin, and malvidin in both intact and artificial saliva, possibly due to structural differences in the B ring (see Figure 7.1) (Kamonpantana *et al.*, 2012).

The fact that several kinetic studies in both animal and humans reported the appearance of intact and metabolized anthocyanins in blood just a few minutes after intake (Braga *et al.*, 2017) is compatible with the involvement of the oral cavity in the absorption of anthocyanins. After exposure of the oral cavity to anthocyanins, saliva showed the highest levels of anthocyanins, followed by buccal mucosa and then oral cells (Ugalde *et al.*, 2009). The uptake of the compounds by the oral cells only happens when they cross the oral mucous layer. More than 97% of the anthocyanins from chokeberries were found to be associated with mucus (Kamonpantana *et al.*, 2014), which may imply a generally low absorption of the anthocyanins at the oral level.

Another significant factor that leads to different absorption rates in the oral cavity is the structure of the anthocyanins. The amounts of different anthocyanins from chokeberries present in the oral mucosa differed according to their structure. Cyanidin-3-*O*-glucoside and cyanidin-3-*O*-arabinoside were found in the highest concentrations, while cyanidin-3-*O*-xyloside was found at lower levels (Kamonpantana *et al.*, 2014). The relative amount of cyanidin-3-*O*-glucoside in washed buccal epithelial cells was significantly higher than those of the other cyanidin-3-monoglucosides, suggesting preferential transport of cyanidin-3-O-glucoside. This may be related to the affinity of the anthocyanin sugar moieties for specific mechanisms of transport. In the mouth, sodium-dependent glucose co-transporters were suggested as being essential for anthocyanin absorption (Mallery *et al.*, 2011).

The aglycone structure has also been shown to be important. Anthocyanins from red grape were retained within the buccal mucous layer after oral retention of the juice for five minutes. However, significant differences were found between them: malvidin-3-*O*-glucoside, cyanidin-3-*O*-glucoside, and peonidin-3-*O*-glucoside were found in higher amounts than delphinidin-3-*O*-glucoside and petunidin-3-*O*-glucoside, possibly due to structural differences in the B ring (see Figure 7.2) (Kamonpantana *et al.*, 2014).

The intestines (and more recently stomach) is usually considered as the leading absorption site, due to the high rate of anthocyanin metabolites that are normally detected in the bloodstream, urine, and feces. However, the enzymatic machinery for phase II metabolism and enteric recycling was also identified in the oral cavity of humans (Kamiloglu *et al.*, 2015; Mallery *et al.*, 2011, 2014; Ugalde *et al.*, 2009). Two critical enzymes for the metabolism of anthocyanins were found in human saliva: β-glucosidases and arylsulfatases (Mallery *et al.*, 2011). It thus becomes clear that the mouth may have an interesting and still far from fully explored role in anthocyanin absorption and bioavailability.

7.1.2 Gastric Absorption

The stomach was for a long time ignored as a potential site of absorption of anthocyanins. This fact was driven by the observation of several metabolites in blood, urine, and feces of humans and animals upon ingestion of these compounds. Such considerations are compatible with an intestinal absorption due to the presence of the enzymatic machinery necessary for the biotransformation of anthocyanins in their metabolites.

However, this perception started to change when the kinetics of anthocyanin absorption in humans and animals were evaluated, and it was observed that anthocyanins appeared in the bloodstream either intact or in their metabolized form just a few minutes after ingestion. The first studies in rats using different anthocyanins showed the presence of these compounds in blood and urine. Bilberry anthocyanins were detected in the bloodstream in a concentration of 3 μg/mL after 15 minutes following ingestion (Morazzoni *et al.*, 1991). Cyanidin-3-*O*-glucoside was detected intact in plasma after 30 minutes upon oral administration (Miyazawa *et al.*, 1999; Tsuda *et al.*, 1999). Anthocyanins from blackcurrants were isolated (delphinidin-3-*O*-rutinoside, cyanidin-3-*O*-glucoside and cyanidin-3-*O*-rutinoside) and orally administered to rats. These compounds were detected in the bloodstream after 30 minutes and then excreted in urine with no detection of potential degradation products (Matsumoto *et al.*, 2001).

Nevertheless, degradation is known to occur, and more recent studies reported the presence of anthocyanin metabolites in urine and plasma of rats, pigs, and rabbits. When rats were dosed with 100 mg of delphinidin-3-*O*-glucoside/kg body weight, the maximum concentration in plasma was reached within 15 minutes, and the methylated metabolites reached a maximum plasma concentration after one hour (Ichiyanagi *et al.*, 2004). These metabolites were suggested to be produced in the liver as a result of the enteric recycling paths, rather than in the intestinal flora, evidencing absorption of intact anthocyanins and further metabolization (Ichiyanagi *et al.*, 2005a,b). Glucuronidated and methylated anthocyanin-derived metabolites from pelargonidin-3-*O*-glucoside and cyanidin-3-*O*-glucoside were also reported as the major types of metabolites present in the urine of pigs after oral administration (Wu *et al.*, 2004). There are also numerous studies showing that in humans, anthocyanins are absorbed from the digestive tract, transit in the circulatory system, and are excreted in the urine and feces.

Blackcurrant anthocyanins were directly absorbed and detected in the blood and the urine as intact forms (Matsumoto *et al.*, 2001). Similar results were found for anthocyanins from boysenberry and blueberry (McGhie *et al.*, 2003). The intact glycosides and their metabolites have been detected in urine after the consumption of plum anthocyanins (Netzel *et al.*, 2012). After the ingestion of red wine containing 136 mg of malvidin-3-*O*-glucoside/L, or red grape juice containing 234 mg of the same anthocyanin/L, malvidin-3-*O*-glucoside was found in plasma 20 minutes after intake of red wine and 180 minutes after the intake of red grape juice, which clearly shows the role of the food matrix in their pharmacokinetics. Neither aglycones nor glucuronide or sulfate conjugates were found in the plasma or urine samples, indicating that this anthocyanin may be absorbed in the glycosylated form with no further significant metabolism (Bub *et al.*, 2001). Intact cyanidin-base anthocyanins were detected in the urine after ingestion of boysenberry anthocyanins, although methylated and glucuronidated metabolites were also found (Cooney *et al.*, 2004). The total bioavailability of anthocyanins becomes an issue when considering all metabolites. Only a few studies have tried to assess and quantify all the possible compounds resulting from anthocyanin absorption (Czank *et al.*, 2013; de Ferrars *et al.*, 2014b).

Interestingly, among all polyphenols of blueberries, 23 phenolic acids were quantified in plasma of healthy volunteers after intake. However, these amounts were not directly related to the quantities of anthocyanin intakes, revealing a far more complex metabolic fate for these compounds (Rodriguez-Mateos *et al.*, 2016). Comparison of the bioavailability of anthocyanins in healthy subjects versus ileostomists revealed higher amounts of anthocyanins and degradants in the plasma/urine of subjects with an intact gut (Mueller *et al.*, 2017). In spite of demonstrating the role of the intestine in the absorption and metabolization of anthocyanins, this study also showed that even without a colon, anthocyanins can still be absorbed (C_{max} in plasma of 27 nmol/L for no colon versus 43 nmol/L for colon existence, for peonidin-3-*O*-glucoside), prompting the potential role of the stomach.

In vitro studies have also brought essential insights about the potential mechanisms in the absorption of anthocyanins at the gastric level. The first studies reported an interaction between anthocyanins and bilitranslocase, which is expressed in gastric cells (Passamonti and Vanzo, 2005; Passamonti *et al.*, 2002). This study showed that the aglycone structure played a vital role in the interactions between bilitranslocase and anthocyanins, which is essential for understanding potential protein–ligand phenomena. More recent studies focused on the role of the glycosyl moiety in the absorption efficacy of anthocyanins at the gastric level. MKN-28 cells were used as an *in vitro* model for a gastric barrier, and competition between glucose and red wine anthocyanins for the transport across this barrier was reported, implying the potential involvement of glucose transporters in the absorption of these anthocyanins (Oliveira *et al.*, 2015). Computational studies verified that the glucose moiety was essential for the docking of cyanidin-3-*O*-glucoside, delphinidin-3-*O*-glucoside, and malvidin-3-*O*-glucoside in the three tested structures: flavylium cation, hemiacetal, and anionic quinoidal base (Oliveira *et al.*, 2015). The mechanism of transport through gastric MKN-28 cells was evaluated using a nano-approach with gold nanoparticles functionalized against hGLUT-1 and hGLUT-3 (Oliveira *et al.*, 2019c). Upon the silencing of these transporters, the cells were used to form an *in vitro* gastric barrier model, and the transport efficiency of anthocyanins from red wine (3-*O*-glucosides) and purple-fleshed sweet potato (3-*O*-sophoroside-5-*O*-glucosides) was evaluated. In both cases, the transport was considerably decreased in the silenced cells, with a higher decrease for the

3-*O*-sophoroside-5-*O*-glucosides (Oliveira *et al.*, 2019c). This type of anthocyanins was reported to be absorbed at both gastric and intestinal level (Oliveira *et al.*, 2019b). Interestingly, computational studies showed that the more complex anthocyanins (from purple-fleshed sweet potato) have three-dimensional arrangements that may allow the glucose to interact with the transporters, suggesting a similar translocation mechanism for these and red wine anthocyanins.

However, this work also reported the potential involvement of other classes of transporters such as OATs or OCTs. In fact, at the pH of the experiments, anthocyanins occur in equilibrium under different structures, and these may explain the observed phenomena.

These recent studies provided insights about the importance of these glucose transporters for our knowledge of how anthocyanins may be absorbed not only at the gastric level but at other levels. Figure 7.3 shows a summarized scheme of these recent findings of the role of glucose transporters in anthocyanin absorption.

7.1.3 Intestinal Absorption

The anthocyanin fraction that is not absorbed in the stomach reaches the small intestine. Once anthocyanins encounter less acidic conditions, in the small intestine, the anionic quinoidal base is likely to predominate. Anthocyanidin glycosides are rapidly and efficiently absorbed in the small intestine (Fang, 2014; Talavera *et al.*, 2004). Furthermore, anthocyanins are quickly metabolized and appear in the circulation or are excreted through urine as both intact and metabolized forms (glucuronidated, sulfated or methylated derivatives) (Czank *et al.*, 2013).

The potential mechanisms of anthocyanin glycoside absorption in the small intestine may involve specific glucose transporters, as previously suggested for other flavonoids. The most recent works point to the involvement of SGLT1 and GLUT2 transporters in anthocyanin absorption at the intestinal level (Faria *et al.*, 2009a; Kamiloglu *et al.*, 2015; Zou *et al.*, 2014).

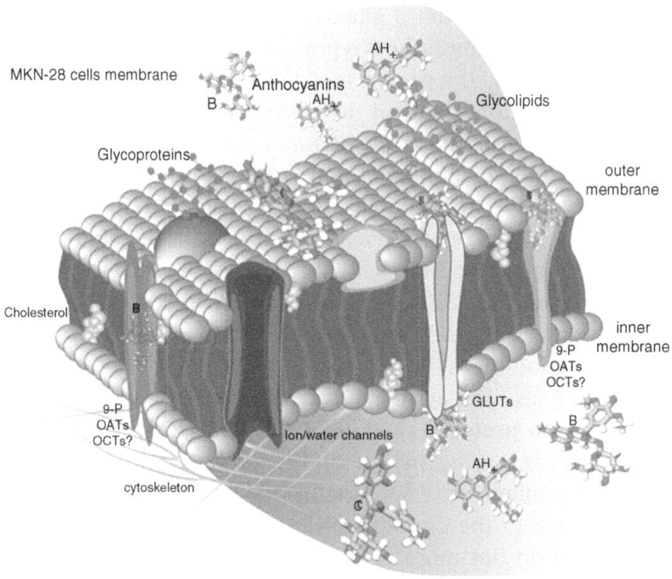

Figure 7.3 Proposed mechanism for anthocyanin transport through MKN-28 cell membrane with the involvement of glucose transporters. Anthocyanins exist in a dynamic equilibrium network between different chemical structures at the pH conditions of the stomach environment. In this figure, the neutral hemiacetal form is represented as being transported by glucose transporters (GLUTs). However, other types may also be transported. Transport of the anthocyanins can also occur by different mechanisms with the involvement of carriers such as efflux transporters and organic cation/anion transporters. AH_+, flavylium cation form; B, hemiacetal form; C, chalcone form.

7.1.4 Microbial Metabolism

Anthocyanins that are not absorbed in the small intestine can be metabolized by the gut microbiota on reaching the colon (Faria *et al.*, 2014a). Microbial metabolism of anthocyanins alters the bioactivity of these compounds and, at the same time, anthocyanins and their metabolites modulate the gut microbiota composition (Marques *et al.*, 2018).

Colonic metabolism has long been speculated to be a major contributor to the overall metabolism of anthocyanins. In fact, in the ^{13}C-labeled cyanidin-3-glucose study conducted by de Ferrars *et al.* (2014b), the authors reported 32 ± 6% of the recovered ^{13}C-label in feces. The methyl metabolite of caffeic acid, ferulic acid, was the most abundant metabolite excreted in feces after anthocyanin consumption (de Ferrars *et al.*, 2014b).

Study of the metabolism of anthocyanins by the gut microbiota has led to the conclusion that bacterial metabolism involves the cleavage of glycosidic linkages and breakdown of anthocyanidin heterocycle, which is rapidly converted to phenolic acids (Aura *et al.*, 2005; Keppler and Humpf, 2005). Cytosolic β-glucosidase and lactase-phlorizin hydrolase are enzymes responsible for the hydrolysis of anthocyanin glycosides into their aglycones. The aglycone cyanidin can be spontaneously degraded into protocatechuic acid (PCA, originated from the B ring during C-ring fission of the cyanidin skeleton) and phloroglucinaldehyde (PGA, originated from the A ring during C-ring fission of the cyanidin skeleton), in the small intestine and circulation but more extensively in the colon (de Ferrars *et al.*, 2014b; Gonzalez-Barrio *et al.*, 2011).

Gonzalez-Barrio *et al.* (2011) suggested that dehydroxylation of PCA by colonic bacteria to form hydroxybenzoic acid, followed by conjugation with glycine to form hippuric acid, is the major colonic pathway followed by cyanidin-3-glucose. This is in accordance with de Ferrars *et al.*, since hippuric acid is the primary metabolite found in serum and urine after the consumption of 500 mg ^{13}C-labeled cyanidin-3-glucose (de Ferrars *et al.*, 2014b). However, in a different study using nonlabeled anthocyanins, hippuric acid was not increased in the serum or urine of women after elderberry anthocyanin consumption comparative to baseline values (before the consumption of anthocyanin-rich foods) (de Ferrars *et al.*, 2014a). This could be explained by the fact that hippuric acid only reaches its maximum concentration in serum and urine at 6–24 hours or because it was present in the background diet of women (de Ferrars *et al.*, 2014a, b).

It should be highlighted that the majority of anthocyanin catabolites formed by the action of gut microbiota can also result from the breakdown of other flavonoids (Figure 7.4) (Braune and Blaut, 2016) or even be ingested as part of the composition of other food sources. In fact, without radioactivity

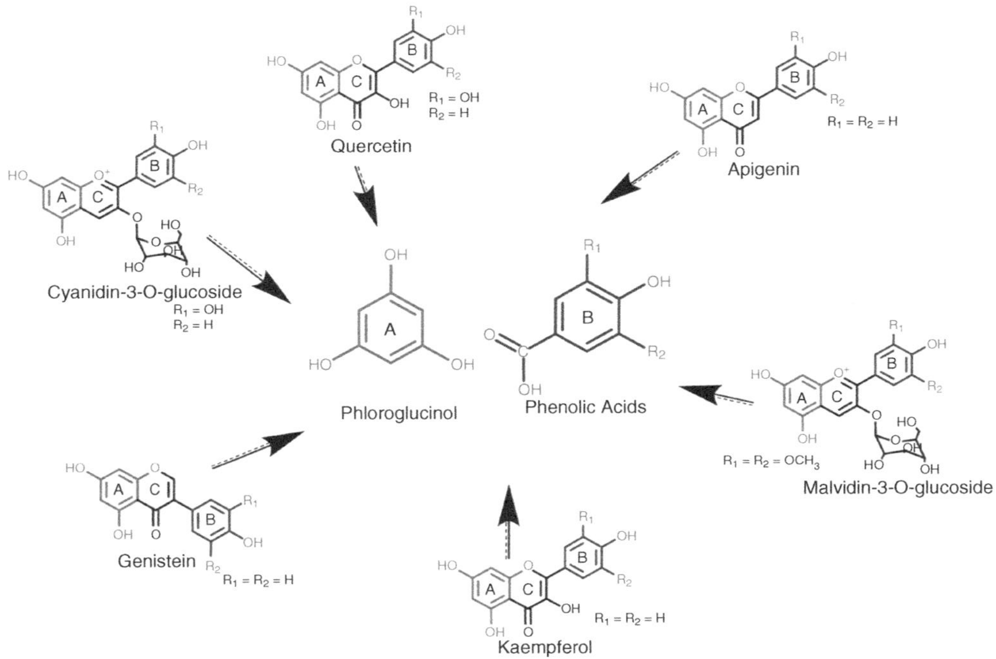

Figure 7.4 Summary of the breakdown of flavonols, isoflavones, flavan-3-ols, and flavones by human intestinal bacteria.

labeling, it is almost impossible to associate the presence of these simpler phenolics with the ingestion of anthocyanins.

After absorption, phenolic acids can also be metabolized into glucuronidated, sulfated or methylated derivatives. Given the enterohepatic recirculation, these compounds may prevail in the human body for several days (Kay *et al.*, 2017). Health benefits associated with anthocyanin-rich foods may, hence, be explained by a slow and continuous release of phenolic compounds through the gut into the bloodstream. Moreover, they can accumulate in several organs where they can exert their effects (El Mohsen *et al.*, 2006).

7.2 Pharmacokinetics of Anthocyanins

The pharmacokinetics of bilberry anthocyanins were evaluated in rats. The plasma concentration peak (C_{max}) of 11.1 ng/mL was reached after 30 minutes. Animals were subjected to fasting, and the bioavailability was significantly increased by more than seven-fold comparative to fed rats. Interestingly, concomitant administration of glucose did not affect anthocyanin bioavailability. However, the bioavailability of particular anthocyanins was influenced by the aglycone and the sugar moiety (Baron *et al.*, 2017).

The bioavailability of anthocyanins from calafate berry was studied in gerbils, but no parental anthocyanins were detected in plasma (Bustamante *et al.*, 2018). The detection of small phenolics was proposed to be due to the contribution of anthocyanin degradation products.

There is a remarkable lack of studies that describe both anthocyanin tissue bioavailability and the pharmacodynamics explaining anthocyanin effects (Sandoval-Ramírez *et al.*, 2018). Nevertheless, regarding the pharmacokinetics of anthocyanin conjugates, some human studies have reported remarkable concentrations of methylated and glucuronidated derivatives(de Ferrars *et al.*, 2014b; Fernandes *et al.*, 2017; Marques *et al.*, 2016). These results validate the assumption that anthocyanins are extensively metabolized and that their putative health effects may be mediated by the originated metabolites instead of the parent anthocyanins (Table 7.1).

Table 7.1 Concentration of major methylated and glucuronidated anthocyanin derivatives found in plasma/serum and urine after anthocyanin consumption (de Ferrars *et al.*, 2014b; Marques *et al.*, 2016).

	Blackberry anthocyanins	^{13}C-labeled cyanidin-3-glucose
Plasma/serum		
Me-Cy-Glucr	272 ± 39 ng/mL	ND
3′-Me-Cy3glc	88 ± 11 ng/mL	ND
Me-Cy3glc-Glucr	78 ± 21 ng/mL	ND
Urine		
Me-Cy-Glucr	3.0 ± 0.8 (ng/mg creatinine)	206 ± 105 nM
3′-Me-Cy3glc	1.2 ± 0.3 (ng/mg creatinine)	76 ± 43 nM
Cy-Glucr (a)	0.9 ± 0.2 (ng/mg creatinine)	88 ± 42 nM

Cy, cyanidin; Cy3glc, cyanidin-3-glucoside; Glucr, glucuronide; Me, methyl; ND, not detected.

7.3 Factors Affecting Anthocyanin Bioavailability

Bioavailability, the degree to which an ingested anthocyanin is available to a target tissue, is a complex issue that despite significant advances in recent years is still not fully understood.

The nonbiological aspects such as the dose, anthocyanin chemical structure, source (purified or a mixture of anthocyanins), food/beverage matrix or processing and the analytical methodology have a strong effect on the analyzed parameters, particularly in anthocyanin availability to absorption, meaning bioaccessibility (Dangles and Fenger, 2018). On the other hand, biological aspects such as age and gender of the subjects, body composition and individual characteristics such as microbiota and biotransformation capacity may have an impact on anthocyanin disposition to target tissues (Lila *et al.*, 2016; Marques *et al.*, 2016). Of course, gastrointestinal (GI) tract characteristics such as pH, concomitant ingested food, digestive enzymes, biliary acids, the motility and permeability of the GI tract as well as microbiota will also have a significant influence on anthocyanin stability.

Anthocyanins' chemical structure is determinant to their rate of absorption (Baron *et al.*, 2017; Lila *et al.*, 2016; Mazza *et al.*, 2002; Talavera *et al.*, 2004; Wu *et al.*, 2002; Yi *et al.*, 2006). The transport/absorption efficiency of anthocyanins varied with the type of anthocyanin, explained by the presence of more free hydroxyl groups and fewer OCH_3 groups which decrease the anthocyanins' bioavailability, which may be associated with the lower lipophilicity (Weiguang *et al.*, 2006). Talavéra *et al.* (2004) demonstrated that the aglycone structure had an impact on anthocyanin intestinal absorption, caused by the presence of methoxylated groups that reduced intestinal absorption.

The passage through the GI tract is critical concerning anthocyanin absorption since these compounds have complex chemistry responsible for their attractive colors but also for their instability, which will influence all the biokinetic processes, and, thus, the bioactivity (McGhie and Walton, 2007). The most common methodologies to detect anthocyanins are centered on their color in a flavylium cationic form and are based on the total conversion of anthocyanin forms in one way in acidic medium. However, even analyzed at acidic conditions, samples with no preacidification could lead to underestimated anthocyanin quantification, since the conversion of anthocyanins to a cation form may not be complete (Fernandes *et al.*, 2012a).

In addition, it is usually not considered that anthocyanins may be metabolized/biotransformed and, thus, conversion back to the cationic form after acidification is no longer a possibility, contributing to this underestimation. Furthermore, whether the biological effects attributed to anthocyanins are due to their cationic form, their hemiacetal form or a metabolite of one of these two forms is still uncertain. The difficulty in overcoming these analytical problems may contribute significantly to the reported low bioavailability of anthocyanins, which does not justify all the biological activities previously associated with the consumption of this flavonoid class. However, Czank *et al.*, making use of isotopically labeled cyanidin-3-glucoside, reported a recovery of 12.4% (Czank *et al.*, 2013).

Another factor affecting anthocyanin bioavailability is their possible ingestion as pigments (anthocyanin derivatives), especially when considering red wine consumption. One report indicated that anthocyanin pyruvic acid adducts could rapidly reach rat plasma 15 minutes after oral administration of 400 mg/kg bw (~21.1 nM/kg/μmol and ~28.8 nM/kg/μmol of the malvidin-3-glucoside-pyruvic acid adduct and malvidin-3-glucoside, respectively) (Faria *et al.*, 2009b). Also, flavanol-anthocyanin pigments presented a higher absorption efficiency in the Caco-2 cell model than procyanidin B3 (Fernandes *et al.*, 2012b).

The study design used to evaluate anthocyanin bioavailability is an important variable, as the food source, food matrix, intake duration, and dosing will affect the results. The effect of intake duration was reported showing that after long-term blueberry juice intake, anthocyanin content in urine declined over time (Kalt *et al.*, 2017). A previous *in vitro* study has pointed out the importance of consumption frequency since cells exposed long term to anthocyanins were shown to be more prone to their transport (Faria *et al.*, 2009a). Also, dosing factor was studied; when the same amount of anthocyanins were ingested in different portions during the day, this resulted in a minor anthocyanin urinary excretion (Kalt *et al.*, 2017), suggesting that a larger volume would lead to higher anthocyanin absorption.

The food matrix would also be a factor, where a more lipophilic environment may facilitate flavonoid solubilization and absorption as well as the presence of ethanol that could determine the extent of anthocyanin absorption (Faria *et al.*, 2009a), promoting their transport across intestinal epithelia.

Anthocyanins are subject to extensive metabolism in the human organism (Czank *et al.*, 2013; Kay *et al.*, 2004), and most of the bioavailability studies do not detect or quantify these metabolites, leading to underestimation. The complexity of the metabolites generated by the chemical and physiological conditions in the GI tract, the human xenobiotic processes and the microbiota make its determination very difficult. Human phenotypic variation in xenobiotic mechanisms of transport and enzymatic biotransformation responsible for modification, conjugation, transport, and excretion will affect this pool of metabolites formed (Zhang *et al.*, 2007) and consequently its biological effect.

This also raises the question about simultaneous ingestion of other substances; polyphenols use the same transport mechanisms as other nutrients, interfering with absorption of these molecules, e.g., organic cations, glucose (Faria *et al.*, 2006, 2009a; Keating *et al.*, 2008).

Despite not directly affecting anthocyanin bioavailability, the microbiota has a great influence on the biological effects reported for anthocyanins. Anthocyanins that are not absorbed, distributed, metabolized, and excreted in the classic pharmacological definition reach the colon. Being microbiota substrates, this can have an impact on human health by promoting the growth of specific bacterial strains (prebiotic effect) and also result in bacterial metabolites that may have a positive biological effect (Faria *et al.*, 2014a; Williamson and Clifford, 2017).

Even though several studies have explored anthocyanin bioavailability, there is no consistency in the study design and the information produced is very heterogeneous. When taken together, there are significant interindividual and intraindividual variations in bioavailability, but the knowledge of the disposition of anthocyanins and their metabolites is a critical determinant to understand its biological effect. Nevertheless, it is agreed that the lower levels determined for anthocyanin bioavailability do not justify their health beneficial effects.

7.4 Biological Activity of Anthocyanin Metabolites

The biological activity of anthocyanin metabolites may include a panoply of compounds including degradation products, phase II metabolites, and colonic catabolites (Figure 7.5).

If the positive health effects associated with anthocyanin consumption resulted only from the gut metabolites formed after colon metabolism, the nutrition allegation should not promote anthocyanin consumption but instead encourage the ingestion of sources of phenolic acids and phloroglucinol. However, since these were the primary metabolites detected after a ^{13}C-tracer study with Cy3glc, they were identified as degradation products, namely phenolic, hippuric, phenylacetic, and phenylpropanoic acids (Czank *et al.*, 2013).

As can be observed in the literature, considerable effort has been expended in the assessment of the neuroprotective and cardiovascular health effects associated with these catabolites.

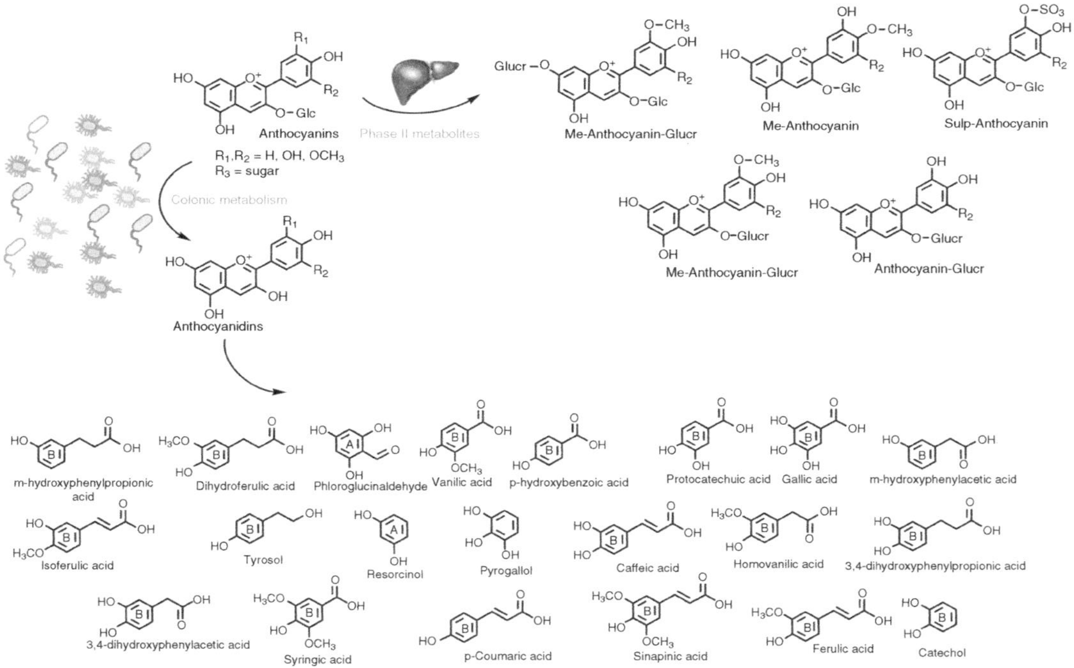

Figure 7.5 Main phase II metabolites and microbial metabolites of monoglucoside anthocyanins. Me, methyl group; Glucr; glucuronyl group; SO_3, sulfate group; Glc, sugar moiety.

Both 4-hydroxybenzoic acid (HBA) and PCA are capable of mitigating oxidative stress induced by hydrogen peroxide, which is thought to contribute to neuronal cell death in neurodegeneration (Winter *et al.*, 2017). These data indicate that phenolic acids could be suitable for the treatment of neurodegeneration.

In addition, accumulating evidence suggests that anthocyanins play a vital role in the cardioprotective effects associated with consumption of anthocyanin-rich foods. Anthocyanin metabolites (PCA and vanillic acid) may modulate vascular reactivity by inducing Heme Oxygenase-1 (HO-1) and modulating NADPH oxidases (NOX) activity, resulting in reduced superoxide production and improved nitric oxide (NO) bioavailability (Edwards *et al.*, 2015). Using human umbilical vein endothelial cells (HUVECs) anthocyanin gut metabolites, protocatechuic, vanillic, ferulic and hippuric acids were able to modulate the adhesion of monocytes to endothelial cells, the initial step in atherosclerosis development (Krga *et al.*, 2016).

Blood samples from human volunteers were treated with gut metabolites at physiologically relevant concentrations, including 4-hydroxybenzaldehyde, protocatechuic, vanillic, ferulic, and hippuric acids, and the impact of these metabolites on platelet activation and platelet–leukocyte aggregation was determined (Krga *et al.*, 2018). Hippuric and protocatechuic acids inhibited P-selectin expression, ferulic acid reduced platelet–monocyte aggregation, while 4-hydroxybenzaldehyde affected P-selectin expression, platelet–neutrophil and –monocyte aggregation (Krga *et al.*, 2018). Altogether, these results support the beneficial effects of frequent consumption of anthocyanin-rich foods for cardiovascular health.

As described earlier, if on the one hand, anthocyanins are subjected to metabolism by microbiota, on the other hand, they and their metabolites may modulate the growth of specific bacteria from microbiota. There is evidence that the composition of human intestinal microbiota influences health and the incidence of disease and that gut health is largely determined by the complex interaction between host and gastrointestinal microbiota (Sekirov *et al.*, 2010). In this regard, anthocyanins have been shown to have anti-inflammatory (Jayarathne *et al.*, 2019; Peng *et al.*, 2019), antidiabetic (Cremonini *et al.*, 2019; Gowd *et al.*, 2018; Petersen *et al.*, 2019; Su *et al.*, 2019) and neuroprotective (Marques *et al.*, 2018) properties though gut microbiota modulation (Hester *et al.*, 2018; Kilua *et al.*, 2019; Mayta-Apaza *et al.*, 2018; Petersen *et al.*, 2019; Rodriguez-Morato *et al.*, 2018; Wankhade *et al.*, 2019), for example by preventing an increase in the *Firmicutes/Bacteroidetes* ratio and a decrease in *Akkermansia* typically observed in animal models of high-fat diet obesity. Table 7.2 summarizes the most recent studies where the health benefits caused by dietary supplementation with anthocyanins were associated with gut microbiota modulation.

7.4.1 Phase II Metabolites

The identity and concentration of anthocyanin metabolites that are the most relevant forms *in vivo* are practically unknown because of the lack of standards. It is only possible to determine the type of conjugation, but it is not possible to ascertain the position of the conjugation. Since this lack of information will compromise the evaluation of conjugate bioactivity (several isomers may be assayed) and metabolic mechanisms followed by anthocyanins after absorption (according to our knowledge there is one main route, but secondary ones are also observed), this is a significant problem.

The references in the literature concerning evaluation of the biological activity of phase II metabolites of anthocyanins are quite scarce since these compounds are not commercially available. On the other hand, synthesis in large amounts and purification of these compounds are difficult. Indeed, previous *in vivo* studies have concluded that Cy3glc, Dp3glc, and Pt3glc are rapidly absorbed and metabolized extensively, following oral consumption in humans and rats, to 3′-Me-Cy3glc, 4′-Me-Cy3glc, 4′-Me-Dp3glc, and 4′-Pt3glc, respectively (Felgines *et al.*, 2005; Ichiyanagi *et al.*, 2006, 2008; Kay *et al.*, 2004, 2005; Marczylo *et al.*, 2009; Tian *et al.*, 2006).

Based on some of these *in vivo* studies, an enzymatic approach was conducted and pure standards of methylated anthocyanins were obtained (Fernandes *et al.*, 2009, 2013). The antioxidant and antiproliferative activity against several cancer cell lines (MKN-28, Caco-2, and MCF-7) and the normal human foreskin fibroblasts (HFF-1) of these metabolites in comparison with native anthocyanins was evaluated (Fernandes *et al.*, 2013). The methylated metabolites were found to retain significant radical

Table 7.2 Gut microbiota modulation and gut microbiota-mediated health benefits induced by berry anthocyanins

Animal model	Dose/duration	Methods	Changes in gut microbiota composition	Biological activity	Reference
Dahl salt-sensitive rats	Lyophilized berry mixture (2 g/day); 9 weeks	16S rRNA gene sequencing	↑ Microbiota richness ↑ *Bacteroidetes* ↓ *Firmicutes*	Attenuate the development of hypertension induced by high-salt diets	(Gomes *et al.*, 2019)
Wistar rats	Cyanidin-3-*O*-glucoside (500 and 1000 mg/kg diet, w/w); 8 weeks	16S rRNA gene sequencing	Ameliorate gut microbial dysbiosis caused by 3-MCPD (food contaminant) ↑ Actinobacteria ↑ Lachnospiraceae NK4A136 group	Protect against the damage of intestinal villus, microvillus, and mucosa caused by 3-MCPD (food contaminant)	(Chen *et al.*, 2019)
C57BL/6J mice High-fat diet (HFD)-induced obesity model	Anthocyanins (40 mg /kg body weight/day); 14 weeks	16S rRNA gene sequencing	Ameliorate gut microbial dysbiosis caused by HFD ↑ *Akkermasia* ↓ *Firmicutes/Bacteroidetes* ratio	Mitigate HFD-induced development of obesity, dyslipidemia, insulin resistance, and steatosis ↑ GLP-2 plasma concentration ↓ Intestinal permeability	(Cremonini *et al.*, 2019)

Diabetic (*db/db*) mice	Pelargonidin-3-*O*-glucoside (150 mg/kg body weight/day); 8 weeks	16S rRNA gene sequencing	↑ *Bacteroidetes/Firmicutes* ratio ↑ *Prevotella*	↓ HOMA-IR ↑ Intestinal barrier integrity	(Su *et al.*, 2019)
C57BL/6J mice Dextran sodium sulfate (DSS)-induced colitis mice model	*L. ruthenicum* anthocyanins and petunidin 3-*O*-[rhamnopyranosyl-(*trans-p*-coumaroyl)]-5-*O*-[β-*d*-glucopyranoside] (200 mg/kg body weight/day); 7 days	16S rRNA gene sequencing	↓ *Firmicutes/Bacteroidetes* ratio ↑ Actinobacteria	Ameliorate DSS-induced colitis ↓ Inflammatory cytokines ↑ Tight junctions (ZO-1, occludin, and claudin-1) ↑ SCFA production	(Peng *et al.*, 2019)
Diabetic (*db/db*) mice	Freeze-dried strawberry powder (2.35%, w/w); 10 weeks	16S rRNA gene sequencing	↓ Verrucomicrobia ↑ *Bifidobacterium*	Ameliorate vascular dysfunction ↓ Vascular inflammation	(Petersen *et al.*, 2019)
Male C57BL/6 × FVB F1 mice	Freeze-dried black raspberries powder (10%, w/w); 6 weeks	16S rRNA gene sequencing	↑ Microbial diversity ↑ *Bacteroidetes* ↓ *Firmicutes* ↓ *Clostridium* ↑ *Barnesiella*	NA	(Gu *et al.*, 2019)

(Continued)

Table 7.2 (Continued)

Animal model	Dose/duration	Methods	Changes in gut microbiota composition	Biological activity	Reference
Male and female C57BL/6J mice	Freeze-dried blueberry powder (5%, w/w); 8 weeks	16S rRNA gene sequencing	↑ *Bacteroidetes* ↓ *Firmicutes* ↓ *Firmicutes/Bacteroidetes* ratio ↑ *Tenericutes* ↓ *Deferribacteres* Sex-specific differences in the changes induced	NA	
Young and old female Sprague–Dawley rats	Bilberry anthocyanin extract (20 mg/kg body weight/day); 10 weeks	16S rRNA gene sequencing	↑ *Aspergillus oryzae* ↑ *Lactobacillus* ↑ *Bacteroides* ↑ Clostridiaceae-1 ↑ Bacteroidales-S24-7-group ↑ Lachnospiraceae NK4A136 group ↓ Verrucomicrobia ↓ Euryarchaeota	Promote intestinal barrier function; assist healthy aging	(Li *et al.*, 2019)

Healthy subjects	Freeze-dried whole cranberry powder (30 g/day); 5 days	16S rRNA gene sequencing	↑ *Bacteroidetes* ↓ *Firmicutes*	Attenuate the impact of animal-based diets on microbiota composition, bile acids, and SCFA	(Rodriguez-Morato *et al.*, 2018)
Healthy subjects	Montmorency tart cherry concentrate juice (237 mL juice/day); 5 days	16S rRNA gene sequencing	High-*Bacteroides* individuals: ↓ *Bacteroides* ↓ *Bifidobacterium* ↑ Lachnospiraceae ↑ *Ruminococcus* ↑ *Collinsella* Low-*Bacteroides* individuals: ↑ *Bacteroides* ↑ *Prevotella* ↑ *Bifidobacterium* ↓ Lachnospiraceae ↓ *Ruminococcus* ↓ *Collinsella*	Individuals consuming a more Western diet may have lower ability to metabolize polyphenols, thereby reducing bioavailability and any potential health benefits	(Mayta-Apaza *et al.*, 2018)

(Continued)

Table 7.2 (Continued)

Animal model	Dose/duration	Methods	Changes in gut microbiota composition	Biological activity	Reference
Wistar rats HFD-induced obesity model	Blackberry anthocyanin-rich extract (25 mg/kg body weight/day); 17 weeks	16S rRNA gene sequencing	↑ *Oscillobacter* vs HF diet: ↓ *Rumminococcus* ↑ *Sporobacter*	Attenuate neuroinflammation (↓ TCK-1 and ↑ fractalkine)	(Marques *et al.*, 2018)
Wistar rats HFD-induced obesity model	Blueberry powder (10%, w/w); 8 weeks	16S rRNA gene sequencing	↓ *Bacteroidetes* ↓ *Firmicutes* ↑ Proteobacteria ↑ Fusobacteria ↑ Gammaproteobacteria ↑ Lactobacillales	Restore gastrointestinal integrity Reduce systemic inflammmation Improve markers of insulin sensitivity	(Lee *et al.*, 2018)
Sprague–Dawley rats	Blackcurrant extract (4%, w/w); 6 weeks	Real-time polymerase chain reaction (PCR)	↑ *Bacteroides-Prevotella-Porphyromonas* group ↑ *Lactobacillus* spp. ↓ *Bifidobacterium* spp. ↓ *C. perfringens*	Improve biomarkers of large intestinal health: ↓ Crypt depth ↑ Goblet cells/crypt	(Paturi *et al.*, 2018)

NA, not applicable; SCFA, short-chain fatty acids.

scavenging activity and reducing activity, suggesting that they could act as potent antioxidants *in vivo*. The conjugation with methyl groups decreased or did not alter the antiproliferative effect of the original anthocyanin. Neither the metabolites nor the natural forms were toxic against the standard human cell line.

Curiously, the methylated forms, 4′-methylated Dp-3-glc and a mixture of 3′ and 4′-*O*-methylcyanidin-3-*O*-glucoside, were able to cross a blood–brain barrier (BBB) model monolayer in a time-dependent manner. The methylated forms were able to cross the BBB more efficiently than the unconjugated forms, with significant differences after 18 hours of incubation. The insertion of a methyl group in the anthocyanin structure positively affected its transport (Faria *et al.*, 2014b).

Several anthocyanin phase II metabolites (methylated, glucuronidated, and sulfated) were identified, in both plasma and urine samples, in a study where volunteers consumed a blackberry purée with or without ethanol (Marques *et al.*, 2016). In this study, the plasma concentration of anthocyanin metabolites was about 10 times higher than the parent anthocyanin's level (even without reaching the C_{max} of anthocyanin metabolites during the time of the protocol).

The enzymatic and chemical hemisynthesis strategies that were developed to obtain the standards for phase II anthocyanin metabolites allowed the *in vivo* identification of metabolites observed after consumption of foodstuffs containing anthocyanins (Cruz *et al.*, 2013, 2015, 2016; Fernandes *et al.*, 2009). Some of these standards were evaluated for their ability to attenuate the negative impact of high-fat (HF) diets on neuroinflammation. Cyanidin-3-glucoside (Cy3glc; the main anthocyanin present in blackberry anthocyanin extract), a mixture of two of its phase II metabolites (3′ and 4′Me-Cy3glc) and the bacterial metabolite PCA were able to modulate fractalkine expression in SH-SY5Y neurons (Meireles *et al.*, 2015). This indicates that the chronic blackberry intake of rats fed a standard or HF diet may have a considerable impact on the neuroinflammatory status.

7.5 Conclusion

It is accepted by all the scientific community that anthocyanins are a class of compounds with proper features, which implies a multidisciplinary approach to give an exact scenario of their bioavailability, which is beginning to solve the old paradox between their low bioavailability and their reported human health effects.

References

Aura, A.M., Martin-Lopez, P., O'Leary, K.A., *et al.* (2005) In vitro metabolism of anthocyanins by human gut microflora. *European Journal of Nutrition*, **44**, 133–42.

Baron, G., Altomare, A., Regazzoni, L., *et al.* (2017) Pharmacokinetic profile of bilberry anthocyanins in rats and the role of glucose transporters: LC–MS/MS and computational studies. *Journal of Pharmaceutical and Biomedical Analysis*, **144**, 112–21.

Braga, A., Murador, D., de Souza, L., and de Rosso, V.V. (2017) Bioavailability of anthocyanins: gaps in knowledge, challenges and future research. *Journal of Food Composition and Analysis*, **68**, 31–40.

Braune, A. and Blaut, M. (2016) Bacterial species involved in the conversion of dietary flavonoids in the human gut. *Gut Microbes*, **7**, 216–34.

Bub, A., Watzl, B., Heeb, D., Reachkemmer, G., and Briviba, K. (2001) Malvidin-3-glucoside bioavailability in humans after ingestion of red wine, dealcoholised red wine and red grape juice. *Molecular Nutrition & Food Research*, **51**, 714–25.

Bustamante, L., Pastene, E., Duran-Sandoval, D., Vergara, C., von Baer, D., and Mardones, C. (2018) Pharmacokinetics of low molecular weight phenolic compounds in gerbil plasma after the consumption of calafate berry (Berberis microphylla) extract. *Food Chemistry*, **268**, 347–54.

Cooney, J.M., Jensen, D.J., and McGhie, T.K. (2004) LC-MS identification of anthocyanins in boysenberry extract and anthocyanin metabolites in human urine following dosing. *Journal of the Science of Food and Agriculture*, **84**, 237–45.

Cremoni, E., Daveri, E., Mastaloudis, A., *et al.* (2019) Anthocyanins protect the gastrointestinal tract from high fat diet–induced alterations in redox signaling, barrier integrity and dysbiosis. *Redox Biology*, **26**, 101269.

Cruz, L., Mateus, N., and de Freitas, V.D. (2013) First chemical synthesis report of an anthocyanin metabolite with in vivo occurrence: cyanidin-4′-O-methyl-3-glucoside. *Tetrahedron Letters*, **54**, 2865–9.

Cruz, L., Basílio, N., Mateus, N., Pina, F., and de Freitas, V.D. (2015) Characterization of kinetic and thermodynamic parameters of cyanidin-3-glucoside methyl and glucuronyl metabolite conjugates. *Journal of Physical Chemistry B*, **119**, 2010–18.

Cruz, L., Feranades, I., Évora, A., De Freitas, V.D., and Mateus, N. (2016) Synthesis of the main red wine anthocyanin metabolite: malvidin-3-O-b-glucuronide. *Synlett*, **28**.

Czank, C., Cassidy, A., Zhang, Q., *et al.* (2013) Human metabolism and elimination of the anthocyanin, cyanidin-3-glucoside: a (13)C-tracer study. *American Journal of Clinical Nutrition*, **97**, 995–1003.

Chen, G., Wang, G., Zhu, C., *et al.* (2019) Effects of cyanidin-3-O-glucoside on 3-chloro-1,2-propanediol induced intestinal microbiota dysbiosis in rats. *Food and Chemical Toxicology*, **133**, 110767.

Dangles, O. and Fenger, J.A. (2018) The chemical reactivity of anthocyanins and its consequences in food science and nutrition. *Molecules*, **23**, e1970.

De Ferrars, R.M., Cassidy, A., Curtis, P., and Kay, C.D. (2014a) Phenolic metabolites of anthocyanins following a dietary intervention study in post-menopausal women. *Molecular Nutrition & Food Research*, **58**, 490–502.

De Ferrars, R.M., Czank, C., Zhang, Q., Botting, N.P., and Kroon, P.A. (2014b) The pharmacokinetics of anthocyanins and their metabolites in humans. *British Journal of Pharmacology*, **171**, 3268–82.

Edwards, M., Czank, C., Woodward, G.M., Cassidy, A., and Kay, C.D. (2015) Phenolic metabolites of anthocyanins modulate mechanisms of endothelial function. *Journal of Agricultural and Food Chemistry*, **63**, 2423–31.

El Mohsen, M.A., Marks, J., Kuhnle, G., *et al.* (2006) Absorption, tissue distribution and excretion of pelargonidin and its metabolites following oral administration to rats. *British Journal of Nutrition*, **95**, 51–8.

Fang, J. (2014) Some anthocyanins could be efficiently absorbed across the gastrointestinal mucosa: extensive presystemic metabolism reduces apparent bioavailability. *Journal of Agricultural and Food Chemistry*, **62**, 3904–11.

Faria, A., Mateus, N., de Freitas, V.D., and Calhau, C. (2006) Modulation of MPP+ uptake by procyanidins in Caco-2 cells: involvement of oxidation/reduction reactions. *FEBS Letters*, **580**, 155–60.

Faria, A., Pestana, D., Azevedo, J., *et al.* (2009a) Absorption of anthocyanins through intestinal epithelial cells – putative involvement of GLUT2. *Molecular Nutrition & Food Research*, **53**, 1430–7.

Faria, A., Pestana, D., Monteiro, R., *et al.* (2009b) Bioavailability of anthocyanin-pyruvic acid adducts in rat. Presented at the International Conference on Polyphenols and Health, Yorkshire, Leeds.

Faria, A., Fernandes, I., Norberto, S., Mateus, N., and Calhau, C. (2014a) Interplay between anthocyanins and gut microbiota. *Journal of Agricultural and Food Chemistry*, **62**, 6898–902.

Faria, A., Meireles, M., Fernandes, I., *et al.* (2014b) Flavonoid metabolites transport across a human BBB model. *Food Chemistry*, **149**, 190–6.

Felgines, C., Talavera, S., Texier, O., Gil-Izquierdo, A., Lamaison, J.L., and Remesy, C. (2005) Blackberry anthocyanins are mainly recovered from urine as methylated and glucuronidated conjugates in humans. *Journal of Agricultural and Food Chemistry*, **53**, 7721–7.

Fernandes, I., Azevedo, J., Faria, A., Calhau, C., de Freitas, V.D., and Mateus, N. (2009) Enzymatic hemisynthesis of metabolites and conjugates of anthocyanins. *Journal of Agricultural and Food Chemistry*, **57**, 735–45.

Fernandes, I., de Freitas, V.D., Reis, C., and Mateus, N. (2012a) A new approach on the gastric absorption of anthocyanins. *Food & Function*, **3**, 508–16.

Fernandes, I., Nave, F., Gonçalves, R., de Freitas, V.D., and Mateus, N. (2012b) On the bioavailability of flavanols and anthocyanins: flavanol-anthocyanin dimers. *Food Chemistry*, **135**, 812–18.

Fernandes, I., Marques, F., de Freitas, V.D., and Mateus, N. (2013) Antioxidant and antiproliferative properties of methylated metabolites of anthocyanins. *Food Chemistry*, **141**, 2923–33.

Fernandes, I., Marques, C., Évora, A., *et al.* (2017) Pharmacokinetics of table and Port red wine anthocyanins: a crossover trial in healthy men. *Food & Function*, **8**, 2030–7.

Ferrer-Gallego, R., Soares, S., Mateus, N., Rivas-Gonzalo, J., Escribano-Bailon, M.T., and de Freitas, V.D. (2015) New anthocyanin-human salivary protein complexes. *Langmuir*, **31**, 8392–401.

Garcia-Estévez, I., Cruz, L., Oliveira, J., Mateus, N., de Freitas, V.D., and Soares, S. (2017) First evidences of interaction between pyranoanthocyanins and salivary proline-rich proteins. *Food Chemistry*, **228**, 574–81.

Gomes, A., Oudot, C., Macia, A., *et al.* (2019) Berry-enriched diet in salt-sensitive hypertensive rats: metabolic fate of (poly)phenols and the role of gut microbiota. *Nutrients*, **11**, e2634.

González-Barrio, R., Edwards, C.A., and Crozier, A. (2011) Colonic catabolism of ellagitannins, ellagic acid, and raspberry anthocyanins: in vivo and in vitro studies. *Drug Metabolism and Disposition*, **39**, 1680–8.

Gowd, V., Bao, T., Wang, L., *et al.* (2018) Antioxidant and antidiabetic activity of blackberry after gastrointestinal digestion and human gut microbiota fermentation. *Food Chemistry*, **269**, 618–27.

Gu, J., Thomas-Ahner, J.M., Riedl, K.M., *et al.* (2019) Dietary black raspberries impact the colonic microbiome and phytochemical metabolites in mice. *Molecular Nutrition & Food Research*, **63**, e1800636.

He, J. and Giusti, M. (2010) Anthocyanins: natural colorants with health-promoting properties. *Annual Review of Food Science and Technology*, **1**, 163–87.

Hester, S.N., Mastaloudis, A., Gray, R., Antony, J.M., Evans, M., and Wood, S M. (2018) Efficacy of an anthocyanin and prebiotic blend on intestinal environment in obese male and female subjects. *Journal of Nutrition and Metabolism*, **2018**, 7497260.

Ichiyanagi, T., Rahman, M.M., Kashiwada, Y., *et al.* (2004) Absorption and metabolism of delphinidin-3-O-(beta)-d-glucopyranoside in rats. *Free Radical Biology and Medicine*, **36**, 930–7.

Ichiyanagi, T., Shida, Y., Rahman, M.M., Hatano, Y., and Konishi, T. (2005a) Extended glucuronidation is another major path of cyanidin-3-O-(beta)-d-glucopyranoside metabolism in rats. *Journal of Agricultural and Food Chemistry*, **53**, 7312–19.

Ichiyanagi, T., Shida, Y., Rahman, M.M., Hatano, Y., Matsumoto, H., and Hirayama, M. (2005b) Metabolic pathway of cyanidin-3-O-(beta)-d-glucopyranoside in rats. *Journal of Agricultural and Food Chemistry*, **53**, 145–50.

Ichiyanagi, T., Shida, Y., Rahman, M.M., Hatano, Y., and Konishi, T. (2006) Bioavailability and tissue distribution of anthocyanins in bilberry (Vaccinium myrtillus L.) extract in rats. *Journal of Agricultural and Food Chemistry*, **54**, 6578–87.

Ichiyanagi, T., Shida, Y., Rahman, M.M., *et al.* (2008) Effect on both aglycone and sugar moiety towards Phase II metabolism of anthocyanins. *Food Chemistry*, **110**, 493–500.

Jayarathne, S., Stull, A.J., Park, O.H., Kim, J.H., Thompson, L., and Moustaid-Moussa, N. (2019) Protective effects of anthocyanins in obesity-associated inflammation and changes in gut microbiome. *Molecular Nutrition & Food Research*, **63**, e1900149.

Kalt, W., McDonald, J.E., Vinqvist-Tymchuk, M.R., Liu, Y., and Fillmore, S.A.E. (2017) Human anthocyanin bioavailability: effect of intake duration and dosing. *Food & Function*, **8**, 4563–9.

Kamiloglu, S., Capanoglu, E., Grootaert, C., and van Camp, J. (2015) Anthocyanin absorption and metabolism by human Caco-2 cells: a review. *International Journal of Molecular Sciences*, **16**, 21555–74.

Kamonpantana, K., Giusti, M., Chitchumroonchokchai, C., *et al.* (2012) Susceptibility of anthocyanins to ex vivo degradation in human saliva. *Food Chemistry*, **135**, 738–47.

Kamonpantana, K., Failla, M., Kumar, P., and Giusti, M. (2014) Anthocyanin structure determines susceptibility to microbial degradation and bioavailability to the buccal mucosa. *Journal of Agricultural and Food Chemistry*, **62**, 6903–10.

Kay, C.D., Mazza, G., Holub, B.J., and Wang, J. (2004) Anthocyanin metabolites in human urine and serum. *British Journal of Nutrition*, **91**, 933–42.

Kay, C. D., Mazza, G.J., and Holub, B.J. (2005) Anthocyanins exist in the circulation primarily as metabolites in adult men. *Journal of Nutrition*, **135**, 2582–8.

Kay, C.D., Kroon, P.A., and Cassidy, A. (2009) The bioactivity of dietary anthocyanins is likely to be mediated by their degradation products. *Molecular Nutrition & Food Research*, **53**, S92–S101.

Kay, C.D., Pereira-Caro, G., Ludwig, I.A., Clifford, M.N., and Crozier, A. (2017) Anthocyanins and flavanones are more bioavailable than previously perceived: a review of recent evidence. *Annual Review of Food Science and Technology*, **8**, 155–80.

Keating, E., Lemos, C., Goncalvez, P., and Martel, F. (2008) Acute and chronic effects of some dietary bioactive compounds on folic acid uptake and on the expression of folic acid transporters by the human trophoblast cell line BeWo. *Journal of Nutritional Biochemistry*, **19**, 91–100.

Keppler, K. and Humpf, H.U. (2005) Metabolism of anthocyanins and their phenolic degradation products by the intestinal microflora. *Bioorganic & Medicinal Chemistry*, **13**, 5195–205.

Kilua, A., Nomata, R., Nagata, R., *et al.* (2019) Purple sweet potato polyphenols differentially influence the microbial composition depending on the fermentability of dietary fiber in a mixed culture of swine fecal bacteria. *Nutrients*, **11**, e1495.

Krga, I., Monfoulet, L.-E., Konic-Ristic, A., *et al.* (2016) Anthocyanins and their gut metabolites reduce the adhesion of monocyte to TNFα-activated endothelial cells at physiologically relevant concentrations. *Archives of Biochemistry and Biophysics*, **599**, 51–9.

Krga, I., Vidovic, N., Milenkovic, D., *et al.* (2018) Effects of anthocyanins and their gut metabolites on adenosine diphosphate-induced platelet activation and their aggregation with monocytes and neutrophils. *Archives of Biochemistry and Biophysics*, **645**, 34–41.

Lee, S., Keirsey, K.I., Kirkland, R., Grunewald, Z.I., Fischer, J.G., and De La Serre, C.B. (2018) Blueberry supplementation influences the gut microbiota, inflammation, and insulin resistance in high-fat-diet-fed rats. *Journal of Nutrition*, **148**, 209–19.

Li, J., Wu, T., Li, N., Wang, X., Chen, G., and Lyu, X. (2019) Bilberry anthocyanin extract promotes intestinal barrier function and inhibits digestive enzyme activity by regulating the gut microbiota in aging rats. *Food & Function*, **10**, 333–43.

Lila, M.A., Burton-Freeman, B., Grace, M., and Kalt, W. (2016) Unraveling anthocyanin bioavailability for human health. *Annual Review of Food Science and Technology*, **7**, 375–93.

Mallery, S.R., Stoner, G.D., Larsen, P.E., *et al.* (2007) Formulation and in-vitro and in-vivo evaluation of a mucoadhesive gel containing freeze dried black raspberries: implications for oral cancer chemoprevention. *Pharmaceutical Research*, **24**, 728–37.

Mallery, S.R., Budenford, D.E., Larsen, M.P., *et al.* (2011) Effects of human oral mucosal tissue, saliva, and oral microflora on intraoral metabolism and bioactivation of black raspberry anthocyanins. *Cancer Prevention Research*, **4**, 1209–21.

Mallery, S.R., Tong, M., Shumway, B.S., *et al.* (2014) Topical application of a mucoadhesive freeze-dried black raspberry gel induces clinical and histologic regression and reduces loss of heterozygosity events in premalignant oral intraepithelial lesions: results from a multicentered, placebo-controlled clinical trial. *Clinical Cancer Research*, **20**, 1910–24.

Marczylo, T., Cooke, D., Brown, K., Steward, W., and Gescher, A. (2009) Pharmacokinetics and metabolism of the putative cancer chemopreventive agent cyanidin-3-glucoside in mice. *Cancer Chemotherapy and Pharmacology*, **64**, 1261–8.

Marques, C., Fernandes, I., Norberto, S., *et al.* (2016) Pharmacokinetics of blackberry anthocyanins consumed with or without ethanol: a randomized and crossover trial. *Molecular Nutrition & Food Research*, **60**, 2319–30.

Marques, C., Fernandes, I., Meireles, M., *et al.* (2018) Gut microbiota modulation accounts for the neuroprotective properties of anthocyanins. *Scientific Reports*, **8**, 11341.

Matsumoto, H., Inaba, H., Kishi, M., Tominaga, S., Hirayama, M., and Tsuda, T. (2001) Orally administered delphinidin-3-rutinoside and cyanidin-3-rutinosideare directly absorbed in rats and humans and apper in the blood as the intact form. *Food Chemistry*, **49**, 1546–51.

Mayta-Apaza, A.C., Pottgen, E., de Bodt, J., *et al.* (2018) Impact of tart cherries polyphenols on the human gut microbiota and phenolic metabolites in vitro and in vivo. *Journal of Nutritional Biochemistry*, **59**, 160–72.

Mazza, G., Kay, C.D., Cottrell, T., and Holub, B.J. (2002) Absorption of anthocyanins from blueberries and serum antioxidant status in human subjects. *Journal of Agricultural and Food Chemistry*, **50**, 7731–7.

McGhie, T.K. and Walton, M.C. (2007) The bioavailability and absorption of anthocyanins: towards a better understanding. *Molecular Nutrition & Food Research*, **51**, 702–13.

McGhie, T.K., Ainge, G.D., Barnett, L.E., Cooney, J.M., and Jensen, D.J. (2003) Anthocyanin glycosides from berry fruit are absorbed and excreted unmetabolized by both humans and rats. *Journal of Agricultural and Food Chemistry*, **51**, 4539–48.

Meireles, M., Marques, C., Norberto, S., *et al.* (2015) The impact of chronic blackberry intake on the neuroinflammatory status of rats fed a standard or high-fat diet. *Journal of Nutritional Biochemistry*, **26**, 1166–73.

Miyazawa, T., Nakagawa, K., Kudo, M., Muraishi, K., and Someya, K. (1999) Direct intestinal absorption of red fruit anthocyanins, cyanidin-3-glucoside and cyanidin-3, 5-diglucoside, into rats and humans. *Journal of Agricultural and Food Chemistry*, **47**, 1083–91.

Morazzoni, P., Livio, S., Scilingo, A., and Malandrino, S. (1991) Vaccinium myrtillus anthocyanosides pharmacokinetics in rats. *Drug Research (Arzneimittel-Forschung)*, **41**, 128–31.

Mueller, D., Jung, K., Winter, M., Rogoll, D., Melcher, R., and Richling, E. (2017) Human intervention study to investigate the intestinal accessibility and bioavailability of anthocyanins from bilberries. *Food Chemistry*, **231**, 275–86.

Netzel, M., Fanninng, K., Netzel, G., *et al.* (2012) Urinary excretion of antioxidants in healthy humans following Queen Garnet plum juice ingestion: a new plum variety rich in antioxidant compounds. *Journal of Food Biochemistry*, **20**, 159–70.

Oliveira, H., Fernandes, I., Bras, N.F., *et al.* (2015) Experimental and theoretical data on the mechanism by which red wine anthocyanins are transported through a human mkn-28 gastric cell model. *Journal of Agricultural and Food Chemistry*, **63**, 7685–92.

Oliveira, H., Basílio, N., Pina, F., Fernandes, I., de Freitas, V.D., and Mateus, N. (2019a) Purple-fleshed sweet potato acylated anthocyanins: equilibrium network and photophysical properties. *Food Chemistry*, **288**, 386–94.

Oliveira, H., Perez-Gregorio, R., de Freitas, V.D., Mateus, N., and Fernandes, I. (2019b) Comparison of the in vitro gastrointestinal bioavailability of acylated and non-acylated anthocyanins: purple-fleshed sweet potato vs red wine. *Food Chemistry*, **276**, 410–18.

Oliveira, H., Roma-Rodrigues, C., Santos, A., *et al.* (2019c) GLUT1 and GLUT3 involvement in anthocyanin gastric transport – nanobased targeted approach. *Scientific Reports*, **9**, 789.

Oliveira, J., Fernandes, V., Miranda, C., *et al.* (2006) Color properties of four cyanidin-pyruvic acid adducts. *Journal of Agricultural and Food Chemistry*, **54**, 6894–903.

Passamonti, S., and Vanzo, A. (2005) Hepatic uptake of grape anthocyanins and the role of bilitranslocase. *Food Chemistry and Toxicology*, **38**, 953–60.

Passamonti, S., Vrhovsek, U., and Mattivi, F. (2002) The interaction of anthocyanins with bilitranslocase. *Biochemical and Biophysical Research Communications*, **296**, 631–6.

Paturi, G., Butts, C.A., Monro, J.A., and Hedderley, D. (2018) Effects of blackcurrant and dietary fibers on large intestinal health biomarkers in rats. *Plant Foods for Human Nutrition*, **73**, 54–60.

Peng, Y., Yan, Y., Wan, P., *et al.* (2019) Gut microbiota modulation and anti-inflammatory properties of anthocyanins from the fruits of Lycium ruthenicum Murray in dextran sodium sulfate-induced colitis in mice. *Free Radical Biology & Medicine*, **136**, 96–108.

Petersen, C., Wankhade, U.D., Bharat, D., *et al.* (2019) Dietary supplementation with strawberry induces marked changes in the composition and functional potential of the gut microbiome in diabetic mice. *Journal of Nutritional Biochemistry*, **66**, 63–9.

Pina, F., Oliveira, J., and de Freitas, V.D. (2015) Anthocyanins and derivatives are more than flavylium cations. *Tetrahedron*, **71**, 14.

Rodriguez-Mateos, A., Feliciano, R., Cifuentes-Gomez, T., and Spencer, J. (2016) Bioavailability of wild blueberry (poly)phenols at different levels of intake. *Journal of Berry Research*, **6**, 137–48.

Rodriguez-Morato, J., Matthan, N.R., Liu, J., De La Torre, R., and Chen, C.O. (2018) Cranberries attenuate animal-based diet-induced changes in microbiota composition and functionality: a randomized crossover controlled feeding trial. *Journal of Nutritional Biochemistry*, **62**, 76–86.

Sandoval-Ramírez, B.A., Catalán, Ú., Fernández-Castillejo, S., Rubió, L., Macià, A., and Solà, R. (2018) Anthocyanin tissue bioavailability in animals: possible implications for human health. A systematic review. *Journal of Agricultural and Food Chemistry*, **66**, 11531–43.

Sekirov, I., Russell, S.L., Antunes, L.C., and Finlay, B.B. (2010) Gut microbiota in health and disease. *Physiological Reviews*, **90**, 859–904.

Su, H., Xie, L., Xu, Y., *et al.* (2019) Pelargonidin-3-O-glucoside derived from wild raspberry exerts antihyperglycemic effect by inducing autophagy and modulating gut microbiota. *Journal of Agricultural and Food Chemistry*. https://doi.org/10.1021/acs.jafc.9b03338

Talavera, S., Felgines, C., Texier, O., *et al.* (2004) Anthocyanins are efficiently absorbed from the small intestine in rats. *Journal of Nutrition*, **134**, 2275–9.

Tian, Q., Giusti, M.M., Stoner, G.D., and Schwartz, S.J. (2006) Urinary excretion of black raspberry (Rubus occidentalis) anthocyanins and their metabolites. *Journal of Agricultural and Food Chemistry*, **54**, 1467–72.

Tsuda, T., Horio, F., and Osawa, T. (1999) Absorption and metabolism of cyanidin-3-O-(beta)-glucoside in rats. *FEBS Letters*, **449**, 179–82.

Ugalde, C.M., Liu, Z., Ren, C., *et al.* (2009) Distribution of anthocyanins delivered from a bioadhesive black raspberry gel following topical intraoral application in normal healthy volunteers. *Pharmaceutical Research*, **26**, 977–86.

Vieira, E.F., Soares, C., Machado, S., *et al.* (2018) Seaweeds from the Portuguese coast as a source of proteinaceous material: total and free amino acid composition profile. *Food Chemistry*, **269**, 264–75.

Wankhade, U.D., Zhong, Y., Lazarenko, O.P., *et al.* (2019) Sex-specific changes in gut microbiome composition following blueberry consumption in C57BL/6J mice. *Nutrients*, **11**, e313.

Weiguang, Y., Akoh, C.C., Fischer, J., and Krewer, G. (2006) Absorption of anthocyanins from blueberry extracts by Caco-2 human intestinal cell monolayers. *Journal of Agricultural and Food Chemistry*, **54**, 5651–8.

Williamson, G. and Clifford, M.N. (2017) Role of the small intestine, colon and microbiota in determining the metabolic fate of polyphenols. *Biochemical Pharmacology*, **139**, 24–39.

Winter, A.N., Brenner, M.C., Punessen, N., *et al.* (2017) Comparison of the neuroprotective and anti-inflammatory effects of the anthocyanin metabolites, protocatechuic acid and 4-hydroxybenzoic acid. *Oxidative Medicine and Cell Longevity*, **2017**, 6297080.

Woodward, G., Kroon, P.A., Cassidy, A., and Kay, C.D. (2009) Anthocyanin stability and recovery: implications for the analysis of clinical and experimental samples. *Journal of Agricultural and Food Chemistry*, **57**, 5271–8.

Wu, X., Cao, G., and Prior, R.L. (2002) Absorption and metabolism of anthocyanins in elderly women after consumption of elderberry or blueberry. *Journal of Nutrition*, **132**, 1865–71.

Wu, X., Pittman, H.E.I., and Prior, R.L. (2004) Pelargonidin is absorbed and metabolized differently than cyanidin after marionberry consumption in pigs. *Journal of Nutrition*, **134**, 2603–10.

Yi, W., Akoh, C C., Fischer, J., and Krewer, G. (2006) Absorption of anthocyanins from blueberry extracts by Caco-2 human intestinal cell monolayers. *Journal of Agricultural and Food Chemistry*, **54**, 5651–8.

Zhang, L., Zuo, Z., and Lin, G. (2007) Intestinal and hepatic glucuronidation of flavonoids. *Molecular Pharmaceutics*, **4**, 833–45.

Zou, T.B., Feng, D., Song, G., Li, H.W., Tang, H.W., and Ling, W.H. (2014) The role of sodium-dependent glucose transporter 1 and glucose transporter 2 in the absorption of cyanidin-3-O-beta-glucoside in Caco-2 cells. *Nutrients*, **6**, 4165–77.

8

Flavan-3-ols: Catechins and Proanthocyanidins

Claudia Favari[1], Pedro Mena[1], Claudio Curti[2], Daniele Del Rio[3], and Donato Angelino[3]*

[1] *Human Nutrition Unit, Department of Food & Drug, University of Parma, Parma, Italy*
[2] *Department of Food & Drug, University of Parma, Parma, Italy*
[3] *Human Nutrition Unit, Department of Veterinary Science, University of Parma, Parma, Italy*

8.1 Introduction: Chemistry and Main Dietary Sources

Flavan-3-ols are the most complex subclass of flavonoids, with structures of various molecular weight ranging from simple monomers to oligomers and polymers of up to 190 units. The monomeric flavan-3-ol presents two stereogenic centers at C2 and C3 that produce four possible stereoisomers (two diastereomeric couples of enantiomers) for each level of B-ring hydroxylation and shows two hydroxyl group in C5 and C7. The hydroxyl groups in the B ring can range from one to three, yielding different structures: (epi)afzelechin, with one hydroxyl at C4′; (epi)catechin, two hydroxyls at C3′ and C4′; and (epi)gallocatechin, three hydroxyls at C3′, C4′, and C5′ (Figure 8.1). (+)-Catechin and (−)-epicatechin are the most common flavan-3-ol monomers, being widespread in nature. Monomers can also undergo esterification with gallic acid, forming (epi)(gallo)catechin derivatives (Crozier *et al.*, 2009; Del Rio *et al.*, 2013). Differently from other flavonoids,

*Corresponding author.

Dietary Polyphenols: Metabolism and Health Effects, First Edition.
Edited by Francisco A. Tomás-Barberán, Antonio González-Sarrías, and Rocío García-Villalba.

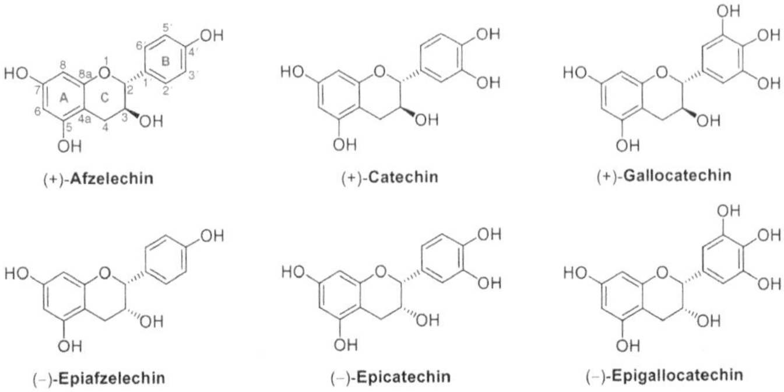

Figure 8.1 Stereoisomers of monomeric flavan-3-ols with different patterns of hydroxylation.

flavan-3-ols exist *in planta* predominantly as aglycones rather than glycosides.

Proanthocyanidins (PACs) are the oligomeric and polymeric structures of flavan-3-ols, also known as condensed tannins. Two types of PACs are distinguished: type B and type A. Type B PACs are formed from flavan-3-ol monomers by oxidative coupling between the C4 of the upper unit and the C6 or C8 of the adjacent lower or extension unit (Figure 8.2). Type A PACs have an additional ether bond between C2 and C7 (see Figure 8.2) and are less widespread in the plant kingdom than type B PACs.

Oligomeric and polymeric structures have an additional chiral center at C4 of every additional flavan-3-ol unit. Condensed tannins made up exclusively of (epi)catechin units are called procyanidins (PCs), whereas propelargonidins (PPs) and prodelphinidins (PDs) are the oligomeric and polymeric structures consisting of (epi)afzelechin or (epi)gallocatechin subunits, respectively. PCs are the most abundant type of PACs in plants, while PPs and PDs are comparatively rare. Many PACs contain more than one type of monomer, and also linkages with anthocyanins and flavonols can occur (Crozier *et al.*, 2009; Del Rio *et al.*, 2013; Rodriguez-Mateos *et al.*, 2014b).

Flavan-3-ols are characteristic polyphenols of tea, cocoa, wine, pome fruits (such as apple and pear), berries, and nuts, but they are also found in stone fruits and legumes (Table 8.1) (Zamora-Ros *et al.*, 2013; Zanotti *et al.*, 2015). This subclass

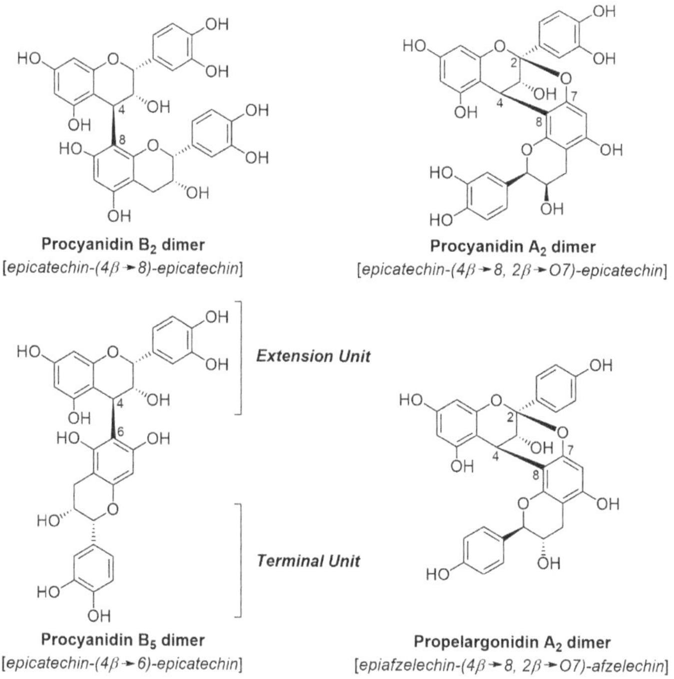

Figure 8.2 Dimeric type B and type A proanthocyanidins.

of compounds is the main source of flavonoids in Western diets (Ferrucci *et al.*, 1991; Zamora-Ros *et al.*, 2016). Different dietary sources contain distinct types of flavan-3-ols (see Table 8.1). Green tea, the beverage prepared from the leaves of *Camellia* spp., contains very high levels of monomers, with the main components being (−)-epigallocatechin-3-*O*-gallate, (−)-epicatechin-3-*O*-gallate, and (−)-epigallocatechin (Figure 8.3). Black tea, the product of the fermentation of green leaves, presents lower amounts of monomers, as a result of the action of polyphenol oxidase, and a concomitant accumulation of theaflavins and thearubigins. Theaflavins and thearubigins are dimer- and polymer-like structures, that can also be esterified with gallic acid (see Figure 8.3) (Crozier *et al.*, 2009; Del Rio *et al.*, 2013). These compounds, that are the result of the monomer transformation during tea leaf fermentation, also belong to the class of flavan-3-ols. Cocoa (*Theobroma*

Table 8.1 Examples of dietary sources of flavan-3-ols and content of their main flavan-3-ols (PhenolExplorer, Neveu *et al*., 2010)

Dietary source	Main flavan-3-ols	Content[a] (mean ± SD)
Green tea, infusion (*Camellia sinensis* L. Kuntze)	(–)-Epigallocatechin-3-*O*-gallate	27.16 ± 39.91 mg/100 mL
	(–)-Epigallocatechin	19.68 ± 25.11 mg/100 mL
	(–)-Epicatechin-3-*O*-gallate	7.50 ± 10.19 mg/100 mL
Black tea, infusion (*Camellia sinensis* L. Kuntze)	(–)-Epigallocatechin-3-*O*-gallate	9.12 ± 12.67 mg/100 mL
	(–)-Epicatechin-3-*O*-gallate	7.34 ± 7.10 mg/100 mL
	(–)-Epigallocatechin	7.19 ± 10.87 mg/100 mL
Cocoa, powder (*Theobroma cacao*)	(–)-Epicatechin	158.30 ± 86.33 mg/100 g FW
	(+)-Catechin	107.75 ± 63.95 mg/100 g FW
	Procyanidin dimer B1	112.00 ± 0.00 mg/100 g FW
Chocolate (dark)	(–)-Epicatechin	70.36 ± 29.54 mg/100 g FW
	Procyanidin tetramer D	53.83 ± 20.09 mg/100 g FW
	Procyanidin dimer B2	36.50 ± 11.69 mg/100 g FW
Red wine (*Vitis vinifera* L.)	Procyanidin dimer B3	9.47 ± 4.29 mg/100 mL
	Procyanidin dimer B4	7.29 ± 3.78 mg/100 mL
	(+)-Catechin	6.81 ± 6.24 mg/100 mL
Grape, black (*Vitis vinifera* L.)	(+)-Catechin	5.46 ± 5.74 mg/100 g FW
	(–)-Epicatechin	5.24 ± 5.61 mg/100 g FW
	(–)-Epicatechin-3-*O*-gallate	1.68 ± 1.87 mg/100 g FW

Apple (*Malus domestica*)	Procyanidin dimer B2	14.56 ± 9.19 mg/100 g FW
	(−)-Epicatechin	8.37 ± 3.67 mg/100 g FW
	(+)-Catechin	1.22 ± 0.83 mg/100 g FW
Pear (*Pyrus communis*)	(−)-Epicatechin	3.77 ± 2.65 mg/100 g FW
	(+)-Catechin	0.28 ± 0.35 mg/100 g FW
Apricot (*Prunus armeniaca* L.)	(−)-Epicatechin	3.47 ± 4.27 mg/100 g FW
	(+)-Catechin	2.96 ± 3.28 mg/100 g FW
	Procyanidin dimer B1	0.09 ± 0.00 mg/100 g FW
Peach (*Prunus persica* L.)	(+)-Catechin	2.33 ± 0.00 mg/100 g FW
Plum (*Prunus domestica* L.)	Procyanidin trimer C1	10.01 ± 0.00 mg/100 g FW
	Procyanidin dimer B1	8.84 ± 0.00 mg/100 g FW
	Procyanidin trimer isomer	7.73 ± 0.00 mg/100 g FW
Hazelnut (*Corylus* L.)	(−)-Epigallocatechin	2.80 ± 0.00 mg/100 g FW
	(+)-Catechin	1.20 ± 0.00 mg/100 g FW
	(−)-Epigallocatechin-3-*O*-gallate	1.10 ± 0.00 mg/100 g FW
Almond (*Prunus dulcis*)	(−)-Epigallocatechin	2.60 ± 0.00 mg/100 g FW
	(+)-Catechin	1.28 ± 1.04 mg/100 g FW
	(−)-Epicatechin	0.59 ± 0.35 mg/100 g FW
Lentil (*Lens culinaris*)	(+)-Catechin-3-*O*-glucose	3.15 ± 0.00 mg/100 g FW
	Procyanidin dimer B3	0.71 ± 0.33 mg/100 g FW
	Prodelphinidin dimer B3	0.45 ± 0.00 mg/100 g FW

a) Results expressed as mg/100 g FW are relative to the fresh weight of the product.

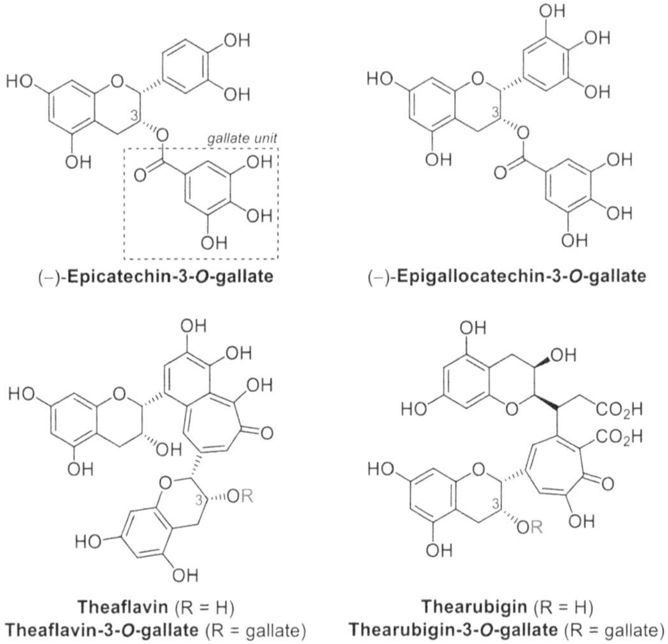

Figure 8.3 Primary green and black tea flavan-3-ol derivatives.

cacao) beans and their derived products (cocoa powder and chocolate) are rich sources of (+)-catechin, (−)-epicatechin, and oligomeric B-type PCs (from dimers to decamers) (Crozier *et al.*, 2009). Wine, the alcoholic fermented juice of *Vitis vinifera* grapes, as well as apples (*Malus domestica*), pears (*Pyrus communis*), and berries are also characterized by high contents of oligomeric and polymeric PACs (Crozier *et al.*, 2009; Rodriguez-Mateos *et al.*, 2014b).

8.2 Bioavailability of Flavan-3-ols

The bioavailability of a dietary compound is a crucial factor in understanding to what extent and in which forms it is available for individual internal compartments. Over recent decades, significant results have been accomplished in elucidation of the mechanisms of absorption, metabolism, distribution, and excretion (ADME) of (poly)phenolic compounds, and among them flavan-3-ols, after dietary intake (Mena *et al.*, 2019). This information is fundamental to unravel the biological effects of these compounds in human health.

In the first instance, it has to be considered that many factors affect the bioavailability of flavan-3-ols: stereochemical configuration, degree of polymerization, flavanol-containing food matrix ingested, as well as interindividual variations (including sex, genetic background, gut microbiota composition, ethnicity, age, dietary habits, etc.) (Ottaviani *et al.*, 2012).

8.2.1 Absorption and Metabolism: Native and Colonic Phase II Metabolites

Flavan-3-ols undergo an extensive metabolism once introduced into the gastrointestinal (GI) tract. After ingestion, they pass through the oral cavity and the stomach almost unchanged, thus reaching the small intestine. Here, some monomers are absorbed in the enterocytes by passive diffusion and subjected to some degree of phase II enzymatic metabolism. Sulfotransferases (SULT), uridine-5′-diphosphate-glucuronosyl-transferases, (UGT) and catechol-*O*-methyltransferases (COMT) originate sulfated, glucuronidated, and *O*-methylated metabolites respectively (Figure 8.4), which pass through the portal vein to the liver or efflux back into the lumen mediated via members of the adenosine-binding cassette (ABC) family of transporters (Del Rio *et al.*, 2013; Rodriguez-Mateos *et al.*, 2014b). Unlike simple monomers, flavan-3-ols with a 3-gallate moiety do not necessarily undergo conjugation by phase II enzymes since they have been detected in the circulation unmetabolized. Besides, a very small percentage of oligomers (<1%, mainly dimers) seems to be absorbed in the small intestine, getting into the portal bloodstream in unconjugated forms (Mena *et al.*, 2019).

Absorbed flavan-3-ols rapidly reach the liver, where they may be further subjected to phase II metabolism before entering the systemic circulation and eventually undergoing renal excretion. A small amount of phase II metabolites may also return from the liver to the small intestine via enterohepatic recirculation in the bile (Del Rio *et al.*, 2013; Rodriguez-Mateos *et al.*, 2014b). However, most ingested flavan-3-ols (> 70%) are not

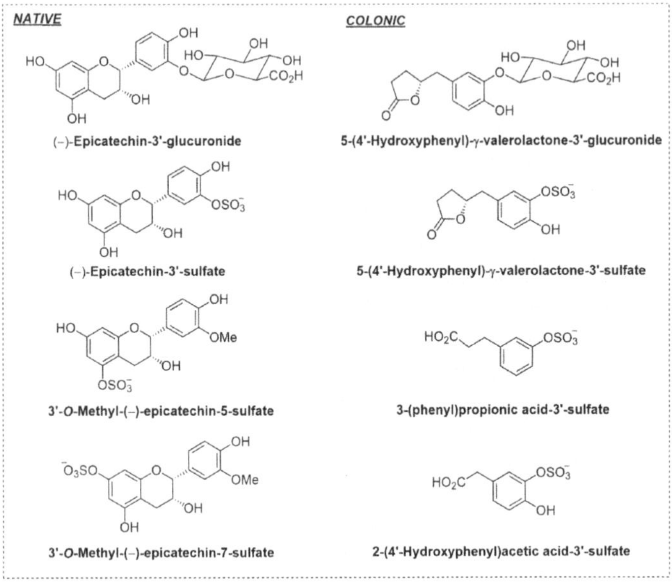

Figure 8.4 Structures of leading native and colonic flavan-3-ol metabolites.

absorbed in the upper part of the GI tract and reach the colon, where they are extensively metabolized by the host microbiota (Mena *et al.*, 2019).

Regarding the microbial metabolism of flavan-3-ol monomers, free (epi)(gallo)catechins present in the large intestine undergo: (1) C-ring fission by the action of specific bacterial species yielding a di- or trihydroxyphenylpropan-2-ol derivative, (2) subsequent conversion into phenyl-γ-valerolactones (PVLs) and 4-hydroxy-phenylvaleric acids by the action of *Flavonifractor plautii*, (3) possible dehydroxylations, and (4) further transformation into low molecular weight phenolics (Appeldoorn *et al.*, 2009; Kutschera *et al.*, 2011; Stoupi *et al.*, 2010) (Figure 8.5). Free (epi)(gallo)catechins subject to these microbial metabolic pathways can derive from: (a) undigested simple monomers, (b) galloyl-moiety removal from 3-gallate monomers by microbial esterases, (c) PAC interflavan link cleavage operated by specific microorganisms.

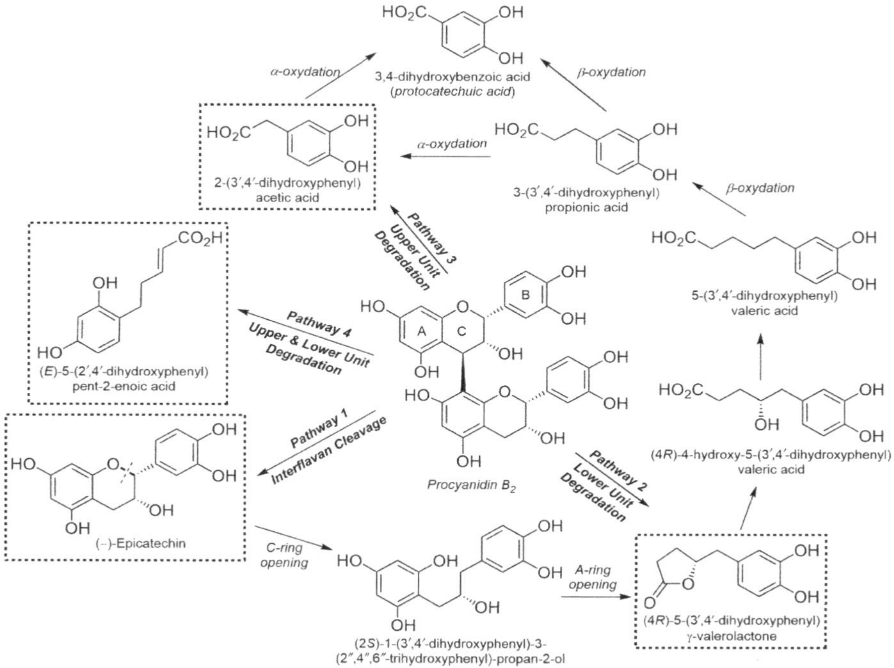

Figure 8.5 Exemplified catabolism of procyanidin B2 and (−)-epicatechin by gut microbiota. Pathways proposed according to Appeldoorn *et al.* (2009) and Stoupi *et al.* (2010).

The number and the position of hydroxyls in the phenyl group of these microbial-derived catabolites depend on the characteristics of the (epi)(gallo)catechin precursor. In particular, free (epi)catechin is catabolized into a 1-(3′,4′-dihydroxyphenyl)-3-(2″,4″,6″-trihydroxyphenyl)propan-2-ol, before being converted into 5-(3′,4′-dihydroxyphenyl)-γ-valerolactone and/or 4-hydroxy-5-(3′,4′-dihydroxyphenyl)valeric acid (Appeldoorn *et al.*, 2009; Kutschera *et al.*, 2011; Stoupi *et al.*, 2010) (see Figure 8.5). The γ-valerolactone ring can also be opened to 4-hydroxy-5-(3′,4′-dihydroxyphenyl)valeric acid and/or be later dehydroxylated to 5-(3′,4′-dihydroxyphenyl)valeric acid (Appeldoorn *et al.*, 2009) (see Figure 8.5). When (epi)gallocatechin is the colonic precursor, 1-(3′,4′,5′-trihydroxyphenyl)-3-(2″,4″,6″-trihydroxyphenyl)propan-2-ol is produced and converted into 5-(3′,4′,5′-trihydroxyphenyl)-γ-valerolactone and 4-hydroxy-5-(3′,4′,5′-trihydroxyphenyl)valeric acid. These trihydroxyphenyl catabolites can be dehydroxylated to the analogous 3′,5′-dihydroxyphenyl and 3′,4′-dihydroxyphenyl derivatives, both of which can subsequently yield the 3′-hydroxyphenyl derivative, and the 3′,4′-dihydroxyphenyl derivative potentially also the 4′-hydroxyphenyl derivative (Mena *et al.*, 2019).

PVLs and their related phenylvaleric acids (PVAs) are the main catabolites produced, but these compounds can be further converted by the intestinal microbiota into other low molecular weight phenolics, such as phenylpropionic, benzoic and cinnamic acid derivatives, by successive loss of carbon atoms from the side chain through β-oxidation (Stoupi *et al.*, 2010). In this sense, a single monomeric precursor can lead to a broad array of phenolic metabolites.

Concerning B-type PAC dimers and oligomers, besides the possible depolymerization into monomeric units previously described (which represents a minor catabolic route; see Figure 8.5 pathway 1) (Appeldoorn *et al.*, 2009; Stoupi *et al.*, 2010), the main degradation pathways involve the direct production of PVLs (see Figure 8.5 pathway 2) and other low molecular weight phenolics (see Figure 8.5 pathways 3 and 4) (Appeldoorn *et al.*, 2009; Stoupi *et al.*, 2010). For example, for procyanidin B2, 5-(3′,4′-dihydroxyphenyl)-γ-valerolactone could result from the direct degradation of its lower unit (see Figure 8.5 pathway 2), while 2-(3′,4′-dihydroxyphenyl)acetic acid could derive from the cleavage of its upper unit (see Figure 8.5 pathway 3). Moreover, other microbial metabolites, such as 5-(2′,4′-dihydroxyphenyl)-2-ene-valeric acid, could arise from the simultaneous degradation of the upper and lower units (see Figure 8.5 pathway 4) (Stoupi *et al.*, 2010).

Less is known about the microbial catabolism of A-type PACs, which are more resistant to microbial breakdown than the B-type. This resistance could be due to their more rigid interflavan ether bonds (Engemann *et al.*, 2012; Ou *et al.*, 2014). Like B-type dimer catabolism, degradation of A-type PCs starts with the cleavage of monomeric unit C-rings, followed by the production of various phenolic acids (Engemann *et al.*, 2012). The occurrence of PVA derivatives after incubation of A-type dimers with colonic microbiota has been observed, while the formation of PVLs from A-type PACs has not been reported to date (Ou *et al.*, 2014).

The products of the colonic microbial catabolism of flavan-3-ols (mainly PVLs, PVAs, and phenolic acids) can be absorbed in colonocytes by passive transport and be subjected to enzymatic phase II metabolism, before getting into the portal bloodstream to reach the liver. In hepatocytes, these catabolites can be further conjugated by phase II enzymes before passing to the systemic circulation or being expelled back into the intestinal lumen via enterohepatic recirculation. The circulating fraction is then excreted in urine (Mena *et al.*, 2019).

Lastly, undigested flavan-3-ols (mostly high molecular weight PACs with a degree of polymerization [DP] >4) and unabsorbed catabolites are voided in feces. Feces may also contain conjugated metabolites released from enterocytes or excreted through the bile (Borges *et al.*, 2018; Stalmach *et al.*, 2010).

8.2.2 Pharmacokinetics and Urinary Excretion of Circulating Metabolites: Interindividual Differences

Over the years, numerous human intervention studies have been carried out to evaluate the profiles of absorption and excretion of flavan-3-ols to understand to what extent and in which forms these compounds are bioavailable for the human organism. Ottaviani and colleagues (Ottaviani *et al.*, 2016) investigated the ADME of (–)-epicatechin, one of the most widely consumed flavan-3-ols, using radiolabeled and stereochemically pure [2-^{14}C](–)-epicatechin ([2-^{14}C] (EC). This fundamental study, that will be followed as an example in this section, revealed that levels of recovered radioactivity were substantially higher in urine than in plasma, as determined by liquid scintillation counting, with mean total radioactivity in plasma never exceeding 2% of intake, while reaching 82% ± 5% of ingested [2-^{14}C]EC in urine, over the 0–48-hour period after consumption. Interindividual differences were observed, especially in the recovery of radioactivity in urine, with individual values ranging from 49% to 90% (Borges *et al.*, 2018).

Analysis of blood samples showed that radioactivity was associated almost exclusively with plasma rather than blood cellular components. The pharmacokinetic profile was biphasic with maximum peaks at ~1 h and 6 h after [2-^{14}C]EC intake. A T_{max} of 1.0 ± 0.1 h is indicative of absorption in the small intestine and was related to the presence of 12 native phase II metabolites, mainly in the forms of (–)-epicatechin-3′-glucuronide, 3′-*O*-methyl-(–)-epicatechin-5-sulfate, and (–)-epicatechin-3′-sulfate. Native phase II (–)-epicatechin metabolites reached an overall C_{max} of 1223 ± 104 nmoL/L, before declining rapidly with an apparent elimination half-life ($AT_{1/2}$) of 1.9 ± 0.1 h and disappearing from the circulatory system within eight hours. On the other hand, a T_{max} of 5.8 ± 0.4 h is indicative of absorption in the distal GI tract and it was characterized by the occurrence of colonic phase II catabolites, namely the PVL and PVA derivatives. The main ones were 5-(4′-hydroxyphenyl)-γ-valerolactone-3′-sulfate and 5-(4′-hydroxyphenyl)-γ-valerolactone-3′-glucuronide. Microbial-derived metabolites attained a combined C_{max} of 588 ± 102 nmoL/L, but they were retained in the circulation longer than the native phase II metabolites, showing an $AT_{1/2}$ of 5.7 ± 0.7 h and area under the curve (AUC) concentration ~3-fold higher. In fact, ~0.2% of [2-^{14}C]EC intake was still present in plasma 24 hours after consumption.

Radioactivity associated with phenolic acid derivatives in plasma was low and this precluded their detection by HPLC-RC, making it impossible to quantify in plasma how much of the ingested (−)-epicatechin was converted into these low molecular weight phenolics. Interindividual variations in plasma C_{max} of metabolites were reported, with higher differences observed for colonic catabolite concentrations (Borges *et al.*, 2018; Ottaviani *et al.*, 2016).

Urinary excretion reflected the plasma pharmacokinetic profiles. Native phase II (−)-epicatechin metabolites were mostly excreted in the first 0–4 h collection period and represented 20% ± 2% of the ingested [2-^{14}C]EC, absorbed in the proximal GI tract. As in plasma, the major urinary structurally related (−)-epicatechin metabolites (SREM) were (−)-epicatechin-3′-glucuronide, 3′-*O*-methyl-(−)-epicatechin-5-sulfate, and (−)-epicatechin-3′-sulfate. PVL and PVA derivatives were excreted later, mainly over the 4–8 h, 8–12 h, and 12–24 h collection periods. Their urinary excretion corresponded to 42 ± 5% of the consumed [2-^{14}C]EC and, as in plasma, the main representatives were 5-(4′-hydroxyphenyl)-γ-valerolactone-3′-sulfate and 5-(4′-hydroxyphenyl)-γ-valerolactone-3′-glucuronide.

Concerning low molecular weight phenolics, their urinary excretion continued over 48 hours post [2-^{14}C]EC intake and represented 28 ± 3% of the ingested radioactivity. Thus, recovery of microbial-derived catabolites absorbed in the colon was about 70% of the consumed [2-^{14}C]EC. Variability among individuals in the levels of urinary excretion was 2.9-fold for native phase II (−)-epicatechin metabolites and 3.2-fold for PVL and PVA derivatives, while there was a much more substantial variation with low molecular weight phenolics, but these were relatively minor metabolites (Borges *et al.*, 2018; Ottaviani *et al.*, 2016).

When considering dimeric flavan-3-ols, native procyanidin B1 can be rapidly absorbed in enterocytes, though in small quantities, and appear in circulation mainly unconjugated. It is also quickly excreted, after undergoing enzymatic methylation and then glucuronidation or sulfation. Colonic-derived 5-(3′,4′-dihydroxyphenyl)-γ-valerolactone conjugates were the most important metabolites detected later after procyanidin B1 intake, exhibiting high plasma levels and significant urine excretion. High interindividual variability was also reported (Wiese *et al.*, 2015).

It must be taken into account that, unless consumed as food supplements, (–)-epicatechin and other flavan-3-ols are not consumed as pure compounds, but as part of more complex food matrices, which may affect their bioavailability and pharmacokinetic profiles. Plenty of human intervention studies have tried to assess the ADME of flavan-3-ols using their primary dietary sources, in particular tea and cocoa products, but also grape and grape by-products, cranberry, apples and almonds (Boto-Ordóñez *et al.*, 2013; Calani *et al.*, 2012; Castello *et al.*, 2018; Del Rio *et al.*, 2010; Feliciano *et al.*, 2016; Garrido *et al.*, 2010; Khymenets *et al.*, 2015; Rios *et al.*, 2003; Roowi *et al.*, 2010; Sang *et al.*, 2008; Trost *et al.*, 2018; Urpi-Sarda *et al.*, 2009a,b; van der Hooft *et al.*, 2012; van Duynhoven *et al.*, 2014; Vitaglione *et al.*, 2013).

In general, native phase II metabolites show shorter plasma T_{max} than colonic phase II metabolites, reflecting a proximal GI tract absorption versus a distal GI tract absorption, as well as higher plasma C_{max} but lower $AT_{1/2}$. Cumulative urinary excretion of colonic phase II metabolites is instead more elevated than that of native phase II metabolites. Monoglucuronide and monosulfate forms of flavan-3-ol monomers are the dominant native phase II metabolites, and monoglucuronide and monosulfate derivatives of PVLs are the main colonic phase II catabolites. In particular, glucuronide and sulfate conjugates of 5-(3′,4′-dihydroxyphenyl)-γ-valerolactone are the most important flavan-3-ol metabolites, regardless of the food source, and have therefore been proposed as biomarkers of flavan-3-ol intake (Ottaviani *et al.*, 2018a). High interindividual variability is always recorded, especially in the production of PVLs and PVAs, likely due to individual differences in gut microbiota composition. Thus, the possible existence of different metabotypes (metabolic phenotypes) in the production of flavan-3-ol colonic metabolites has been proposed. They were observed in a free-living study in which volunteers were daily supplemented with very high amounts of green tea flavan-3-ols (Mena *et al.*, 2018). Three putative metabotypes related to green tea flavan-3-ols were distinguished and characterized by: (1) elevated excretion of tri- and dihydroxyphenyl-γ-valerolactones and reduced excretion of 3-(hydroxyphenyl)propionic acids (metabotype 1);

(2) medium excretion of dihydroxyphenyl-γ-valerolactones and reduced excretion of trihydroxyphenyl-γ-valerolactones and 3-(hydroxyphenyl)propionic acids (metabotype 2); and (3) high excretion of 3-(hydroxyphenyl)propionic acids and limited production of phenyl-γ-valerolactones (metabotype 3).

Recently, a study on this topic has denied the existence of metabotypes coming from nut PACs (Cortés-Martín *et al.*, 2019). Differences in the flavan-3-ols consumed in each study (green tea monomers vs nut oligomers) might account for these differences. However, it is important to emphasize that the results between both studies are not contradictory. Actually, the work by Cortés-Martín and colleagues (2019) observed clusters of individuals using the same metabolites that were considered in the tea study by Mena *et al.* (2018). In particular, Cortés-Martín *et al.* (2019) observed four clusters representing groups of volunteers with different concentrations of PVLs and hydroxyphenylpropionic acids, fully in line with our preliminary study (Mena *et al.*, 2018). A relevant aspect supporting the robustness of clear interindividual differences in the production of flavan-3-ol colonic catabolites might be related to a limitation of the work carried out with nut PACs, since the amount of PACs provided under free living conditions was very low (54 mg per day) and nut PACs cannot be considered a good source of flavan-3-ol catabolites (80% of the nut flavan-3-ols presented a mean degree of polymerization ranging from 2 to 9). In this sense, it could be considered that Cortés-Martín *et al.* (2019) confirmed the presence of different metabolic profiles in a work closer to a free diet setting than to an intervention with sources of flavan-3-ols able to yield colonic metabolites.

The only contradictory point between these studies is related to the concept of metabotypes that both groups have considered. Specifically, while the study by Mena *et al.* (2018) considered the definition of metabotyping as the classification of individuals into subgroups according to their metabolic profile (as usually done in the nutrition field) (Hillesheim and Brennan, 2020), Cortés-Martín *et al.* adhered to a more restrictive definition considering "gut microbiota polyphenol metabotypes," which are characterized by the presence–absence of specific metabolites for specific individuals. Therefore, beyond definitions, it is important to highlight that Cortés-Martín *et al.* (2019) confirmed the existence of a high interindividual variability in the production of colonic metabolites of flavan-3-ols and that volunteers can be grouped according to different metabolic profiles.

Nevertheless, further research is needed to confirm the existence and stability of these metabotypes and to understand how they may vary based on the flavan-3-ol profile of the food source consumed. The elucidation of metabotypes is of particular interest to fully understand the potential health benefits associated with flavan-3-ol consumption and their microbiota-derived metabolites on an individual basis.

8.3 Health Benefits of Flavan-3-ols and Their Derived Circulating Metabolites

During the writing of this chapter, the number of manuscripts focusing on "polyphenol" AND "health" on Medline Database is close to 7000, almost 5000 of which were written in the year range 2009–2019. After an initial focus on their chemical structure and their purification and/or synthesis processes, most of the research has concentrated on the health effects of these bioactives. Flavan-3-ols had (and have) a leading interest because of their extensive presence, the abundant concentration in the most consumed plant-based foods and beverages (i.e., cocoa, tea, some berries, wine, etc.), and their high intake (Williamson, 2017; Ziauddeen *et al.*, 2018). Being categorized as "natural antioxidants," the first studies focused on the flavan-3-ol role in oxidative stress and redox state of the cellular environment (Rice-Evans *et al.*, 1995; Serafini *et al.*, 2003). However, several other possible biochemical and molecular mechanisms have been identified, including various effects within intra- and intercellular signaling pathways, such as regulating nuclear transcription factors and fat metabolism and modulating the synthesis of inflammatory mediators (Fraga *et al.*, 2019; Mena *et al.*, 2014; Potì *et al.*, 2019).

8.3.1 Cognitive

Several observational and intervention studies suggest that diets rich in (poly)phenols beneficially affect nervous system functions (Vauzour, 2017). Among these, one of the most critical observational studies identified neuroprotective dietary components based on the change in cognition over 4.7 years among 960 participants, aged 58–98 years, recruited within the Mediterranean-DASH diet Intervention for Neurodegenerative Delay (MIND) study (Morris *et al.*, 2015). A MIND diet score was developed and assigned to food categories based on their association with cognitive decline, with the highest MIND scores being assigned to "healthy brain" foods such as green leafy vegetables, nuts, berries, beans, whole grains, olive oil, and wine. All these food items were positively associated with a slower decline in global cognitive score (Morris *et al.*, 2015). These foods are rich sources of (poly)phenols and most of them have a high PAC content, a subclass of flavan-3-ols associated with improvements in cognitive function (Lamport *et al.*, 2014).

Despite the formulation of several hypotheses to explain the positive impact of (poly)phenols on the central nervous system, including free radical scavenging, metal chelation, modulation of detoxification enzyme activities, etc. (Schaffer and Halliwell, 2012; Spencer, 2008, 2009), the scarce bioavailability and profound rearrangement of their structures leading to new secondary metabolites make a final conclusion complicated. Schaffer and Halliwell (2012) suggested that (poly)phenols might alter brain function at three locations: (1) outside the central nervous system (i.e., by improving cerebral blood flow), (2) at the blood–brain barrier (i.e., by altering multidrug-resistant protein-dependent influx and efflux mechanisms of various biomolecules) and (3) inside the central nervous system (i.e., by directly modifying the activity of neurons and glial cells).

The first hypothesized mechanism has been proven in a recent randomized controlled trial (RCT) involving sedentary older adults consuming, for three months, a high (900 mg/d) or low (45 mg/d) dose of cocoa flavan-3-ol supplements (Brickman *et al.*, 2014). Results highlighted that high flavan-3-ol supplementation significantly enhanced the hippocampal dentate gyrus function by increasing cerebral blood volume (measured using functional magnetic resonance imaging) compared to those who were supplemented with low-flavan-3-ol cocoa. This study confirmed a previous one from the same group, where young adults, after the consumption of pure (−)-epicatechin (up to 3 mg/kg) or a flavan-3-ol-rich cocoa drink (917 mg/d), showed a significant transient increase in the flow-mediated vasodilation response at 1–4 h post consumption compared to baseline and a low-flavan-3-ol cocoa beverage (37 mg/d) (Schroeter *et al.*, 2006). The authors demonstrated that the increase of (−)-epicatechin and its metabolites, such as 4′-*O*-methyl-epicatechin-glucuronide, 4′-*O*-methyl-epicatechin, and epicatechin-7-glucuronide, significantly correlated with increasing concentrations of plasma and urinary nitric oxide (NO), as well as with the induction of NO synthesis. They concluded that this might be the mechanism behind amelioration of flow-mediated vasodilation after flavan-3-ol consumption (Schroeter *et al.*, 2006).

The close link between the cerebral blood flow (including oxygen and nutrient delivery) and brain cell activities is defined as "neurovascular coupling" (NVC). A direct association has been underlined between impaired NVC and cognitive diseases (Girouard and Iadecola, 2006). Sorond *et al.* (2013) investigated whether flavan-3-ol-rich cocoa could affect NVC through the improvement in endothelial and cognitive functions in a parallel-arm, double-blind clinical trial, where 60 older volunteers, with or without impaired NCV, consumed, for 30 days, up to 1200 mg/d of cocoa flavan-3-ols. Cocoa consumption was associated with significantly increased NVC for impaired individuals and amelioration of their cognitive performance after the intervention, compared to healthy seniors. Other than the above-mentioned biological effects, the authors also hypothesized cholinergic, antiinflammatory, antioxidant, and β-amyloid-reducing properties for cocoa flavan-3-ols (Sorond *et al.*, 2013).

Flavan-3-ols have been consistently studied for their preventive role in mild and severe cognitive decline models, such as dementia and Alzheimer's disease (AD). Concerning the former, one of the leading studies in the field is the Cocoa, Cognition, and Aging (CoCoA) Study, a RCT in which 90 seniors with mild cognitive impairment were supplemented daily, for eight weeks, with a drink containing high, intermediate or low cocoa flavan-3-ols (990, 520, or 45 mg/day, respectively) (Desideri *et al.*, 2012; Mastroiacovo *et al.*, 2015). Not only cognitive functions, assessed by Mini-Mental State Examination, Trail Making Test A and B, and verbal fluency test, but also insulin resistance, blood pressure, and lipid peroxidation markers improved at eight weeks following consumption of the high flavan-3-ol beverage in comparison to the intermediate and lower ones. Consequently, it was hypothesized that circulating flavan-3-ol metabolites might have improved vascular reactivity, such as NO-dependent endothelial function, or markers of oxidative stress and, in turn, they promoted neuronal functions (Mastroiacovo *et al.*, 2015).

Oxidative damage has been indicated as one of the most dangerous enhancers of neuronal degeneration, caused by the accumulation of iron or radical species – mainly derived by oxygen or nitrogen – which trigger neuroinflammation and depletion of endogenous antioxidant enzymes and compounds (Weinreb *et al.*, 2004). Green tea catechins have been considered among the most effective polyphenols counteracting neuroinflammation and oxidative processes. EGCG is regarded as a "multimodal acting molecule" as it seems to be involved in various cellular neuroprotection mechanisms including iron chelation, scavenging of oxygen and nitrogen radical species, and activation of protein kinase-C signaling pathways (Weinreb *et al.*, 2009). The biological activity of EGCG also shows a characteristic biphasic pattern, where higher doses exhibit prooxidant and proapoptotic effects, supposedly involved in its antitumorigenic role, whereas low doses have neuroprotective effects (Ortiz-López *et al.*, 2016). Other potential mechanisms involve EGCG binding to peroxisome proliferator-activated receptors (PPARs), which pleiotropically act as transcription factors for a large number of downstream target genes (Lee and Jia, 2015).

The protective effects of green tea catechins on AD have also been considered in several human intervention studies. Three Japanese studies focusing on green tea beverages confirmed the improvement in cognitive functions (by means of Mini-Mental State Examination or Dementia Scales) of volunteers with different shades of cognitive impairment and dysfunction, after the consumption of more than 200 mg/day green tea catechins, up to 12 months, compared to placebo (Ide *et al.*, 2018).

Very few studies have so far focused on the putative biological role of flavan-3-ol gut microbial metabolites, PVLs and PVAs, in neuronal improvement. Unno *et al.* (2017) treated human SH-SY5Y neuroblastoma cells with 5-(3′,5′-dihydroxyphenyl)-γ-valerolactone, 5-(5′-hydroxyphenyl)-γ-valerolactone-3′-sulfate, and 5-(5′-hydroxyphenyl)-γ-valerolactone-3′-glucuronide, in the range of 0.01–1.0 mM for 48 hours. These colonic metabolites induced an increase in neurite numbers (Unno *et al.*, 2017). Recent work carried out by the authors of this chapter in the frame of an international collaboration used five experimental models with increased complexity and experimental realism (including *in silico, in vitro,* and *in vivo* animal studies) to understand whether PVLs and PVAs are able to cross the blood–brain barrier (Angelino *et al.*, 2019). Results showed the blood–brain barrier permeability of one of the main flavan-3-ol microbial metabolites, 5-(hydroxyphenyl)-γ-valerolactone-sulfate, supporting the hypothesis that those microbial derivatives might be present in brain tissues. Therefore, further work with this metabolite may shed light on the prospects of flavan-3-ol colonic metabolites for the prevention of neurological disorders.

8.3.2 Inflammation and Cardiometabolic Diseases

Several different flavan-3-ol-rich foods have been considered for their beneficial effects on inflammation, as observational studies reported an association between flavan-3-ol intake and positive impact on cardiometabolic outcomes, such as cholesterol levels, blood pressure, and myocardial infarction (Fraga *et al.*, 2019). Among these, a prospective study on the Cancer Prevention Study II Nutrition cohort showed that flavan-3-ol intake was positively associated with a lower risk of fatal cardiovascular disease (McCullough *et al.*, 2012), while a meta-analysis of cohort, case–control, and cross-sectional studies reported more moderate risk of any cardiovascular disease and diabetes and a reduced risk of stroke in individuals who consumed higher levels of cocoa and chocolate (Buitrago-Lopez *et al.*, 2011). Similarly, a recent meta-analysis focused on interindividual variability also indicated that flavan-3-ols derived from green tea, apple, and cocoa products might improve blood lipid levels, in particular in overweight adults (González-Sarrías *et al.*, 2017).

A meta-analysis by Berends *et al.* (2015) summarized findings from RCTs focused on both short- and long-term consumption of chocolate and cocoa-based products; it evidenced significant decrease in blood pressure (both systolic and diastolic), improvements in insulin resistance, flow-mediated dilation (FMD), HDL-C, and LDL-C. Again, another review which considered RCTs on the protective role of berry (poly)phenols, among them flavan-3-ols, towards markers of cardiovascular disease pointed out that the most active positive effects were shown on blood pressure, lipid-related markers, and FMD (Rodriguez-Mateos *et al.*, 2014a).

Exciting and contrasting results have been highlighted when not only flavan-3-ols but also their colonic metabolites have been considered as biological effectors. A recent RCT found that consumption of cocoa extract, containing 130 mg (−)-epicatechin and 560 mg procyanidins, by 45 healthy men for four weeks significantly improved FMD and decreased blood pressure, arterial stiffness markers (pulse wave velocity and augmentation index), and total cholesterol, compared to those who consumed 20 mg (−)-epicatechin and 560 mg procyanidins and controls (Rodriguez-Mateos *et al.*, 2018). The authors hypothesized that flavan-3-ol monomers and their derived SREMs could be more effective in the improvement of vascular function than procyanidins and circulating PVLs, while reduction in total cholesterol was related to the procyanidin content rather than to the monomers.

Nevertheless, plasma 5-(3′-hydroxyphenyl)-γ-valerolactone-4′-sulfate was found to be positively correlated with an increase in FMD four hours and eight hours after cranberry juice consumption (containing up to 1910 mg of total (poly)phenols), compared to control. Some improvement in blood pressure after cranberry juice intake was observed compared to baseline in the intervention group, but no significant results were observed when comparing changes in blood pressure, pulse wave velocity or augmentation index to the control (Rodriguez-Mateos *et al.*, 2016). Contrasting results on endothelial function markers have been found when considering the health status of the participants and the chronic/acute consumption of the supplement. While a crossover RCT involving volunteers with coronary artery disease showed a decrease in carotid–femoral pulse wave velocity after consumption of cranberry juice for four weeks (Dohadwala *et al.*, 2011), these results were not confirmed when healthy men consumed a single supplementation of a blueberry drink (Rodriguez-Mateos *et al.*, 2013).

Both cocoa- and berry-containing products have been found to significantly improve platelet function using a platelet function analyzer. A significant increase in adenosine-5′-diphosphate (ADP)/collagen closure time has been found (1) after two hour and six hour consumption of dark chocolate, by healthy volunteers (Holt *et al.*, 2002), (2) after eight-week supplementation of 100 g berries in subjects with metabolic syndrome (Erlund *et al.*, 2008) and (3) after four hour dark chocolate consumption, by healthy male volunteers (Montagnana *et al.*, 2018). Particularly in this last study, the focus was on the putative role of SREMs and PVLs in increasing ADP/collagen closure time, revealing a significantly positive correlation with single and total SREMs but not with PVLs.

Several molecular mechanisms underlying the biological effects of flavan-3-ols and their derivatives have been hypothesized. Flavan-3-ols have been found to increase NO availability through the activation of several different phosphorylation cascades, leading to activation of endothelial NO synthase (eNOS), the major enzyme responsible for the production of NO (Jiménez *et al.*, 2012). As mentioned for cognitive dysfunction, ROS have a key role in the oxidative stress process, with the production of prooxidants which are not adequately scavenged by the oxidant defense enzymes, such as catalase or superoxide dismutase (Al-Dashti *et al.*, 2018). Angiotensin-converting enzyme (ACE) plays a key role in the renin-angiotensin system, as it cleaves angiotensin-I to produce angiotensin-II, which is a potent vasoconstrictor and stimulates the secretion of aldosterone and vasopressin. In humans, some data support the concept that flavan-3-ols can competitively inhibit ACE activity and then reduce blood pressure by lowering peripheral vasoconstriction and plasma volume (Actis-Goretta *et al.*, 2006).

An intriguing hypothesis by Rodriguez-Mateos *et al.* (2014a) claims that some phenolic metabolites have a structure similar to the nicotinamide adenine dinucleotide phosphate (NADPH) oxidase inhibitor apocynin (acetovanillone), which has been shown to act as a potent NAPDH oxidase inhibitor in *in vitro* models of endothelial cells. A decrease in NADPH oxidase activity has been proposed as the primary mechanism mediating the short-term improvements in FMD observed after consumption of flavan-3-ol-rich foods. Among these, Bernatoniene *et al.* focused on sources of catechins, such as tea, cocoa and wine, and their potential beneficial effects on cardiovascular diseases (Bernatoniene and Kopustinskiene, 2018). Catechins from green tea decreased blood pressure, reduced the risk of stroke and coronary heart disease, and alleviated conditions associated with vascular dysfunction, including vascular inflammation and smooth muscle cell proliferation, blood platelet aggregation, lipoprotein oxidation, altered lipid profile, and vascular reactivity. EGCG enhanced expression of p53, p21, and NF-κB, induced the apoptosis of vascular smooth muscle cells and prevented the development of atherosclerosis. In addition, catechins reduced the accumulation of cholesterol and its oxidation products in artery walls *in vivo*, thus improving blood circulation (Bernatoniene and Kopustinskiene, 2018).

8.3.3 Urinary Tract Infections

Urinary tract infections (UTIs) are caused by bacteria and certain fungi, with uropathogenic *Escherichia coli* (UPEC) as the most common cause. UTIs can be uncomplicated (e.g., cystitis and pyelonephritis affecting the lower and upper urinary tract, respectively, but without structural abnormalities in the tract) or complicated, when able to compromise the affected parts (e.g., obstruction of the urinary tract) (Terlizzi *et al.*, 2017). UTIs are among the most pervasive bacterial infections and represent a significant economic and medical burden worldwide. They account for several millions of outpatient hospital visits and millions of emergency room visits, with a vast annual direct cost for national healthcare systems (Flores-Mireles *et al.*, 2015).

Cranberry consumption has been reported to be effective in decreasing the occurrence and severity of UTI in women

(Nicolle, 2016). The preventive effect of cranberry on UTIs has been attributed to its bioactives and/or their metabolites during the phase of bacterial adherence to the uroepithelial cells, disabling or inhibiting the adhesion of UPEC and preventing bacterial colonization and progression of UTI. Importantly, cranberries are supposed to reduce UTI-related symptoms by suppressing inflammatory cascades as an immunological response to bacterial invasion (González de Llano *et al.*, 2015). Another interesting theory, which considers the large intestine as a reservoir for uropathogenic bacteria, claims that A-type PACs might specifically decrease the transient intestinal colonization by UPEC, consequently reducing the risk of UTI incidence (González de Llano *et al.*, 2015). Despite several studies suggesting that the beneficial effects for UTI-related diseases and recurrence might be driven by type A PACs (Kaspar *et al.*, 2015), Peron *et al.* (2017) pointed out that type A PACs might not be the sole agent responsible for the antiadhesive activity, as these compounds present in the original food have been detected in very low concentrations (<nM) in urine samples of women who consumed cranberry products.

There is still debate in the literature concerning which are the compound(s) present in urine after cranberry consumption that prevent(s)/inhibit(s) the effective adherence of uropathogens to uroepithelial cells. Three recent *in vitro* investigations focused on the major gut microbial derivatives from cranberry flavan-3-ols and their antiadhesive activity against UPEC on human bladder cells. Gonzalez de Llano *et al.* (2015) found that catechol, benzoic acid, vanillic acid, phenylacetic acid, and 3,4-dihydroxyphenylacetic acid showed antiadhesive activity against UPEC on human bladder T24 cells in a concentration-dependent manner from 100 to 500 μM, whereas procyanidin A2, widely reported as an inhibitor of UPEC adherence on uroepithelium, was only statistically significant at 500 μM. Mena *et al.* (2017) investigated the role of 5-(3′,4′-dihydroxyphenyl)-γ-valerolactone derivatives in the inhibition of UPEC adherence to bladder epithelial cells, showing that all the compounds significantly decreased UPEC adherence at 100 μM, up to 50 μM for 5-(3′-hydroxyphenyl)-γ-valerolactone-4′-sulfate. Gonzalez de Llano *et al.* (2019) found that 3′,4′-dihydroxyphenylacetic acid (but not its methyl derivative) inhibited UPEC adherence to both Gram-positive and Gram-negative bacteria up to 250 μM.

Despite the *in vitro* findings supporting the biological effects of flavan-3-ol gut microbial derivatives, *in vivo* studies have reported conflicting results. One of the primary limits worth considering is the wide interindividual variability observed in the human urinary/plasmatic phenolic metabolite profile. As previously described, interindividual differences in the production of flavan-3-ol metabolites might lead to the existence of urinary "metabotypes" which group subjects with similar capacity to metabolize flavan-3-ols (Mena *et al.*, 2018). It remains to be proven whether the intake of cranberry polyphenols would also cluster subjects according to their phenolic metabolite profile in physiological fluids (urine, plasma, and feces) and whether these hypothetical metabotypes could be one of the factors putatively associated with the considerable variability observed in the efficacy of cranberry products.

8.4 Conclusions and Future Perspectives

The existing literature is consistent with the evidence that flavan-3-ols are bioactive compounds with beneficial effects on health. Despite this, discrepancies have been highlighted when encouraging results obtained using observational and *in vivo* animal studies are compared with more wary results from *in vivo* human intervention studies (Potì *et al.*, 2019). Most of the reasons for this concern the profound structural modifications that these compounds undergo once in contact with the human body, and the interindividual variability that defines every single organism as "unique."

Recently, essential viewpoints and position papers have been published to fill this evidence gap. Ottaviani *et al.* (2018b) produced an interesting handbook of rules worth considering when studying the health effects of flavan-3-ols, including an in-depth characterization of the food/supplement flavan-3-ol profile, the study of ADME of the compounds, with the identification of both flavan-3-ol conjugates and gut microbial derivatives, the choice of outcomes/endpoints and of adequate statistical methods. A recent viewpoint from the European COST Action

"POSITIVe" (www6.inra.fr/cost-positive) pointed out that the failing results in *in vivo* studies might be linked to the high variability in ADME among volunteers, particularly because of the different gut microbiota metabolism or the presence of polymorphisms in phase I and II metabolism enzymes (Morand and Tomás-Barberán, 2019).

To fully understand the effects of flavan-3-ol intake on human health at the individual level, further research is needed. This research should be conducted taking into account the evidence collected to date and considering real-life situations of flavan-3-ol intake (Mena and Del Rio, 2018).

References

Actis-Goretta, L., Ottaviani, J.I., and Fraga, C.G. (2006) Inhibition of angiotensin converting enzyme activity by flavanol-rich foods. *Journal of Agricultural and Food Chemistry*, **54**, 229–34.

Al-Dashti, Y.A., Holt, R.R., Stebbins, C.L., Keen, C.L., and Hackman, R.M. (2018) Dietary Flavanols: a review of select effects on vascular function, blood pressure, and exercise performance. *Journal of the American College of Nutrition*, **37**, 553–67.

Angelino, D., Carregosa, D., Domenech-Coca, C., *et al.* (2019) 5-(Hydroxyphenyl)-γ-valerolactone-sulfate, a key microbial metabolite of flavan-3-ols, is able to reach the brain: evidence from different in silico, in vitro and in vivo experimental models. *Nutrients*, **11**, 2678.

Appeldoorn, M.M., Vincken, J.-P., Aura, A.-M., Hollman, P.C.H., and Gruppen, H. (2009) Procyanidin dimers are metabolized by human microbiota with 2-(3,4-dihydroxyphenyl)acetic acid and 5-(3,4-dihydroxyphenyl)-γ-valerolactone as the major metabolites. *Journal of Agricultural and Food Chemistry*, **57**, 1084–92.

Berends, L.M., van der Velpen, V., and Cassidy, A. (2015) Flavan-3-ols, theobromine, and the effects of cocoa and chocolate on cardiometabolic risk factors. *Current Opinion in Lipidology*, **26**, 10–19.

Bernatoniene, J. and Kopustinskiene, D. (2018) The role of catechins in cellular responses to oxidative stress. *Molecules*, **23**, 965.

Borges, G., Ottaviani, J.I., van der Hooft, J.J.J., Schroeter, H., and Crozier, A. (2018) Absorption, metabolism, distribution and excretion of (−)-epicatechin: a review of recent findings. *Molecular Aspects of Medicine*, **61**, 18–30.

Boto-Ordóñez, M., Urpi-Sarda, M., Queipo-Ortuño, M.I., *et al.* (2013) Microbial metabolomic fingerprinting in urine after regular dealcoholized red wine consumption in humans. *Journal of Agricultural and Food Chemistry*, **61**, 9166–75.

Brickman, A.M., Khan, U.A., Provenzano, F.A., *et al.* (2014) Enhancing dentate gyrus function with dietary flavanols improves cognition in older adults. *Nature Neuroscience*, **17**, 1798–803.

Buitrago-Lopez, A., Sanderson, J., Johnson, L., *et al.* (2011) Chocolate consumption and cardiometabolic disorders: systematic review and meta-analysis. *BMJ*, **343**, 1–8.

Calani, L., Del Rio, D., Luisa Callegari, M.L., Morelli, L., and Brighenti, F. (2012) Updated bioavailability and 48 h excretion profile of flavan-3-ols from green tea in humans. *International Journal of Food Sciences and Nutrition*, **63**, 513–21.

Castello, F., Costabile, G., Bresciani, L., *et al.* (2018) Bioavailability and pharmacokinetic profile of grape pomace phenolic compounds in humans. *Archives of Biochemistry and Biophysics*, **646**, 1–9.

Cortés-Martín, A., Selma, M.V., Espín, J.C., and García-Villalba, R. (2019) The human metabolism of nuts proanthocyanidins does not reveal urinary metabolites consistent with distinctive gut microbiota metabotypes. *Molecular Nutrition & Food Research*, **63**, e1800819.

Crozier, A., Jaganath, I.B., and Clifford, M.N. (2009) Dietary phenolics: chemistry, bioavailability and effects on health. *Natural Product Reports*, **26**, 1001–43.

Del Rio, D., Calani, L., Cordero, C., Salvatore, S., Pellegrini, N., and Brighenti, F. (2010) Bioavailability and catabolism of green tea flavan-3-ols in humans. *Nutrition*, **26**, 1110–16.

Del Rio, D., Rodriguez-Mateos, A., Spencer, J.P.E., Tognolini, M., Borges, G., and Crozier, A. (2013) Dietary (poly)phenolics in human health: structures, bioavailability, and evidence of protective effects against chronic diseases. *Antioxidants & Redox Signaling*. **18**, 1818–92.

Desideri, G., Kwik-Uribe, C., Grassi, D., *et al.* (2012) Benefits in cognitive function, blood pressure, and insulin resistance through cocoa flavanol consumption in elderly subjects with mild cognitive impairment: the cocoa, cognition, and aging (CoCoA) study. *Hypertension (Dallas, Texas)*, **60**, 794–801.

Dohadwala, M.M., Holbrook, M., Hamburg, N.M., *et al.* (2011) Effects of cranberry juice consumption on vascular function in patients with coronary artery disease. *American Journal of Clinical Nutrition*, 934–40.

Engemann, A., Hübner, F., Rzeppa, S., and Humpf, H.U. (2012) Intestinal metabolism of two A-type procyanidins using the pig cecum model: detailed structure elucidation of unknown catabolites with Fourier transform mass spectrometry (FTMS). *Journal of Agricultural and Food Chemistry*, **60**, 749–57.

Erlund, I., Koli, R., Alfthan, G., *et al.* (2008) Favorable effects of berry consumption on platelet function, blood pressure, and HDL cholesterol. *American Journal of Clinical Nutrition*, **87**, 323–31.

Feliciano, R.P., Boeres, A., Massacessi, L., *et al.* (2016) Identification and quantification of novel cranberry-derived plasma and urinary (poly)phenols. *Archives of Biochemistry and Biophysics*, **599**, 31–41.

Ferrucci, F., Gaudiosi, G., Hirn, A., and Nicolich, R. (1991) Ionian basin and calabria arc: some new elements from DSS data. *Tectonophysics*, **195**, 411–19.

Flores-Mireles, A.L., Walker, J.N., Caparon, M., and Hultgren, S.J. (2015) Urinary tract infections: epidemiology, mechanisms of infection and treatment options. *Nature Reviews. Microbiology*, **13**, 269–84.

Fraga, C.G., Croft, K.D., Kennedy, D.O., and Tomás-Barberán, F.A. (2019) The effects of polyphenols and other bioactives on human health. *Food & Function*, **10**, 514–28.

Garrido, I., Urpi-Sarda, M., Monagas, M., *et al.* (2010) Targeted analysis of conjugated and microbial-derived phenolic metabolites in human urine after consumption of an almond skin phenolic extract. *Journal of Nutrition*, **140**, 1799–807.

Girouard, H. and Iadecola, C. (2006) Neurovascular coupling in the normal brain and in hypertension, stroke, and Alzheimer disease. *Journal of Applied Physiology*, **100**, 328–35.

González de Llano, D., Esteban-Fernández, A., Sánchez-Patán, F., Martínlvarez, P., Moreno-Arribas, M., and Bartolomé, B. (2015) Anti-adhesive activity of cranberry phenolic compounds and their microbial-derived metabolites against uropathogenic escherichia coli in bladder epithelial cell cultures. *International Journal of Molecular Sciences*, **16**, 12119–30.

González de Llano, D., Liu, H., Khoo, C., Moreno-Arribas, M.V., and Bartolomé, B. (2019) Some new findings regarding the antiadhesive activity of cranberry phenolic compounds and their microbial-derived metabolites against uropathogenic bacteria. *Journal of Agricultural and Food Chemistry*, **67**, 2166–74.

González-Sarrías, A., Combet, E., Pinto, P., *et al.* (2017) A systematic review and meta-analysis of the effects of flavanol-containing tea, cocoa and apple products on body composition and blood lipids: exploring the factors responsible for variability in their efficacy. *Nutrients*, **9**, 746.

Hillesheim, E. and Brennan, L. (2020) Metabotyping and its role in nutrition research. *Nutrition Research Reviews*, **33**, 33–42. doi.org/10.1017/S0954422419000179.

Holt, R.R., Schramm, D.D., Keen, C.L., Lazarus, S.A., and Schmitz, H.H. (2002) Chocolate consumption and platelet function. *JAMA*, **287**, 2212–13.

Ide, K., Matsuoka, N., Yamada, H., *et al.* (2018) Effects of tea catechins on Alzheimer's disease: recent updates and perspectives. *Molecules*, **23**, 2357.

Jiménez, R., Duarte, J., and Perez-Vizcaino, F. (2012) Epicatechin: endothelial function and blood pressure. *Journal of Agricultural and Food Chemistry*, **60**, 8823–30.

Kaspar, K.L., Howell, A.B., and Khoo, C. (2015) A randomized, double-blind, placebo-controlled trial to assess the bacterial anti-adhesion effects of cranberry extract beverages. *Food & Function*, **6**, 1212–17.

Khymenets, O., Andres-Lacueva, C., Urpi-Sarda, M., *et al.* (2015) Metabolic fingerprint after acute and under sustained consumption of a functional beverage based on grape skin extract in healthy human subjects. *Food & Function*, **6**, 1288–98.

Kutschera, M., Engst, W., Blaut, M., and Braune, A. (2011) Isolation of catechin-converting human intestinal bacteria. *Journal of Applied Microbiology*, **111**, 165–75.

Lamport, D.J., Saunders, C., Butler, L.T., and Spencer, J.P. (2014) Fruits, vegetables, 100% juices, and cognitive function. *Nutrition Reviews*, **72**, 774–89.

Lee, S.-J. and Jia, Y. (2015) The effect of bioactive compounds in tea on lipid metabolism and obesity through regulation of peroxisome proliferator-activated receptors. *Current Opinion in Lipidology*, **26**, 3–9.

Mastroiacovo, D., Kwik-Uribe, C., Grassi, D., *et al.* (2015) Cocoa flavanol consumption improves cognitive function, blood pressure control, and metabolic profile in elderly subjects: the cocoa, cognition, and aging (CoCoA) study – a randomized controlled trial. *American Journal of Clinical Nutrition*, **101**, 538–48.

McCullough, M.L., Peterson, J.J., Patel, R., Jacques, P.F., Shah, R., and Dwyer, J.T. (2012) Flavonoid intake and cardiovascular disease mortality in a prospective cohort of US adults. *American Journal of Clinical Nutrition*, **95**, 454–64.

Mena, P. and Del Rio, D. (2018) Gold standards for realistic (poly)phenol research. *Journal of Agricultural and Food Chemistry*, **66**, 8221–3.

Mena, P., Domínguez-Perles, R., Gironés-Vilaplana, A., Baenas, N., García-Viguera, C., and Villaño, D. (2014) Flavan-3-ols, anthocyanins, and inflammation. *IUBMB Life*, **66**, 745–58.

Mena, P., González de Llano, D., Brindani, N., *et al.* (2017) 5-(3′,4′-Dihydroxyphenyl)-γ-valerolactone and its sulphate conjugates, representative circulating metabolites of flavan-3-ols, exhibit anti-adhesive activity against uropathogenic Escherichia coli in bladder epithelial cells. *Journal of Functional Foods*, **29**, 275–80.

Mena, P., Ludwig, I.A., Tomatis, V.B., *et al.* (2018) Inter-individual variability in the production of flavan-3-ol colonic metabolites: preliminary elucidation of urinary metabotypes. *European Journal of Nutrition*, **58**, 1529-43.

Mena, P., Bresciani, L., Brindani, N., *et al.* (2019) Phenyl-γ-valerolactones and phenylvaleric acids, the main colonic metabolites of flavan-3-ols: synthesis, analysis, bioavailability, and bioactivity. *Natural Product Reports*, **36**, 714–52.

Montagnana, M., Danese, E., Angelino, D., *et al.* (2018) Dark chocolate modulates platelet function with a mechanism mediated by flavan-3-ol metabolites. *Medicine*, **97**, e13432.

Morand, C. and Tomás-Barberán, F.A. (2019) Interindividual variability in absorption, distribution, metabolism, and excretion of food phytochemicals should be reported. *Journal of Agricultural and Food Chemistry*, **67**, 3843–4.

Morris, M.C., Tangney, C.C., Wang, Y., Sacks, F.M., Bennett, D.A., and Aggarwal, N.T. (2015) MIND diet associated with reduced incidence of Alzheimer's disease. *Alzheimer's & Dementia*, **11**, 1007–14.

Neveu, V., Perez-Jimenez, J., Vos, F., *et al.* (2010) Phenol-Explorer: an online comprehensive database on polyphenol contents in foods. *Database*, **2010**, bap024.

Nicolle, L.E. (2016) Cranberry for prevention of urinary tract infection? Time to move on. *JAMA*, **316**, 1873–4.

Ortiz-López, L., Márquez-Valadez, B., Gómez-Sánchez, A., *et al.* (2016) Green tea compound epigallo-catechin-3-gallate (EGCG) increases neuronal survival in adult hippocampal neurogenesis in vivo and in vitro. *Neuroscience*, **322**, 208–20.

Ottaviani, J.I., Momma, T.Y., Kuhnle, G.K., Keen, C.L., and Schroeter, H. (2012) Structurally related (-)-epicatechin metabolites in humans: assessment using de novo chemically synthesized authentic standards. *Free Radical Biology & Medicine*, **52**, 1403–12.

Ottaviani, J.I., Borges, G., Momma, T.Y., *et al.* (2016) The metabolome of [2-14C](-)-epicatechin in humans: implications for the assessment of efficacy, safety, and mechanisms of action of polyphenolic bioactives. *Scientific Reports*, **6**, 29034.

Ottaviani, Javier I., Fong, R., *et al.* (2018a) Evaluation at scale of microbiome-derived metabolites as biomarker of flavan-3-ol intake in epidemiological studies. *Scientific Reports*, **8**, 9859.

Ottaviani, Javier I, Heiss, C., Spencer, J.P.E., Kelm, M., and Schroeter, H. (2018b) Recommending flavanols and procyanidins for cardiovascular health: revisited. *Molecular Aspects of Medicine*, **61**, 63–75.

Ou, K., Sarnoski, P., Schneider, K.R., Song, K., Khoo, C., and Gu, L. (2014) Microbial catabolism of procyanidins by human gut microbiota. *Molecular Nutrition & Food Research*, **58**, 2196–205.

Peron, G., Sut, S., Pellizzaro, A., *et al.* (2017) The antiadhesive activity of cranberry phytocomplex studied by metabolomics: intestinal PAC-A metabolites but not intact PAC-A are identified as markers in active urines against uropathogenic Escherichia coli. *Fitoterapia*, **122**, 67–75.

Potì, F., Santi, D., Spaggiari, G., Zimetti, F., and Zanotti, I. (2019) Polyphenol health effects on cardiovascular and neurodegenerative disorders: a review and meta-analysis. *International Journal of Molecular Sciences*, **20**, 351.

Rice-Evans, C.A., Miller, N.J., Bolwell, P.G., Bramley, P.M., and Pridham, J.B. (1995) The relative antioxidant activities of plant-derived polyphenolic flavonoids. *Free Radical Research*, **22**, 375–83.

Rios, L.Y., Gonthier, M.-P., Rémésy, C., *et al.* (2003) Chocolate intake increases urinary excretion of polyphenol-derived phenolic acids in healthy human subjects. *American Journal of Clinical Nutrition*, **77**, 912–18.

Rodriguez-Mateos, A., Rendeiro, C., Bergillos-Meca, T., *et al.* (2013) Intake and time dependence of blueberry flavonoid-induced improvements in vascular function: a randomized, controlled, double-blind, crossover intervention study with mechanistic insights into biological activity. *American Journal of Clinical Nutrition*, **98**, 1179–91.

Rodriguez-Mateos, A., Heiss, C., Borges, G., and Crozier, A. (2014a) Berry (poly)phenols and cardiovascular health. *Journal of Agricultural and Food Chemistry*, **62**, 3842–51.

Rodriguez-Mateos, A., Vauzour, D., Krueger, C.G., *et al.* (2014b) Bioavailability, bioactivity and impact on health of dietary flavonoids and related compounds: an update. *Archives of Toxicology*, **88**, 1803–53.

Rodriguez-Mateos, A., Feliciano, R.P., Boeres, A., *et al.* (2016) Cranberry (poly)phenol metabolites correlate with improvements in vascular function: a double-blind, randomized, controlled, dose-response, crossover study. *Molecular Nutrition & Food Research*, **60**, 2130–40.

Rodriguez-Mateos, A., Weber, T., Skene, S.S., *et al.* (2018) Assessing the respective contributions of dietary flavanol monomers and procyanidins in mediating cardiovascular effects in humans: randomized, controlled, double-masked intervention trial. *American Journal of Clinical Nutrition*, **108**, 1229–37.

Roowi, S., Stalmach, A., Mullen, W., Lean, M.E.J., Edwards, C.A., and Crozier, A. (2010) Green tea flavan-3-ols: colonic degradation and urinary excretion of catabolites by humans. *Journal of Agricultural and Food Chemistry*, **58**, 1296–304.

Sang, S., Lee, M.-J., Yang, I., Buckley, B., and Yang, C.S. (2008) Human urinary metabolite profile of tea polyphenols analyzed by liquid chromatography/electrospray ionization tandem mass spectrometry with data-dependent acquisition. *Rapid Communication in Mass Spectrometry*, **22**, 1567–78.

Schaffer, S. and Halliwell, B. (2012) Do polyphenols enter the brain and does it matter? Some theoretical and practical considerations. *Genes & Nutrition*, **7**, 99–109.

Schroeter, H., Heiss, C., Balzer, J., *et al.* (2006) (-)-Epicatechin mediates beneficial effects of flavanol-rich cocoa on vascular function in humans. *Proceedings of the National Academy of Sciences of the USA*, **103**, 1024–9.

Serafini, M., Bugianesi, R., Maiani, G., Valtuena, S., de Santis, S., and Crozier, A. (2003) Plasma antioxidants from chocolate. *Nature*, **424**, 1013.

Sorond, F.A., Hurwitz, S., Salat, D.H., Greve, D.N., and Fisher, N.D.L. (2013) Neurovascular coupling, cerebral white matter integrity, and response to cocoa in older people. *Neurology*, **81**, 904–9.

Spencer, J.P.E. (2008) Flavonoids: modulators of brain function? *British Journal of Nutrition*, **99**, ES60–ES77.

Spencer, J.P.E. (2009) Flavonoids and brain health: multiple effects underpinned by common mechanisms. *Genes & Nutrition*, **4**, 243–50.

Stalmach, A., Mullen, W., Steiling, H., Williamson, G., Lean, M.E.J., and Crozier, A. (2010) Absorption, metabolism, and excretion of green tea flavan-3-ols in humans with an ileostomy. *Molecular Nutrition & Food Research*, **54**, 323–34.

Stoupi, S., Williamson, G., Drynan, J.W., Barron, D., and Clifford, M.N. (2010) A comparison of the in vitro biotransformation of (-)-epicatechin and procyanidin B2 by human faecal microbiota. *Molecular Nutrition & Food Research*, **54**, 747–59.

Terlizzi, M.E., Gribaudo, G., and Maffei, M.E. (2017) UroPathogenic Escherichia Coli (UPEC) infections: virulence factors, bladder responses, antibiotic, and non-antibiotic antimicrobial strategies. *Frontiers in Microbiology*, **8**, 1566.

Trost, K., Ulaszewska, M.M., Stanstrup, J., *et al.* (2018) Host: microbiome co-metabolic processing of dietary polyphenols – an acute, single blinded, cross-over study with different doses of apple polyphenols in healthy subjects. *Food Research International*, **112**, 108–28.

Unno, K., Pervin, M., Nakagawa, A., *et al.* (2017) Blood-brain barrier permeability of green tea catechin metabolites and their neuritogenic activity in human neuroblastoma SH-SY5Y cells. *Molecular Nutrition & Food Research*, **5**, 1700294.

Urpi-Sarda, M., Garrido, I., Monagas, M., *et al.* (2009a) Profile of plasma and urine metabolites after the intake of almond [prunus dulcis (Mill.) D.A. Webb] polyphenols in humans. *Journal of Agricultural and Food Chemistry*, **57**, 10134–42.

Urpi-Sarda, M., Monagas, M., Khan, N., *et al.* (2009b) Epicatechin, procyanidins, and phenolic microbial metabolites after cocoa intake in humans and rats. *Analytical and Bioanalytical Chemistry*, **394**, 1545–56.

Van der Hooft, J.J.J., de Vos, R.C.H., Mihaleva, V., *et al.* (2012) Structural elucidation and quantification of phenolic conjugates present in human urine after tea intake. *Analytical Chemistry*, **84**, 7263–71.

Van Duynhoven, J., van der Hooft, J.J.J., van Dorsten, F.A., *et al.* (2014) Rapid and sustained systemic circulation of conjugated gut microbial catabolites after single-dose black tea extract consumption. *Journal of Proteome Research*, **13**, 2668–78.

Vauzour, D. (2017) Polyphenols and brain health. *OCL*, **24**, A202.

Vitaglione, P., Barone Lumaga, R., Ferracane, R., *et al.* (2013) Human bioavailability of flavanols and phenolic acids from cocoa-nut creams enriched with free or microencapsulated cocoa polyphenols. *British Journal of Nutrition*, **109**, 1832–43.

Weinreb, O., Mandel, S., Amit, T., and Youdim, M.B.H. (2004) Neurological mechanisms of green tea polyphenols in Alzheimer's and Parkinson's diseases. *Journal of Nutritional Biochemistry*, **15**, 506–16.

Weinreb, O., Amit, T., Mandel, S., and Youdim, M.B.H. (2009) Neuroprotective molecular mechanisms of

(-)-epigallocatechin-3-gallate: a reflective outcome of its antioxidant, iron chelating and neuritogenic properties. *Genes & Nutrition*, **4**, 283–96.

Wiese, S., Esatbeyoglu, T., Winterhalter, P., *et al.* (2015) Comparative biokinetics and metabolism of pure monomeric, dimeric, and polymeric flavan-3-ols: a randomized cross-over study in humans. *Molecular Nutrition & Food Research*, **59**, 610–21.

Williamson, G. (2017) The role of polyphenols in modern nutrition. *Nutrition Bulletin*, **42**, 226–35.

Zamora-Ros, R., Forouhi, N.G., Sharp, S.J., *et al.* (2013) The association between dietary flavonoid and lignan intakes and incident type 2 diabetes in european populations: the EPIC-InterAct study. *Diabetes Care*, **36**, 3961–70.

Zamora-Ros, R., Knaze, V., Rothwell, J.A., *et al.* (2016) Dietary polyphenol intake in europe: the european prospective investigation into cancer and nutrition (EPIC) study. *European Journal of Nutrition*, **55**, 1359–75.

Zanotti, I., Dall'Asta, M., Mena, P., *et al.* (2015) Atheroprotective effects of (poly)phenols: a focus on cell cholesterol metabolism. *Food & Function*, **6**, 13–31.

Ziauddeen, N., Rosi, A., Del Rio, D., *et al.* (2018) Dietary intake of (poly)phenols in children and adults: cross-sectional analysis of UK National Diet and Nutrition Survey Rolling Programme (2008–2014). *European Journal of Nutrition*, **8**, 3183–98.

9

Ellagitannins and Their Gut Microbiota-Derived Metabolites: Urolithins

Rocío García-Villalba, Juan A. Giménez-Bastida, María A. Ávila-Gálvez, Francisco A. Tomás-Barberán, Juan C. Espín, and Antonio González-Sarrías*

Laboratory of Food and Health, Research Group on Quality, Safety, and Bioactivity of Plant Foods, Department of Food Science and Technology, CEBAS-CSIC, Murcia, Spain

9.1 Chemistry and Sources of Ellagitannins and Ellagic Acid

Tannins are a heterogeneous group of polyphenols of high molecular weight ranging from 500 to 30 000 Da, widely present in plant foods, predominantly fruits, beverages and vegetables such as tea, red wine, juices, nuts, cocoa, etc. (Khanbabaee and van Ree, 2001; Serrano *et al.*, 2009; Smeriglio *et al.*, 2017). Besides their involvement in protection against different biotic and abiotic stressors in plants, they also contribute substantially to the astringency and bitterness of plant foods (Serrano *et al.*, 2009; Smeriglio *et al.*, 2017). They can be classified into two groups: condensed tannins (also called proanthocyanidins) and hydrolyzable tannins that can be further categorized into gallotannins and ellagitannins (ET). Structurally, the hydrolyzable tannins are composed of a carbohydrate moiety, usually d-glucose, partially or totally esterified in the hydroxyl groups with gallic acid units (gallotannins) or hexahydroxydiphenic acid (HHDP) units, resulting from the intramolecular oxidative C-C

*Corresponding author.

Dietary Polyphenols: Metabolism and Health Effects, First Edition.
Edited by Francisco A. Tomás-Barberán, Antonio González-Sarrías, and Rocío García-Villalba.

coupling between galloyl groups (ellagitannins) (Khanbabaee and van Ree, 2001; Tomás-Barberán *et al.*, 2008).

Monomeric ETs refer to those containing only one glucose core with different coupling patterns of HHDP groups. Upon hydrolysis, they release HHDP units that undergo spontaneous conversion to form the dilactone, ellagic acid (EA) (Figure 9.1a) (Khanbabaee and van Ree, 2001; Niemetz and Gross, 2005; Tomás-Barberán *et al.*, 2008). Besides HHDP, other groups such as valoneoyl (present in rugosin A), dehydrohexahydroxy-diphenoyl (DHHDP) (present in geraniin), chebuloyl (present in chebulagic acid) or gallagyl (specific for punicalagin and punicalin) can be coupled to the glucose core (Figure 9.1b). Variations in the number and position of these groups provide a wide variety of compounds in this family. Another group of monomeric ellagitannins includes those with acyclic glycosides, such as vescalagin and castalagin (Figure 9.1c) (Yoshida *et al.*, 2010).

In addition, monomeric ETs can be polymerized to form oligomeric or polymeric ellagitannins. The polymerization can be through the most common C-O couplings (e.g., coriariin A, rugosin D, sanguiin H-6) or C-C couplings (e.g., roburin A, roburin D) (Figure 9.1d).

Ellagitannins and ellagic acid (EA) (rarely free but usually glycosylated) are found in many fruits and beverages (pomegranates, muscadine grapes, raspberries, black raspberries, strawberries, camu-camu, jaboticaba, green and black teas), nuts (walnuts, pecans, chestnuts, almonds), and oak-aged beverages (wine, whisky, cognac) (Crozier *et al.*, 2009; Tomás-Barberán *et al.*, 2008). Although there is a wide variability of ETs, some of them are specific to particular foods. The most representative monomeric ETs include punicalagin found in pomegranate, casuarictin in blackberries, pedunculagin in walnuts, corilagin found in longan seed, and castalagin and vescalagin in chestnuts and oak-aged beverages (wine, whiskey, cognac), respectively. Regarding dimeric or oligomeric ETs, sanguiin H-6 and agrimoniin found in strawberries are the most representative. Table 9.1 shows familiar food sources of the most representative ETs and an estimation of their content in each food. This content may vary within the same food due to different reasons, including fruit cultivars (genotype),

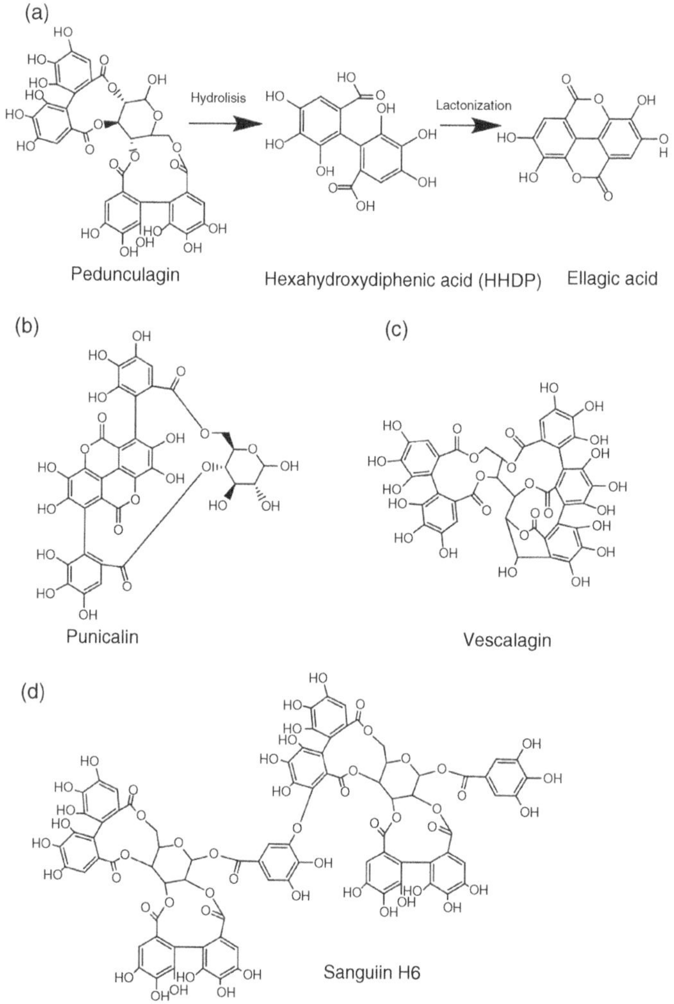

Figure 9.1 Representative dietary ellagitannins: pedunculagin (a), punicalin (b), vescalagin (c), and sanguiin H6 (d).

environmental and agronomic conditions, and food-processing factors (Häkkinen *et al.*, 2000; Zafrilla *et al.*, 2001).

The correct identification and quantification of ETs are difficult due to their structural complexity, their diversity, and

Table 9.1 Representative ellagitannins present in some food products

Ellagitannin	Food sources	Content (mg/g or mg/L)	Reference
Agrimoniin	Strawberry	0.77–0.85	(Koponen *et al.*, 2007)
Casuarictin	Blackberry	0.69	(Hager *et al.*, 2008)
Castalagin	Chestnut	0.0016–0.025	(Gonçalves *et al.*, 2010)
Castalin	Chestnut bark	3.1–10	(Comandini *et al.*, 2014)
Corilagin	Longan (seed)	3.7–8.6	(Rangkadilok *et al.*, 2005)
Geraniin	Rambutan (peel)	140.02	(Li *et al.*, 2018)
Lambertianin C	Raspberry jam	0.76	(Koponen *et al.*, 2007)
Pedunculagin	Walnut	16.04	(Anderson *et al.*, 2001)
Punicalagin	Pomegranate Pomegranate juice	0.35–0.75 2020–2660	(Gil *et al.*, 2000) (Cerdá *et al.*, 2006)
Punicalin	Pomegranate (peel, juice)	0.13	(Fischer *et al.*, 2011)
Sanguiin H-5	Muscadine grapes and juice	8–84	(Talcott and Lee *et al.*, 2002)
Sanguiin H-6	Strawberry Cloudberry Raspberry	3.15 2.63–3.30	(Koponen *et al.*, 2007)
Vescalagin	Whisky Cognac Oak-aged red wine	1–2 31–55 50	(Glabasnia and Hofmann, 2006) (Clifford and Scalbert, 2000)

the lack of commercially available reference standards. Acid hydrolysis has been applied for the analysis of ETs based on the quantification of total ellagic acid and other hydrolysis products (García-Villalba *et al.*, 2015). Due to the lack of accurate quantification, the estimated mean total daily intake of ellagitannins and ellagic acid remains unclear, although it was postulated that could reach hundreds of mg per day depending on the type of ET-rich foods regularly consumed, mainly as beverages and juices (Tomás-Barberán *et al.*, 2009).

9.2 Bioavailability of Ellagitannins and Ellagic Acid

The metabolism of ETs has been evaluated *in vitro* by simulation of gastrointestinal digestion using human gastrointestinal cell lines, and *in vivo* studies with humans and animal models. The release of free EA from ETs under physiological conditions has not been unequivocally proven so far (González-Sarrías *et al.*, 2015a). The low bioavailability of ETs and EA is well known as they are poorly absorbed. They are usually further hydrolyzed and/or metabolized by the gut microbiota into hydroxy-6H-dibenzopyran-6-one derivatives known as urolithins. Since 2005, urolithins (initially urolithin A) were suggested as potential biomarkers for exposure to ET-rich foods in humans (Cerdá *et al.*, 2005).

No ETs are detected in body tissues apart from the gastrointestinal tract. However, several human and animal pharmacokinetic studies reported that low quantities of free EA (rarely exceeding 30 ng/mL) may be absorbed within 30–60 minutes after the intake of ET-rich foods, reporting a rapid elimination (Lei *et al.*, 2003; Mertens-Talcott *et al.*, 2006; Seeram *et al.*, 2004, 2006; Stoner *et al.*, 2005). The occurrence of free EA in plasma has suggested that its absorption mainly occurs in the stomach, whereas the later presence of phase II conjugates suggested its absorption in the intestine. During phase II metabolism, the absorbed EA is converted mostly to methyl and dimethyl ellagic acid glucuronides that are identified in plasma and urine (Mertens-Talcott *et al.*, 2006; Seeram *et al.*, 2006). A pharmacokinetic study in healthy humans who consumed two different extracts with different ET/EA ratios showed that a higher free EA intake does not enhance EA bioavailability due to the limited EA absorption, although it does promote a higher urolithin production (González-Sarrías *et al.*, 2015a). Human EA bioavailability also showed large interindividual variability.

Regarding their distribution in the body, both free EA and its conjugates have been detected in different systemic tissues such as kidneys, liver, heart, and prostate of animals (Cerdá *et al.*, 2003; Espín *et al.*, 2007; Yan *et al.*, 2014) as well as in the prostate and colon tissues of humans after ingestion of ET-rich foods (González-Sarrías *et al.*, 2010a; Nuñez-Sánchez *et al.*, 2014).

9.3 The Microbial Metabolism of Ellagitannins and Ellagic Acid: Urolithins

9.3.1 Urolithin Production and Bioavailability

As indicated above, the bioavailability of ETs and EA is low, and therefore they reach the colon where the gut microbiota metabolizes them through different catabolic reactions. These include the lactone ring cleavage, decarboxylations, and dehydroxylations to produce a series of metabolites known as urolithins (Uro) (Cerdá *et al.*, 2004; Espín *et al.*, 2013). Unlike urolithins, in some cases, EA can suffer bacterial dehydroxylation without the opening of the lactone ring and leading to metabolites known as nasutins (González-Barrio *et al.*, 2011). These were only detected in animal models, particularly as free form (nonconjugated) in the luminal content of the stomach, intestine, and cecum of rats (Kosmala *et al.*, 2015), as well as in feces of beavers and pigs (Gonzalez-Barrio *et al.*, 2011) and as nasutin glucuronide in plasma of pigs (Espín *et al.*, 2007) and rats (Kosmala *et al.*, 2015).

The pathway for urolithin formation seems to be as follows: HHDP is released from ETs, which is quickly converted into 3,4,8,9,10-pentahydroxy-6H-dibenzo[b,d]pyran-6-one (Uro-M5). Then a sequential dehydroxylation takes place producing tetrahydroxy- (Uro-M6 and Uro-D), trihydroxy- (Uro-M7 and Uro-C), dihydroxy- (Uro-A and IsoUro-A), and monohydroxy- (Uro-B) dibenzopyranones (Figure 9.2) (Garcia-Villalba *et al.*, 2013). All urolithins share a nucleus of dibenzo-pyran-one, with different hydroxyl substitutions. Different studies *in vitro* (Bialonska *et al.*, 2010; García-Villalba *et al.*, 2013) and in animal models (Espín *et al.*, 2007) have shown in detail the sequential transformation of EA leading to different urolithin intermediates to end in the final catabolic products Uro-A, IsoUro-A, and Uro-B.

Urolithins are partially excreted in feces as aglycones. The most common urolithins, Uro-A, Uro-B and IsoUro-A, were

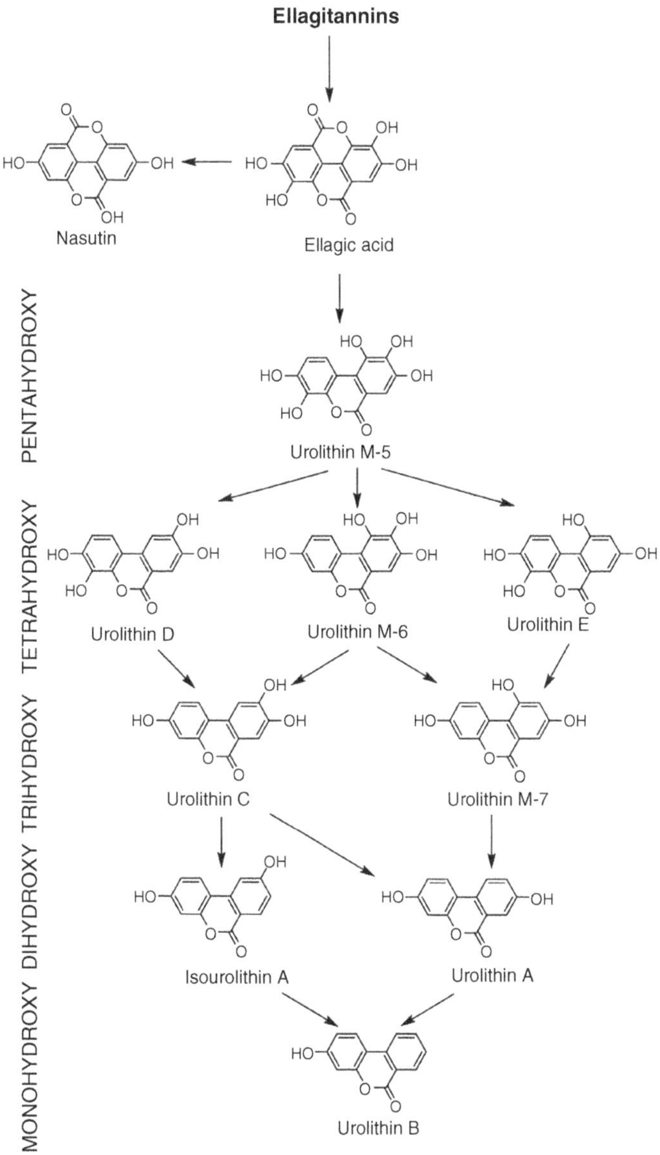

Figure 9.2 Main catabolic pathway from EA to urolithins.

identified and quantified in rat fecal samples (Ito *et al.*, 2012), and also in other animals such as beavers, squirrels (González-Barrio *et al.*, 2011), ruminants (Gonzalez-Barrio *et al.*, 2012), and pigs (Espín *et al.*, 2007). Diverse urolithin patterns were observed in different animals. Thus, the main urolithins produced in cattle were IsoUro-A and Uro-B, which are different metabolites from those found in the monogastric mammals in which Uro-A was the primary metabolite produced. In addition, intermediate metabolites (Uro-M5, Uro-D, Uro-M6, Uro-C, and Uro-M7) were also identified in feces of different animal models (González-Barrio *et al.*, 2011; Ito *et al.*, 2012).

Regarding humans, different bioavailability studies reported the presence of terminal nonconjugated urolithins (Uro-A, IsoUro-A, and Uro-B) in feces (Li *et al.*, 2015; Mosele *et al.*, 2015; Pfundstein *et al.*, 2014; Romo-Vaquero *et al.*, 2015). Recently, it has been reported that the fecal concentration for these metabolites is in the range 7–7180 μg/g of Uro-A, 13–1894 μg/g of Uro-B, and 36–207 μg/g of IsoUro-A (Tomás-Barberán *et al.*, 2017). The identification of the pathway of urolithin formation in humans was first described *in vitro* (García-Villalba *et al.*, 2013). Next, it was confirmed in an intestinal simulator model (TWIN-SHIME) composed of three compartments that simulated ascending, transverse, and descending colon regions. The highest urolithin production occurred in the descending colon (García-Villalba *et al.*, 2017). However, urolithins can be detected in the three colon segments due to the enterohepatic circulation, as reported in colon cancer patients (Núñez-Sánchez *et al.*, 2014). *In vivo*, some of these intermediate metabolites were identified in human fecal samples after consumption of walnut (Pfundstein *et al.*, 2014; Romo-Vaquero *et al.*, 2015) and pomegranate products (Mosele *et al.*, 2015; Romo-Vaquero *et al.*, 2015).

Therefore, currently, it is well established that the presence of urolithins may have local impact on the colonic mucosa as well as being absorbed, appearing in the circulatory system and reaching systemic tissues up to 3–4 days after the intake where they may trigger biological effects (González-Sarrías *et al.*, 2017a; Tomas-Barberán *et al.*, 2017). As widely reported, urolithins, unlike ETs or EA, are much better absorbed in the colon and undergo extensive phase II metabolism, yielding glucuronide and sulfate conjugates that reach plasma and systemic tissues, and are finally excreted in urine.

In this regard, the presence of urolithin conjugates in plasma and urine has been widely described in animal models such as rodents (Cerdá *et al.*, 2003), pigs (Espín *et al.*, 2007), ruminants (González-Barrio *et al.*, 2012), and other mammals (González-Barrio *et al.*, 2011). Urolithin metabolites have been identified in the bile from pigs that suggests extensive hepatic metabolism and active enterohepatic circulation (Espín *et al.*, 2007).

Regarding humans, urolithin conjugates have been identified in plasma and urine after consumption of different ET-rich foods (Cerdá *et al.*, 2005), such as strawberries (Sandhu *et al.*, 2018; Truchado *et al.*, 2012), raspberries (Ludwig *et al.*, 2015; Puupponen-Pimiä *et al.*, 2013), nuts (Tulipani *et al.*, 2012), oak-aged wine, blackberries (García-Muñoz *et al.*, 2014), and pomegranates (Cerdá *et al.*, 2004; Mertens-Talcott *et al.*, 2006; Seeram *et al.*, 2008). In most of the human studies, mainly Uro-A and Uro-B conjugates reached maximum concentrations in plasma and urine between 24 and 48 hours and were detected in these biological fluids for up to 72 hours, evidencing enterohepatic recirculation. In this regard, Tomas-Barberán *et al.* (2017) determined the concentrations of urolithin metabolites in biological fluids and tissues of animals and humans. According to this review, the main metabolites reach concentrations in human plasma in the micromolar range (up to 35 μM for Uro-A glucuronide, 0.745 μM for IsoUro-A and 7.3 μM for Uro-B glucuronide), whereas the concentrations in human urine reached up to 100 μM for these urolithin conjugates. On the other hand, in a few human studies, lower levels of free urolithins have also been detected, as well as other conjugates of intermediate metabolites such as Uro-C glucuronide (García-Muñoz *et al.*, 2014; Puupponen-Pimiä *et al.*, 2013; Nuñez-Sánchez *et al.*, 2014; Teixeira *et al.*, 2017).

An essential factor to be considered is the impact of the food matrix and type of ET on bioavailability and metabolism. Different studies have shown that the nature of ingested ET could determine the rate of urolithin production in humans (Cerdá *et al.*, 2005), rodents (Milala *et al.*, 2017), and *ex vivo* cultures of human gut microbiota (Piwowarski *et al.*, 2016).

Regarding the food matrix, several studies have been focused on this factor. A human study reported that urolithin formation was not affected when pomegranate juice was mixed with soy protein isolate in healthy humans (Yang *et al.*, 2016). In the same way, strawberry processing (strawberry purée) did not affect the production and urinary excretion of urolithins in humans compared to fresh strawberries (Truchado *et al.*, 2012). Finally, a previous animal study with rats fed a diet containing fructooligosaccharides (FOS) and monomeric or dimeric ellagitannins reported a better intestinal breakdown with the monomeric diet, and therefore higher concentrations of EA and urolithins, as well as an increase in the absorption of urolithins with the simultaneous ingestion of FOS (Jurgonski *et al.*, 2017).

9.3.2 Tissue Distribution of Urolithins after Consumption of Ellagitannins

Understanding the molecular form and concentration of urolithins that target different tissues is necessary to correlate the biological effects of ET-containing foods and identifying the actual molecules responsible for the beneficial effects. This information is very useful for designing *in vitro* mechanistic studies using physiologically relevant metabolites at realistic concentrations.

Recently, an extensive review reported the levels of urolithins that may be reached in different tissues (gut and systemic) of animals and humans (Tomás-Barberán *et al.*, 2017). Briefly, according to animal studies, up to 25 urolithin metabolites have been identified in plasma, intestine (jejunum and colon), and bile from pigs (Espín *et al.*, 2007). Urolithin conjugates were also identified in different systemic tissues such as liver, prostate, brain, uterus, kidney, small intestine, and colon in rodents after ingestion of a standard diet enriched in ETs or synthetic urolithin A (Cerdá *et al.*, 2003; Gonzalez-Sarrias *et al.*, 2010a; Gu *et al.*, 2019; Seeram *et al.*, 2007). Intravenous administration of both urolithin A and B to rats showed that these compounds had a fast distribution among target organs, reaching the brain and the heart (Gasperotti *et al.*, 2015).

Regarding human studies, high levels of urolithins (Uro-A, B, C, D and IsoUro-A, and their glucuronide and sulfate conjugates) between 42 and 1671 ng/g tissue were identified in normal and malignant colon tissues after consumption of a pomegranate extract (Núñez-Sánchez *et al.*, 2014). Urolithin conjugates have been identified to reach systemic tissues such as prostate and breast tissues. In human prostate, Uro-A glucuronide (concentration range 0.5–2 ng/g) and traces of Uro-B glucuronide were identified after consumption of walnuts or pomegranate (González-Sarrías *et al.*, 2010a). Recently, Uro-A glucuronide (29.4 ± 23.9 pmol/g), IsoUro-A glucuronide (9.6 ± 5.4 pmol/g), Uro-A sulfate (4.8 ± 3.5 pmol/g) and Uro-B sulfate (1.0 ± 0.6 pmol/g) were identified in malignant and normal mammary tissues from breast cancer patients consuming a mixture of polyphenol-containing extracts (Ávila-Gálvez *et al.*, 2019a).

9.3.3 Interaction of ETs and Urolithins with the Gut Microbiota

A well-known two-way interaction occurs between ETs and the gut microbiota. As has been described, the gut microbiota can metabolize ETs to urolithins, but also ETs and even urolithins may modulate the microbial population, stimulating the growth of some strains while inhibiting others.

To date, there is limited information regarding the ability of phenolic compounds in genera, and ETs in particular to influence the gut microbiota. In this regard, different studies (*in vitro*, animals, and humans) have evaluated the impact of ET-rich foods on gut microbial composition. Pomegranate extracts and juices showed an increase in the mean counts of *Bifidobacterium* and *Lactobacillus* spp. in human studies *in vitro* (Bialonska *et al.*, 2010) and *in vivo* (Li *et al.*, 2015) and a significant decrease of *B. fragilis* group, *Clostridia*, and Enterobacteriaceae (Li *et al.*, 2015). In another human study, consumption of a pomegranate extract for four weeks reduced the ratio of *Firmicutes* to *Bacteroidetes* and increased *Akkermansia* in the gut microflora (Li *et al.*, 2015).

Contrary to these human studies, the daily ingestion for four weeks of 200 mL of pomegranate juice showed nonsignificant modification in the composition of gut microbiota (Mosele *et al.*, 2015). Other studies also reported differences in gut microbiota composition, showing an increase of *Akkermansia* and *Gordonibacter* using a gastrointestinal simulation model (TWIN-SHIME) with gut microbiota from two volunteers with different urolithin metabotypes fed with a pomegranate extract (García-Villalba *et al.*, 2017). This effect of pomegranate extracts on *Akkermansia muciniphila* was also reported in a human intervention trial, and in addition, the Uro-A formation was associated with a high proportion of *A. muciniphila* (Henning *et al.*, 2017).

Similar modulation in gut microbial communities has been observed with the intake of other ET-rich foods. Walnut consumption increased the abundance of *Firmicutes*, *Lactobacillus*, Ruminococcaceae, and *Roseburia* and reduces the abundance of *Bacteroidetes*, *Bacteroides*, and *Anaerotruncus* in rats (Byerley *et al.*, 2017). Regarding blackberry consumption, a proportional decrease of *Firmicutes* and *Clostridium* and an increase of *Bacteroidetes* were reported in mice (Gu *et al.*, 2019).

In general, the objective of most studies was to determine the impact of the whole foods and not of individual constituents. In this regard, since punicalagins did not increase the growth of fecal microbiota in batch culture (Bialonska *et al.*, 2010), the effects of pomegranate extracts or juice on fecal bacteria are probably attributed to the abundance of other components (other phytochemicals, fiber, polysaccharides, etc.).

Finally, studies on the effect of gut microbiota modulation by the microbiota metabolites urolithins are still limited. There is only one study on a rat intestinal inflammation model in which orally administered urolithins showed high prebiotic effects promoting the growth of *Lactobacillus* and *Bifidobacterium* (Larrosa *et al.*, 2010).

On the other hand, in recent years the identification of the microorganism able to transform the ETs and EA into urolithins has gained great interest, although this remains at an initial stage. First, one urolithin-producing strain, belonging to a new species described as *Gordonibacter urolithinfaciens* sp. nov., was isolated fromthe feces of a healthy human volunteer (Selma *et al.*, 2014a) and characterized by phenotypic, biochemical, and molecular methods (Selma *et al.*, 2014b). After checking the closest relatives of this species from the Coriobacteriaceae family, only *Gordonibacter pamelae* DSM 19378T was able to convert EA to urolithins. To date, *Gordonibacter* species are able to produce some intermediate urolithins under anaerobic conditions (Uro-M5, Uro-M6, and Uro-C), suggesting that other bacteria from the gut are necessary to complete the metabolism to reach the final urolithins (Uro-A, IsoUro-A, and Uro- B). However, a correlation between *Gordonibacter* and Uro-A production capacity was shown *in vivo* and also confirmed *in vitro* (Romo-Vaquero *et al.*, 2015).

Recently, the new bacterium *Ellagibacter isourolithinifaciens*, capable of metabolizing EA into IsoUro-A, was isolated from the feces of a healthy human adult (Selma *et al.*, 2017). This bacterium, called strain CEBAS 4A, belongs to a new genus of the Eggerthellaceae family and differs from other genera of this family both phylogenetically and phenotypically (Beltrán *et al.*, 2018).

Two different approaches have been used to find potential novel probiotic strains able to produce Uro-A and/or Uro-B. Thus, 48 strains of bifidobacteria and 1070 bacterial isolates from the feces of a healthy woman were screened. *Bifidobacterium pseudocatenulatum* INIA P815 was the only strain able to produce Uro-A and B from EA (Gaya *et al.*, 2018).

In conclusion, urolithin-producing bacteria could potentially be applied either as probiotics or in the industrial production of urolithins because production of the metabolites directly by microbial fermentation is considered more convenient (sustainable and leaving less residue in the final metabolite) than using synthetic chemistry. However, further studies should be carried out to demonstrate beneficial effects and safety.

9.3.4 Interindividual Variability: Metabotypes

A large interindividual variability in the production of urolithins has been characterized associated with gut microbiota composition which could explain, at least partly, the large variability in health effects observed *in vivo* associated with ET-rich food intake. Thus, the metabolism of ETs by the gut microbiota is not the same in all individuals and three urolithin-producing groups (metabotypes) have been observed in different human intervention trials (Tomás-Barberán *et al.*, 2014): "Metabotype A" produced only Uro-A conjugates, "metabotype B" produced IsoUro-A and/ or Uro-B in addition to Uro-A, and in "metabotype 0" these urolithins were not detected. Thus, these metabotypes were related to differences in the composition of the gut microbiota in healthy humans (Romo-Vaquero *et al.*, 2015).

The distribution of urolithin-related metabotypes mainly depends on aging, as recently reported (Cortés-Martín *et al.*, 2018). Metabotype A represents around 70% and metabotype B around 20% of the healthy population from 5–30 years of age, whereas from that age onward, metabotype A decreases and metabotype B increases to reach 55% and 45%, respectively. Remarkably, the proportion of metabotype 0 (10%) remained constant from age 5 to 92 years (Cortés-Martín *et al.*, 2018).

Further studies have characterized the gut microbiota associated with these human metabotypes and a dysbiosis-prone gut microbiota composition functionality has been attributed to metabotypes B and 0, and a more "protector" microbial profile to metabotype A. This has recently been supported by the normalization of the dysbiosis and anthropometric values of metabotype A women one year postpartum compared to women with metabotype B (Cortés-Martín *et al.*, 2019). These results supported previous observations that associated metabotype B with a higher cardiovascular risk in overweight-obese subjects (González-Sarrías *et al.*, 2017a; Selma *et al.*, 2016).

9.3.5 Analysis of Urolithins

Considering the importance of these metabolites and their implications for human health, the development of sensitive and accurate methods for their identification and quantification is a critical factor. In general, reverse-phase liquid chromatography coupled to photodiode array and/or mass spectrometry (MS) detection has been used to determine urolithins in biological samples. C18 columns and mobile phases consisted of water plus formic or acetic acid (Phase A) and acetonitrile or methanol (Phase B) have been traditionally used for separation.

Due to the characteristic UV spectra of urolithins (Figure 9.3), most researchers have used UV chromatograms for identification and quantification, especially in feces and urine samples (Cerdá *et al.*, 2003, 2004; García-Muñoz *et al.*, 2014; Kosmala *et al.*, 2015; Truchado *et al.*, 2012). The UV spectra of urolithins show two major absorption bands, referred to as band I (300–380 nm) and band II (240–280 nm), and in many cases an additional band II (280–300 nm) is observed. Although different wavelengths have been used in literature (360, 330, 305, 280 nm, etc.), the most characteristic and specific absorption band was that with a maximum around 305 nm, with the exception of Uro-M7 that was better determined at 360 nm.

The need for higher sensitivity and selectivity methods to detect these metabolites in biological samples such as plasma and tissues led to the application of high-resolution mass spectrometry triple quadrupole (QQQ) (Ludwig *et al.*, 2015; Tulipani *et al.*, 2012) and quadrupole time-of-flight (QTOF) (González-Sarrías *et al.*, 2015a; Nuñez-Sanchez *et al.*, 2014). However, due to the absence of available standards for most compounds (especially conjugates), these methodologies have been little used for quantification or used inadequately carrying out a relative quantification with the most available compounds Uro-A and Uro-B, which provides inaccurate results, assuming similar detector response for conjugates and aglycones. Also, in other studies, enzymatic hydrolysis of glucuronide and sulfate conjugates has been used for quantification, although relevant information about the naturally occurring metabolites is missing in this case (Ito *et al.*, 2012; Li *et al.*, 2015; Seeram *et al.*, 2006).

To date, only one recent study has reported the most important LC, UV, and MS features of a wide variety of urolithins using synthetized and purified standards (García-Villalba *et al.*, 2016). Table 9.2 shows the most useful chromatographic (elution order, retention time) and spectroscopic characteristics (maximum absorption, exact mass, MS/MS fragments, transitions on QQQ) for the identification and quantification of urolithins. This information could be used in future work where no commercial standards are available.

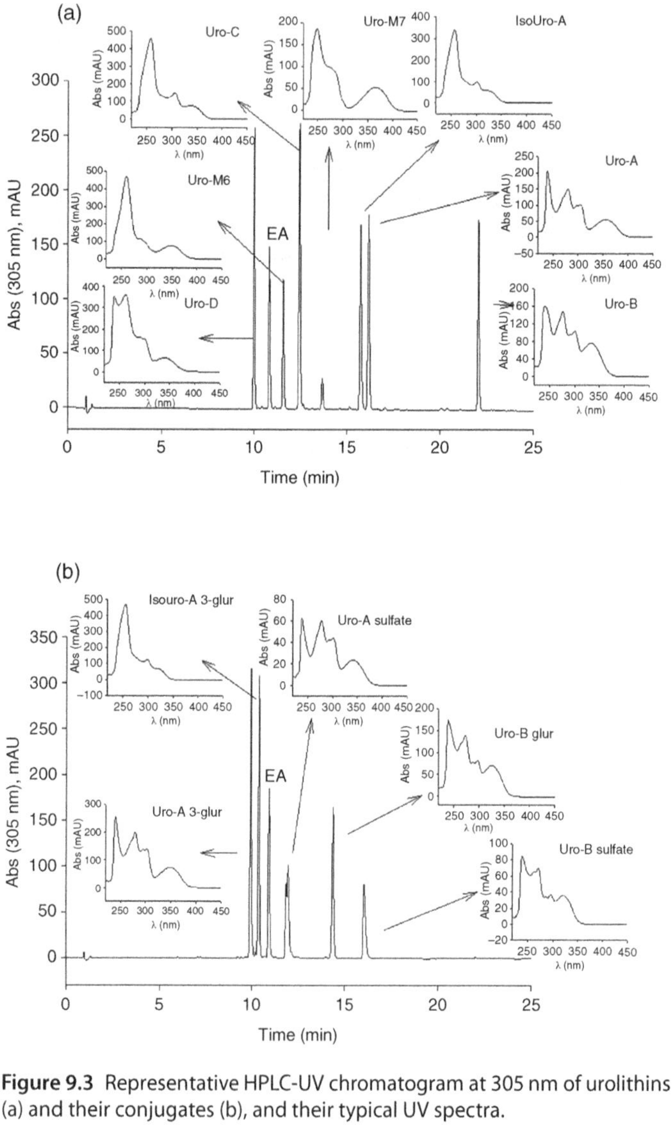

Figure 9.3 Representative HPLC-UV chromatogram at 305 nm of urolithins (a) and their conjugates (b), and their typical UV spectra.

Table 9.2 Main spectroscopic characteristics of the available urolithins (García-Villalba *et al.*, 2016)

Compounds	RRF	Accurate mass	Molecular formulae	MS/MS	QQQ transitions
Uro-D	0.95	259.0248	$C_{13}H_8O_6$	213, 242	259>213; 259>242
Uro-M6	0.56	259.0248	$C_{13}H_8O_6$	213,187,159	259>227; 259>198
Uro-C	1.20	243.0299	$C_{13}H_8O_5$	199,187,171	243>187;243>171
Uro-M7	0.13	243.0299	$C_{13}H_8O_5$	198,147	243>198;243>147
IsoUro-A	0.89	227.0350	$C_{13}H_8O_4$	171,159,183	227>171; 227>159
Uro-A	1.00	227.0350	$C_{13}H_8O_4$	198,182,171	227>198;227>182
Uro-B	0.71	211.0401	$C_{13}H_8O_3$	167,139	211>167;211>139
Uro-A 3-glur	0.85	403.0671	$C_{19}H_{16}O_{10}$	227,113	403>227;403>113
IsoUro-A 3-glur	0.79	403.0671	$C_{19}H_{16}O_{10}$	227,113	403>227;403>113
Uro-A sulfate	0.60	306.9918	$C_{13}H_8O_7S$	227	307>227;307>198
Uro-B glur	0.56	387.0722	$C_{19}H_{16}O_9$	211,113	387>211;387>113
Uro-B sulfate	0.40	290.9969	$C_{13}H_8O_6S$	211	291>211;291>167

RRF, relative response factor in UV, calculated with respect to Uro-A at 305 nm.

9.4 Significance of Ellagitannins, Ellagic Acid, and Urolithins for Human Health

There is now a consensus that a diet rich in plant-based foods is associated with reduced risk of chronic conditions such as cardiometabolic diseases, inflammation, cancer, and neurodegenerative diseases. Regarding ETs and EA-rich foods, there is much evidence that pomegranate, berries (strawberries, blackberries, and raspberries), and walnuts are effective for several chronic diseases. Nevertheless, according to clinical trials and recent metaanalyses, these protective effects remain controversial (García-Conesa *et al.*, 2018; Núñez-Sánchez *et al.*, 2015). Our current level of knowledge of ET bioavailability and metabolism makes it difficult to unambiguously determine if ETs are responsible for the observed effects. Furthermore, we cannot forget that interindividual variability in ET metabolism could also be behind the lack of effect in the response of ET-rich foods in some individuals.

Since the identification of urolithins in humans and the current knowledge that they reach significant concentrations in colon, plasma, and systemic tissues, they have received increasing attention and were suggested to be responsible for the health effects attributed to the consumption of ET-rich foods (Cerdá *et al.*, 2005; Tomás-Barberán *et al.*, 2017). In this regard, in the last decade, many *in vitro* but few *in vivo* studies have reported that urolithins can potentially display a wide range of biological functions including antioxidant, estrogenic, antiinflammatory, and anticarcinogenic properties, among others. Previous comprehensive reviews have provided a critical and updated vision of the potential role of these gut microbiota-derived metabolites in human health (Espín *et al.*, 2013; Tomás-Barberán *et al.*, 2017). As covered and criticized in these reviews, many of the *in vitro* studies have been carried out using both nonphysiological concentrations and metabolites that have not been demonstrated to occur *in vivo* in the specific cells and systemic tissues investigated.

Fortunately, although some *in vitro* studies persist in the use of irrelevant assay conditions, an increasing number use the conjugated phase II metabolites of urolithins or a physiological mixture of them that are present in the bloodstream and can reach different systemic tissues to evaluate their involvement in the beneficial effects attributed to the intake of ET-rich foods. Additionally, new *in vivo* studies in animal models and human trials have also been conducted, although their number is still limited.

Therefore, here we summarize the current evidence on the potential biological activities attributed to ETs such as antioxidant, antiinflammatory, anticarcinogenic, estrogenic, and neuroprotective effects, with potential significance against chronic diseases. Following the statement "First *in vivo* and then *in vitro*" (Ávila-Gálvez *et al.*, 2018), studies should focus on physiologically relevant *in vitro* and *in vivo* studies conducted with urolithins, to critically elucidate if urolithins are responsible for the effects attributed to ET-rich foods.

9.4.1 Antioxidant Effects

The beneficial effects associated with the consumption of ET-containing foods were initially related to their inherent antioxidant activity (AOX). This was mainly supported by *in vitro* studies pointing out the strong AOX exhibited by ET-containing foods like pomegranate (Gil *et al.*, 2000) as well as animal and human studies describing an improvement in antioxidant status (Aviram *et al.*, 2000). Based on their bioavailability and metabolism, EA, but mainly urolithins, are the candidates to exert their biological effects through the AOX instead of the ETs present in the foodstuff such as punicalagin and pedunculagin (Cerdá *et al.*, 2005).

Nevertheless, the *in vitro* studies using different analytical methods to assay the AOX of urolithins can be defined as contradictory. The oxygen radical absorbance capacity (ORAC) assay has provided evidence of the urolithins as antioxidants (Ishimoto *et al.*, 2012), whereas different methods such as DPPH or ABTS show nearly no radical scavenging activity of these metabolites (Cerdá *et al.*, 2004; Rosenblat *et al.*, 2015). In line with this, Uro-A exhibited high (Ito *et al.*, 2011; Kallio *et al.*, 2013) or low (Kallio *et al.*, 2013) AOX depending on the methodology used (i.e., ORAC assay vs cyclic voltammetry).

On the other hand, alternative methodologies such as cell models have been used to establish the relationship between "urolithins–AOX–beneficial effects." Several *in vitro* studies conducted using human leukocytes, bladder cancer, hepatocytes, and BV2 microglia cells in the presence of different urolithins (Uro-A, Uro-B, Uro-C, and Uro-D), showed a decrease of oxidative stress (i.e., reducing ROS and MDA formation) induced by strong oxidants (Bialonska *et al.*, 2009; Lee *et al.*, 2019; Olennikov *et al.*, 2015; Piwowarski *et al.*, 2014a; Qiu *et al.*, 2013). Additional mechanisms of oxidative protection exerted by urolithins are related to the modulation of endogenous defensive systems, including NADPH oxidase subunit expression reduction, hemeoxygenase-1 upregulation, as well as superoxide dismutase (SOD) and glutathione peroxidase (GPx) (Ito *et al.*, 2011; Lee *et al.*, 2019; Qiu *et al.*, 2013).

However, all these studies have not considered the bioavailability of urolithins nor their tissue disposition as free urolithins do not reach systemic tissues at relevant concentrations to exert antioxidant effects. Indeed, to the best of our knowledge, there is no study evaluating the antioxidant effect of physiologically relevant urolithin phase II conjugates (glucuronides and sulfates), with the exception of a slight protective effect after the pretreatment, but not by co-treatment, of Uro-A and IsoUro-A glucuronides on H_2O_2-induced oxidative stress in human neuroblastoma SH-SY5Y cells (González-Sarrías *et al.*, 2017b).

Regarding *in vivo* studies, the urolithins' AOX is further supported by animal studies, which found associations between the plasma level of Uro-A and the maximum ORAC activity in plasma (Ishimoto *et al.*, 2011) and the protective effects against cisplatin-induced oxidative damage in mouse kidneys by increasing the SOD activity and increasing the ratio of reduced/oxidized GSH (Jing *et al.*, 2019) after oral direct and intraperitoneal administration of Uro-A, respectively. However, the AOX in plasma and colon mucosa was unaffected (compared to animals fed a normal diet) in a DSS-intestinal inflammation animal model after oral administration of Uro-A (Larrosa *et al.*, 2010).

In conclusion, to date, according to the contradictory *in vitro* studies and current *in vivo* findings, the evidence is still weak and additional research is needed to endorse urolithins as antioxidant compounds.

9.4.2 Antiinflammatory Properties

An important number of studies support the antiinflammatory effects of ET-rich foods. A PubMed search for "pomegranate and inflammation" or "walnuts and inflammation" will give more than 100 and 200 hits, respectively, describing the beneficial effects associated with the consumption of these food sources. In the past few years, identification of the active components, as well as the mechanistic pathways regulated, has been one of the main goals of the research groups involved in the study of ETs.

In 2010, Uro-A was placed in the spotlight as the main EA-derived microbial metabolite responsible for the antiinflammatory effects in an *in vivo* model of intestinal inflammation (Larrosa *et al.*, 2010). The *in vivo* benefits associated with Uro-A include preservation of the colonic architecture, prebiotic effect, gene expression modulation (related to colon cancer), and downregulation of iNOS, COX-2, and PTGES (Larrosa *et al.*, 2010). Additionally, the mechanism of antiinflammatory action for physiologically relevant concentrations of Uro-A and

Uro-B, as well as a representative mixture of them both, was confirmed *in vitro* using an inflammatory colonic cell model (Giménez-Bastida *et al.*, 2012a; González-Sarrías *et al.*, 2010b). An interesting *in vivo* study has recently provided new insights into the protective role of Uro-A in an intestinal inflammation animal model. The authors demonstrated that Uro-A protected the intestinal barrier of mice through activation of the aryl hydrocarbon receptor (AhR)-nuclear factor erythroid 2-related factor 2 (NRF2)-dependent pathways, which in turn upregulated the expression of tight junction proteins (Singh *et al.*, 2019).

Therefore, regarding colonic environment, strong evidence on the antiinflammatory properties of urolithins has been reported from the concordance between *in vivo* and *in vitro* results that assayed conditions physiologically relevant underlying molecular mechanisms involved. However, to date, there are no human studies to confirm this potential effect.

On the other hand, the interesting effects described in the colonic environment raised the curiosity of researchers to enlarge our knowledge of the antiinflammatory effects of urolithins in systemic tissues. Thus, several antiinflammatory *in vitro* studies have been performed in different cell models from systemic tissues as well as immune system cells. However, the authors of most of these studies focused mainly on investigation using free urolithins, mainly Uro-A, but not their relevant conjugates, and even using nonphysiological concentrations (ranging from 1 to 60 μM), on circulating cells of the immune system (Boakye *et al.*, 2018; Kiss *et al.*, 2012; Komatsu *et al.*, 2018; Piwowarski *et al.*, 2014a, b, 2015) as well as human umbilical vein endothelial cells (Mele *et al.*, 2016).

Therefore, despite the antiinflammatory effects and the molecular mechanisms reported for nonconjugated urolithins, it is difficult to draw conclusions from these studies as the metabolites/concentrations tested and cell models used were unrealistic. Regarding this, to date, only two studies have reported the antiinflammatory effect of the circulating conjugated metabolites, but lower than their aglycones, against cardiomyocytes for Uro-B glucuronide (Sala *et al.*, 2015), and against human aortic endothelial cells for Uro-A and Uro-B glucuronides (Giménez-Bastida *et al.*, 2012b). Regarding the *in vivo* systemic antiinflammatory effect, only one recent study conducted in mice fed Uro-A reported the decrease of high fat-induced adipocyte hypertrophy, inflammatory markers, and adipose tissue macrophages recruitment in the adipose tissue, as well as the increase of M2 macrophage polarization and augmenting mitochondrial function (Toney and Chung, 2019).

In conclusion, evidence remains inconclusive on whether these molecules are responsible for the antiinflammatory benefits associated with the consumption of ET-rich foods in systemic tissues. More *in vitro* studies conducted with phaseII urolithin metabolites as well as *in vivo* studies are needed. However, notwithstanding the numerous references that suggest avoiding the use of unconjugated metabolites and systemic cells (Ávila-Gálvez *et al.*, 2018; Tomás-Barberán *et al.*, 2017), some authors hold the "deconjugation theory" (Kunihiro *et al.*, 2019; Shimoi *et al.*, 2001; Terao *et al.*, 2011) as the main argument to keep performing these investigations. In this regard, a recent *in vivo* study has reported the tissue deconjugation of Uro-A glucuronide to Uro-A under inflammatory conditions, thus bringing new evidence to support this theory (Ávila-Gálvez *et al.*, 2019b). However, this hypothesis should be considered with caution in the design of *in vitro* studies, with special emphasis on the concentrations assayed, since more investigation is needed in order to shed light on this intriguing process.

9.4.3 Anticarcinogenic Effects

There is epidemiological evidence supporting the potential benefits of polyphenols, including ETs, in decreasing the risk of different types of cancer (mainly colorectal), although the link between phenolics consumption and anticarcinogenic properties is currently inconclusive (Núñez-Sánchez *et al.*, 2015). Thus, the interest in ETs for chemoprevention and cancer therapy is stated in different reviews (Heber 2008; Ismail *et al.*, 2016). Early investigations reported the *in vitro* anticarcinogenic effect via apoptosis induction and cell cycle arrest of dietary ETs and EA against many types of cancer cell lines, although according to their bioavailability, only those studies conducted against colon cancer cells should be considered (Larrosa *et al.*, 2006). In the last decade, urolithins have received increasing attention as reflected by the increasing number of publications investigating their potential anticancer effects.

Initially, the *in vitro* anticancer activity of urolithins was mainly investigated against different colon cancer cells due to their presence at higher concentrations (micromolar range) in colon tissues and surrounding environment, reported in several clinical trials after pomegranate extract intake (Núñez-Sánchez *et al.*, 2014; Romo-Vaquero *et al.*, 2015). In 2009, urolithins were identified for the first time to exert *in vitro* antiproliferative activities, particularly Uro-A, and B and representative mixtures with EA against colon cancer cells (Gonzalez-Sarrias *et al.*, 2009a). Subsequent *in vitro* studies corroborated the anticarcinogenic effect exerted by urolithins, including others such as Uro-C, Uro-D and IsoUro-A, as well as mixtures mimicking the profile of metabolites found in human colon tissues, on a wide range of colon cancer cells. These *in vitro* studies delved into the mechanisms of action involved, indicating that urolithins can modulate the expression of genes and proteins that affect the inhibition of cell proliferation by cell cycle arrest and induce apoptosis, as well as the modulation of key cancer markers and signaling events associated with cancer development such as p21 (Cho *et al.*, 2015; González-Sarrías *et al.*, 2009b, 2014, 2016, 2017c; Kasimsetty *et al.*, 2010; Ramírez de Molina *et al.*, 2015). Additionally, Uro-A was also reported to potentiate the anticancer effects of 5-fluorouracil chemotherapy on human colon cancer cells (González-Sarrías *et al.*, 2015b).

Recently, other anticarcinogenic mechanisms have been reported for Uro-A on colon cancer cells exerting autophagy and inhibition of metastasis (Zhao *et al.*, 2018) as well as induction of a senescence-like phenotype upon long-term exposure via the p53/TIGAR axis and showing synergism with the chemotherapeutic oxaliplatin (Norden and Heiss, 2019). On the other hand, physiologically relevant mixtures of urolithins and ellagic acid were reported to inhibit phenotypic and molecular colon cancer stem cell features, suggesting a potential role for the control of cancer metastasis and modulation of chemotherapy resistance (Núñez-Sánchez *et al.*, 2016).

Therefore, these previous *in vitro* studies have suggested strong evidence supported by the design of *in vitro* studies (regarding exposure time and concentrations) that mimic *in vivo* conditions at the intestinal level and the invaluable information about the molecular mechanisms of action of urolithins. However, considering the big gap between *in vitro* and *in vivo* studies, to date, there is no *in vivo* study that corroborates this potential. In this regard, only one study reported the modulation of a number of markers (p53, Rb1, Akt, BclXL, etc.) in the colonic mucosa of rodents fed Uro-A, involved in signaling pathways, transcription regulation, apoptosis, cell proliferation, and cell cycle implicated in cancer development (Larrosa *et al.*, 2010).

On the other hand, the study of the anticancer activity of urolithins has gone beyond the *in vitro* intestinal cell models. Numerous *in vitro* studies have reported anticarcinogenic properties for urolithins against different human cancer cell lines from systemic tissues, including bladder, liver, prostate, and kidney (studies reviewed in Tomás-Barberán *et al.*, 2017). However, almost all these studies were conducted with the free form of urolithins rather than their *in vivo* relevant phase II conjugates that reach the systemic tissues, and therefore, these studies have provided interesting mechanistic information but poor *in vivo* extrapolation.

In this regard, similar to other phenolic compounds, phase II conjugates of urolithins (glucuronides and sulfates) have been reported to be less anticarcinogenic than their aglycone counterparts in different *in vitro* models of cancer such as colon (González-Sarrías *et al.*, 2014, 2017c), prostate (Giorgio *et al.*, 2015), and breast (Ávila-Gálvez *et al.*, 2018). Therefore, whether the phase II conjugated metabolites of urolithins can maintain certain anticarcinogenic activity remains elusive. With respect to this fact, these sulfated and glucuronidated metabolites could exert anticarcinogenic effects upon long-term exposure involving other mechanisms of action such as senescence of cancer cells, as well as possible *in situ* deconjugation processes that could convert the conjugates into their more bioactive free forms, as recently reported for Uro-A mediated by an inflammatory status (Ávila-Gálvez *et al.*, 2019b).

Overall, to date, the anticarcinogenic effect attributed to ETs mediated by the action of the urolithins remains unclear, mainly against noncolorectal types of cancer. Future *in vitro* studies need to be meticulously approached (considering cell model, molecular form, concentrations, deconjugation) in order to understand the effects and molecular mechanisms by which urolithins act against systemic cancer cells. This information is indispensable in the design of future clinical trials to prove their therapeutic effect.

9.4.4 Neuroprotective Effects

The improvement in symptoms in neurodegenerative diseases such as Parkinson's or Alzheimer's in rodent models consuming a pomegranate-enriched diet (Ahmed *et al.*, 2014; Essa *et al.*, 2015; Rojanathammanee *et al.*, 2013; Subash *et al.*, 2014, 2015) has encouraged the interest of food scientists in the identification of the molecule(s) responsible for the effects described. Nevertheless, considering those unrealistic *in vitro* studies conducted with the direct exposure of ET-rich products (plant or food extracts) as well as complex polyphenols or derivatives with a limited physiological relevance, the potential neuroprotective effect of ETs remains unclear. Fortunately, in recent years, the question "Is the brain the landing place of ET-derived metabolites?" has been addressed by different methodologies. The ability of urolithins to reach the brain of rats after intravascular administration has been reported (Gasperotti *et al.*, 2015). Seeram and co-workers predicted by computational studies the ability of urolithins to cross the blood–brain barrier (Yuan *et al.*, 2016). On the other hand, EA has also been detected in the rat brain after oral administration (Yan *et al.*, 2014). However, there is still a need to design improved *in vivo* studies that better reflect the bioavailability of ETs to identify if urolithin phase II conjugates can really pass through the blood–brain barrier.

Regarding neuroprotective effects, a study reported a positive effect of Uro-A, Uro-B, and some of their methyl derivatives in ameliorating $A\beta_{1\text{-}42}$-amyloid fibrillation *in vitro* and in an *in vivo* model of amyloid β_{1-42} induced neurotoxicity and paralysis (*Caenorhabditis elegans*) (Yuan *et al.*, 2016). Recent studies have described new neuroprotective mechanisms of action for Uro-A, through activation of sirtuin-1 (SIRT-1) and induction of autophagy (Velagapudi *et al.*, 2019), and for Uro-B (at supraphysiological concentrations, 30–100 μM) by inhibition of NO and proinflammatory cytokine biosynthesis, increase of

IL-10 level, blockage of NF-κB activation, suppression of MAPK (ERK and JNK) and Akt phosphorylation, as well as enhanced AMPK phosphorylation (Lee *et al.*, 2019) in inflamed BVL2 microglia cells. Additionally, Uro-B was able to attenuate the *in vivo* microglia activation induced by LPS injection in mouse brains (Lee *et al.*, 2019).

Despite the interesting results described in these studies, evidence is still weak because the authors focused mainly on nonconjugated urolithins, which in the bloodstream are not detected or achieve much lower concentrations than those used in *in vitro* assays. Exploration of the effect of the urolithins' conjugated metabolites is essential to prove which molecule(s) could be responsible for the beneficial effects related to the consumption of ET-rich foods. Recent elegant examples of this approach have shown, for the first time, a protective neuronal effect of urolithins and their conjugates (glucuronides, sulfates, and methylated derivatives) in SH-SY5Y cells via apoptosis prevention (through reduction of ROS and caspase-3 activation) and elevated REDOX activity (Gonzalez-Sarrias *et al.*, 2017b), as well as in LPS-stimulated BV-2 through modulation of NO, IL-6, TNF-α, PGE_2 modulation, and ROS formation inhibition (DaSilva *et al.*, 2019).

In conclusion, although new approaches provide further evidence regarding the neuroprotective activity of physiologically relevant ET-derived metabolites, the evidence is still weak to confirm their contribution in the protection of neurodegenerative diseases. Further studies, mainly *in vivo*, are needed to investigate the detailed molecular mechanisms underlying the neuroprotective effects of these metabolites.

9.4.5 Estrogenic Modulation

In recent decades, phytoestrogens have captured research and clinical attention due to their effectiveness in the prevention and treatment of different diseases related to hormones such as cancer, osteoporosis, cardiovascular diseases, and menopausal symptoms, among others (Sirotkin and Harrath, 2014). Since the identification of urolithins, the structural parallelism between urolithins A and B and estradiol made these ET-derived metabolites excellent candidates to act as phytoestrogens. A good example of this approach comes from a study that evidenced for the first time that Uro-A and Uro-B can interact with the ERα and ERβ exhibiting estrogenic (absence of estradiol) and antiestrogenic (presence of estradiol) effects (Larrosa *et al.*, 2006). Further studies provided additional evidence of the interaction of Uro-A with estrogenic receptors, mainly ERα, in cellular models (Zhang *et al.*, 2016) and *in silico* approaches (Dellafiora *et al.*, 2013). The *in silico* data also suggested the affinity of other urolithins such as Uro-C and IsoUro-A, but not for conjugated urolithins (Dellafiora *et al.*, 2013). Another study reported the capacity of Uro-B, but not its sulfated conjugate or Uro-A, to inhibit aromatase activity and diminish testosterone-induced cell proliferation in MCF-7 breast cancer cells (Adams *et al.*, 2010).

Despite the initial enthusiasm, these studies did not evaluate the potential estrogenic effect of conjugated urolithins and there is no *in vivo* study to support this effect. A recent trial gave a step forward in this field and described, for the first time, the occurrence of urolithin conjugates (mainly glucuronides and sulfates), rather than their nonconjugated counterparts, in normal and cancerous mammary tissues at concentrations of 4.8–29.4 pmol/g (Ávila-Gálvez *et al.*, 2019b). However, a recent *in vitro* study has reported for the first time that the glucuronide and sulfate conjugated metabolites (even at supraphysiological concentrations, 10–50 μM) lacked the *in vitro* estrogenic and antiestrogenic activities on ER+ breast cancer cells, rather than urolithin aglycones (Ávila-Gálvez *et al.*, 2018).

In conclusion, the lack of *in vivo* studies that provide information together with the limited effect of conjugated urolithins make it necessary to continue with this research field to confirm or reverse the lack of evidence.

9.4.6 Urolithins, Clinical Trials, and Interindividual Variability–Health Relationship

The number of human intervention studies on the benefits associated with the consumption of ET-containing products is limited and mainly regarding pomegranate and walnuts. However, the link between health effects and urolithins remains elusive. In recent years, the growing knowledge of urolithins as potential bioactive metabolites has prompted *in vivo* studies to demonstrate unequivocally their involvement in beneficial effects. However, among other factors, the difficulty and high cost of producing and administering urolithins to humans have limited their achievement. Thus, to date, only a few trials have tried to correlate the urolithins with possible beneficial effects in humans.

The first trial was conducted to prove the anticarcinogenic effects of urolithins in the normal and malignant colon mucosa of 35 colorectal cancer patients who consumed a pomegranate extract. However, despite the occurrence of high concentrations of urolithins in the colonic mucosa (Núñez-Sánchez *et al.*, 2014), no correlation was observed with a moderate modulation of several colorectal cancer genes and microRNAs related to colon cancer, evidencing the difficulty of confirming *in vivo* the *in vitro* results (Núñez-Sánchez *et al.*, 2015, 2017). Later, in 49 overweight-obese subjects, no correlation was found between urolithin occurrence in plasma, urine or feces and effects observed on serobiochemical variables, mainly blood lipids, after consuming a pomegranate extract (González-Sarrías *et al.*, 2017a). Overall, the lack of clear association is plausible due to the turnover of microbial metabolites and their occurrence in the bloodstream and other biological reservoirs. In contrast, and surprinsingly, in a recent trial conducted in 10 healthy males who consumed red raspberries, the occurrence in plasma of EA (at 2 h), and urolithin A glucuronide and urolithin A sulfate (at 24 h) was correlated with improvements in flow-mediated dilation, suggesting that ETs might be responsible for the observed effects (Istas *et al.*, 2018). Taking into account the low number of subjects, the high interindividual variability and the somewhat unpredictable turnover of microbial metabolites, this association seems to be casual rather than causal.

Additionally, until recent years, the safety of urolithins in humans had not been evaluated. Currently, the safety of direct oral exposure to synthetic Uro-A in rat models (Heilman *et al.*, 2017) and in humans (Andreux *et al.*, 2019) has been proved. In this first-in-human clinical trial, Uro-A at doses of 500 mg and 1000 mg was administered over a four-week period to healthy sedentary elderly individuals, reporting a beneficial impact on mitochondrial and cellular health (Andreux *et al.*, 2019). More recently, according to these investigations, the US Food and Drug Administration (FDA) has designated Uro-A as GRAS (generally recognized as safe) for use as an ingredient in food products such as protein shakes, meal replacement drinks, instant oatmeal, nutrition bars, and yogurt drinks. Therefore, this opens a new door to investigate the role of urolithins using new strategies in clinical trials.

Finally, it is important to consider the potential relationship in human studies between interindividual variability in urolithin production and the observed health effects as it could explain the controversial benefits seen after ET-rich food intake. As stated before, a potential link between urolithin metabotypes and gut dysbiosis has been hypothesized (Tomás-Barberán *et al.*, 2014). Therefore, stratification or clustering according to urolithin metabotype should be considered in intervention trials to identify and understand the differential response to ETs, and therefore, to support associations between intake and health benefits. In this regard, to date, one trial conducted with 49 healthy overweight-obese individuals who consumed a pomegranate extract reported a significant decrease in a panel of serum cardiovascular disease risk markers, but only in those individuals with urolithin metabotype B (which, in turn, also presented a higher cardiovascular risk at baseline) rather than those with urolithin metabotype A (González-Sarrías *et al.*, 2016). In conclusion, more human studies, including stratification based on relevant gut microbiota metabolites, are needed to increase our understanding of the exact role of different urolithins in the beneficial effects attributed to ETs. However, we cannot overlook that perhaps the beneficial effects could be mediated by the specific gut microbial community, as recent reports suggest (Cortés-Martín *et al.*, 2019; Romo-Vaquero *et al.*, 2019), and considering urolithins only as biomarkers, or perhaps by a synergic or additive effect (Espín *et al.*, 2017; Tomás-Barberán *et al.*, 2018). Therefore, this recent hypothesis should be explored in more depth in the coming years.

9.5 Conclusion

In this chapter, we have presented an overview of the current knowledge on ETs and EA, mainly focusing on their *in vivo* derived metabolites, urolithins, and the major challenges facing this research. There is growing evidence that urolithins can exert both local effects in the colon as free forms as well as in systemic tissues as phase II metabolism conjugates. However, despite many preclinical studies reporting the beneficial effects of urolithins including antioxidant, estrogenic, antiinflammatory and anticarcinogenic properties, clinical evidence of most of these effects remains unclear or inconclusive and, therefore, further studies, mainly *in vivo*, are needed. Interindividual variability in urolithin production could explain the controversial health benefits of ET-rich foods, and therefore, the stratification of individuals by urolithin metabotypes should be considered in intervention trials. Nevertheless, the improved design of new *in vitro* studies, the recent approval of Uro-A for use in human trials, and the recent identification of microorganisms able to transform ETs and EA into urolithins open exciting possibilities for the future research of ETs to demonstrate unequivocally their involvement in health beneficial effects.

Acknowledgments

This research was supported by the Projects Fundación Séneca de la Región de Murcia, Ayudas a Grupos de Excelencia 19900/GERM/15, AGL-2015-73106-EXP, AGL2015-64124-R, AGL2015-73744-JIN (MINECO, Spain), 201870E014, 201770E081 and 201870I028 (CSIC, Spain).

References

Adams, L.S., Zhang, Y., Seeram, N.P., Heber, D., and Chen, S. (2010) Pomegranate ellagitannin-derived compounds exhibit antiproliferative and antiaromatase activity in breast cancer cells in vitro. *Cancer Prevention Research*, **3**, 108–13.

Ahmed, A.H., Subaiea, G.M., Eid, A., Li, L., Seeram, N.P., and Zawia, N.H. (2014) Pomegranate extract modulates processing of amyloid-β precursor protein in an aged Alzheimer's disease animal model. *Current Alzheimer Research*, **11**, 834–43.

Anderson, K.J., Teuber, S.S., Gobeille, A., Cremin, P., Waterhouse, A.L., and Steinberg, F.M. (2001) Walnut polyphenolics inhibit in

vitro human plasma and LDL oxidation. *Journal of Nutrition*, **131**, 2837–42.
Andreux, P.A., Blanco-Bose, W., Ryu, D., *et al.* (2019) The mitophagy activator urolithin A is safe and induces a molecular signature of improved mitochondrial and cellular health in humans. *Nature Metabolism*, **1**, 595–603.
Ávila-Gálvez, M.A., González-Sarrías, A., and Espín, J.C. (2018) In vitro research on dietary polyphenols and health: a call of caution and a guide on how to proceed. *Journal of Agricultural and Food Chemistry*, **66**, 7857–8.
Ávila-Gálvez, M.A., García-Villalba, R., Martínez-Díaz, F., *et al.* (2019a) Metabolic profiling of dietary polyphenols and methylxanthines in normal and malignant mammary tissues from breast cancer patients. *Molecular Nutrition & Food Research*, **63**, e1801239.
Ávila-Gálvez, M.A., Giménez-Bastida, J.A., González-Sarrías, A., and Espín, J.C. (2019b) Tissue deconjugation of urolithin A glucuronide to free urolithin A in systemic inflammation. *Food & Function*, **10**, 3135–41.
Aviram, M., Dornfeld, L., Rosenblat, M., *et al.* (2000) Pomegranate juice consumption reduces oxidative stress, atherogenic modifications to LDL, and platelet aggregation: studies in humans and in atherosclerotic apolipoprotein E-deficient mice. *American Journal of Clinical Nutrition*, **71**, 1062–76.
Beltrán, D., Romo-Vaquero, M., Espín, J.C., Tomás-Barberán, F.A., and Selma, M.V. (2018) Ellagibacter isourolithinifaciens gen. nov., sp. nov., a new member of the family Eggerthellaceae, isolated from human gut. *International Journal of Systematic and Evolutionary Microbiology*, **68**, 1707–12.
Bialonska, D., Kasimsetty, S.G., Khan, S.I., and Ferreira, D. (2009) Urolithins, intestinal microbial metabolites of pomegranate ellagitannins, exhibit potent antioxidant activity in a cell-based assay. *Journal of Agricultural and Food Chemistry*, **57**, 10181–6.
Bialonska, D., Ramnani, P., Kasimsetty, S.G., Muntha, K.R., Gibson, G.R., and Ferreira, D. (2010) The influence of pomegranate by-product and punicalagins on selected groups of human intestinal microbiota. *International Journal of Food Microbiology*, **140**, 175–82.
Boakye, Y.D., Groyer, L., and Heiss, E.H. (2018) An increased autophagic flux contributes to the anti-inflammatory potential

of urolithin A in macrophages. *Biochimica et Biophysica Acta. General Subjects*, **1862**, 61–70.

Byerley, L.O., Samuelson, D., Blanchard, E., *et al.* (2017) Changes in the gut microbial communities following addition of walnuts to the diet. *Journal of Nutritional Biochemistry*, **48**, 94–102.

Cerdá, B., Llorach, R., Cerón, J.J., Espín, J.C., and Tomás-Barberán, F.A. (2003) Evaluation of the bioavailability and metabolism in the rat of punicalagin, an antioxidant polyphenols from pomegranate juice. *European Journal of Nutrition*, **42**, 18–28.

Cerdá, B., Espín, J.C., Parra, S., Martínez, P., and Tomás-Barberán, F.A. (2004) The potent in vitro antioxidant ellagitannins from pomegranate juice are metabolised into bioavailable but por antioxidant hydroxy-6H-dibenzopyran-6-one derivatives by the colonic microflora of healthy humans. *European Journal of Nutrition*, **43**, 205–20.

Cerdá, B., Tomás-Barberán, F.A., and Espín, J.C. (2005) Metabolism of antioxidant and chemopreventive ellagitannins from strawberries, raspberries, walnuts, and oak-aged wine in humans: identification of biomarkers and individual variability. *Journal of Agricultural and Food Chemistry*, **53**, 227–35.

Cerdá, B., Soto, M.C., Albaladejo, M.D., *et al.* (2006) Pomegranate juice supplementation in COPD: a 5-week randomised, double blind, placebo-controlled trial. *European Journal of Clinical Nutrition*, **60**, 245–53.

Cho, H., Jung, H., Lee, H., Yi, H. C., Kwak, H.K., and Hwang, K.T. (2015) Chemopreventive activity of ellagitannins and their derivatives from black raspberry seeds on HT-29 colon cancer cells. *Food & Function*, **6**, 1675–83.

Clifford, M.N. and Scalbert, A. (2000) Ellagitannins – nature, occurrence and dietary burden. *Journal of the Science of Food and Agricultural*, **80**, 1118–25.

Comandini, P., Lerma-García, M.J., Simó-Alfonso, E.F., and Toschi, T.G. (2014) Tannin analysis of chestnut bark samples (Castanea sativa Mill.) by HPLC-DAD-MS. *Food Chemistry*, **157**, 290–5.

Cortés-Martín, A., García-Villalba, R., González-Sarrías, A., *et al.* (2018) The gut microbiota urolithin metabotypes revisited: the human metabolism of ellagic acid is mainly determined by aging. *Food & Function*, **9**, 4100.

Cortés-Martín, A., Romo-Vaquero, M., García-Mantrana, I., *et al.* (2019) Urolithin metabotypes can anticipate the different

restoration of the gut microbiota and anthropometric profiles during the first year postpartum. *Nutrients,* **11**, e2079.
Crozier, A., Jaganath, I.B., and Clifford, M.N. (2009) Dietary phenolics: chemistry, bioavailability and effects on health. *Natural Product Reports,* **26**, 1001–43.
DaSilva, N.A., Nahar, P.P., Ma, H., *et al.* (2019) Pomegranate ellagitannin-gut microbial-derived metabolites, urolithins, inhibit neuroinflammation in vitro. *Nutritional Neuroscience,* **22**, 185–95.
Dellafiora, L., Mena, P., Cozzini, P., Brighenti, F., and del Rio, D. (2013) Modelling the possible bioactivity of ellagitannin-derived metabolites. In silico tools to evaluate their potential xenoestrogenic behavior. *Food & Function,* **4**, 1442–51.
Espín, J.C., González-Barrio, R., Cerdá, B., López-Bote, C., Rey, A.I., and Tomás-Barberán, F.A. (2007) Iberian pig as a model to clarify obscure points in the bioavailability and metabolism of ellagitannins in humans. *Journal of Agricultural and Food Chemistry,* **55**, 10476–85.
Espín, J.C., Larrosa, M., García-Conesa, M.T., and Tomás-Barberán, F.A. (2013) Biological significance of urolithins, the gut microbial ellagic acid-derived metabolites: the evidence so far. *Evidence-Based Complementary and Alternative Medicine,* **2013**, 270418.
Espín, J.C., González-Sarrías, A., and Tomás-Barberán, F.A. (2017) The gut microbiota: a key factor in the therapeutic effects of (poly)phenols. *Biochemical Pharmacology,* **139**, 82–93.
Essa, M.M., Subash, S., Akbar, M., Al-Adawi, S., and Guillemin, G.J. (2015) Long-term dietary supplementation of pomegranates, figs and dates alleviate neuroinflammation in a transgenic mouse model of Alzheimer's disease. *PLoS One,* **10**, e0120964.
Fischer, U.A., Carle, R., and Kammerer, D.R. (2011) Identification and quantification of phenolic compounds from pomegranate (Punica granatum L.) peel, mesocarp, aril and differently produced juices by HPLC-DAD-ESI/MSn. *Food Chemistry,* **127**, 807–21.
García-Conesa, M.T., Chambers, K., Combet, E., *et al.* (2018) Meta-analysis of the effects of foods and derived products containing ellagitannins and anthocyanins on cardiometabolic biomarkers: analysis of factors influencing variability of the

individual responses. *International Journal of Molecular Sciences*, **19**, 694.

García-Muñoz, C., Hernández, L., Pérez, A., and Vaillant, F. (2014) Diversity of urinary excretion patterns of main ellagitannins' colonic metabolites after ingestion of tropical highland blackberry (Rubus adenotrichus) juice. *Food Research International*, **55**, 161–9.

García-Villalba, R., Beltrán, D., Espín, J.C, Selma, M.V, and Tomás-Barberán, F.A. (2013) Time course production of urolithins from ellagic acid by human gut microbiota. *Journal of Agricultural and Food Chemistry*, **61**, 8787–806.

García-Villalba, R., Espín, J.C., Aaby, K., *et al.* (2015) Validated method for the characterization and quantification of extractable and nonextractable ellagitannins after acid hydrolysis in pomegranate fruits, juices and extracts. *Journal of Agricultural and Food Chemistry*, **63**, 6555–66.

García-Villalba, R., Espín, J.C., and Tomás-Barberán, F.A. (2016) Chromatographic and spectroscopic characterization of urolithins for their determination in biological samples after the intake of foods containing ellagitannins and ellagic acid. *Journal of Chromatography A*, **1428**, 162–75.

García-Villalba, R., Vissenaekens, H., Pitart, J., *et al.* (2017) Gastrointestinal simulation model TWIN-SHIME shows differences between human urolithin-metabotypes in gut microbiota composition, pomegranate polyphenol metabolism, and transport along the intestinal tract. *Journal of Agricultural and Food Chemistry*, **65**, 5480–93.

Gasperotti, M., Passamonti, S., Tramer, F., *et al.* (2015) Fate of microbial metabolites of dietary polyphenols in rats: is the brain their target destination? *ACS Chemical Neuroscience*, **6**, 1341–52.

Gaya, P., Peirotén, A., Medina, M., Álvarez, I, and Landete, J.M. (2018) *Bifidobacterium pseudocatenulatum* INIA P815: the first bacterium able to produce urolithins A and B from ellagic acid. *Journal of Functional Foods*, **45**, 95–9.

Gil, M. I., Tomás-Barberán, F.A., Hess-Pierce, B., Holcroft, D.M., and Kader, A.A. (2000) Antioxidant activity of pomegranate juice and its relationship with phenolic composition and processing. *Journal of Agricultural and Food Chemistry*, **48**, 4581–9.

Giménez-Bastida, J.A., Larrosa, M., González-Sarrías, A., Tomás-Barberán, F.A., Espín, J.C., and García-Conesa, M.T. (2012a) Intestinal ellagitannin metabolites ameliorate cytokine-induced inflammation and associated molecular markers in human colon fibroblasts. *Journal of Agricultural and Food Chemistry*, **60**, 8866–76.

Giménez-Bastida, J.A., González-Sarrías, A., Larrosa, M., Tomás-Barberán, F.A., Espín, J.C., and García-Conesa, M.T. (2012b) Ellagitannin metabolites, urolithin A glucuronide and its aglycone urolithin A, ameliorate TNF-alpha-induced inflammation and associated molecular markers in human aortic endothelial cells. *Molecular Nutrition & Food Research*, **56**, 784–96.

Giorgio, C., Mena, P., del Rio, D., *et al.* (2015) The ellagitannin colonic metabolite urolithin D selectively inhibits EphA2 phosphorylation in prostate cancer cells. *Molecular Nutrition & Food Research*, **59**, 2155–67.

Glabasnia, A. and Hofmann, T. (2006) Sensory directed identification of taste-active ellagitannins in American (Quercus alba L.) and European oak wood (Quercus robur L.) and quantitative analysis in bourbon whiskey and oak-matured red wines. *Journal of Agricultural and Food Chemistry*, **54**, 3380–90.

Gonçalves, B., Borges, O., Soares-Costa, H., Bennett, R., Santos, M., and Silva, A.P. (2010) Metabolite composition of chestnut (Castanea sativa Mill.) upon cooking: proximate analysis, fibre, organic acids and phenolics. *Food Chemistry*, **122**, 154–60.

González-Barrio, R., Truchado, P., Ito, H., Espín, J.C., and Tomás-Barberán, F.A. (2011) UV and MS identification of urolithins and nasutins, the bioavailable metabolites of ellagitannins and ellagic acid in different mammals. *Journal of Agricultural and Food Chemistry*, **59**, 1152–62.

González-Barrio, R., Truchado, P., García-Villalba, R., *et al.* (2012) Metabolism of oak leaf ellagitannins and urolithin production in beef cattle. *Journal of Agricultural and Food Chemistry*, **60**, 3068–77.

González-Sarrías, A., Espín, J.C., Tomás-Barberán, F.A., and García-Conesa, M.T. (2009a) Gene expression, cell cycle arrest and MAPK signalling regulation in Caco-2 cells exposed to ellagic acid and its metabolites, urolithins. *Molecular Nutrition & Food Research*, **53**, 686–98.

González-Sarrías, A., Azorín-Ortuño, M., Yáñez-Gascón, M.J., Tomás-Barberán, F.A., García-Conesa, M.T., and Espín, J.C. (2009b) Dissimilar in vitro and in vivo effects of ellagic acid and its microbiota-derived metabolites, urolithins, on the cytochrome P450 1A1. *Journal of Agricultural and Food Chemistry*, **57**, 5623–32.

González-Sarrías, A., Giménez-Bastida, J.A., García-Conesa, M.T., *et al.* (2010a) Occurrence of urolithins, gut microbiota ellagic acid metabolites and proliferation markers expression response in the human prostate gland upon consumption of walnuts and pomegranate juice. *Molecular Nutrition & Food Research*, **54**, 311–22.

González-Sarrías, A., Larrosa, M., Tomás-Barberán, F.A., Dolara, P., and Espín, J.C. (2010b) NF-kappaB-dependent anti-inflammatory activity of urolithins, gut microbiota ellagic acid-derived metabolites, in human colonic fibroblasts. *British Journal of Nutrition*, **104**, 503–12.

González-Sarrías, A., Giménez-Bastida, J.A., Núñez-Sánchez, M.A., *et al.* (2014) Phase-II metabolism limits the antiproliferative activity of urolithins in human colon cancer cells. *European Journal of Nutrition*, **53**, 853–64.

González-Sarrías, A., García-Villalba, R., Nuñez-Sánchez, M.A., *et al.* (2015a) Identifying the limits for ellagic acid bioavailability: a crossover pharmacokinetic study in healthy volunteers after consumption of pomegranate extracts. *Journal of Functional Foods*, **19**, 225–35.

González-Sarrías, A., Tomé-Carneiro, J., Bellesia, A., Tomás-Barberán, F.A., and Espín, J.C. (2015b) The ellagic acid-derived gut microbiota metabolite, urolithin A, potentiates the anticancer effects of 5-fluorouracil chemotherapy on human colon cancer cells. *Food & Function*, **6**, 1460–9.

González-Sarrías, A., Núñez-Sánchez, M.A., Tomé-Carneiro, J., Tomás-Barberán, F.A., García-Conesa, M.T., and Espín, J.C. (2016) Comprehensive characterization of the effects of ellagic acid and urolithins on colorectal cancer and key-associated molecular hallmarks: microRNA cell specific induction of CDKN1A (p21) as a common mechanism involved. *Molecular Nutrition & Food Research*, **60**, 701–16.

González-Sarrías, A., García-Villalba, R., Romo-Vaquero, M., *et al.* (2017a) Clustering according to urolithin metabotype explains the interindividual variability in the improvement of cardiovascular risk biomarkers in overweight-obese individuals consuming pomegranate: a randomized clinical trial. *Molecular Nutrition & Food Research*, **61**, e1600830.

González-Sarrías, A., Núñez-Sánchez, M.A., Tomás-Barberán, F.A., and Espín, J.C. (2017b) Neuroprotective effects of bioavailable polyphenol-derived metabolites against oxidative stress-induced cytotoxicity in human neuroblastoma SH-SY5Y cells. *Journal of Agricultural and Food Chemistry*, **65**, 752–8.

González-Sarrías, A., Núñez-Sánchez, M.A., García-Villalba, R., Tomás-Barberán, F.A., and Espín, J.C. (2017c) Antiproliferative activity of the ellagic acid-derived gut microbiota isourolithin A and comparison with its urolithin A isomer: the role of cell metabolism. *European Journal of Nutrition*, **56**, 831–41.

Gu, J., Thomas-Ahner, J.M., Riedl, K.M., *et al.* (2019) Dietary black rapsberries impact the colonic microbiome and phytochemical metabolites in mice. *Molecular Nutrition & Food Research*, **63**, e1800636.

Hager, T.J., Howard L.R., Liyanage, R., Lay, J.O., and Prior, R.L. (2008) Ellagitannin composition of blackberry as determined by HPLC-ESI-MS and MALDI-TOF-MS. *Journal of Agricultural and Food Chemistry*, **13**, 661–9.

Häkkinen, S.H., Karenlampi, S.O., Mykkanen, H.M., Heinonen, I.M., and Torronen, A.R. (2000) Ellagic acid in berries: influence of domestic processing and storage. *European Food Research and Technology*, **212**, 75–80.

Heber, D. (2008) Multitargeted therapy of cancer by ellagitannins. *Cancer Letters*, **269**, 262–8.

Heilman, J., Andreux, P., Tran, N., Rinsch, C., and Blanco-Bose, W. (2017) Safety assessment of Urolithin A, a metabolite produced by the human gut microbiota upon dietary intake of plant derived ellagitannins and ellagic acid. *Food and Chemical Toxicology*, **108**, 289–97.

Henning, S.M., Summanen, P.H., Lee, R.P., *et al.* (2017) Pomegranate ellagitannins stimulate the growth of Akkermansia muciniphila in vivo. *Anaerobe*, **43**, 56–60.

Ishimoto, H., Shibata, M., Myojin, Y., *et al.* (2011) In vivo anti-inflammatory and antioxidant properties of ellagitannin metabolite urolithin A. *Bioorganic & Medicinal Chemistry Letters*, **21**, 5901–4.

Ishimoto, H., Tai, A., Yoshimura, M., *et al.* (2012) Antioxidative properties of functional polyphenols and their metabolites assessed by an ORAC assay. *Bioscience, Biotechnology, and Biochemistry*, **76**, 395–9.

Ismail, T., Calcabrini, C., Diaz, A.R., *et al.* (2016) Ellagitannins in cancer chemoprevention and therapy. *Toxins*, **8**, e151.

Istas, G., Feliciano, R.P., Weber, T., *et al.* (2018) Plasma urolithin metabolites correlate with improvements in endothelial function after red raspberry consumption: a double-blind randomized controlled trial. *Archives of Biochemistry and Biophysics*, **651**, 43–51.

Ito, H. (2011) Metabolites of the ellagitannin geraniin and their antioxidant activities. *Planta Medica*, **77**, 1110–15.

Ito, H., Iguchi, A., and Hatano, T. (2012) Identification of urinary and intestinal bacterial metabolites of ellagitannin geraniin in rats. *Journal of Agricultural and Food Chemistry*, **60**, 3068–77.

Jing, T., Liao, J., Shen, K., *et al.* (2019) Protective effect of urolithin a on cisplatin-induced nephrotoxicity in mice via modulation of inflammation and oxidative stress. *Food and Chemical Toxicology*, **129**, 108–14.

Jurgonski, A., Juskiewicz, J., Fotschki, B., *et al.* (2017) Metabolism of strawberry mono- and dimeric ellagitannins in rats fed a diet containing fructo-oligosaccharides. *European Journal of Nutrition*, **56**, 853–64.

Kallio, T., Kallio, J., Jaakkola, M., Mäki, M., Kilpeläinen, P., and Virtanen, V. (2013) Urolithins display both antioxidant and pro-oxidant activities depending on assay system and conditions. *Journal of Agricultural and Food Chemistry*, **61**, 10720–9.

Kasimsetty, S. G., Bialonska, D., Reddy, M. K., Ma, G., Khan, S.I., and Ferreira, D. (2010) Colon cancer chemopreventive activities of pomegranate ellagitannins and urolithins. *Journal of Agricultural and Food Chemistry*, **58**, 2180–7.

Khanbabaee, K. and van Ree, T. (2001) Tannins: classification and definition. *Natural Product Reports*, **18**, 641–9.

Kiss, A.K., Granica, S., Stolarczyk, M., and Melzig, M.F. (2012) Epigenetic modulation of mechanisms involved in inflammation: influence of selected polyphenolic substances on histone acetylation state. *Food Chemistry*, **131**, 1015–20.

Komatsu, W., Kishi, H., Yagasaki, K., and Ohhira, S. (2018) Urolithin A attenuates pro-inflammatory mediator production by suppressing PI3-K/Akt/NF-kappaB and JNK/AP-1 signaling pathways in lipopolysaccharide-stimulated RAW264 macrophages: possible involvement of NADPH oxidase-derived reactive oxygen species. *European Journal of Pharmacology*, **833**, 411–24.

Koponen, J.M., Happonen, A.M., Mattila, P.H., and Törrönen, A.R. (2007) Contents of anthocyanins and ellagitannins in selected foods consumed in Finland. *Journal of Agricultural and Food Chemistry*, **55**, 1612–19.

Kosmala, M., Zdunczyk, Z., Juskiewicz, J., *et al.* (2015) Chemical composition of defatted strawberry and raspberry seeds and the effect of these dietary ingredients on polyphenol metabolites, intestinal function, and selected serum parameters in rats. *Journal of Agricultural and Food Chemistry*, **63**, 2989–96.

Kunihiro, A.G., Brickey, J.A., Frye, J.B., Luis, P.B., Schneider, C., and Funk, J.L. (2019) Curcumin, but not curcumin-glucuronide, inhibits Smad signaling in TGFbeta-dependent bone metastatic breast cancer cells and is enriched in bone compared to other tissues. *Journal of Nutritional Biochemistry*, **63**, 150–6.

Larrosa, M., González-Sarrías, A., García-Conesa, M.T., Tomás-Barberán, F.A., and Espín, J.C. (2006) Urolithins, ellagic acid-derived metabolites produced by human colonic microflora, exhibit estrogenic and antiestrogenic activities. *Journal of Agricultural and Food Chemistry*, **54**, 1611–20.

Larrosa, M., González-Sarrías, A., Yáñez-Gascón, M.J., *et al.* (2010) Anti-inflammatory properties of a pomegranate extract and its metabolite urolithin-A in a colitis rat model and the effect of colon inflammation on phenolic metabolism. *Journal of Nutritional Biochemistry*, **21**, 717–25.

Lee, G., Park, J.S., Lee, E.J., Ahn, J.H., and Kim, H.S. (2019) Anti-inflammatory and antioxidant mechanisms of urolithin B in activated microglia. *Phytomedicine*, **55**, 50–7.

Lei, F., Xing, D.M., Xiang, L., *et al.* (2003) Pharmacokinetic study of ellagic acid in rat after oral administration of pomegranate leaf extract. *Journal of Chromatography B*, **796**, 189–94.

Li, Y., Li, Z., Hou, H., Zhuang, Y., and Sun, L. (2018) Metal chelating, inhibitory DNA damage, and anti-inflammatory activities of phenolics from rambutan (*Nephelium lappaceum*) peel and the quantifications of geraniin and corilagin. *Molecules*, **23**, e2263.

Li, Z., Henning, S.M., Lee, R.P., *et al.* (2015) Pomegranate extract induces ellagitannin metabolite formation and changes stool microbiota in healthy volunteers. *Food & Function*, **6**, 2487–95.

Ludwig, I.A., Mena, P., Calani, L., *et al.* (2015) New insights into the bioavailability of red raspberry anthocyanins and ellagitannins. *Free Radical Biology and Medicine*, **89**, 758–69.

Mele, L., Mena, P., Piemontese, A., *et al.* (2016) Antiatherogenic effects of ellagic acid and urolithins in vitro. *Archives of Biochemistry and Biophysics*, **599**, 42–50.

Mertens-Talcott, S.U., Jilma-Stohlawetz, P., Rios, J., Hingorani, L., and Derendorf, H. (2006) Absorption, metabolism, and antioxidant effects of pomegranate (*Punica granatum L.*) polyphenols after ingestion of a standardized extract in healthy human volunteers. *Journal of Agricultural and Food Chemistry*, **54**, 8956–61.

Milala, J., Kosmala, M., Karlinska, E., Juskiewicz, J., Zdunczyk, Z., and Fotschki, B. (2017) Ellagitannins from strawberries with different degrees of polymerization showed different metabolism through gastrointestinal tract of rats. *Journal of Agricultural and Food Chemistry*, **65**, 10738–48.

Mosele, J.I., Gosalbes, M.J., Maciá, A., *et al.* (2015) Effect of daily intake of pomegranate juice on fecal microbiota and feces metabolites from healthy volunteers. *Molecular Nutrition & Food Research*, **59**, 1942–53.

Niemetz, R. and Gross, G.G. (2005) Enzymology of gallotannin and ellagitannin biosynthesis. *Phytochemistry*, **66**, 2001–11.

Norden, E. and Heiss, E.H. (2019) Urolithin A gains in antiproliferative capacity by reducing the glycolytic potential via the p53/TIGAR axis in colon cancer cells. *Carcinogenesis*, **40**, 93–101.

Núñez-Sánchez, M.A., García-Villalba, R., Monedero-Saiz, T., *et al.* (2014) Targeted metabolic profiling of pomegranate polyphenols and urolithins in plasma, urine and colon tissues from colorectal cancer patients. *Molecular Nutrition & Food Research*, **58**, 1199–211.

Núñez-Sánchez, M.A., González-Sarrías, A., Romo-Vaquero, M., *et al.* (2015) Dietary phenolics against colorectal cancer – from promising preclinical results to poor translation into clinical trials: pitfalls and future needs. *Molecular Nutrition & Food Research*, **59**, 1274–91.

Núñez-Sánchez, M.T., Karmokar, A., González-Sarrías, A., *et al.* (2016) In vivo relevant mixed urolithins and ellagic acid inhibit phenotypic and molecular colon cancer stem cell features: a new potentially for ellagitannin metabolites against cancer. *Food and Chemical Toxicology*, **92**, 8–16.

Nuñez-Sánchez, M.A., González-Sarrías, A., García-Villalba, R., *et al.* (2017) Gene expression changes in colon tissues from colorectal cancer patients following the intake of an ellagitannin-containing pomegranate extract: a randomized clinical trial. *Journal of Nutritional Biochemistry*, **42**, 126–33.

Olennikov, D.N., Kashchenko, N.I., and Chirikova, N.K. (2015) In vitro bioaccessibility, human gut microbiota metabolites and hepatoprotective potential of chebulic ellagitannins: a case of Padma Hepaten(R) formulation. *Nutrients*, **7**, 8456–77.

Pfundstein, B., Haubner, R., Würtele, G., Gehres, N., Ulrich, C.M., and Owen, R.W. (2014) Pilot walnut intervention study of urolithin bioavaiability in human volunteers. *Journal of Agricultural and Food Chemistry*, **62**, 10264–73.

Piwowarski, J.P., Granica, S., and Kiss, A.K. (2014a) Influence of gut microbiota-derived ellagitannins' metabolites urolithins on pro-inflammatory activities of human neutrophils. *Planta Medica*, **80**, 887–95.

Piwowarski, J.P., Granica, S., Zwierzynska, M., *et al.* (2014b) Role of human gut microbiota metabolism in the anti-inflammatory effect of traditionally used ellagitannin-rich plant materials. *Journal of Ethnopharmacology*, **155**, 801–9.

Piwowarski, J.P., Kiss, A.K., Granica, S., and Moeslinger, T. (2015) Urolithins, gut microbiota-derived metabolites of ellagitannins, inhibit LPS-induced inflammation in RAW 264.7 murine macrophages. *Molecular Nutrition & Food Research*, **59**, 2168–77.

Piwowarski, J.P., Granica, S., Stefanska, J., and Kiss, A.K. (2016) Differences in metabolism of ellagitannins by human gut microbiota ex vivo cultures. *Journal of Natural Products*, **79**, 3022–30.

Puupponen-Pimiä, R., Seppánen-Laakso, T., Kankainen, M., *et al.* (2013) Effects of ellagitannin-rich berries on blood lipids, gut microbiota, and urolithins production in human subjects with symptoms of metabolic syndrome. *Molecular Nutrition & Food Research*, **57**, 2258–63.

Qiu, Z., Zhou, B., Jin, L., *et al.* (2013) In vitro antioxidant and antiproliferative effects of ellagic acid and its colonic metabolite, urolithins, on human bladder cancer T24 cells. *Food and Chemical Toxicology*, **59**, 428–37.

Ramírez de Molina, A., Vargas, T., Molina, S., *et al.* (2015) The ellagic acid derivative 4,4'-di-O-methylellagic acid efficiently inhibits colon cancer cell growth through a mechanism involving WNT16. *Journal of Pharmacology and Experimental Therapeutics*, **353**, 433–44.

Rangkadilok, N., Worasuttayangkurn, L., Bennett, R.N., and Satayavivad, J. (2005) Identification and quantification of polyphenolic compounds in longan (Euphoria longanaLam.) fruit. *Journal of Agricultural and Food Chemistry*, **53**, 1387–92.

Rojanathammanee, L., Puig, K.L., and Combs, C.K. (2013) Pomegranate polyphenols and extract inhibit nuclear factor of activated T-cell activity and microglial activation in vitro and in a transgenic mouse model of Alzheimer disease. *Journal of Nutrition*, **143**, 597– 605.

Romo-Vaquero, M., García-Villalba, R., González-Sarrías, A., *et al.* (2015) Interindividual variability in the human metabolism of ellagic acid: contribution of Gordonibacter to urolithin production. *Journal of Functional Foods*, **17**, 785–91.

Romo-Vaquero, M., Cortés-Martín, A., Loria-Kohen, V., *et al.* (2019) Deciphering the human gut microbiome of urolithin metabotypes: association with enterotypes and potential cardiometabolic health implications. *Molecular Nutrition & Food Research*, **63**, e1800958.

Rosenblat, M., Volkova, N., Borochov-Neori, H., Judeinstein, S., and Aviram, M. (2015) Anti-atherogenic properties of date vs. pomegranate polyphenols: the benefits of the combination. *Food & Function*, **6**, 1496–509.

Sala, R., Mena, P., Savi, M., *et al.* (2015) Urolithins at physiological concentrations affect the levels of pro-inflammatory cytokines and growth factor in cultured cardiac cells in hyperglucidic conditions. *Journal of Functional Foods*, **15**, 97–105.

Sandhu, A.K., Miller, M.G., Thangthaeng, N., *et al.* (2018) Metabolic fate of strawberry polyphenols after chronic intake in healthy older adults. *Food & Function*, **9**, 96–106.

Seeram, N.P., Lee, R. and Heber, D. (2004) Bioavailability of ellagic acid in human plasma after consumption of ellagitannins from pomegranate (Punica granatum L.) juice. *Clinica Chimica Acta*, **348**, 63–8.

Seeram, N.P., Henning, S.M., Zhang, Y., Suchard, M., Li, Z., and Heber, D. (2006) Pomegranate juice ellagitannin metabolites are present in human plasma and some persist in urine for up to 48 hours. *Journal of Nutrition*, **136**, 2481–5.

Seeram, N.P., Aronson, W.J., Zhang, Y., *et al.* (2007) Pomegranate ellagitannin-derived metabolites inhibit prostate cancer growth and localize to the mouse prostate gland. *Journal of Agricultural and Food Chemistry*, **55**, 7732–7.

Seeram, N.P., Zhang, Y., McKeever, R., *et al.* (2008) Pomegranate juice and extracts provide similar levels of plasma and urinary ellagitannin metabolites in human subjects. *Journal of Medicinal Food*, **11**, 390–4.

Selma, M.V., Beltrán, D., García-Villalba, R., Espín, J.C., and Tomás-Barberán, F.A. (2014a) Description of urolithin production capacity from ellagic acid of two human intestinal *Gordonibacter* species. *Food & Function*, **5**, 1779–84.

Selma, M.V., Tomás-Barberán, F.A., Beltrán, D., García-Villalba, R., and Espín, J.C. (2014b) Gordonibacter urolithinfaciens sp. Nov., a urolithin-producing bacterium isolated from the human gut. *International Journal of Systematic and Evolutionary Microbiology*, **64**, 2346–52.

Selma, M.V., Romo-Vaquero, M., García-Villalba, R., González-Sarrías, A., Tomás-Barberán, F.A., and Espín, J.C. (2016) The human gut microbial ecology associated with overweight and obesity determines ellagic acid metabolism. *Food & Function*, **7**, 1769–74.

Selma, M.V., Beltrán, D., Luna, M.C., *et al.* (2017) Isolation of human intestinal bacteria capable of producing the bioactive metabolite isourolithin A from elleagic acid. *Frontiers in Microbiology*, **8**, 1521.

Serrano, J., Puupponen-Pimiä, R., Dauer, A., Aura, A.M., and Saura-Calixto, F. (2009) Tannins: current knowledge of food sources, intake, bioavailability and biological effects. *Molecular Nutrition & Food Research*, **53**, S310–S329.

Shimoi, K., Saka, N., Nozawa, R., *et al.* (2001) Deglucuronidation of a flavonoid, luteolin monoglucuronide, during inflammation. *Drug Metabolism and Disposition*, **29**, 1521–4.

Singh, R., Chandrashekharappa, S., Bodduluri, S.R., *et al.* (2019) Enhancement of the gut barrier integrity by a microbial metabolite through the Nrf2 pathway. *Nature Communications*, **10**, 89.

Sirotkin, A.V. and Harrath, A.H. (2014) Phytoestrogens and their effects. *European Journal of Pharmacology*, **741**, 230–6.

Smeriglio, A., Barreca, D., Belloco, D., Bellocco, E., and Trombetta, D. (2017) Proanthocyanidins and hydrolysable tannins: occurrence, dietary intake and pharmacological effects. *British Journal of Pharmacology*, **174**, 1244–62.

Stoner, G.D., Sardo, C., Apseloff, G., *et al.* (2005) Pharmacokinetics of anthocyanins and ellagic acid in healthy volunteers fed freeze-dried black raspberries daily for 7 days. *Journal of Clinical Pharmacology*, **45**, 1153–64.

Subash, S., Essa, M.M., Al-Asmi, A., *et al.* (2014) Pomegranate from Oman alleviates the brain oxidative damage in transgenic mouse model of Alzheimer's disease. *Journal of Traditional and Complementary Medicine*, **4**, 232–8.

Subash, S., Braidy, N., Essa, M.M., *et al.* (2015) Long-term (15 mo) dietary supplementation with pomegranates from Oman attenuates cognitive and behavioral deficits in a transgenic mice model of Alzheimer's disease. *Nutrition*, **31**, 223–9.

Talcott, S.T. and Lee, J.H. (2002) Ellagic acid and flavonoid antioxidant content of muscadine wine and juice. *Journal of Agricultural and Food Chemistry*, **50**, 3186–92.

Teixeira, L.L., Costa, G.R., Dörr, F.A., *et al.* (2017) Potential antiproliferative activity of polyphenol metabolites against human breast cancer cells and their urine excretion pattern in healthy subjects following acute intake of a polyphenol-rich juice of grumixama (Eugenia brasiliensis Lam.). *Food & Function*, **8**, 2266–74.

Terao, J., Murota, K., and Kawai, Y. (2011) Conjugated quercetin glucuronides as bioactive metabolites and precursors of aglycone in vivo. *Food & Function*, **2**, 11–17.

Tomás-Barberán, F.A., García-Conesa, M.T., Larrosa, M., *et al.* (2008) Bioavailability, metabolism, and bioactivity of food ellagic acid and related polyphenols. In F. Daayf and V. Lattanzio (Eds.),

Recent Advances in Polyphenol Research, Volume 1. Oxford: Blackwell Publishing.
Tomás-Barberan, F.A., Espín, J.C., and García-Conesa, M.T. (2009) Bioavailability and metabolism of ellagic acid and ellagitannins. In *Chemistry and Biology of Ellagitannins* (pp. 273–97). Singapore: World Scientific.
Tomás-Barberán, F.A., García-Villalba, R., González-Sarrías, A., Selma, M.V., and Espín, J.C. (2014) Ellagic acid metabolism by human gut microbiota: consistent observation of three urolithin phenotypes in intervention trials, independent of food source, age and health status. *Journal of Agricultural and Food Chemistry*, **62**, 6535–8.
Tomás-Barberán, F.A., González-Sarrías, A., García-Villalba, R., *et al.* (2017) Urolithins, the rescue of "old" metabolites to understand a "new" concept: metabotypes as a nexus among phenolic metabolism, microbiota dysbiosis, and host health status. *Molecular Nutrition & Food Research*, **61**, e1500901.
Tomás-Barberán, F.A., Selma, M.V., and Espín, J.C. (2018) Polyphenols'gut microbiota metabolites: bioactives or biomarkers? *Journal of Agricultural and Food Chemistry*, **66**, 3593–4.
Toney, A.M. and Chung, S. (2019) Urolithin A, a gut metabolite, induces metabolic reprogramming of adipose tissue by promoting M2 macrophage polarization and mitochondrial function (OR12-02-19). *Current Developments in Nutrition*, **3** (suppl 1). https://academic.oup.com/cdn/article/3/Supplement_1/nzz049.OR12-02-19/5518039
Truchado, P., Larrosa, M., García-Conesa, M.T., *et al.* (2012) Strawberry processing does not affect the production and urinary excretion of urolithins, ellagic acid metabolites, in humans. *Journal of Agricultural and Food Chemistry*, **60**, 5749–54.
Tulipani, S., Urpi-Sarda, M., García-Villalba, R., *et al.* (2012) Urolithins are the main urinary microbial-derived phenolic metabolites discriminating a moderate consumption of nuts in free-living subjects with diagnosed metabolic syndrome. *Journal of Agricultural and Food Chemistry*, **60**, 8930–40.
Velagapudi, R., Lepiarz, I., El-Bakoush, A., *et al.* (2019) Induction of autophagy and activation of SIRT-1 deacetylation mechanisms mediate neuroprotection by the pomegranate

metabolite urolithin A in BV2 microglia and differentiated 3D human neural progenitor cells. *Molecular Nutrition & Food Research*, **63**, e1801237.

Yan, L., Yin, P., Ma, C., and Liu, Y. (2014) Method development and validation for pharmacokinetic and tissue distributions of ellagic acid using ultrahigh performance liquid chromatography-tandem mass spectrometry (UPLC-MS/MS). *Molecules*, **19**, 18923–35.

Yang, J., Lee, R., Henning, S.M., *et al.* (2016) Soy protein isolate does not affect ellagitannin bioavailability and urolithin formation when mixed with pomegranate juice in humans. *Food Chemistry*, **194**, 1300–03.

Yoshida, T., Amakura, Y., and Yoshimura, M. (2010) Structural features and biological properties of ellagitannins in some plant families of the order Myrtales. *International Journal of Molecular Sciences*, **11**, 79–106.

Yuan, T., Ma, H., Liu, W., *et al.* (2016) Pomegranate's neuroprotective effects against Alzheimer's disease are mediated by urolithins, its ellagitannin-gut microbial derived metabolites. *ACS Chemical Neuroscience*, **7**, 26–33.

Zafrilla, P., Ferreres, F., and Tomás-Barberán, F.A. (2001) Effect of processing and storage on the antioxidant ellagic acid derivatives and flavonoids of red raspberry (Rubus idaeus) jams. *Journal of Agricultural and Food Chemistry*, **49**, 3651–5.

Zhang, W., Chen, J. H., Aguilera-Barrantes, I., *et al.* (2016) Urolithin A suppresses the proliferation of endometrial cancer cells by mediating estrogen receptor-alpha-dependent gene expression. *Molecular Nutrition & Food Research*, **60**, 2387–95.

Zhao, W., Shi, F., Guo, Z., Zhao, J., Song, X., and Yang, H. (2018) Metabolite of ellagitannins, urolithin A induces autophagy and inhibits metastasis in human sw620 colorectal cancer cells. *Molecular Carcinogenesis*, **57**, 193–200.

10

Lignans

Knud E. Bach Knudsen, Natalja Nørskov, Anne K. Bolvig, Mette Skou Hedemann, and Helle Nygaard Lærke*

Aarhus University, Department of Animal Science, Tjele, Denmark

10.1 Introduction

Plant lignans constitute a group of diphenolic compounds widely distributed in the plant kingdom. Plant lignans occur as glycosides in whole grain, seeds, nuts, vegetables, berries, and beverages such as tea and coffee (Milder *et al.*, 2005; Penalvo *et al.*, 2008; Smeds *et al.*, 2007) and due to structural similarities they have been suggested to be intermediates or by-products of the pathway of lignin formation (Begum *et al.*, 2004; Davin *et al.*, 2008). The most common plant lignans in foods are matairesinol (Mat), pinoresinol (Pino), medioresinol (Med), lariciresinol (Lari), sesamin (Ses), syringaresinol (Syr), secoisolariciresinol (Seco), the glycosylated form of Seco – secoisolariciresinol diglucoside (SDG), and hydroxymatairesinol (hMat) (Figure 10.1) (Milder *et al.*, 2005; Penalvo *et al.*, 2008; Nørskov and Bach Knudsen, 2016; Smeds *et al.*, 2007) but several other plant lignans may also occur. Plant lignans are believed to play a role in the protection of the plant by having antifungal, antimicrobial, antiviral, and insecticidal properties (Cunha *et al.*, 2012; Harmatha and Dinan, 2003; Pan *et al.*, 2009).

Most plant lignans are converted to the enterolignans enterolactone (Enl) and enterodiol (End) by the intestinal microbiota

*Corresponding author.

Dietary Polyphenols: Metabolism and Health Effects, First Edition.
Edited by Francisco A. Tomás-Barberán, Antonio González-Sarrías, and Rocío García-Villalba.

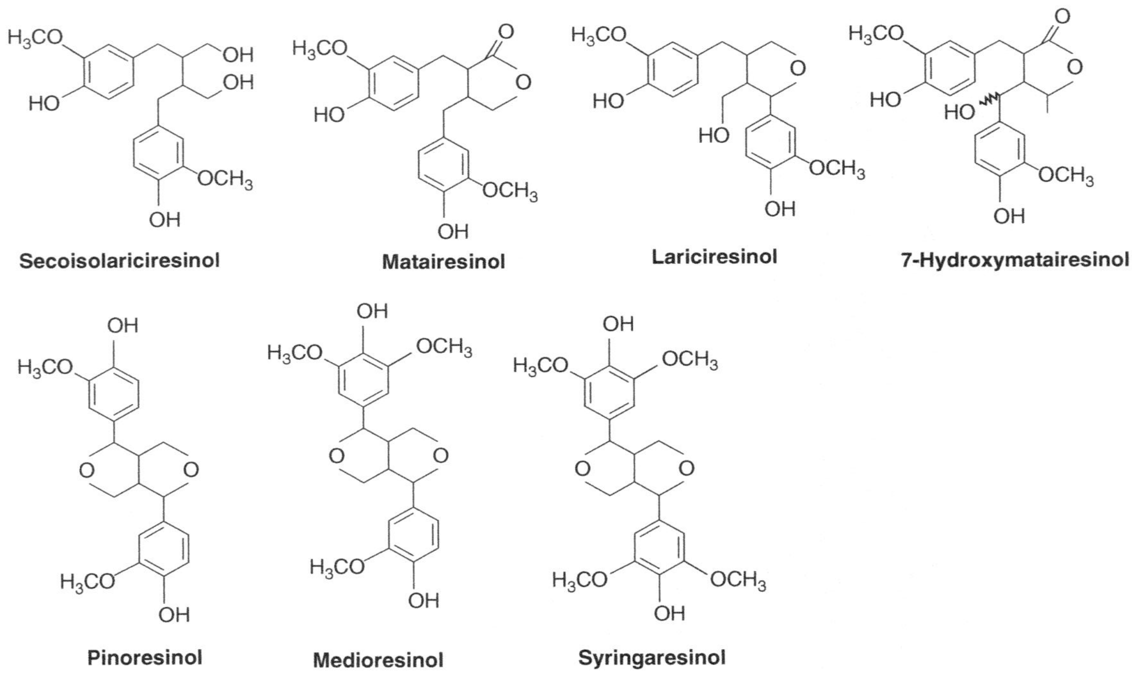

Figure 10.1 Molecular structure of the main plant lignans in cereals.

(Axelson and Setchell, 1981; Clavel *et al.*, 2005, 2006a; Nose *et al.*, 1992; Setchell *et al.*, 1980; Woting *et al.*, 2010). Although not conclusive, epidemiological studies have suggested that high plasma enterolignan concentrations are associated with a decreased risk of several cancers such as breast, prostate, and colorectal cancer but also with cardiovascular diseases and type 2 diabetes (Aarestrup *et al.*, 2013; Adlercreutz, 2007; Eriksen *et al.*, 2017, 2019; Hollaender *et al.*, 2015; Johnsen *et al.*, 2010; Kyrø *et al.*, 2018a, 2018b; Olsen *et al.*, 2004; Sun *et al.*, 2014; Tuomisto *et al.*, 2019). The cancer effects are thought to be related to the estrogen-like structure brought about by the binding of enterolignans to estrogen receptors (Penttinen *et al.*, 2007). The affinity of enterolignans for the estrogen receptors, however, is approximately 1000–10 000 times lower than for estradiol (Rice and Whitehead, 2006). Enterolignans may also interact with the activity of sex hormone-binding globulins (SHBG) (Schottner *et al.*, 1997) and Adlercreutz *et al.* (1987) identified a significantly positive correlation between the intake of fiber and urinary excretion of lignans and phytoestrogens, and the concentration of plasma SHBG. Lignans have further been shown to inhibit 5-α-reductase (Evans *et al.*, 1995), reduce insulin-like growth factor 1 (Rickard *et al.*, 2000) and suppress fatty acid synthase protein expression (Menendez *et al.*, 2008).

Finally, enterolignans have been shown to modulate angiogenesis and insulin-like growth factor 1, indicating that they may affect breast carcinogenesis (Brooks and Thompson, 2005). Studies in animal models have shown that the same factors may be of importance for prevention of early-stage cancer leading to lower incidence as well as inhibition of the progression of already established tumors (Chen *et al.*, 2003, 2006; Wang *et al.*, 2005, 2006). Although the epidemiological studies in general are inconclusive, studies on breast cancer patients have shown reduced mortality rates in patients with high plant lignan intake (Buck *et al.*, 2011; Guglielmini *et al.*, 2012; McCann *et al.*, 2010; Olsen *et al.*, 2011).

The intake of plant lignans varies considerably among the different European countries. Estimation of lignan intake based on the Dutch lignan database showed that the major sources of lignans in Europe are from the food groups cereals and grain products, vegetables, fruit and berries, and beverages

(Tetens *et al.*, 2013). In Scandinavia, whole-grain cereals and vegetables are the main sources of plant lignans (Johnsen *et al.*, 2004; Kilkkinen *et al.*, 2001), whereas fruits and berries are the relatively most important lignan contributors in Italy and nonalcoholic beverages such as tea and coffee are the main contributors in the UK (Tetens *et al.*, 2013).

The primary purpose of this chapter is to provide an overview of the main plant lignans in foods and their metabolism and biological activity.

10.2 Lignans in Foods

The concentration of lignans varies widely among the various groups of plant foods (Table 10.1); for more details see Landete (2012). Flaxseed (linseed) is by far the richest source of plant lignans not only in the oilseed and nuts group but in plant foods in general (Landete, 2012; Mazur, 1998; Milder *et al.*, 2005; Smeds *et al.*, 2007; Thompson *et al.*, 2006). Flaxseed contains 75–800 times more lignans than cereal grains, legumes, fruits, and vegetables. The major lignan in flaxseed is SDG (Landete, 2012; Milder *et al.*, 2005), along with lower content of Lari, Pin, Mat, and isolariciresinol (I-Lar) (Landete, 2012; Milder *et al.*, 2005). SDG is present in a complex-bound form in the outer layers of the seed (Kajla *et al.*, 2015). Sesame (*Sesamum indicum*) is also a rich source of lignans, particularly in the form of sesaminol (Ses) (not present in Table 10.1) (Landete, 2012; Penalvo *et al.*, 2005b; Smeds *et al.*, 2007). Mono-, di-, and triglucoside forms of sesaminol, sesamolinol, and Pin can be present in the oil-free meal (Dar and Arumugam, 2013). Sunflower seeds and cashew nuts also have high lignan concentrations, especially of Seco and Lar, and I-Lar was also found in cashew nuts (Landete, 2012; Smeds *et al.*, 2007).

Cereals are significant contributors of lignans in countries with a high intake of whole-grain cereals, such as the Nordic countries (Johnsen *et al.*, 2004; Kilkkinen *et al.*, 2001; Tetens *et al.*, 2013). Among cereals, the highest lignan contents are found in rye and wheat (Smeds *et al.*, 2007). Cereals, in particular rye, are rich in Syr but also Seco and Mat are found in high concentrations along with hMat (Penalvo *et al.*, 2005a; Smeds *et al.*, 2012).

Table 10.1 Lignan content (μg/100 g as is basis) of selected foods

	Plant lignans							
	Seco	Pino	Lari	Mat	Syr	Med	Total	Reference
Oilseed and nuts								
Flaxseed	294210	3324	3041	553			301 129	(Milder *et al.*, 2005)
	379012	730	2808	153			379 012	(Thompson *et al.*, 2006)
Sunflower seeds	53	167	671	0			891	(Milder *et al.*, 2005)
Sesame seed	66	29331	9470	481			379 012	(Milder *et al.*, 2005)
Cashew nuts	133	0	496	0			629	(Milder *et al.*, 2005)
Cereals								
Whole-grain rye	30	55	278	23	2954	129	3515	(Nørksov, 2019)
Rye bran	462	1547	1505	729	3540	858		(Smeds *et al.*, 2007)
	46	53	495	34	4247	1219	6146	(Nørksov, 2019)
Whole-grain wheat	23	4	93	4	1263	199	1592	(Nørksov, 2019)
Wheat flour	6	1	19	<1	174	17	219	(Nørksov, 2019)
Wheat bran	868	138	672	410	882	232		(Smeds *et al.*, 2007)
	31	36	200	10	3993	429	4721	(Nørksov, 2019)
Oat	90	567	766	440	297	112		(Smeds *et al.*, 2007)
Corn bran	125	33	69	21	2			(Smeds *et al.*, 2007)

Table 10.1 (Continued)

	Plant lignans							
	Seco	Pino	Lari	Mat	Syr	Med	Total	Reference
Vegetables								
Asparagus	743	122	92	14	58	3	1034	(Penalvo *et al.*, 2008)
Brassica vegetables	19	1691	599	12			2321	(Milder *et al.*, 2005, Thompson *et al.*, 2006)
Broccoli	6	6	82				94	(Thompson *et al.*, 2006)
Cabbage	3	44	32				79	(Thompson *et al.*, 2006)
Carrot	93	19	60				171	(Milder *et al.*, 2005)
Garlic	42	482	54	5			583	(Thompson *et al.*, 2006)
Spinach	8	1	59	0			67	(Thompson *et al.*, 2006)
Sweet peppers (green)	7	1	164	0			172	(Milder *et al.*, 2005)
Tomatoes (cherry)	16	19	38	–	3	3	80	(Penalvo *et al.*, 2008)
Fruits								
Apricot	31	105	314	0			450	(Horn-Ross *et al.*, 2000)

Dried raisins	9	<1	12	<1		22	(Thompson *et al.*, 2006)
Strawberries	5	21	23	<1		49	(Thompson *et al.*, 2006)
Nonalcoholic beverages							
Coffee	9–16	<1–2	9–13	<1		19–31	(Milder *et al.*, 2005)
Orange juice	8	<1	<1	0		8	(Thompson *et al.*, 2006)
Milk, cow	<1	<1	<1	<1		1	(Thompson *et al.*, 2006)
Milk, soya	1	30	7	0		38	(Milder *et al.*, 2005)
Tea, black	5–6	27–41	29–31	1–2		64–77	(Milder *et al.*, 2005)
Tea, green	13	6	19	2		39	(Milder *et al.*, 2005)
Alcoholic beverages							
Beer	0–1	13–22	6–9	0		19–32	(Milder *et al.*, 2005)
Wine, red	42–61	6-–2	9–16	6–8		69–91	(Milder *et al.*, 2005)
Wine, white	5–8	2–3	5–12	3–3		16–24	(Milder *et al.*, 2005)

Seco, secoisolariciresinol; Pino, pinoresinol; Lari, larirecinol; Mat, matairesinol; Syr, syringaresinol; Med, medioresinol.

The lignans in cereals are mainly located in the outer bran layer (pericarp/testa and aleurone) (Glitsø and Bach Knudsen, 1999; Nilsson *et al.*, 1997; Nørskov and Bach Knudsen, 2016) where it can be found as part of the lignin polymers or in free form as glycosides (Begum *et al.*, 2004; Nørskov and Bach Knudsen, 2016).

Legumes (bean, lentil, soybean) and vegetables (broccoli, garlic, asparagus, carrots) contain relatively high levels of lignans, although the concentration is generally lower than those of cereals, and oilseed and nuts (Landete, 2012; Mazur, 1998; Milder *et al.*, 2005; Smeds *et al.*, 2007). In fruits, lower contents of lignans are found in dried apricots, dates, and prunes as well as yuzu, a citrus fruit originating in East Asia (Landete, 2012).

Lignans are also found in several beverages (Landete, 2012; Mazur, 1998; Mazur *et al.*, 1998; Milder *et al.*, 2005). Among the nonalcoholic beverages, the highest lignan concentrations have been found in tea, followed by coffee and juices such as orange and pomegranate (Landete, 2012; Milder *et al.*, 2005). Due to the presence of lignans in legumes, soy-based drinks contain substantially more lignans than cows' milk (Milder *et al.*, 2005). Several alcoholic beverages such as red wine also contain lignans, with Seco and isoLar as the most abundant ones, while lower amounts are found in beer and white wine (Landete, 2012; Mazur, 1998; Milder *et al.*, 2005).

Our knowledge of how plant lignans are bound to macromolecules in the plant cell walls is somewhat limited. Smeds *et al.* (2007) studied different extraction methods and found that accelerated solvent extraction gave the highest yield of Syr, whereas a combination of alkaline and acid extraction gave the highest contents of Pino, Med, Lari, Seco, and Mat. It was suggested that the difference in extractability might be related to the esterification of the plant lignans. Since the accelerated solvent extraction method only detected aglycones and unesterified lignan glycosides, the data of Smeds *et al.* (2007) indicate that the majority of Syr was not in esterified form possibly due to steric hindrance of the phenolic hydroxyl groups. This contrasts with Seco that has been found as diglucoside-hydroxymethyl glutaryl ester-linked oligomers in flax seeds (Ford *et al.*, 2001; Kamal-Eldin *et al.*, 2001) and indications for ester-linked

Table 10.2 Total and free plant lignans (μg/100 g dry matter) in wheat and rye bran before and after treatment with cell wall degrading enzymes

		Enz	Seco	Pino	Lari	I-Lari	Mat	Syr	Med	Total
Wheat bran	Total	–	181	42	370	158	6	1658	691	3106
	Free		13	5	38	12	<1	263	37	369
	%		7.1	11.9	10.2	7.6		15.9	5.4	11.9
Wheat bran	Total	+	200	36	290	204	6	1346	822	2905
	Free		22	4	36	28	<1	585	55	731
	%		11.0	11.1	12.4	13.7		43.5	6.7	25.2
Rye bran	Total	–	47	44	407	72	52	3807	391	4820
	Free		12	12	177	16	15	1496	75	1803
	%		25.5	27	53.5	22.2	28.8	39.3	19.2	37.5
Rye bran	Total	+	57	44	302	93	51	4456	447	5450
	Free		10	11	96	45	15	2797	73	3047
	%		17.5	25.0	31.8	48.4	29.4	62.7	16.3	55.9

Enz., enzymes; Seco, secoisolariciresinol; Pino, pinoresinol; Lari, larirecinol; Mat, matairesinol; Syr, syringaresinol; Med, medioresinol.

lignans in other food items have also been observed (Milder *et al.*, 2004).

Although the majority of lignans are bound to plant cell wall macromolecules, a proportion of lignans is present as free glycosides, bound to one or two carbohydrate moieties (Bondia-Pons *et al.*, 2009; Schwartz and Sontag, 2006). The ratio of free lignans varies between plant foods and within groups of foods. For instance, the proportion of free lignans is higher in rye bran than wheat bran (Table 10.2) and when treated with cell wall degrading enzymes targeting arabinoxylan, cellulose, and β-glucan, the proportion of free lignans, particularly Syr, increased substantially and more in the case of rye than wheat (Figure 10.2) (Bolvig *et al.*, 2017b; Nørksov, 2019).

10.3 Metabolism of Lignans

In order to be absorbed, ingested lignans have to be hydrolyzed by endogenous or bacterial enzymes in the gastrointestinal tract

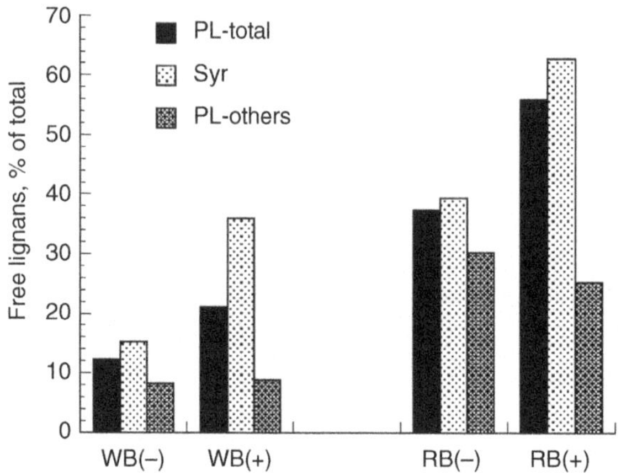

Figure 10.2 Total free plant lignans (PL-total), free syringaresinol (Syr) and free others plant lignans (PL-others) in wheat (W) and rye (R) brans (B) before (-) and after (+) treatment with cell wall degrading enzymes.

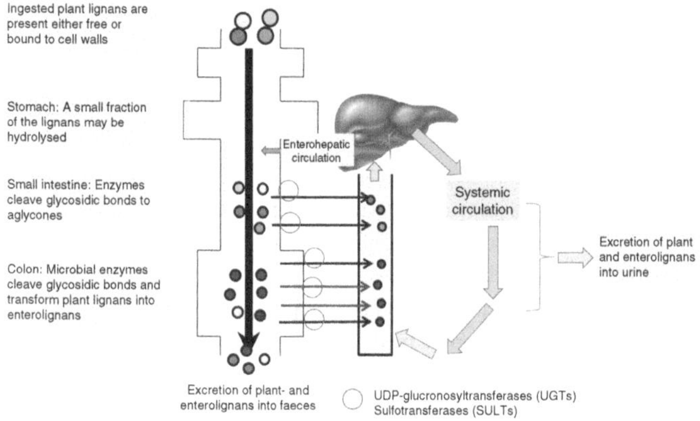

Figure 10.3 Proposed metabolism of plant lignans and enterolignans (*red circle*) in monogastric animals.

to lignan aglycones, which can be absorbed directly (Bolvig *et al.*, 2016a, 2017b) or converted by colonic bacteria to the enterolignans Enl and End (Axelson and Setchell, 1981; Nose *et al.*, 1992) (Figure 10.3). The hydrolysis of lignans is, therefore, an essential step for bioaccessibility and bioavailability in the body.

The uptake of unconjugated lignans occurs most likely by passive diffusion. This was indicated by a study with Caco-2 cells showing that while SDG was not able to permeate the epithelium, Seco and the enterolignans were able to undergo passive permeation and conjugation with glucuronic acid or, to a lesser extent, sulfonic acid by phase II enzymes in the Caco-2 cells (Jansen *et al.*, 2005; Mukker *et al.*, 2014). There is a growing body of evidence that when passing the intestinal epithelium, lignans are conjugated with glucuronic and sulfonic acids and circulate in plasma mainly as conjugates. The analyses of 2913 human blood samples revealed the conjugation pattern for enterolactone: 95.3/4.7% glucuronide/sulfate for men and 94.8/5.2% for women (Nørskov *et al.*, 2016). A previous study by Adlercreutz *et al.* (1995) on human urine has also shown similar patterns of conjugation for enterolactone with enterolactone glucuronide as the major conjugate, followed by sulfate and further by free enterolactone, and the corresponding sulfoglucuronide, disulfate, and diglucuronide.

The conjugation reactions are facilitated by a large family of membrane-bound enzymes, sulfotransferases (SULTs) and uridine diphosphate-glucuronosyltransferases (UGTs), which are found primarily in liver tissue but also in the intestine and kidney (Jancova *et al.*, 2010). Most UGTs have been shown to exhibit overlapping in substrate specificities, such as is known for the UGT1A1 type. This isoform is responsible for bilirubin, phenols, flavones, steroids, and coumarins glucuronidation. Phase II metabolic processes are common detoxification steps in the formation of hydrophilic compounds for excretion into urine or bile (Jancova *et al.*, 2010). The conjugation of lignans with glucuronic and sulfonic acids makes the molecules more water soluble and promotes their urinary and biliary excretion. When excreted in the bile, lignans undergo second enterohepatic circulation.

Multiple animal and human intervention studies have shown large interindividual variability in absorption, distribution, metabolism, and excretion (ADME) of lignans and phytochemicals in general (Morand and Tomas-Barberan, 2019). This interindividual variation is often the reason why it is difficult to demonstrate the biological effects of food phytochemicals. The interindividual variation in ADME reflects multiple factors, for example genetic polymorphisms, microbial composition and functionality, phase I and phase II metabolism, and intestinal and kidney transporters. The stratification of interindividual variability in ADME before or after interventional trials can be a sensible approach which will improve the statistical power of the study (Morand and Tomas-Barberan, 2019).

10.3.1 Kinetics of Absorption of Plant Lignans

Studies with plant lignans provided as SDG and sesamin have shown that a certain fraction of the plant lignans is rapidly absorbed and appears in the systemic circulation within hours after ingestion (Penalvo *et al.*, 2005b; Setchell *et al.*, 2014) (Figure 10.4). In a study on sesamin intake, a total of 10 different lignans were quantified in postprandial plasma, and most of these reached their maximum concentration less than 2h after the meal; Pino, Mat, 7'-hydroxymatairesinol (Hmr), and Ses were the most abundant. Neither Enl nor End appeared in plasma until 8 h after the intake of sesame seeds (Penalvo *et al.*, 2005b). The elimination half-life varied from 2.1 hours for Ses to 5.9 hours for Hmr.

A study by Setchell *et al.* (2014) investigated the pharmacokinetics and bioavailability of SDG in healthy postmenopausal women after oral intake of graded levels of SDG. Extracts of SDG differing in purity were compared, and steady-state lignan concentrations measured after daily intake for one week after which blood and urine samples were collected at time intervals for the analysis of Seco, End, and Enl concentrations by mass spectroscopy. It was found that SDG was efficiently hydrolyzed and converted to Seco. Serum concentrations increased rapidly after oral intake, peaking after 5–7 hours and with a plasma elimination half-life of 4.8 hours (Figure 10.5).

The same absorption pattern as seen for SDG and Ses has been shown to occur with cereals provided as breads (Bolvig *et al.*, 2016a) or in raw form (Bolvig *et al.*, 2017b) in a pig model (Figures 10.6 and 10.7). Portal vein catheterized pigs were fed whole-grain wheat or wheat and rye aleurone-enriched breads at breakfast (0 h) and lunch (5 h) and samples were collected and pooled in intervals of 2.5 hours up to 10 hours. After ingestion of the breads, a substantial rise was seen in the concentration

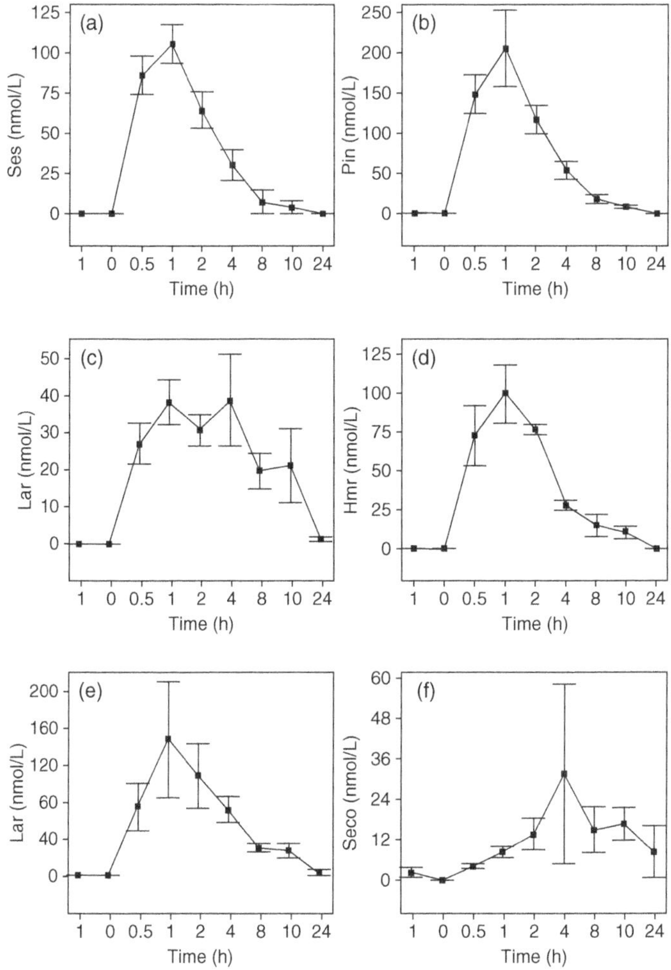

Figure 10.4 Postprandial absorption/elimination curves of (a) Ses, (b) Pin, (c) Lar (as the sum of Lar and IsoL), (d) Hmr, (e) Mat, (f), Seco (as the sum of Seco and Anse), (g) Enl, and (h) End in four humans. Values are means ± SEM, n=4. Source: Data used with permission from Penalvo *et al.* (2005b).

of plant lignans in the portal vein and the mesenteric artery at 0–2.5 hours after breakfast and with a continuously increasing concentration over the following phases (see Figure 10.7) (Bolvig *et al.*, 2016a). The uptake is presumably occurring by

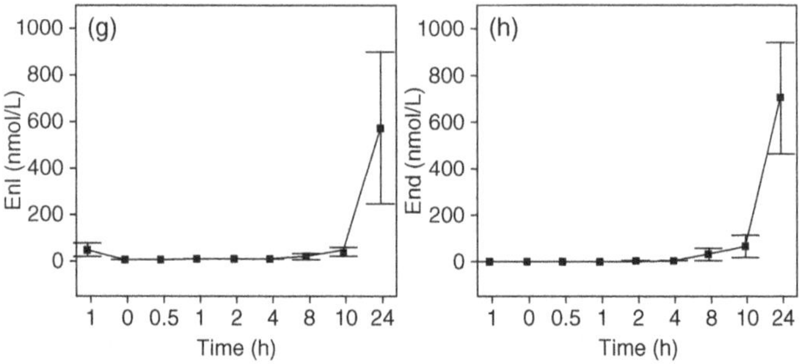

Figure 10.4 *(Continued)*

passive diffusion as indicated by an *in vitro* study (Mukker *et al.*, 2014). The fraction of enterolignans versus plant lignans changed from 2% plant lignans in the postabsorptive phase before ingesting the breakfast, to after breakfast and lunch where the plant lignans represented 21–32% of the total lignans and with a concomitant decrease in the concentration of Enl (Bolvig *et al.*, 2016a). Over the 10-hour sampling period, the cumulative net absorption of plant lignans corresponded to 7% of ingested plant lignans (Bolvig *et al.*, 2016a).

Treating rye bran with cell wall degrading enzymes increased the proportion of free lignans particularly Syr but not for the other plant lignans (Seco, Mat, Lari, I-Lari, Pino, and Hydr) (see Figure 10.2) (Bolvig *et al.*, 2017b). The enzymatic treatment resulted in a very rapid postprandial uptake of Syr with peak levels at 49 min postprandially (see Figure 10.7). For the other plant lignans (PL_{Other}), the uptake was also higher after enzymatic treatment, but the plateau concentration was not reached until 118 minutes after consumption. For untreated rye bran, there was also a more rapid rise in the plasma concentrations of Syr than PL_{Total}, but not nearly as fast as for enzymatically treated bran.

The half-lives in the study of Bolvig *et al.* (2017b) were measured postprandially or during the following wash-out phase with two different outcomes; the half-lives were substantially longer when estimated postprandially than during the wash-out. This most likely reflects an ongoing absorption of plant lignans from the gut in the postprandial phase. In the study there was a four-hour difference (11.4 h and 7.5 h, respectively) in the

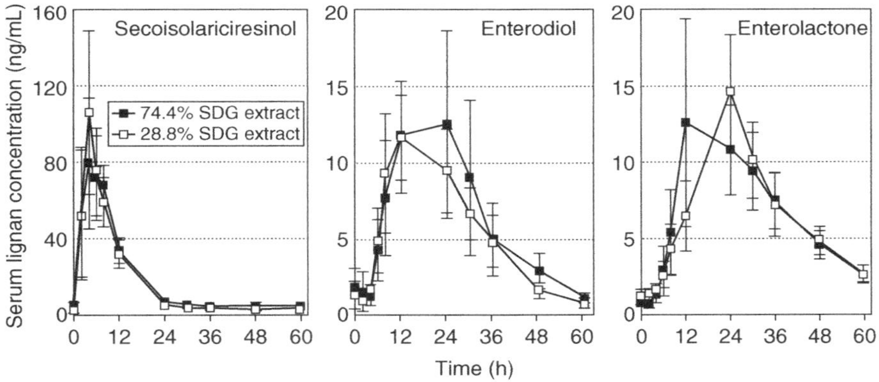

Figure 10.5 Serum concentrations (mean ± SEM) of secoisolariciresinol, enterodiol, and enterolactone in 10 healthy postmenopausal women following a single oral bolus of 50 mg of a 28.8% and 74.4% pure SDG extract given in a randomized double-blinded cross-over design. Source: Data used with permission from Setchell *et al*. (2014).

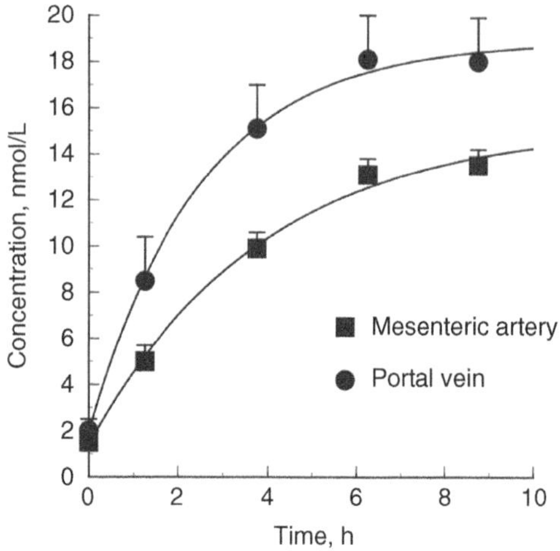

Figure 10.6 Portal and arterial plant lignan plasma concentrations after consumption of lignan-rich diets (WWG, WAF, RAF) at breakfast and lunch. Plasma concentrations are presented as means resulting from intake of WWG, WAF or RAF. The variation in plasma concentration in the mesenteric artery (MA) could be described by: Plant lignanMA = 15.3 × (1-0.911 × EXP(-0.260 × T), R2 = 0.99; and in the portal vein (PV) by: Plant lignanPV =18.9 × (1-0.898 × EXP(-0.404 × T), R2 = 0.997, where T is the time after breakfast. SEM are indicated with error bars. Source: Data used with permission from Bolvig *et al.* (2016a).

half-lives of Syr compared to the other plant lignans. This is most likely driven by the occurrence of other plant lignans in the diet such as Lari and I-Lari. This indicates that while Syr is being rapidly absorbed and eliminated, Lari and I-Lari might contribute to a longer half-life of $PL_{Total.}$ This was also supported by the half-life measurements in the wash-out period where the half-lives of PL_{Total} and Syr were approximately three times longer when feeding nonenzyme-treated rye bran compared to the enzyme-treated rye bran in the wash-out phase, while the half-life of PL_{Other} was only 1.5 times longer.

Elimination of the absorbed plant lignans from cereal occurs most likely via the bile. While approximately 7% of the ingested plant lignans were absorbed, urinary excretion accounted only for 1.7–3.1% of the ingested plant lignans for comparable diets (Bolvig *et al.*, 2016a). This corroborates the findings of Lærke

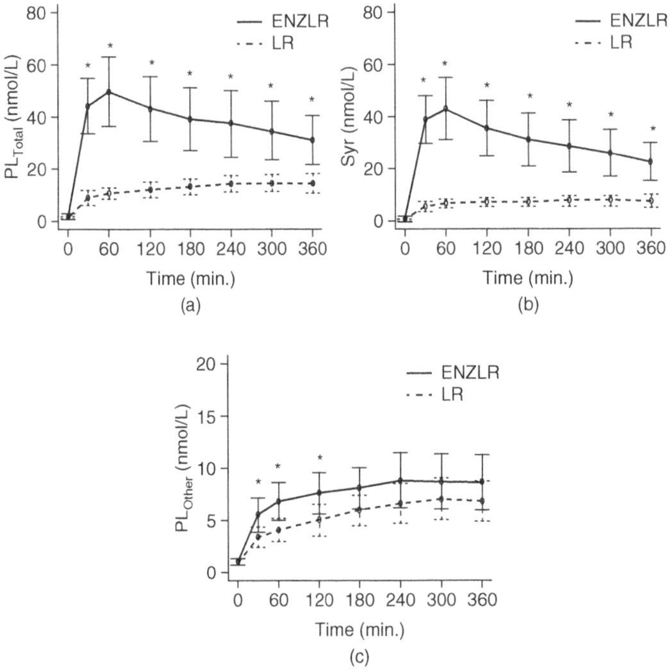

Figure 10.7 Plasma concentrations of (A) PL_{Total}, (B) Syr, and (C) PL_{Other} collected from the mesenteric artery of pigs 0–360 min after a single dose of LR or ENZLR. Values are least square means ± SD, n=10 per diet. Asterisks indicate different from LR at that time, $P<0.05$. PL_{Total}, total plant lignan; Syr, syringaresinol; PL_{Other}, other plant lignans than Syr; LR, lignan-rich diet; ENZLR, enzymatically treated lignan-rich diet.Source: Data used with permission from Bolvig *et al.* (2017b).

et al. (2009) who found a much higher proportion of plant lignans compared to total lignan (sum of plant and enterolignans) in bile relative to the portal vein and other blood vessels, suggesting that plant lignans are diverted towards systemic elimination via the bile. From these data, it appears that most of the Enl formed may pass the liver for immediate excretion in urine as is the case for estriol-3-glucuronide (Aubertin-Leheudre *et al.*, 2008) while the plant lignans may be selectively taken up by the hepatocytes, conjugated and excreted in the bile.

Before conjugation, the plant lignans are biologically active in the liver cells or in any other cell in the body. However, after conjugation in the liver, the lignans will not be biologically active again before they reach the intestine, are deconjugated by enzymes, and continue in a second enterohepatic circulation.

10.3.2 Conversion of Plant Lignans to Enterolignans

A common feature of the colonic bacteria is their diverse repertoire of catabolizing enzymes and metabolic capabilities that by all measures is much higher than those of their host (Flint *et al.*, 2008, 2012; Rowland *et al.*, 2017). Combined with the long retention time and moist environment in the large intestine, these conditions are favorable for microbial degradation of complex cell wall polysaccharides (Bach Knudsen, 2005) and associated compounds, i.e., plant lignans (Flint *et al.*, 2008, 2012; Rowland *et al.*, 2017). Axelson and Setchell (1981) were the first to show that the Enl present in urine was of microbial origin as conventional, but not germ-free, rats excreted Enl in urine (Axelson and Setchell, 1981; Setchell *et al.*, 1980). This finding has been confirmed *in vitro* using human stool samples, and in germ-free and human flora associated rats (Borriello *et al.*, 1985; Bowey *et al.*, 2003). Studies conducted on ileostomy subjects further showed that while Mat was higher after consuming a lignan-rich high-fiber diet compared to a low-fiber diet, plasma and urinary Enl were below the detection limit (Pettersson *et al.*, 1996).

An artificial stomach and small intestinal system has been used to study the gastrointestinal fate of SDG (Eeckhaut *et al.*, 2008). It was found that SDG remained intact through the stomach and small intestine and that human digestive enzymes did not initiate hydrolysis of the glycosides on the lignans (Eeckhaut *et al.*, 2008). In a study with ileal cannulated pigs, the majority (95%) of the lignans present in the ileal effluent were plant lignans (Glitsø *et al.*, 2000). Influenced by type and level of cell wall materials, this proportion changed dramatically in the cecum where there was a significant increase in the content of enterolignans, primarily as Enl, and with a graduate increase along the large intestine (Glitsø *et al.*, 2000). From *in vitro* fermentation in a system simulating the human microbial ecosystem, it appeared that the transverse colon is the main site for production of enterolignans (Eeckhaut *et al.*, 2008).

The individual plant lignans, however, are not equally metabolized to enterolignans (Heinonen *et al.*, 2001). When incubating isolated plant lignans with human fecal inoculum *in vitro*, it was found that the conversion rate of Syr to enterolignans was very low, while a higher fraction of Pino, Mat, Seco, and Lari was converted to enterolignans (Heinonen *et al.*, 2001). The conversion pathway of Syr is not fully elucidated but it has been suggested that the metabolism of Syr occurs in a similar way to that of Pino (Heinonen *et al.*, 2001).

Impact of Intestinal Microbiota

It appears evident from several studies that the complete metabolism of plant lignans requires interaction of a consortium of anaerobic bacteria that are distant both functionally and phylogenetically (Clavel *et al.*, 2006a, 2006b; Yoder *et al.*, 2015). For example, the metabolism of SDG involves a total of four reactions (Figure 10.8); the initial deglycosylation is catalyzed by three *Bacteroides* species (*B. distasonis*, *B. fragilis*, and *B. ovatus*) and two strains of *Clostridium* (*C. cocleatum*, *C saccharogumia*) resulting in Seco. In the next step, Seco is converted to the intermediate dehydroxyenterodiol (DHEnd) by strains of *Butyribacterium methylotrophicum*, *Eubacterium callanderi*, *E. limosum*, *Blautia producta*, and *Peptostreptococcus productus*. The intermediate DHEnd is dehydroxylated by *Clostridium scindens* and *Eggerthella lenta* to form End and the final step, dehydrogenation of End to Enl, and closure of the lactone ring is catalyzed by subdominant populations of Clostridiales, in particular *Lactonifactor longoviformis*. *Lactonifactor longoviformis* is a strict anaerobic Gram-positive bacterium that does not grow in the presence of penicillin or rifampicin (Clavel *et al.*, 2007). Jin *et al.* (2007a, 2007b) isolated *Ruminococcus* sp. END-1, Clostridiaceae bacterium END-2, *Eggerthella* sp. SDG-2, and *Eubacterium* sp. ARC-2, all capable of metabolizing Enl in an enantiomeric selective manner. In a recent study, it was also found that strains of *Bifidobacterium* and *Lactobacillus* are capable of carrying out the complete enterolignan metabolism, transforming a purified lignan (Seco) into End and Enl (Peiroten *et al.*, 2019).

The bacterial strains capable of producing Enl have been shown to constitute less than 1% of the total microbial

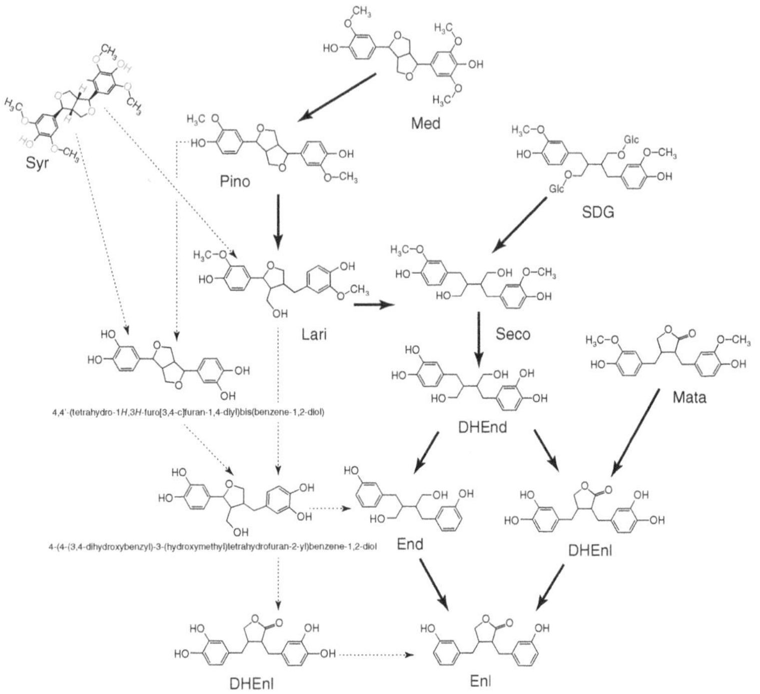

Figure 10.8 Conversion of plant lignans to enterolignans by human intestinal bacteria. Solid arrows indicate described pathways, dashed arrows indicate theoretical pathways. Mat, Matairesinol; Pino, pinoresinol; Med, medioresinol; Lari, lariciresinol ; Ses, sesamin; Syr, syringaresinol; Seco, secoisolariciresinol; SDG, secoisolariciresinol diglucoside; DHEND; dihydroxyenterodiol, DHENL; dihydroxyenterolactone. Source: Adapted from Yoder *et al.* (2015).

community (Clavel *et al.*, 2007; Hullar *et al.*, 2015). In a cross-sectional study including 84 men, the fecal microbiota of the participants with high (n=10) and low (n=10) serum Enl were investigated. It was found that men with low Enl concentrations had a lower diversity of faecal bacteria, particularly bacteria belonging to the lactobacilli-enterococci group (Clavel *et al.*, 2005). Thus, high diversity of the gut microbial community and high abundance and presence of specific microbial strains seem to be associated with high Enl levels in humans.

Impact of Antibiotic Use

Antibiotics are known to have a profound effect on the mammalian gut microbiota (Antunes and Finlay, 2011; Dethlefsen *et al.*, 2008; Palleja *et al.*, 2018). At an early stage in the research, it was shown that antibiotics played a significant role in the capacity of the microflora to produce enterolignans. This link was identified by Axelson and Setchell (1981) who showed an immediate and pronounced effect of antibiotic treatment on urinary excretion of both End and Enl. Later, Kilkkinen *et al.* (2002), Adgent and Rogan (2015) and Bolvig *et al.* (2016b) all identified an inverse association between Enl serum, plasma, and urine concentrations and oral intake of antibiotics. In the studies of Kilkkinen *et al.* (2002) and Bolvig *et al.* (2017a) it was found that the number of treatments, the time from the last treatment, and the type of antibiotics influenced serum Enl levels. A follow-up study in pigs fed either a low- or high-lignan diet with or without dosing with controlled levels of antibiotics showed that treatment with antibiotics resulted in substantially lower concentrations of Enl compared with concentrations detected in untreated animals, whereas the plasma concentrations of plant lignans were unchanged (Bolvig *et al.*, 2017a). It was also found that both diet and antibiotic treatment affected the clustering of urinary metabolites and significantly altered the proportions of taxa in the gut microbiota.

Kinetics of Absorption of Enterolignans

From kinetic studies with a single dose of SDG and Ses in humans, it was known that peak plasma concentrations of enterolignans were not obtained until 20–36 hours and 12–24 hours (see Figures 10.4 and 10.5), respectively (Kuijsten *et al.*, 2005b; Penalvo *et al.*, 2005b; Setchell *et al.*, 2014). The associated half-lives have been estimated to be 12–13 hours and 4–10 hours (Kuijsten *et al.*, 2005b; Setchell *et al.*, 2014). In model studies with catheterized pigs, it was also shown that when the pigs were transferred from a low-lignan diet to lignan-rich rye-based diets, it took around six hours (Figure 10.9) until the concentration of Enl started to increase in the portal vein (Bach Knudsen *et al.*, 2003) and an additional 3–4 days (Figure 10.10) to reach plateau level (Bach Knudsen *et al.*, 2003; Bolvig *et al.*, 2017b).

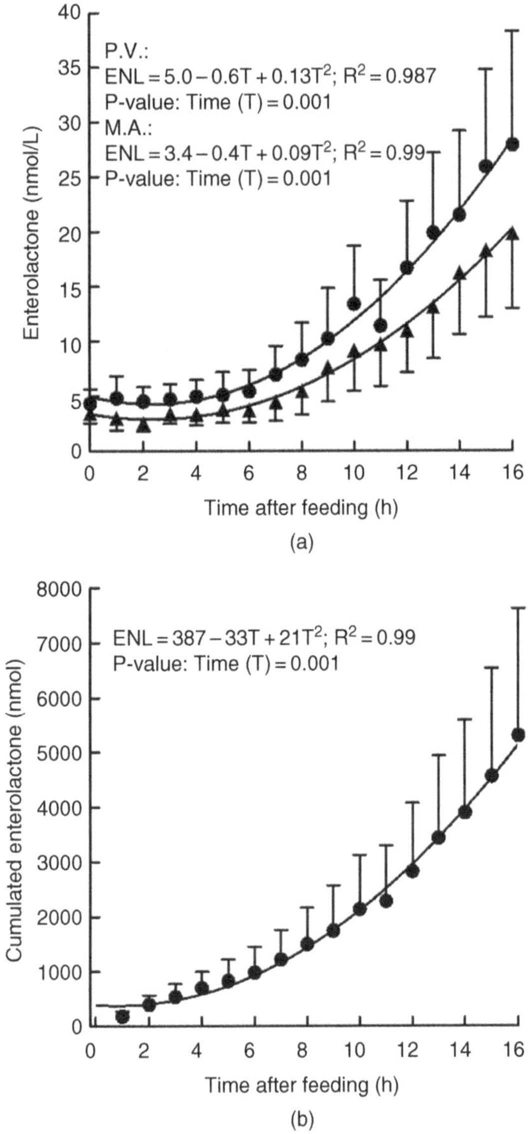

Figure 10.9 Portal (*closed circle*) and arterial (*closed triangle*) blood concentrations (nmol/L) of enterolactone (ENL) (a) and cumulated uptake of Enl (b) following the intake of a single meal of a rye diet. The values are means of four animals. Source: Data used with permission from Bach Knudsen *et al.* (2003).

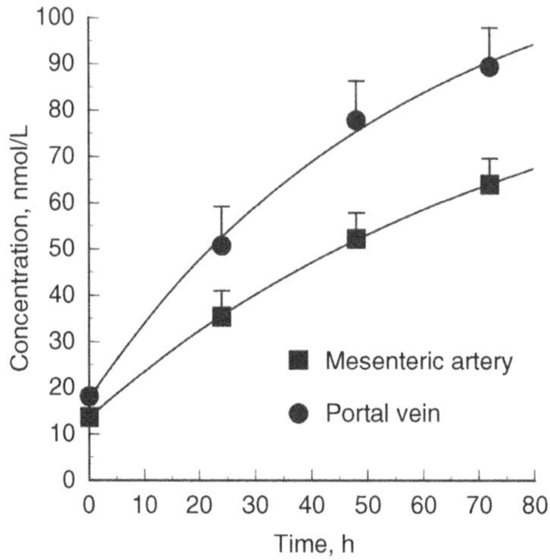

Figure 10.10 Fasting portal and arterial enterolignan plasma concentrations after switching from a lignan-poor diet (time 0) to lignan-rich diets. Plasma concentrations are presented as means resulting from intake of the three lignan-rich diets WWG, WAF or RAF. The variation in plasma concentration in the mesenteric artery (MA) could be described by: $\text{EnterolignanMA} = 99.0 \times (1\text{-}0.863 \times \text{EXP}(\text{-}0.0124 \times T)$, $R2 = 0.999$; and in the portal vein (PV) by: $\text{EnterolignanPV} = 119.2 \times (1\text{-}0.852 \times \text{EXP}(\text{-}0.0175 \times T)$, $R2 = 0.997$, where T is the time after switching to the lignan-rich diets. SEM are indicated with error bars. Source: Data used with permission from Bolvig *et al.* (2016a).

The enterohepatic circulation plays a crucial role in the diurnal variation of plasma levels of Enl (Bach Knudsen *et al.*, 2003). Flow of lignans from the intestine to the liver and reexcretion through the bile duct back to the intestinal tract contribute to a longer residence time for enterolignans in the organism. The enterohepatic circulation will also have a buffering effect on the plasma levels of Enl, which levels out the diurnal fluctuation in microbial production in the gut (Bach Knudsen *et al.*, 2003).

10.4 Blood Levels of Lignans after Dietary Intervention

Numerous human (Di *et al.*, 2017; Jacobs *et al.*, 2002; Juntunen *et al.*, 2000; Kuijsten *et al.*, 2005a; Mazur *et al.*, 2000; McIntosh *et al.*, 2003; Stumpf *et al.*, 2000; Tarpila *et al.*, 2002) and animal (Bach Knudsen *et al.*, 2003; Bolvig *et al.*, 2016a, 2017a, 2017b; Lærke *et al.*, 2009) studies have shown that dietary intervention with lignan-containing foods leads to an increase in blood levels of Enl in almost all subjects (Tables 10.3 and 10.4). The interindividual variation for Enl, however, is far higher than for other metabolites deriving from microbial fermentation in the large intestine such as, for example, short-chain fatty acids (Bach Knudsen, 2005; Bach Knudsen *et al.*, 2003; Halldin *et al.*, 2019).

Most of the studies used amounts of whole grains, oilseeds or fruits that are relevant in terms of daily food consumption and it is likely that moderate changes in dietary habits may significantly alter blood levels of Enl (see Table 10.3). For instance, Mazur *et al.* (2000), Kuijsten *et al.* (2005a), Penalvo *et al.* (2005b) and Setchell *et al.* (2014) all showed that blood levels of Enl in human subjects significantly increased after a single meal of strawberries, flaxseed, and sesamin. Studies with pigs have further explained that it takes some time after the transfer from a lignan-poor to a lignan-rich (cereal-based) diet until a plateau level is established (Bach Knudsen *et al.*, 2003; Bolvig *et al.*, 2016a, 2017b).

However, it is not fully established whether Enl levels in biological matrices reach maximum values caused by a specific ingested dose of dietary precursors or whether there will be a gradual adaptation of the microbiota that will lead to an increase over longer periods. For instance, in a 12-week intervention favoring intake of lignan-containing foods (Stumpf *et al.*, 2000), most increases in the blood concentration of Enl occurred during the first six weeks; median values were 12.2, 17.2 and 19.5 nmol/L at baseline and after six and 12 weeks, respectively (n=85). Conversely, Tarpila *et al.* (2002) observed a continuous increase during four months of intervention with flaxseed (serum Enl concentrations were 33, 52 and 70 nmol/L at baseline and after two and four months, respectively; n=80) (see Table 10.3). In the latter study, the participants also ingested inulin, which potentially could alter the microbial activities and composition towards more *Lactobacillus* and *Bifidobacterium* spp. (Peiroten *et al.*, 2019). Further, the time it takes to reach a plateau may be different depending on whether the lignans are provided from nonlignified plant tissues, such as those

Table 10.3 Influence of intake of plant lignans on plasma, urinary, and feces concentrations of enterolactone in human subjects

	Subjects,		Plasma	Urine		
	n	Intake	nmol/L	µmol/L	Mmol/24 h	Reference
Baseline	7		10.3±2.5	1.5±0.4		(Mazur *et al.*, 2000)
Strawberry meal, 500 g		12.3 mg/d	20.6±6.2	1.4±0.2		
Wheat – males	10	0.06 µmol/d	12.5±1.6	2.7±0.4	4.0±0.7	(Juntunen *et al.*, 2000)
		0.05 µmol/d				
Wheat – females	12		14.8±1.9	2.3±0.5	3.7±0.8	
Rye – males		0.54 µmol/d	25.6±4.7	5.1±1.2	6.8±1.4	
		0.40 µmol/d	39.7±6.5	4.9±0.8	7.8±1.8	
Rye – females						
Baseline	85		12.2 [10.4, 19.3]			(Stumpf *et al.*, 2000)
Interventions with whole grains, vegetables, fruits, and berries			19.5 [16.1, 31.5]			
Switchback			17.6 [14.2, 26.5]			
Baseline	80 (62 females, 18 males)		33			(Tarpila *et al.*, 2002)
Ground flaxseed, flaxseed oil plus 3–4 g inulin and wheat bran per 100 g:						
2 month		8 mg/d	52			
4 month		8 mg/d	70			

Table 10.3 (Continued)

	Subjects, n	Intake	Plasma nmol/L	Urine µmol/L	Urine Mmol/24 h	Reference
Low-fiber diet (19 g DF/d)			18.1±2.7			(McIntosh *et al.*, 2003)
High-fiber wheat diet (32 g DF/d; 18 g from wheat foods)	28 males		21.0±3.8			
High-fiber rye diet (32 g DF/d; 18 g from rye foods)			30.9±3.8			
Refined grain (white bread, refined wheat, rice and corn products)	12 (6 females, 6 males)		11.1			(Jacobs *et al.*, 2002)
Whole-grain diet (whole-grain products 80% wheat, remaining oats, rice, corn, barley, and rye)			19.2			
Baseline providing a diet poor in lignans		<0.15 µmol lignans/ kg body weight	9.5±1.1[a]			(Kuijsten *et al.*, 2005a)
Whole flaxseed		3.0 µmol lignans/ kg body weight	65±16[b]			
Crushed flaxseed			85±17[c]			
Ground flaxseed			167±25[d]			

a) Enterodiol 1.9±0.5, total enterolignans 11.4±1.6.
b) Enterodiol, 20±5, total enterolignans 85±19.
c) #Enterodiol 37±11, total enterolignans 122±21.
d) Enterodiol 103±31, total enterolignans 270±34.
DF, dietary fiber.

Table 10.4 Influence of diets varying in plant lignans on plasma and urine concentrations of enterolactone and urinary excretion of enterolactone in model experiments with pigs

Intervention	Content per kg DM		Plasma Enl concentration, nmol/L		Urine Enl		Reference
	DF (g)	PL (ng)	Portal vein	Mesentery artery	µmol/L	µmol/d	
Wheat flour plus refined wheat fiber	234	3.1	74±6.7	56±4.0	0.2±0.02	2.3±0.3	(Bach Knudsen *et al.*, 2003)
Whole-grain rye plus rye bran	253	35.2	451±51.3	343±53.0	1.6±0.3	16.6±1.4	
Wheat flour plus refined wheat fiber	194	1.2	3.9 [2.9, 5.3]	3.6 [2.5, 5.2]	0.4±0.06	2.0±0.2	(Lærke *et al.*, 2009)
Whole-grain rye plus rye bran	203	41.5	147.3 [110.3, 196.9]	96.1 [66.1, 133.9]	16.3 ±1.4	106±4.9	
Wheat flour plus refined wheat fiber	155	1.9	18.1	13.6	0.8 [0.4, 1.9]	3.4 [2.3, 5.1]	(Bolvig *et al.*, 2016a)
Whole-grain wheat	155	9.2	81.7±9.3	57.2±5.5	4.9 [1.9, 12.5]	16.1 [10.3, 25.3]	
Wheat aleurone flour	151	15.0	69.6	43.0	4.9 [1.9, 12.3]	23.7 [15.1, 37.2]	
Rye aleurone flour	186	13.2	67.0	48.0	5.7 [2.5, 13.1]	23.5 [15.7, 35.1]	
Wheat flour plus refined wheat fiber	162	1.5	5.1	3.5	1.3 [1.0, 1.7]		(Bolvig *et al.*, 2017a)
Rye bran	161	21.1	180.5	123.4	58.7 [45.4, 75.9]		

DF, dietary fiber; DM, dry matter; PL, plant lignans.

found in oilseeds and fruits, or from lignified tissues like in cereals. Juntunen *et al.* (2000) found that a daily intake of more than 90 g rye bread for four weeks did not trigger a further increase in blood levels of Enl. On the other hand, Nesbitt *et al.* (1999) observed a dose-response in daily urinary excretion of lignans after intake of 5, 15 or 25 g flaxseed/d for seven days and with no plateau level at 25 g/d. Therefore, it is likely that after a graduated increase of plasma Enl over some days (Bach Knudsen *et al.*, 2003; Bolvig *et al.*, 2016a, 2017b), there will be a smaller relative increase in Enl levels long term, which may be caused by adaptive responses in microbiota activities, that not necessarily are accompanied by detectable changes in bacterial diversity or proportions.

There are indications from both animal and human studies of a sex difference in Enl plasma concentration. In one of the early studies on Enl, Axelson and Setchell (1981) found that female rats excreted more Enl than male. An intervention study with men and women fed whole-grain rye rich in lignans also showed numerically higher Enl in women than in men in spite of a lower intake (see Table 10.3) (Juntunen *et al.*, 2000). Kilkkinen *et al.* (2001) investigated determinants of serum Enl concentration and found a higher variability between individuals, but with a higher range in women (0–182.6 nmol/L) compared to men (0–95.6 nmol/L). The higher production is possibly related to the higher number of Enl-producing bacteria found in women (Clavel *et al.*, 2005), perhaps in combination with a generally longer gastrointestinal transit (Lampe *et al.*, 1993) that allows the colonic bacteria more time to metabolize the plant lignans. However, as reviewed by Yoder *et al.* (2015), not all studies show the same picture.

The majority of studies on lignans have so far primarily measured blood levels of Enl, whereas our knowledge on the levels of plant lignans is much more limited. Studies with sesamin (Penalvo *et al.*, 2005b), flaxseeds (Kuijsten *et al.*, 2005b; Setchell *et al.*, 2014) and cereals (Bolvig *et al.*, 2016a, 2017b) have all shown that plant lignans are rapidly absorbed and present in the blood shortly after ingestion. These studies, however, also showed that plant lignans have a much shorter half-life than Enl and thus are less suited as biomarkers in epidemiological studies.

10.5 Bioactivity of Plant Lignans and Enterolignans

Most studies on the biological activity of lignans have been related to their anticancer role, with particular emphasis on their estrogen-like structure. Enl has been found to bind weakly to estrogen receptors (Penttinen *et al.*, 2007), to possibly interact with the activity of sex hormone-binding globulins (SHBG) (Schottner *et al.*, 1997) and be positively associated with levels of SHBG (Adlercreutz *et al.*, 1987). Enterolignans have also been shown to have modulating effects on angiogenesis and growth factors, indicating that Enl may affect breast carcinogenesis (Brooks and Thompson, 2005).

Although the biological activity of enterolignans, particularly concerning cancer, has been discussed in several studies (Adlercreutz, 2002, 2007; Landete, 2012), much less is known about the biological function of plant lignans. In an *in vitro* study by Hu *et al.* (2007), it was found that the antioxidant activities of flaxseed lignan SDG and its aglycone Seco were much higher than those of Enl or End. In a recent study with the plant lignan Syr isolated from *Rubia philippinensis*, it was found that Syr mediated antiinflammatory effects by downregulating NF-κB expression through interference with JNK and p38 phosphorylation and by reducing mRNA levels of iNOS, COX-2, TNF-α, IL-1β, and IL-6, pointing towards its significant biological potential (Figure 10.11) (Bajpai *et al.*, 2018). Syr is present in high concentrations in whole-grain rye and bran (see Tables 10.1 and 10.2). Other cell studies have also indicated that Syr has the ability to alter lipid and glucose metabolism in HepG2 cells and C2C12 myotubes (Wang *et al.*, 2017) and cause vasorelaxation by elevating nitric oxide production (Chung *et al.*, 2012).

Although it can be challenging to extrapolate from *in vitro* studies to *in vivo* conditions, the above-mentioned data suggest that plant lignans themselves could have health-promoting properties. In this context, it is interesting that the systemic concentration of plant lignans is high (Bolvig *et al.*, 2016a, 2017b; Kuijsten *et al.*, 2005b; Setchell *et al.*, 2014) during the periods with high influx of nutrients to the body and high cellular activity (Ingerslev *et al.*, 2014).

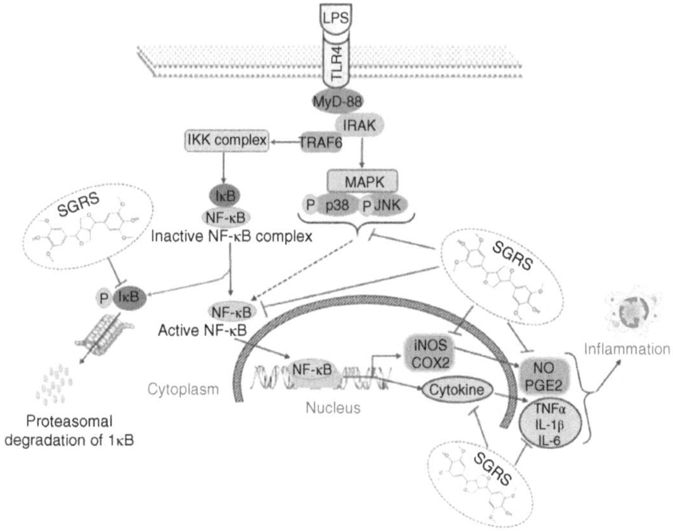

Figure 10.11 Schematic presentation of the mechanisms behind the antiinflammatory effect of (+)-syringaresinol (SGRS). Source: Data used with permission from Bajpai *et al.* (2018).

10.6 Conclusions and Future Perspectives

Plant lignans are diphenolic compounds present in a wide variety of edible plants, of which cereals, nuts, vegetables, berries, and beverages such as tea and coffee are the most important in normal diets. The generally accepted view is that plant lignans are converted to the metabolic active enterolignans End and Ent by the microbiota in the large intestine, absorbed, and taken up in circulation, where they provide protection against several diseases including cancer (breast, prostate, and colorectal), cardiovascular diseases, and type 2 diabetes. Studies with flaxseeds, sesamin, and cereals, however, have also shown that a certain fraction of the plant lignans is absorbed as such but with a half-life much lower than that of the enterolignans. The biological role of the absorbed plant lignans is still virtually unknown but cell studies have indicated that plant lignans could have anti-inflammatory and other cellular effects. To obtain the whole picture of the health properties of lignans, future studies should include the plant lignans as well.

Acknowledgments

This work was supported by Innovation Fund Denmark (Project ELIN: the effects of enterolignans in chronic disease: 0603-00580B) and Aarhus University.

References

Aarestrup, J., Kyrø, C., Bach Knudsen, K.E., *et al.* (2013) Plasma enterolactone and incidence of endometrial cancer in a case–cohort study of Danish women. *British Journal of Nutrition*, **109**, 2269–75.

Adgent, M.A. and Rogan, W.J. (2015) Triclosan and prescription antibiotic exposures and enterolactone production in adults. *Environmental Research*, **142**, 66–71.

Adlercreutz, H. (2002) Phyto-oestrogens and cancer. *Lancet Oncology*, **3**, 364–73.

Adlercreutz, H. (2007) Lignans and human health. *Critical Review in Clinical Laboratory Sciences*, **44**, 483–525.

Adlercreutz, H., Höckerstedt, K., Bannwart, C., *et al.* (1987) Effect of dietary components, including lignans and phytoestrogens, on enterohepatic circulation and liver metabolism of estrogens and on sex hormone binding globulin (SHBG). *Journal of Steroid Biochemistry and Molecular Biology*, **27**, 1135–44.

Adlercreutz, H., Vanderwildt, J., Kinzel, J., *et al.* (1995) Lignan and isoflavonoid conjugates in human urine. *Journal of Steroid Biochemistry and Molecular Biology*, **52**, 97–103.

Antunes, L.C. and Finlay, B.B. (2011) A comparative analysis of the effect of antibiotic treatment and enteric infection on intestinal homeostasis. *Gut Microbes*, **2**, 105–8.

Aubertin-Leheudre, M., Gorbach, S., Woods, M., Dwyer, J.T., Goldin, B., and Adlercreutz, H. (2008) Fat/fiber intakes and sex hormones in healthy premenopausal women in USA. *Journal of Steroid Biochemistry and Molecular Biology*, **112**, 32–9.

Axelson, M. and Setchell, D.R. (1981) The excretion of lignans in rats – evidence for an intestinal source for this new group of compounds. *FEBS Letters*, **123**, 337–43.

Bach Knudsen, K.E. (2005) Effect of dietary non-digestible carbohydrates on the rate of SCFA delivery to peripheral tissues. *Foods and Food Ingredients Journal of Japan*, **210**, 1008–17.

Bach Knudsen, K.E., Serena, A., Kjær, A.K., *et al.* (2003) Rye bread in the diet of pigs enhances the formation of enterolactone and increases its levels in plasma, urine and feces. *Journal of Nutrition*, **133**, 1368–75.

Bach Knudsen, K.E., Serena, A., Kjær, A.B.K., Jørgensen, H., and Engberg, R. (2005) Rye bread enhances the production and plasma concentration of butyrate but not the plasma concentrations of glucose and insulin in pigs. *Journal of Nutrition*, **135**, 1696–704.

Bajpai, V.K., Alam, M.B., Quan, K.T., *et al.* (2018) Attenuation of inflammatory responses by (+)-syringaresinol via MAP-Kinase-mediated suppression of NF-kappaB signaling in vitro and in vivo. *Science Reports*, **8**, 9216.

Begum, A.N., Nicolle, C., Mila, I., *et al.* (2004) Dietary lignins are precursors of mammalian lignans in rats. *Journal of Nutrition*, **134**, 120–7.

Bolvig, A.K., Adlercreutz, H., Theil, P.K., Jørgensen, H., and Bach Knudsen, K.E. (2016a) Absorption of plant lignans from cereals in an experimental pig model. *British Journal of Nutrition*, **115**, 1711–20.

Bolvig, A.K., Kyrø, C., Nørskov, N.P., *et al.* (2016b) Use of antibiotics is associated with lower enterolactone plasma concentration. *Molecular Nutrition & Food Research*, **60**, 2712–21.

Bolvig, A.K., Nørskov, N.P., Hedemann, M.S., *et al.* (2017a) Effect of antibiotics and diet on enterolactone concentration and metabolome studied by targeted and nontargeted LC-MS metabolomics. *Journal of Proteome Research*, **16**, 2135–50.

Bolvig, A.K., Nørskov, N.P., Van Vliet, S., *et al.* (2017b) Rye bran modified with cell wall-degrading enzymes influences the kinetics of plant lignans but not of enterolignans in multicatheterized pigs. *Journal of Nutrition*, **147**, 2220–7.

Bondia-Pons, L., Aura, A.M., Vuorela, S., Kolehmainen, M., Mykkenen, H., and Poutanen, K.S. (2009) Rye phenolics in nutrition and health. *Journal of Cereal Science*, **49**, 323–36.

Borriello, S.P., Setchell, D.R., Axelson, M., and Lawson, A.M. (1985) Production and metabolism of lignans by the human faecal flora. *Journal of Applied Bacteriology*, **58**, 37–43.
Bowey, E., Adlercreutz, H., and Rowland, I. (2003) Metabolism of isoflavones and lignans by the gut microflora: a study in germ-free and human flora associated rats. *Food and Chemical Toxicology*, **41**, 631–6.
Brooks, J.D. and Thompson, L.U. (2005) Mammalian lignans and genistein decrease the activities of aromatase and 17beta-hydroxysteroid dehydrogenase in MCF-7 cells. *Journal of Steroid Biochemistry and Molecular Biology*, **94**, 461–7.
Buck, K., Zaineddin, A.K., Vrieling, A., *et al.* (2011) Estimated enterolignans, lignan-rich foods, and fibre in relation to survival after postmenopausal breast cancer. *British Journal of Cancer*, **105**, 1151–7.
Chen, J., Tan, K.P., Ward, W.E., and Thompson, L.U. (2003) Exposure to flaxseed or its purified lignan during suckling inhibits chemically induced rat mammary tumorigenesis. *Experimental Biology and Medicine*, **228**, 951–8.
Chen, J., Wang, L., and Thompson, L.U. (2006) Flaxseed and its components reduce metastasis after surgical excision of solid human breast tumor in nude mice. *Cancer Letters*, **234**, 168–75.
Chung, B.H., Kim, S., Kim, J.D., *et al.* (2012) Syringaresinol causes vasorelaxation by elevating nitric oxide production through the phosphorylation and dimerization of endothelial nitric oxide synthase. *Experimental & Molecular Medicine*, **44**, 191–201.
Clavel, T., Henderson, G., Alpert, C.A., *et al.* (2005) Intestinal bacterial communities that produce active estrogen-like compounds enterodiol and enterolactone in humans. *Applied and Environmental Microbiology*, **71**, 6077–85.
Clavel, T., Borrmann, D., Braune, A., Dore, J., and Blaut, M. (2006a) Occurrence and activity of human intestinal bacteria involved in the conversion of dietary lignans. *Anaerobe*, **12**, 140–7.
Clavel, T., Henderson, G., Engst, W., Dore, J., and Blaut, M. (2006b) Phylogeny of human intestinal bacteria that activate the dietary lignan secoisolariciresinol diglucoside. *FEMS Microbiology Ecology*, **55**, 471–8.
Clavel, T., Lippman, R., Gavini, F., Dore, J., and Blaut, M. (2007) Clostridium saccharogumia sp. nov. and Lactonifactor longoviformis gen. nov., sp. nov., two novel human faecal

bacteria involved in the conversion of the dietary phytoestrogen secoisolariciresinol diglucoside. *Systemic and Applied Microbiology*, **30**, 16–26.

Cunha, W., E Silva, M., Sola R., Veneziani, S., and Bastos, J. (2012) Lignans: chemical and biological properties. In V. Rao (Ed.), *Phytochemicals – A Global Perspective of Their Role in Nutrition and Health*. Rijeka, Croatia: InTech Europe.

Dar, A.A. and Arumugam, N. (2013) Lignans of sesame: purification methods, biological activities and biosynthesis – a review. *Bioorganic Chemistry*, **50**, 1–10.

Davin, L.B., Jourdes, M., Patten, A.M., Kim, K.W., Vassao, D.G., and Lewis, N.G. (2008) Dissection of lignin macromolecular configuration and assembly: comparison to related biochemical processes in allyl/propenyl phenol and lignan biosynthesis. *Natural Product Reports*, **25**, 1015–90.

Dethlefsen, L., Huse, S., Sogin, M.L., and Relman, D.A. (2008) The pervasive effects of an antibiotic on the human gut microbiota, as revealed by deep 16S rRNA sequencing. *PLoS Biology*, **6**, e280.

Di, Y., Jones, J., Mansell, K., *et al.* (2017) Influence of flaxseed lignan supplementation to older adults on biochemical and functional outcome measures of inflammation. *Journal of the American College of Nutrition*, **36**, 646–53.

Eeckhaut, E., Struijs, K., Possemiers, S., Vincken, J.P., Keukeleire, D.D., and Verstraete, W. (2008) Metabolism of the lignan macromolecule into enterolignans in the gastrointestinal lumen as determined in the simulator of the human intestinal microbial ecosystem. *Journal of Agricultural and Food Chemistry*, **56**, 4806–12.

Eriksen, A.K., Kyrø, C., Nørskov, N.P., *et al.* (2017) Prediagnostic enterolactone concentrations and mortality among Danish men diagnosed with prostate cancer. *European Journal of Clinical Nutrition*, **71**, 1235–40.

Eriksen, A.K., Kyrø, C., Nørskov, N.P., *et al.* (2019) Pre-diagnostic plasma enterolactone concentrations are associated with lower mortality among individuals with type 2 diabetes: a case-cohort study in the Danish Diet, Cancer and Health cohort. *Diabetologia*, **62**, 959–69.

Evans, B.A., Griffiths, K., and Morton, M.S. (1995) Inhibition of 5 alpha-reductase in genital skin fibroblasts and prostate tissue by dietary lignans and isoflavonoids. *Journal of Endocrinology*, **147**, 295–302.

Flint, H.J., Bayer, E.A., Rincon, M.T., Lamed, R., and White, B.A. (2008) Polysaccharide utilization by gut bacteria: potential for new insights from genomic analysis. *Nature Reviews Microbiology*, **6**, 121–31.

Flint, H.J., Scott, K.P., Duncan, S.H., Louis, P., and Forano, E. (2012) Microbial degradation of complex carbohydrates in the gut. *Gut Microbes*, **3**, 289–306.

Ford, J.D., Huang, K.S., Wang, H.B., Davin, L.B., and Lewis, N.G. (2001) Biosynthetic pathway to the cancer chemopreventive secoisolariciresinol diglucoside-hydroxymethyl glutaryl ester-linked lignan oligomers in flax (Linum usitatissimum) seed. *Journal of Natural Products*, **64**, 1388–97.

Glitsø, L.V. and Bach Knudsen, K.E. (1999) Milling of whole grain rye to obtain fractions with different dietary fibre characteristics. *Journal of Cereal Science*, **29**, 89–97.

Glitsø, L.V., Mazur, W.M., Adlercreutz, H., *et al.* (2000) Intestinal metabolism of rye lignans in pigs. *British Journal of Nutrition*, **84**, 429–37.

Guglielmini, P., Rubagotti, A., and Boccardo, F. (2012) Serum enterolactone levels and mortality outcome in women with early breast cancer: a retrospective cohort study. *Breast Cancer Research and Treatment*, **132**, 661–8.

Halldin, E., Eriksen, A.K., Brunius, C., *et al.* (2019) Factors explaining interpersonal variation in plasma enterolactone concentrations in humans. *Molecular Nutrition & Food Research*, **63**, e1801159.

Harmatha, J. and Dinan, L. (2003) Biological activities of lignans and stilbenoids associated with plantinsect chemical interactions. *Phytochemistry Reviews*, **2**, 321–30.

Heinonen, S., Nurmi, T., Liukkonen, K., *et al.* (2001) In vitro metabolism of plant lignans: new precursors of mammalian lignans enterolactone and enterodiol. *Journal of Agricultural and Food Chemistry*, **49**, 3178–86.

Hollaender, P.L., Ross, A.B., and Kristensen, M. (2015) Whole-grain and blood lipid changes in apparently healthy adults: a systematic review and meta-analysis of randomized controlled studies. *American Journal of Clinical Nutrition*, **102**, 556–72.

Horn-Ross, P.L., Lee, M., John, E.M., and Koo, J. (2000) Sources of phytoestrogen exposure among non-Asian women in California, USA. *Cancer Causes & Control*, **11**, 299–302.

Hu, C., Yuan, Y.V., and Kitts, D.D. (2007) Antioxidant activities of the flaxseed lignan secoisolariciresinol diglucoside, its aglycone secoisolariciresinol and the mammalian lignans enterodiol and enterolactone in vitro. *Food and Chemical Toxicology*, **45**, 2219–27.

Hullar, M.A., Lancaster, S.M., Li, F., *et al.* (2015) Enterolignan-producing phenotypes are associated with increased gut microbial diversity and altered composition in premenopausal women in the United States. *Cancer Epidemiology, Biomarkers & Prevention*, **24**, 546–54.

Ingerslev, A.K., Theil, P.K., Hedemann, M.S., Lærke, H.N., and Bach Knudsen, K.E. (2014) Resistant starch and arabinoxylan augment SCFA absorption, but affect postprandial glucose and insulin responses differently. *British Journal of Nutrition*, **111**, 1564–76.

Jacobs, D.R., Pereira, M.A., Stumpf, K., Pins, J.J., and Adlercreutz, H. (2002) Whole grain food intake elevates serum enterolactone. *British Journal of Nutrition*, **88**, 111–16.

Jancova, P., Anzenbacher, P., and Anzenbacherova, E. (2010) Phase II drug metabolizing enzymes. *Biomedical Papers of the Medical Faculty of the University of Palacky, Olomouc, Czech Republic*, **154**, 103–16.

Jansen, G.H., Arts, I.C., Nielen, M.W., Muller, M., Hollman, P.C., and Keijer, J. (2005) Uptake and metabolism of enterolactone and enterodiol by human colon epithelial cells. *Archives of Biochemistry and Biophysics*, **435**, 74–82.

Jin, J.S., Kakiuchi, N., and Hattori, M. (2007a) Enantioselective oxidation of enterodiol to enterolactone by human intestinal bacteria. *Biological & Pharmaceutical Bulletin*, **30**, 2204–6.

Jin, J.S., Zhao, Y.F., Nakamura, N., *et al.* (2007b) Enantioselective dehydroxylation of enterodiol and enterolactone precursors by human intestinal bacteria. *Biological & Pharmaceutical Bulletin*, **30**, 2113–19.

Johnsen, N.F., Hausner, H., Olsen, A., *et al.* (2004) Intake of whole grains and vegetables determines the plasma enterolactone concentration of Danish women. *Journal of Nutrition*, **134**, 2691–7.

Johnsen, N.F., Olsen, A., Thomsen, B.L., *et al.* (2010) Plasma enterolactone and risk of colon and rectal cancer in a case-cohort study of Danish men and women. *Cancer Causes & Control*, **21**, 153–62.

Juntunen, K.S., Mazur, W.M., Liukkonen, K.H., *et al.* (2000) Consumption of wholemeal rye bread increases serum concentrations and urinary excretion of enterolactone compared with consumption of white wheat bread in healthy Finnish men and women. *British Journal of Nutrition*, **84**, 839–46.

Kajla, P., Sharma, A., and Sood, D.R. (2015) Flaxseed – a potential functional food source. *Journal of Food Science and Technology*, **52**, 1857–71.

Kamal-Eldin, A., Peerlkamp, N., Johnsson, P., *et al.* (2001) An oligomer from flaxseed composed of secoisolariciresinoldiglucoside and 3-hydroxy-3-methyl glutaric acid residues. *Phytochemistry*, **58**, 587–90.

Kilkkinen, A., Stumpf, K., Pietinen, P., Valsta, L.M., Tapanaine, H., and Adlercreutz, H. (2001) Deterninants of serum enterolactone concentration. *American Journal of Clinical Nutrition*, **73**, 1094–100.

Kilkkinen, A., Pietinen, P., Klaukka, T., Virtamo, J., Korhonen, P., and Adlercreutz, H. (2002) Use of oral antimicrobials decreases serum enterolactone concentration. *American Journal of Epidemiology*, **155**, 472–77.

Kuijsten, A., Arts, I.C., Van't Veer, P., and Hollman, P.C. (2005a) The relative bioavailability of enterolignans in humans is enhanced by milling and crushing of flaxseed. *Journal of Nutrition*, **135**, 2812–16.

Kuijsten, A., Arts, J.C.W., Vree, T.B., and Hollman, P.C. (2005b) Pharmacokinetics of enterolignans in healthy men and women consuming a single dose of secoisolariciresinol diglucoside. *Journal of Nutrition*, **135**, 795–801.

Kyrø, C., Frederiksen, K., Holm, M., *et al.* (2018a) Prediagnosis plasma concentrations of enterolactone and survival after colorectal cancer: the Danish Diet, Cancer and Health cohort. *British Journal of Nutrition*, **122**, 552–63.

Kyrø, C., Hansen, L., Frederiksen, K., *et al.* (2018b) Prediagnostic plasma enterolactone concentrations and breast cancer prognosis among postmenopausal women – the Danish Diet, Cancer and Health cohort. *Clinical Nutrition*, **37**, 2217–25.

Lærke, H.N., Mortensen, M.A., Hedemann, M.S., Bach Knudsen, K.E., Penalvo, J.L., and Adlercreutz, H. (2009) Quantitative aspects of the metabolism of lignans in pigs fed fibre-enriched rye and wheat bread. *British Journal of Nutrition*, **102**, 985–94.

Lampe, J.W., Fredstrom, S.B., Slavin, J.L., and Potter, J.D. (1993) Sex differences in colonic function: a randomised trial. *Gut*, **34**, 531–6.

Landete, J.M. (2012) Plant and mammalian lignans: a review of source, intake, metabolism, intestinal bacteria and health. *Food Research International*, **46**, 410–24.

Mazur, W.M. (1998) Phytoestrogen content in foods. *Baillière's Clinical Endocrinology and Metabolism*, **12**, 729–43.

Mazur, W.M., Wähälä, K., Rasku, S., Salakka, A., Hase, T., and Adlercreutz, H. (1998) Lignan and isoflavonoid concentrations in tea and coffe. *British Journal of Nutrition*, **79**, 37–45.

Mazur, W.M., Uehara, M., Wahala, K., and Adlercreutz, H. (2000) Phyto-oestrogen content of berries, and plasma concentrationsand urinary excretion of enterolactone after a single strawberry-meal in human subjects. *British Journal of Nutrition*, **83**, 381–7.

McCann, S.E., Thompson, L.U., Nie, J., *et al.* (2010) Dietary lignan intakes in relation to survival among women with breast cancer: the Western New York Exposures and Breast Cancer (WEB) Study. *Breast Cancer Research and Treatment*, **122**, 229–35.

McIntosh, G.H., Noakes, M., Royle, P.J., and Foster, P.R. (2003) Whole-grain rye and wheat foods and markers of bowel health in overweight middle-aged men. *American Journal of Clinical Nutrition*, **77**, 967–74.

Menendez, J.A., Vazquez-Martin, A., Oliveras-Ferraros, C., *et al.* (2008) Analyzing effects of extra-virgin olive oil polyphenols on breast cancer-associated fatty acid synthase protein expression using reverse-phase protein microarrays. *International Journal of Molecular Medicine*, **22**, 433–9.

Milder, I.E., Arts, I.C., Venema, D.P., Lasaroms, J.J., Wahala, K., and Hollman, P.C. (2004) Optimization of a liquid chromatography-tandem mass spectrometry method for quantification of the plant lignans secoisolariciresinol, matairesinol, lariciresinol, and pinoresinol in foods. *Journal of Agricultural and Food Chemistry*, **52**, 4643–51.

Milder, I.E., Arts, I.C., van de Putte, B., Venema, D.P., and Hollman, P.C. (2005) Lignan contents of Dutch plant foods: a database including lariciresinol, pinoresinol, secoisolariciresinol and matairesinol. *British Journal of Nutrition*, **93**, 393–402.

Morand, C. and Tomás-Barberán, F.A. (2019) Interindividual variability in absorption, distribution, metabolism, and excretion of food phytochemicals should be reported. *Journal of Agricultural and Food Chemistry*, **67**, 3843–4.

Mukker, J.K., Michel, D., Muir, A.D., Krol, E.S., and Alcorn, J. (2014) Permeability and conjugative metabolism of flaxseed lignans by Caco-2 human intestinal cells. *Journal of Natural Products*, **77**, 29–34.

Nesbitt, P.D., Lam, Y., and Thompson, L.U. (1999) Human metabolism of mammalian lignan precursors in raw and processed flaxseed. *American Journal of Clinical Nutrition*, **69**, 549–55.

Nilsson, M., Åman, P., Harkonen, H., *et al.* (1997) Content of nutrients and lignans in roller milled fractions of rye. *Journal of the Science of Food and Agriculture*, **73**, 143–8.

Nørskov, N.P. (2019) Total and free plant lignans in whole grain and cereal fractions. Unpublished.

Nørskov, N.P. and Bach Knudsen, K.E. (2016) Validated LC-MS/MS method for the quantification of free and bound lignans in cereal-based diets and feces. *Journal of Agricultural and Food Chemistry*, **64**, 8343–51.

Nørskov, N.P., Kyrø, C., Olsen, A., Tjønneland, A., and Bach Knudsen, K.E. (2016) High-throughput lc-ms/ms method for direct quantification of glucuronidated, sulfated, and free enterolactone in human plasma. *Journal of Proteome Research*, **15**, 1051–8.

Nose, M., Fujimoto, F., Takeda, T., Nishibe, S., and Ogihara, Y. (1992) Structural transformation of lignan compounds in rat gastrointestinal tract. *Planta Medica*, **58**, 520–3.

Olsen, A., Bach Knudsen, K.E., Thomsen, B.L., *et al.* (2004) Plasma enterolactone and breast cancer incidence by estrogen receptor status. *Cancer Epidemiology, Biomarkers & Prevention*, **13**, 2084–9.

Olsen, A., Christensen, J., Bach Knudsen, K.E., Johnsen, N.F., Overvad, K., and Tjønneland, A. (2011) Prediagnostic plasma enterolactone levels and mortality among women with breast cancer. *Breast Cancer Research and Treatment*, **128**, 883–9.

Palleja, A., Mikkelsen, K.H., Forslund, S.K., *et al.* (2018) Recovery of gut microbiota of healthy adults following antibiotic exposure. *Nature Microbiology*, **3**, 1255–65.

Pan, J.Y., Chen, S.L., Yang, M.H., Wu, J., Sinkkonen, J., and Zou, K. (2009) An update on lignans: natural products and synthesis. *Natural Product Reports*, **26**, 1251–92.

Peiroten, A., Gaya, P., Alvarez, I., Bravo, D., and Landete, J.M. (2019) Influence of different lignan compounds on enterolignan production by Bifidobacterium and Lactobacillus strains. *International Journal of Food Microbiology*, **289**, 17–23.

Penalvo, J.L., Haajanen, K.M., Botting, N., and Adlercreutz, H. (2005a) Quantification of lignans in food using isotope dilution gas chromatography/mass spectrometry. *Journal of Agricultural and Food Chemistry*, **53**, 9342–7.

Penalvo, J L., Heinonen, S.M., Aura, A.M., and Adlercreutz, H. (2005b) Dietary sesamin is converted to enterolactone in humans. *Journal of Nutrition*, **135**, 1056–62.

Penalvo, J.L., Adlercreutz, H., Uehara, M., Ristimaki, A., and Watanabe, S. (2008) Lignan content of selected foods from Japan. *Journal of Agricultural and Food Chemistry*, **56**, 401–9.

Penttinen, P., Jaehrling, J., Damdimopoulos, A.E., *et al.* (2007) Diet-derived polyphenol metabolite enterolactone is a tissue-specific estrogen receptor activator. *Endocrinology*, **148**, 4875–86.

Pettersson, D., Åmam, P., Bach Knudsen, K.E., *et al.* (1996) Intake of rye bread ileostomists increases ileal excretion of fiber polysaccharide components and organic acids but does not increase plasma or urine lignans and isoflavonoids. *Journal of Nutrition*, **126**, 1594–600.

Rice, S. and Whitehead, S.A. (2006) Phytoestrogens and breast cancer – promoters or protectors? *Endocrine-Related Cancer*, **13**, 995–1015.

Rickard, S.E., Yuan, Y.V., and Thompson, L.U. (2000) Plasma insulin-like growth factor I levels in rats are reduced by dietary supplementation of flaxseed or its lignan secoisolariciresinol diglycoside. *Cancer Letters*, **161**, 47–55.

Rowland, I., Gibson, G., Heinken, A., *et al.* (2017) Gut microbiota functions: metabolism of nutrients and other food components. *European Journal of Nutrition*, **57**, 1–24.

Schottner, M., Gansser, D., and Spiteller, G. (1997) Interaction of lignans with human sex hormone binding globulin (SHBG). *Zeitschrift fur Naturforschung C*, **52**, 834–43.

Schwartz, H. and Sontag, G. (2006) Determination of secoisolariciresinol, lariciresinol and isolariciresinol in plant foods by high performance liquid chromatography coupled with coulometric electrode array detection. *Journal of Chromatography.*
B, Analytical Technologies in the Biomedical and Life Sciences, **838**, 78–85.

Setchell, K.D., Lawson, A.M., Mitchell, F.L., Adlercreutz, H., Kirk, D.N., and Axelson, M. (1980) Lignans in man and in animal species. *Nature*, **287**, 740–2.

Setchell, K.D., Brown, N.M., Zimmer-Nechemias, L., Wolfe, B., Jha, P., and Heubi, J.E. (2014) Metabolism of secoisolariciresinol-diglycoside the dietary precursor to the intestinally derived lignan enterolactone in humans. *Food & Function*, **5**, 491–501.

Smeds, A.I., Eklund, P.C., Sjoholm, R.E., *et al.* (2007) Quantification of a broad spectrum of lignans in cereals, oilseeds, and nuts. *Journal of Agricultural and Food Chemistry*, **55**, 1337–46.

Smeds, A.I., Eklund, P.C., and Willfor, S.M. (2012) Content, composition, and stereochemical characterisation of lignans in berries and seeds. *Food Chemistry*, **134**, 1991–8.

Stumpf, K., Pietinen, P., Puska, P., and Adlercreutz, H. (2000) Changes in serum enterolactone, genistein, and daidzein in a dietary intervention study in Finland. *Cancer Epidemiology Biomarkers & Prevention*, **9**, 1369–72.

Sun, Q., Wedick, N.M., Pan, A., *et al.* (2014) Gut microbiota metabolites of dietary lignans and risk of type 2 diabetes: a prospective investigation in two cohorts of U.S. women. *Diabetes Care*, **37**, 1287–95.

Tarpila, S., Aro, A., Salminen, I., *et al.* (2002) The effect of flaxseed supplementation in processed foods on serum fatty acids and enterolactone. *European Journal of Clinical Nutrition*, **56**, 157–65.

Tetens, I., Turrini, A., Tapanainen, H., *et al.* (2013) Dietary intake and main sources of plant lignans in five European countries. *Food & Nutrition Research*, **57**, e19805.

Thompson, L.U., Boucher, B.A., Liu, Z., Cotterchio, M., and Kreiger, N. (2006) Phytoestrogen content of foods consumed in Canada, including isoflavones, lignans, and coumestan. *Nutrition and Cancer*, **54**, 184–201.

Tuomisto, A., Norskov, N.P., Sirnio, P., *et al.* (2019) Serum enterolactone concentrations are low in colon but not in rectal cancer patients. *Science Reports*, **9**, 11209.

Wang, S., Wu, C., Li, X., *et al.* (2017) Syringaresinol-4-O-beta-d-glucoside alters lipid and glucose metabolism in HepG2 cells and C2C12 myotubes. *Acta Pharmaceutica Sinica B*, **7**, 453–60.

Wang, X., Wei, Y., Yuan, S., *et al.* (2005) Potential anticancer activity of tanshinone IIA against human breast cancer. *International Journal of Cancer*, **116**, 799–807.

Wang, X., Wei, Y., Yuan, S., Liu, G., Zhang, Y.L., and Wang, W. (2006) Potential anticancer activity of litchi fruit pericarp extract against hepatocellular carcinoma in vitro and in vivo. *Cancer Letters*, **239**, 144–50.

Woting, A., Clavel, T., Loh, G., and Blaut, M. (2010) Bacterial transformation of dietary lignans in gnotobiotic rats. *FEMS Microbiology Ecology*, **72**, 507–14.

Yoder, S.C., Lancaster, S.M., Hullar, M.A.J., and Lampe, J.W. (2015) Gut microbial metabolism of plant lignans: influence on human health. In K. Tuohy and D. del Rio (Eds.) *Diet–Microbe Interactions in the Gut*. Philadelphia: Elsevier.

11

Stilbenes: Beneficial Effects of Resveratrol Metabolites in Obesity, Dyslipidemia, Insulin Resistance, and Inflammation

Itziar Eseberri[1,2], Iñaki Milton-Laskibar[1,2], Alfredo Fernández-Quintela[1,2], Saioa Gómez-Zorita[1,2], and María P. Portillo[1,2]*

[1] Nutrition and Obesity Group, Department of Nutrition and Food Sciences, Faculty of Pharmacy, University of the Basque Country (UPV/EHU) and Lucio Lascaray Research Centre, Vitoria, Spain
[2] CIBER Physiopathology of Obesity and Nutrition (CIBERobn), Institute of Health Carlos III, Madrid, Spain

11.1 Introduction: Occurrence and Intake

Resveratrol (3,4',5-trihydroxystilbene) is a naturally occurring polyphenol which is found in several plants and produced in response to a stressful situation such as ultraviolet radiation or bacterial, viral, and fungal infection (Langcake and Price, 1976). It was isolated for the first time from roots of *Veratrum grandiflorum*, or white hellebore plant (Takaoka, 1939). Its main sources are grapes and their products (wine and grape juice), berries, peanuts, tea, and chocolate, among others (Burns *et al.*, 2002; Pervaiz, 2003). Although it is challenging to estimate resveratrol intake, Zamora-Ros and co-workers made an estimation of stilbene intake in Europe. They calculated that the total stilbene intake was in the range of 2–3 mg per day with the consumption of resveratrol and piceid representing almost half this amount (Zamora-Ros *et al.*, 2016).

Its beneficial health effects were first described in several epidemiological studies, which demonstrated a negative

*Corresponding author.

Dietary Polyphenols: Metabolism and Health Effects, First Edition.
Edited by Francisco A. Tomás-Barberán, Antonio González-Sarrías, and Rocío García-Villalba.

correlation between wine consumption and cardiovascular diseases (Delmas *et al.*, 2005; Lekakis *et al.*, 2005; Wu *et al.*, 2001). This phenomenon has become widely known as "the French paradox," a concept that was formulated by the scientist Serge Renaud (Simini, 2000). Over the years, various beneficial properties have been attributed to this polyphenol, such as antioxidant, antiinflammatory, antiplatelet aggregation, vasodilator, hypolipidemic, antineoplastic, neuroprotective, antiosteoporotic, and estrogenic effects (Cho *et al.*, 2017; Das and Das, 2010; Gescher *et al.*, 2013; Honari *et al.*, 2019; Huang *et al.*, 2019; Ko *et al.*, 2017; Kulkarni and Cantó, 2015; Marchal *et al.*, 2013; Santos *et al.*, 2013; Švajger and Jeras, 2012; Švajger *et al.*, 2010; Xu and Si, 2012; Yu *et al.*, 2012).

Its low bioavailability has been a matter of concern for scientists since its discovery because only a small amount of the parent compound reaches plasma and tissues. However, the beneficial effects established in *in vivo* studies and clinical trials suggest that its metabolites could show similar properties to those described for the parent compound. Moreover, it has been postulated that resveratrol metabolites could act as an intracellular reserve pool of the parent compound because they could be deconjugated in some target cells (Patel *et al.*, 2013). Nevertheless, it is not yet known if this process can occur in all tissues, as enzymes involved in recycling are not ubiquitously expressed (Miksits *et al.*, 2010).

11.2 Absorption, Metabolism, and Excretion of Resveratrol

Regarding resveratrol utilization, around 70% of the phenolic compound is absorbed at the intestinal level (Chaplin *et al.*, 2018). Absorption occurs by passive diffusion or through membrane transporters such as integrins (Gambini *et al.*, 2015; Lin *et al.*, 2006). The resveratrol absorbed is transported through the portal vein to the liver. Resveratrol undergoes phase II xenobiotic metabolism reactions, first in the intestine and then in the liver. Phase II metabolism serves as a detoxification process to reduce its potential toxic effect and to facilitate biliary and urinary excretion (Liu and Hu, 2007).

Uridine-5′-diphosphate glucuronosyltransferase (UGT) enzymes, located in the endoplasmic reticulum of cells, transfer a glucuronic acid from UDP-glucuronic acid to resveratrol, thus generating resveratrol glucuronides. Moreover, sulfation reactions, which occur mainly in the liver (Falany, 1997), are mediated by sulfotransferases (SULT), which transfer a sulfate group from phosphoadenosine phosphosulfate to a hydroxyl group of polyphenols. Consequently, only a small amount of free resveratrol and higher amounts of glucuronide and sulfate forms (conjugated primarily in its third and fourth positions) are found in plasma, indicating that resveratrol shows low bioavailability (Walle *et al.*, 2004) (Figure 11.1). The peak of free plasma resveratrol occurs at 30 minutes, due to its gastric absorption (Chedea *et al.*, 2017).

An important aspect is the fact that as resveratrol supplementation dosage increases, the proportion of glucuronide metabolites in the bloodstream decreases, while that of sulfate metabolites increases (Ardid-Ruiz *et al.*, 2018), due to a saturation phenomenon. In this context, in a previous study from our group, tissue distribution of resveratrol metabolites in Sprague–Dawley rats was analyzed; while sulfate metabolites increased when resveratrol dosage was enhanced, glucuronides increased between 6 and 30 mg/kg/d but remained unchanged when resveratrol dosage reached 60 mg/kg/d. These results

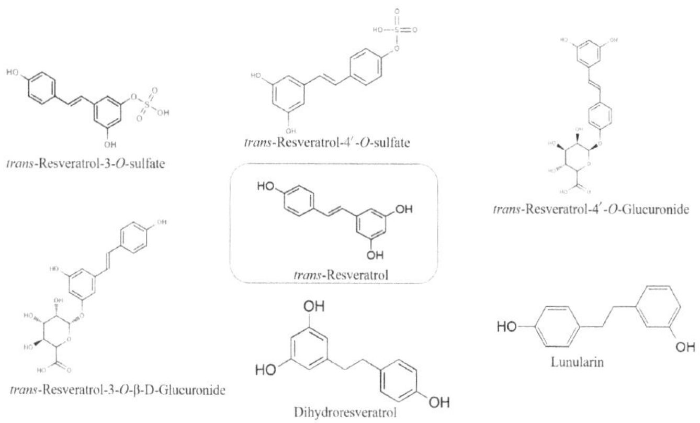

Figure 11.1 Chemical structure of resveratrol and its main metabolites.

suggest that glucuronidation, but not sulfation, is a saturable metabolic pathway, at least in the range of doses used in the study mentioned (Andres-Lacueva *et al.*, 2012). Moreover, Kapetanovic *et al.* (2011) observed a shift from glucuronidation to sulfation of resveratrol due to the saturation of the glucuronide pathway.

It should be underlined that the study of resveratrol metabolism is a complex issue because the types and proportions of the primary metabolites can vary depending on the dosage, the administration procedure, and the species. Several authors have reported that glucuronidation is the main conjugation reaction in rodents and pigs, whereas sulfatation is more abundant in humans (Azorín-Ortuño *et al.*, 2011; Boocock *et al.*, 2007; Marier *et al.*, 2002). Nevertheless, in line with the observations of Ardid-Ruiz and co-workers (2018), it has been stated that conjugation reactions are also dose-dependent in humans. After supplementation with small doses of resveratrol (5 or 50 mg), glucuronidation reactions are the most abundant, whereas more monosulfate metabolites are found in plasma and urine after the administration of higher doses (≥250 mg) (Chow *et al.*, 2010; Kennedy *et al.*, 2010; Poulsen *et al.*, 2013; Zhu *et al.*, 2012).

It is also important to highlight that flavonoids present in the diet can inhibit the sulfation of resveratrol, thus improving its bioavailability. Quercetin is the most potent of these inhibitors. Several studies have been conducted to establish the protective effect of quercetin for resveratrol bioavailability (Arias *et al.*, 2016; de Santi *et al.*, 2000a,b; Wenzel and Somoza, 2005).

The transport of resveratrol in the blood is bound to albumin, blood cells, and low-density lipoproteins (LDL) (Blache *et al.*, 1997; Gambini *et al.*, 2015; Jannin *et al.*, 2004). The complexes generated are dissociated in those cellular membranes that contain specific receptors, thereby allowing free resveratrol to enter the target cells.

Resveratrol and its metabolites undergo a rapid and wide distribution in tissues and organs (Andres-Lacueva *et al.*, 2012; Azorín-Ortuño *et al.*, 2011; Böhmdorfer *et al.*, 2017; Juan *et al.*, 2010; Lin *et al.*, 2012; Wang *et al.*, 2008). Studies carried out in animal models or humans have reported that resveratrol is mainly distributed in abundant blood-supplied tissues, such as liver, lung, and kidney (Juan *et al.*, 2010). Bearing in mind that resveratrol shows high lipid solubility, it can also be expected to be deposited, along with its metabolites, in tissues with high lipid content such as adipose tissue (Soleas *et al.*, 2001). Furthermore, the affinity of resveratrol for its transporters and the presence of specific receptors play a crucial role in its distribution (Delmas and Lin, 2011; Gambini *et al.*, 2015). The different distribution of resveratrol and its metabolites in various tissues is an essential issue because it may be predictive for the effects of resveratrol on those tissues. It is as yet unclear whether resveratrol metabolites found in tissues other than liver are due to direct absorption of the parent compound and/or the absorption of cleaved conjugates. To clarify this issue, studies addressing the presence of phase II metabolizing enzymes in several tissues are needed.

Some of the metabolites generated in the liver enter the enterohepatic circulation. After biliary excretion, these metabolites are either reabsorbed in the small intestine or reach the colon (Marín *et al.*, 2015). Also, unabsorbed resveratrol is metabolized by the gut bacteria in the colon. Dihydroresveratrol, which was the first resveratrol derivative to be identified, is produced by hydrogenation of a double bond on its structure (see Figure 11.1), but others such as 3,4'-dihydroxystilbene and 3,4'-dihydroxybibenzyl (lunularin) were later identified (Chaplin *et al.*, 2018). As occurs with resveratrol in other tissues, resveratrol derivatives synthesized by gut microbiota undergo conjugation reactions. At the same time, gut microbiota hydrolyzes resveratrol glycosides, such as piceid, to an aglycone. Resveratrol can be returned to piceid and be conjugated to a glucuronide form, the main form in which it is absorbed (Andres-Lacueva *et al.*, 2012). The remaining resveratrol is excreted via feces (0.3–38% of human resveratrol intake) (Walle, 2011). Moreover, the excretion of absorbed resveratrol and its metabolites occurs mainly via urine.

More than 20 resveratrol metabolites have been found in human urine after a moderate intake of red wine for 28 days, some of them being by-products generated by gut microbiota (Rotches-Ribalta *et al.*, 2012). In this regard, dihydroresveratrol (DHR) and lunularin have been considered as the main microbial-derived resveratrol metabolites

(Bode *et al.*, 2013; Rotches-Ribalta *et al.*, 2012). It has been reported that DHR can be produced by the hydrogenation of the aliphatic double bond of *trans*-resveratrol, which is mediated by the bacteria *Eggerthella lenta* and *Bacteroides uniformis*, both microorganisms identified in human fecal samples (Juan *et al.* 2013). Lunularin has also been identified in humans, although only some are able to produce it (along with DHR), who have been classified as high or low producers (Bode *et al.*, 2013). This resveratrol metabolite can be produced by both DHR dehydroxylation or through the reduction of 3,4′-dihydroxy-*trans*-stilbene. In this regard, even though information regarding the bacteria involved in its synthesis is currently scarce, it has been suggested that the aforementioned processes could occur in individual microorganisms, as well as relying on the presence of a diversity of bacteria (coriobacteria) (Bode *et al.*, 2013).

11.3 Biological Effects of Resveratrol Metabolites

In spite of extensive resveratrol metabolism, the positive health effects induced by this phenolic compound when it is orally administered, in animals or humans, have led researchers to focus on the potential activity of their metabolites. Resveratrol has been extensively used in preclinical and clinical approaches, but in the case of resveratrol metabolites, only preclinical research has been carried out, mostly in cell cultures. The main limitation for *in vivo* studies is the relatively large amount of each metabolite needed to feed the animals. It is challenging to have these amounts available either by using metabolites synthesized in the laboratory or commercial metabolites, in this case due to their high price.

In this scenario, the vast majority of the reported studies have analyzed the effects of resveratrol metabolites on cancer. However, to date, studies relating to other activities are rarer. The present chapter summarizes the reported biological effects of resveratrol metabolites on adipocyte fat accumulation, as well as on other issues related to obesity.

11.3.1 *In vitro* Studies

In our laboratory, we carried out a survey in maturing and mature 3T3-L1 adipocytes aimed to test the effects of some phase II resveratrol metabolites – resveratrol-3-*O*-glucuronide, resveratrol-4'-*O*-glucuronide and resveratrol-3-*O*-sulfate – on adipogenesis and triglyceride (TG) metabolism (Lasa *et al.*, 2012). For this purpose, maturing preadipocytes were treated over the eight days of the differentiation period, while mature adipocytes were treated for 24 hours on day 12 after differentiation. Cells were treated with resveratrol and its metabolites at doses of 1, 10, and 25 μM.

No changes in TG content were observed in maturing preadipocytes when cells were treated with 1 or 10 μM of the tested molecules. By contrast, resveratrol and the three metabolites reduced this parameter in a similar percentage at 25 μM, although the lowering effect exerted by resveratrol-3-*O*-glucuronide did not reach statistical significance. To elucidate the mechanisms underlying this delipidating effect, the mRNA levels of several transcription factors involved in adipogenesis were studied. All the analyzed compounds effectively inhibited the increase in CCAAT/enhancer-binding protein ß (Cebpß) expression that occurs at the early phase of differentiation. By contrast, gene expression of peroxisome proliferator-activated receptor γ (Pparγ), CCAAT/enhancer-binding protein α (Cebpα) and lipoprotein lipase (Lpl), which are induced in later stages of the differentiation process, were only decreased by resveratrol-3-*O*-sulfate.

Concerning mature 3T3-L1 adipocytes, resveratrol proved to be effective in reducing TG content at the three analyzed concentrations, whereas glucuronide metabolites decreased this parameter at 10 and 25 μM and the sulfate metabolite was not effective at none of the tested doses. Concerning gene expression, resveratrol-4'-*O*-glucuronide, as well as resveratrol, affected the lipolytic pathway, although the mechanism was different. Thus, whereas resveratrol increased adipose triglyceride lipase (Atgl) expression, resveratrol-4'-*O*-glucuronide increased that of hormone-sensitive lipase (Hsl). Furthermore, we observed that resveratrol-3-*O*-glucuronide increased fatty acid synthase (Fasn) expression, suggesting its implication in *de novo* lipogenesis. Only resveratrol was able to significantly increase peroxisome proliferator-activated receptor-γ coactivator 1α (Pgc1α) and carnitine palmitoyltransferase 1b (Cpt1b) expressions. Taken together, these data suggest that glucuronide metabolites showed biological activity in mature adipocytes, although they were less effective than resveratrol.

It has been reported that resveratrol modifies microRNAs (miRNAs) expression, which are involved in several biological functions such as adipogenesis regulation. Thus, in another study by our group, we focused on analyzing the potential role of miRNAs in the antiadipogenic effect of resveratrol and its metabolites (Eseberri *et al.*, 2017). miRNAs are small noncoding RNAs of about 19–24 nucleotides in length that typically inhibit the translation and stability of messenger RNAs (mRNAs), thus controlling genes involved in cellular processes such as cell differentiation (di Leva *et al.*, 2014). Two criteria were established for miRNA selection: they had to be validated or predicted by five algorithms in the miRWalk 2.0 database (Dweep and Gretz, 2015) and reported in the literature after "miR+adipogenesis" search in Pubmed database. Those that fulfilled these criteria were miR-155 for Cebpβ, miR-27a, miR-27b and miR-130b for Pparγ, and miR-31, miR-144, miR-205, miR-224 and miR-326 for Cebpα. After treating maturing 3T3-L1 preadipocytes with 25 μM of resveratrol or its metabolites, total RNA (including large and small RNA) was extracted. Resveratrol and glucuronide metabolites increased miR-155 expression, suggesting that the previously observed inhibitory effect of molecules on Cebpβ expression (Lasa *et al.*, 2012) could be exerted via miRNAs. To confirm this, 3T3-L1 preadipocytes were transfected with an anti-miR-155 in the presence or absence of resveratrol, and it was observed that after miR-155 silencing, resveratrol was not able to reduce Cebpβ gene or protein expression. Thus, we concluded that the Cebpβ downregulation exerted by resveratrol, and presumably by that of its glucuronide metabolites, was carried out via miR-155. Moreover, taking into account that the Cebpβ expression reduction after cell treatment with resveratrol-3-sulfate was not carried out via miR-155, we measured krüppel-like factor 5 (Klf5) and cAMP responding element binding protein 1 (Creb1) mRNA levels, since they are involved in Cebpβ transcriptional regulation. The expression of both genes was reduced after resveratrol-3-*O*-sulfate treatment, and Klf5 expression was also reduced in cells treated with resveratrol-4'-*O*-glucuronide. This observation suggested that Cebpβ downregulation carried out by resveratrol-3-sulfate was exerted at transcriptional level, whereas the effect of resveratrol-4'-*O*-glucuronide was at both transcriptional and posttranscriptional levels.

Taking into account that in our previous study by Lasa and co-workers (2012) resveratrol-3-*O*-suldate treatment reduced Pparγ and Cebpα mRNA levels, the expression of miRNAs that target these genes was measured, although no changes were observed. As in the case of Cebpβ, we measured the expression of other transcription factors involved in Pparγ regulation (sterol regulatory element-binding transcription factor 1 [Srebf1] and liver x receptor α [Lxrα]), and we observed a reduction after resveratrol-3-*O*-sulfate treatment, indicating that the effect of this metabolite on Pparγ was exerted at transcriptional level. Moreover, considering that Pparγ and Cebpα induced the expression of each other, it seemed that the Cebpα reduction observed after resveratrol-3-sulfate treatment could be due to the inhibitory effect of the metabolite on Pparγ expression.

An essential role of adipocytes is adipokine production, which is involved in the regulation of a significant number of metabolic pathways. In addition, some of the beneficial effects of resveratrol are mediated by changes in adipokine production. Because of this situation, we carried out a new experiment to investigate the impact of resveratrol-3-*O*-glucuronide, resveratrol-4'-*O*-glucuronide and resveratrol-3-*O*-sulfate on adipokine production (Eseberri *et al.*, 2013). The experiments were based on our previous study (Lasa *et al.*, 2012), and thus, cells were treated under the same experimental conditions. As far as dose is concerned, we selected 25 μM for maturing preadipocytes and 10 μM for mature adipocytes, since those were effective in TG reduction on each condition. The results obtained showed that, in maturing preadipocytes, all the compounds increased adiponectin mRNA levels, although no changes in adiponectin secretion to culture media were observed. While exposure to resveratrol or the sulfate metabolite reduced leptin expression, glucuronide metabolites did not induce such a change. Surprisingly, only glucuronide metabolites decreased leptin secretion. Concerning visfatin mRNA levels, these were increased by resveratrol and glucuronide metabolites, while their release remained unchanged. As far as apelin gene expression is concerned, this was significantly increased after treatment with resveratrol and glucuronide metabolites, but its expression was reduced in resveratrol-3-*O*-sulfate treated groups. Once again, the release of apelin to the medium remained unchanged.

In mature adipocytes, unlike in preadipocytes, resveratrol but not its metabolites increased adiponectin mRNA levels, although, as in the case of maturing adipocytes, no changes were observed in adiponectin secretion to culture media. Leptin gene expression and secretion were reduced by resveratrol and its metabolites. Visfatin and apelin mRNA levels were increased by the metabolites but not by resveratrol, and their release remained unaltered by the treatments. This unexpected and repeated discrepancy between adipokine gene expression and release has also been reported in the literature (Houshmand-Oeregaard *et al.*, 2017; Zeigerer *et al.*, 2008). In this regard, we hypothesized that several agents used in cell culture studies could have disguised the effect of molecules on adipokine production (Ryan *et al.*, 2010; Wang *et al.*, 2006), which could be explained by polysome activation on mRNA, alternative mRNA splicing or protein turnover. Furthermore, differences in the half-life of the protein and the mRNA could be another explanation.

The results described in these studies carried out in maturing 3T3-L1 preadipocytes and mature 3T3-L1 adipocytes are summarized in Figures 11.2 and 11.3.

It is well known that obesity is accompanied by some degree of inflammation, called chronic low-grade inflammation. The increased size of adipocytes plays a crucial role in this process. Adipocyte hypertrophy occurring in obesity leads to increased production of several proinflammatory cytokines (Hosogai *et al.*, 2007). Moreover, the death of adipocytes, which is very common in obese individuals and has been linked to adipocyte hypoxia, attracts macrophages to adipose tissue to eliminate the

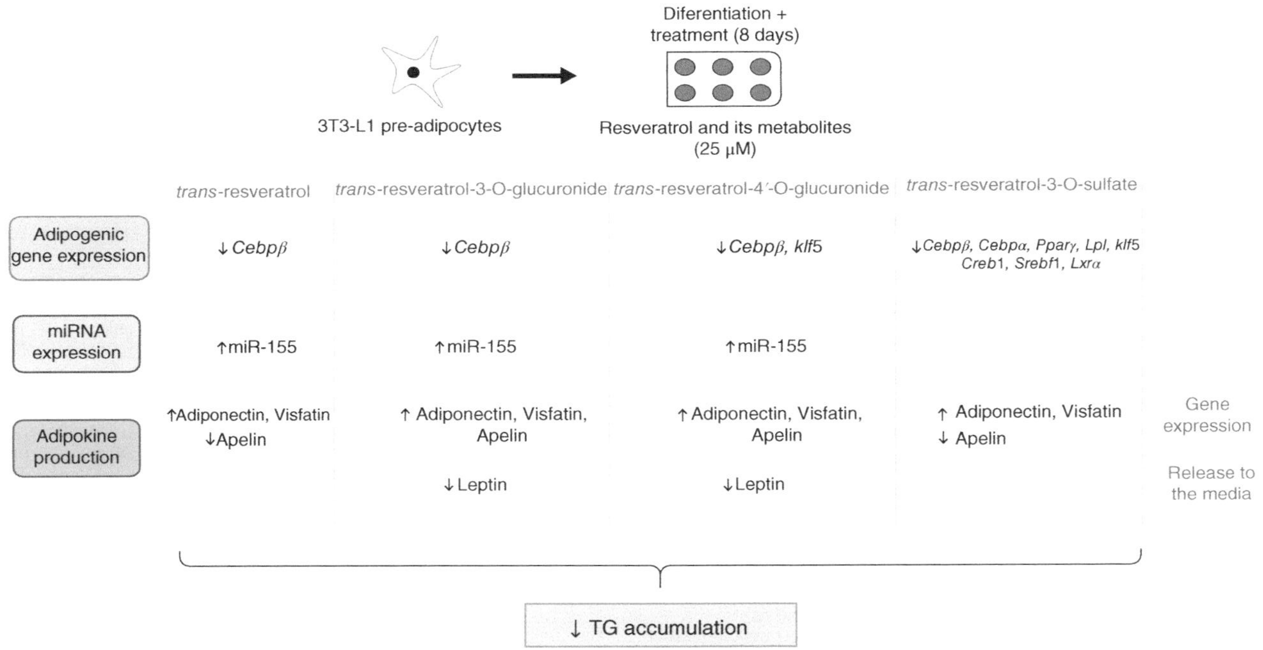

Figure 11.2 Summary of the effects of resveratrol metabolites (resveratrol-3-*O*-glucuronide, resveratrol-4′-*O*-glucuronide or resveratrol-3-*O*-sulfate) and those of the parent compound on 3T3-L1 maturing pre-adipocytes (Eseberri *et al*. 2013, 2017; Lasa *et al*. 2012). Cebpα, CCAAT/enhancer-binding protein α; Cebpβ, CCAAT/enhancer-binding protein β; Creb1, cAMP responding element-binding protein 1; Klf5, krüppel-like factor 5; Lpl, lipoprotein lipase; Lxrα, liver x receptor α; Pparγ, peroxisome proliferator-activated receptor γ; Srebf1, sterol regulatory element-binding transcription factor 1; TG, triglyceride.

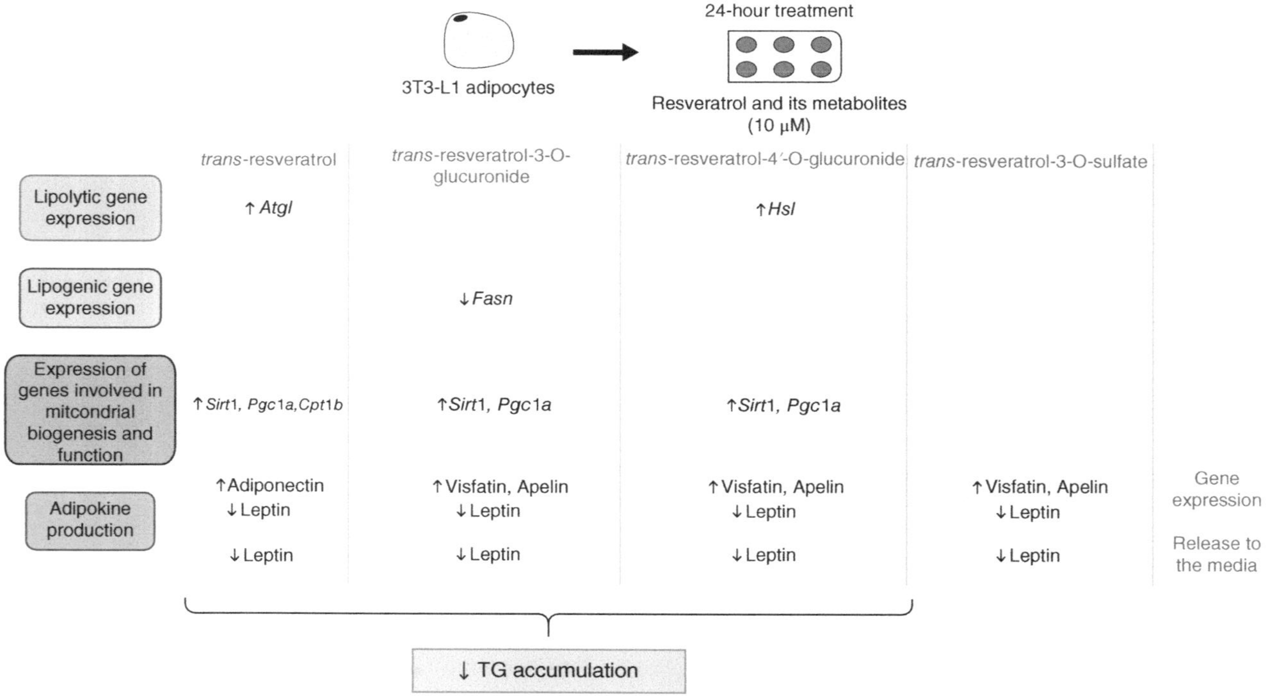

Figure 11.3 Summary of the effects of resveratrol metabolites ((resveratrol-3-*O*-glucuronide, resveratrol-4′-*O*-glucuronide or resveratrol-3-*O*-sulfate) and those of the parent compound on 3T3-L1 mature adipocytes (Eseberri *et al*. 2013; Lasa *et al*. 2012). Atgl, adipose triglyceride lipase; Cpt1b, carnitine palmitoyltransferase 1b; Fasn, fatty acid synthase; Hsl, hormone-sensitive lipase; Pgc1α, peroxisome proliferator-activated receptor γ coactivator 1α; Sirt1, sirtuin 1; TG, triglyceride.

dead cells. These macrophages also produce proinflammatory cytokines (van Gaal *et al.*, 2006).

In this context, Walker *et al.* (2014) compared the effects of resveratrol and some of their metabolites (resveratrol, resveratrol-3-*O*-sulfate, resveratrol disulfates, resveratrol-3-*O*-glucuronide and resveratrol-4'-*O*-glucuronides) on proinflammatory cytokine production in U-937 macrophages stimulated by *Escherichia coli* lipopolysaccharide (LPS). For this purpose, U-937 monocytes were differentiated to macrophages and incubated with each molecule at different concentrations (1 and 10 μM) in combination with 1 μg/mL *E. coli*-LPS for six hours (Figure 11.4a). After this incubation period, the authors studied the effects of resveratrol and its metabolites on gene expression of interleukin IL-6, tumor necrosis factor (TNF)-α, IL-1α, IL-1β, and IL-1R. They observed that the reductions in IL-1α and IL-1β gene expression induced by resveratrol-3-*O*-sulfate and resveratrol disulfates were similar to those induced by resveratrol. By contrast, the glucuronide metabolites enhanced the expression of the genes mentioned above (except resveratrol-3-*O*-glucuronide at a dose of 1 μM). In the case of IL-6, 10 μM of resveratrol and the resveratrol-3-sulfate metabolite showed similar reducing effects. By contrast, at a dose of 1 μM, the reduction in Il-6 gene expression exerted by resveratrol-3-*O*-sulfate was the greatest one. As far as TNF-α is concerned, resveratrol and its sulfate metabolites increased its gene expression, while the glucuronides did not induce any change.

These results suggest that the resveratrol sulfates have an antiinflammatory effect (at least in the case of the genes studied) while resveratrol glucuronides not. Moreover, the combined effects of the studied metabolites were analyzed by including in the incubation media a mixture of all the analyzed metabolites at the tested doses (either 1 or 10 μM). This mixture significantly reduced Il-1α and Il-1β gene expression, while no changes were appreciated in Il-6 and TNF-α.

Data concerning cytokine release were in good accordance with those of cytokine expression. Thus, mono- and disulfate resveratrol metabolites significantly inhibited the IL-6 and TNF-α release, while no effect was observed when the glucuronide metabolites were used. When the metabolite mixture

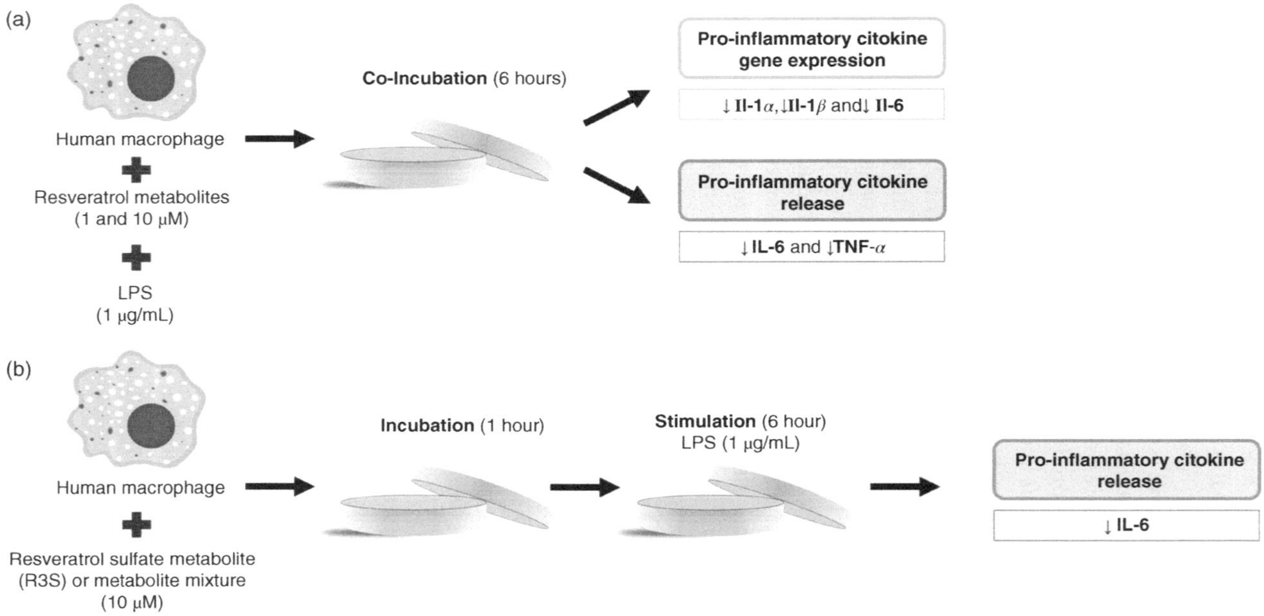

Figure 11.4 Antiinflammatory effects of resveratrol metabolites in human macrophages after co-incubation with metabolites and LPS (a) or after pretreatment with resveratrol metabolites and subsequent stimulation with LPS (b) (Walker *et al*. 2014). IL-1α, interleukin 1α; IL-1β, interleukin 1β; IL-6, interleukin 6; LPS, lipopolysaccharide; R3S, resveratrol-3-sulfate; TNF-α, tumor necrosis factor α.

was tested, IL-6 release was significantly decreased, while that of TNF-α was not affected.

Finally, the authors analyzed the preventive effect of resveratrol and its metabolites on IL-6 and TNF-α release. For this purpose, U-937 macrophages were concomitantly incubated with resveratrol, resveratrol-3-*O*-sulfate and the metabolite mixture (in all cases at a dose of 10 μM) for one hour and then stimulated with *E. coli*-LPS for six hours (Figure 11.4b). In this case, resveratrol-3-*O*-sulfate and the metabolite mixture reduced IL-6 release more significantly than resveratrol did, while no effects regarding TNF-α release were reported.

Based on the results obtained, the authors concluded that resveratrol sulfates exhibit the most antiinflammatory properties. In this line, since resveratrol-3-*O*-sulfate is found in higher concentrations in plasma than those of the parent compound, the authors suggested that it could be mediating the antiinflammatory effects showed by the sulfate metabolites.

Using the same cell line (U-937 macrophages), Schueller *et al.* (2015) carried out a study of the effects of physiological concentrations (1 μM) of the most abundant resveratrol metabolites in the expression of several genes and proteins involved in inflammation and immune cell chemotaxis (Figure 11.5). For this purpose, once differentiated, cells were co-incubated with 1 μg/mL of *Porphyromonas gingivalis* LPS (pgLPS) and 1 μM of resveratrol or 1 μM of its metabolites (resveratrol-3-*O*-sulfate, resveratrol disulfates, resveratrol-3-*O*-glucuronide and resveratrol-4'-*O*-glucuronide), for three and six hours, in the case of gene expression studies, and nine hours when protein release was studied.

After three hours of treatment, similar reductions in monocyte chemotactic protein-1 (Mcp-1) mRNA levels were found in cells treated with resveratrol, resveratrol-3-*O*-sulfate and resveratrol disulfates when compared to the pgLPS control. In the case of the glucuronide metabolites (resveratrol-3-*O*-glucuronide and resveratrol-4'-*O*-glucuronide), no such changes were appreciated. Moreover, no differences in Mcp1-1 mRNA levels were found in any of the groups comparing to the pgLPS control when the treatment was extended to six hours. No changes were observed regarding macrophage inflammatory protein 1 β (Mip-1β) or in Il-8 gene expression in any of the treated groups

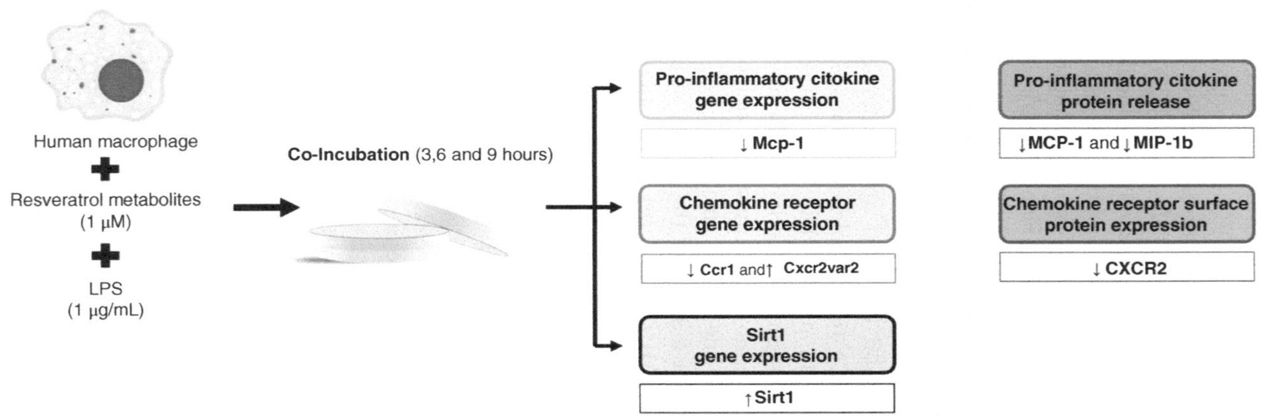

Figure 11.5 Antiinflammatory effects of resveratrol metabolites in human macrophages after co-incubation with metabolites and LPS for different time periods (Schueller *et al*. 2015). Ccr1, C-C motif chemokine receptor 1; Cxcr2var2, C-X-C motif chemokine receptor 2 variant 2; IL-8, interleukin 8; LPS, lipopolysaccharide; Mcp-1, monocyte chemotactic protein-1; MIP-1β, macrophage inflammatory protein 1 β; Sirt1, sirtuin 1.

when compared to the pgLPS control group. Indeed, this lack of change was appreciated in both treatment periods (three and six hours). Furthermore, the six-hour treatment induced a significant increase in Mip-1β mRNA levels in cells treated with resveratrol-4'-*O*-glucuronide.

When the protein release of the genes mentioned above was studied, the authors found that mcp1 was decreased after treating cells with resveratrol-3-*O*-sulfate and resveratrol disulfates for three hours. Moreover, when treatment period was extended to nine hours, lower MCP-1 protein release was found in the groups treated with resveratrol, resveratrol disulfates and resveratrol-3-*O*-glucuronide. In the case of Mip-1β, its protein release was decreased in resveratrol and resveratrol-3-*O*-sulfate groups after a three-hour-treatment. A significant decrease in this parameter was also found in cells treated with resveratrol disulfates after a 6-hour-treatment period, and in cells treated with resveratrol-3-*O*-sulfate, resveratrol disulfates and resveratrol-3-*O*-glucuronide, when the treatment was for nine hours. Of all these changes, the most significant MIP-1β protein release reduction was that induced by resveratrol disulfates after a nine-hour treatment.

After studying the expression and release of these cytokines, the authors analyzed the effects of the proposed treatments on the gene and surface protein expression of chemokine receptors. In this regard, they observed that treatments with resveratrol and resveratrol-3-*O*-sulfate for six hours decreased C-C motif chemokine receptor 1 (Ccr1) gene expression, while it was significantly increased when cells were treated with resveratrol-4'-*O*-glucuronide under the same conditions. Concerning the gene expression of C-X-C motif chemokine receptor 2 variant 2 (Cxcr2var2), resveratrol and all the tested metabolites induced significant increases after three hours of treatment when compared to the pgLPS control group, while no such effect was appreciated when the treatment was extended to six hours. When the surface protein expression was analyzed, no changes were observed in this parameter regarding Ccr1. By contrast, in the case of Cxcr2var2, downregulation was induced by resveratrol disulfates, resveratrol-3-*O*-glucuronide, and resveratrol-4'-*O*-glucuronide.

Based on the results obtained, the authors concluded that the most abundant resveratrol metabolites are effective in modulating genes and proteins involved in inflammatory immune response and immune cell chemotaxis. Moreover, the authors highlighted that these events took place at concentrations of metabolites that could be reached with the usual dietary intakes of resveratrol.

Among the obesity-related co-morbidities, cardiovascular diseases are a matter of concern. High levels of either plasma total cholesterol or low-density lipoprotein cholesterol (LDL-C) are considered risk factors for cardiovascular diseases (Manson and Bassuk, 2015). It has been reported that a diet rich in phenolic compounds could significantly lower total plasma cholesterol and/or LDL-C concentrations (Kamal-Eldin *et al.*, 2000). This health effect can be reached by increasing the excretion of neutral sterols and bile acids from liver to feces, which is considered to be one of the mechanisms for reducing cholesterol (Chen *et al.*, 2012).

Bearing this in mind, Shao *et al.* (2016) studied the effects of resveratrol and resveratrol-4'-*O*-glucuronide on hepatic cholesterol metabolism. Hepatic cholesterol and genes related to cholesterol metabolism in the liver were analyzed in a human hepatoma cell line (HepG2). The construction of a high cholesterol-containing cell model was performed by incubating HepG2 cells with complete Dulbecco's modified Eagle's medium (DMEM) supplemented with 10% fetal bovine serum (10% FBS–DMEM medium; control group) or with 50% fetal bovine serum (50% FBS–DMEM medium; model group) for 48 hours. The model cells obtained were certified in relation to different aspects, including total cholesterol content, accumulation of lipid droplets and gene expression of the enzyme that catalyzes *de novo* cholesterol synthesis in liver, hydroxy-3-methylglutaryl-CoA reductase (HMGR) and the receptor that transfers cholesterol in LDL-C to liver LDL-receptor (LDLR). Later on, cells with the hypercholesterolemic medium (50% FBS–DMEM) were treated with resveratrol and resveratrol-4'-*O*-glucuronide at doses of 6.25, 12.5 and 25 μM.

Resveratrol, but not its glucuronide metabolite, significantly reduced cell viability (up to 80%) at all the tested concentrations.

Once the treatment was completed, all the doses of both compounds decreased total cellular cholesterol and increased bile acid release to the extracellular medium in a dose-dependent manner. Oil red O staining showed that resveratrol and resveratrol-4'-*O*-glucuronide decreased intracellular TG accumulation, but statistical significance was only reached with the highest dose. When gene expressions were measured in cells treated with 25 μM of the compounds studied, upregulation of cholesterol 7a-hydroxylase (Cyp7a1) (an essential gene involved in bile acid biosynthesis), bile salt export pump (Bsep) (involved in the bile acid efflux), ATP-binding cassette half-transporter 5 (Abcg5) (gene encoding the biliary cholesterol transporter), and LDLR was observed. By contrast, both compounds reduced HMGR-CoAR mRNA levels and no significant differences were observed among groups in Na^+-taurocholate cotransport polypeptide (Ntcp), which is involved in the reabsorption of bile acids in the liver.

Taking the results as a whole, the authors concluded that resveratrol and its glucuronide metabolite reduced cholesterol and lipid content due to an increase in the synthesis and efflux of bile acids, and a decrease in the synthesis and an increase in the efflux of cholesterol (Figure 11.6). Moreover, it was stated that due to its lower cytotoxicity, the metabolite was more advisable.

In humans, obesity and insulin resistance are characterized by endothelial dysfunction and impaired vasodilation (Steinberg *et al.*, 1996). Moreover, it has been reported that, in rodents, the inhibition of endothelial nitric oxide synthase (eNOS) decreases insulin-stimulated uptake of glucose by skeletal muscle, and deletion of the eNOS gene induces insulin resistance (Duplain *et al.*, 2001; Shankar *et al.*, 2000) and impairs fatty acid oxidation (Le Gouill *et al.*, 2007).

Ladurner *et al.* (2014) carried out a study devoted to analyzing changes induced by different resveratrol metabolites on eNOS activity, nitric oxide and intracellular reactive oxygen species (ROS) levels in endothelial cells. For this purpose, EA.hy926 human endothelial cells were treated with resveratrol and its main metabolites at concentrations ranging from 1 to 100 μM for 24 hours.

While the treatment of cells with 100 μM of resveratrol significantly increased this parameter, none of the tested resveratrol

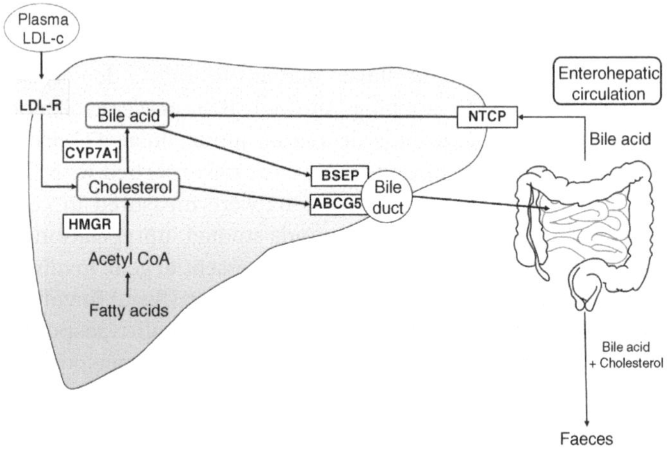

Figure 11.6 Hepatic bile acid and cholesterol metabolism. ABCG5, ATP-binding cassette half-transporter 5; BSEP, bile salt export pump; CYP7A1, cholesterol 7a-hydroxylase; HMGR, hydroxymethyl glutaryl reductase; LDL-c, low-density lipoprotein cholesterol; LDL-R, low-density lipoprotein receptor; NTCP, Na+-taurocholate co-transport polypeptide. Source: Modified from Shao *et al.* (2016).

metabolites induced significant changes when used at doses ranging from 1 to 100 μM. To elucidate whether the tested resveratrol metabolites could enhance endothelial NO bioavailability, without changing eNOS enzyme activity, the authors analyzed NO release. Again, in this case, no changes produced by resveratrol treatment (peaking at 30 μM) were observed in cells treated with the metabolites. Finally, the authors wanted to study if the tested resveratrol metabolites had any effect on the ROS detoxification system of cells, as resveratrol does. In good accordance with the results obtained for eNOS activity and NO release, none of the used metabolites induced significant changes.

Based on these results, the authors concluded that the primary physiological metabolites of resveratrol are not effective in inducing changes in eNOS enzyme activity, NO release or cellular ROS production in endothelial cells.

Some of the effects induced by resveratrol, for instance those related to antiobesity, are mediated by peroxisome proliferator-activated receptors (PPARs), a family of nuclear receptors that act as ligand-dependent transcription factors.

Thus, the reduction in adipogenesis induced by this polyphenol is mediated by PPARγ (Fischer-Posovszky *et al.*, 2010; Kim *et al.*, 2011; Rayalam *et al.*, 2008), whereas the activation in fatty acid oxidation is mediated by PPARα (Ahn *et al.*, 2008; Milton-Laskibar *et al.*, 2018). However, evidence for the direct interaction between this polyphenol and PPARs is lacking.

Working on this issue, Calleri *et al.* analyzed whether resveratrol and some of its metabolites (resveratrol-3-*O*-glucuronide, resveratrol-4'-*O*-glucuronide and resveratrol-3-*O*-sulfate) could interact directly with PPAR proteins (Calleri *et al.*, 2014). For this purpose, bioaffinity studies were carried out to establish the binding of resveratrol and the mentioned metabolites to PPARγ and PPARα. Resveratrol and the three metabolites analyzed effectively bound to PPARγ, but only resveratrol was able to bind to PPARα. Regarding binding to PPARγ, resveratrol and resveratrol-4'-*O*-glucuronide showed similar affinities. In addition, the authors analyzed the pattern of receptor-ligand recognition and observed that resveratrol-3-*O*-glucuronide might have the same resveratrol binding site whereas resveratrol-3-sulfate may bind PPARγ differently.

Another essential mediator in resveratrol actions is SIRT1 (Borra *et al.*, 2005), a protein considered as a sensor for metabolism and oxidation of nutrients and energy state (Li, 2013). SIRT1 is responsible for the regulation of protein activation by deacetylating a variety of proteins that play important roles in the regulation of lipid and glucose metabolism, insulin sensitivity, and inflammation. In this context, Andersen *et al.* (2011) carried out a study to analyze the effects of the main resveratrol metabolites on *Sirt1* gene expression in human HepG2 cells. For this purpose, HepG2 cells were cultured and treated with resveratrol conjugates resveratrol-3-*O*-sulfate, resveratrol-4'-*O*-sulfate, resveratrol disulfates, resveratrol-3-O-glucuronide, resveratrol-4'-*O*-glucuronide and resveratrol diglucuronides (resveratrol-2-C-β-D-/4'-*O*-β-D-diglucuronide and resveratrol-2-C-β-D-/5-*O*-β-D -diglucuronide) at 10 μM/L for one hour. Of all the metabolites tested, resveratrol diglucuronides were the only compounds which induced a significant increase in *Sirt1* gene expression, as resveratrol did.

In the study reported by Schueller *et al.* (2015), previously described in this chapter, where U-937 macrophages were co-incubated with 1 μg/mL of pgLPS and 1 μM of resveratrol or 1 μM of its metabolites (resveratrol-3-*O*-sulfate, resveratrol disulfates, resveratrol-3-*O*-glucuronide and resveratrol-4′-*O*-glucuronide), for three and six hours, *sirt1* gene expression was measured. Upregulation of this parameter was observed in cells co-incubated with pgLPS and resveratrol-3-*O*-glucuronide or resveratrol-4'-*O*-glucuronide. Similarly, in our study carried out in mature 3T3-L1 adipocytes incubated with resveratrol or resveratrol metabolites (resveratrol-3-O-glucuronide, resveratrol-4'-*O*-glucuronide or resveratrol-3-*O*-sulfate), previously described in this chapter, the parent compound and the glucuronide metabolites, but not resveratrol-3-*O*-sulfate, increased *Sirt1* expression (Lasa *et al.*, 2012).

11.3.2 *In vivo* Studies

To date, only one *in vivo* has been reported in the literature. Pallauf *et al.* (2019) carried out a study in C57BL/6 to determine whether resveratrol and the microbial metabolites, dihydroresveratrol and lunularin, act differently on molecular targets implicated in aging. Mice were fed a high-fat, high-sugar diet for 12 weeks. After this period, the animals were administered intraperitoneally (24 mg/kg/d) resveratrol, dihydroresveratrol or lunularin for eight weeks.

At the end of the experimental period, mice receiving lunularin showed lower body weight than the control group, which could be due to a reduction in food intake. The other phenolic compounds did not alter food intake or body weight. To determine insulin sensitivity, oral glucose tolerance tests were carried out. Lunularin decreased blood glucose levels, but resveratrol and dihydroresveratrol did not modify insulin sensitivity when compared to the control mice. Adiponectin, leptin, TG, and total cholesterol levels remained unchanged in all groups.

The expression of different genes was analyzed in liver: *phosphoenolpyruvate carboxykinase 1 (Pck1), heme oxygenase 1 (Hmox1), forkhead box protein O3 (Foxo3), Pgc1*α, *Sirt1*, and *Tnf*α. The only changes observed were increased *Sirt1* and *Hmox1* gene expression induced by lunularin and the increase of *Hmox1* caused by dihydroresveratrol. The activity of AMPK, a well-known mediator of resveratrol, was also analyzed, but no activation was found in mice treated with resveratrol or the microbiota metabolites. Finally, hepatic resveratrol, dihydroresveratrol, and lunularin contents were quantified. In the case of the resveratrol-treated group, slightly more than the 50% of the animals presented resveratrol at low levels, while the vast majority presented dihydroresveratrol at higher levels than resveratrol. This indicates that resveratrol was metabolized to dihydroresveratrol. By contrast, the only compounds detected in mice receiving dihydroresveratrol and lunularin were those which were supplemented in each case.

Based on the results obtained, the authors concluded that lunularin was more beneficial than resveratrol or dihydroresveratrol, although it is essential to point out that the effects observed could be due to decreased food intake.

11.4 Conclusion

Although reported studies are currently scarce, published *in vitro* studies show that some phase II resveratrol metabolites can contribute to the beneficial effects attributed to resveratrol on obesity and its co-morbidities. Sometimes the effectiveness of metabolites is similar to that of resveratrol, and other times it is lower. This means that the low bioavailability of resveratrol is likely to be less important than expected. Regarding the resveratrol metabolites produced by gut microbiota, very few data have been reported, and thus it is difficult to reach definite conclusions. Because of this situation, not only is more research needed in this field but also *in vivo* studies designed to provide resveratrol metabolites are warranted.

Acknowledgments

This study was supported by grants from the Ministerio de Economía y Competitividad (AGL-2015-65719-R), Fondo Europeo de Desarrollo Regional (FEDER), the Instituto de Salud

Carlos III (CIBERobn) under Grant CB12/03/30007 and the University of the Basque Country under Grant GIU18-173.

References

Ahn, J., Lee, H., Kim, S., Park, J., and Ha, T. (2008)The anti-obesity effect of quercetin is mediated by the AMPK and MAPK signaling pathways. *Biochemical and Biophysical Research Communications*, **373**, 545–9.

Andersen, G., Burkon, A., Sulzmaier, F.J., *et al.* (2011) High dose of dietary resveratrol enhances insulin sensitivity in healthy rats but does not lead to metabolite concentrations effective for SIRT1 expression. *Molecular Nutrition & Food Research*, **55**, 1197–206.

Andres-Lacueva, C., Macarulla, M.T., Rotches-Ribalta, M., *et al.* (2012) Distribution of resveratrol metabolites in liver, adipose tissue, and skeletal muscle in rats fed different doses of this polyphenol. *Journal of Agricultural and Food Chemistry*, **60**, 4833–40.

Ardid-Ruiz, A., Ibars, M., Mena, P., *et al.* (2018) Potential Involvement of peripheral leptin/STAT3 signaling in the effects of resveratrol and its metabolites on reducing body fat accumulation. *Nutrients*, **10**, 1757.

Arias, N., Macarulla, M.T., Aguirre, L., Milton, I., and Portillo, M.P. (2016) The combination of resveratrol and quercetin enhances the individual effects of these molecules on triacylglycerol metabolism in white adipose tissue. *European Journal of Nutrition*, **55**, 341–8.

Azorín-Ortuño, M., Yáñez-Gascón, M. J., Vallejo, F., *et al.* (2011) Metabolites and tissue distribution of resveratrol in the pig. *Molecular Nutrition & Food Research*, **55**, 1154–68.

Blache, D., Rustan, I., Durand, P., Lesgards, G., and Loreau, N., (1997) Gas chromatographic analysis of resveratrol in plasma, lipoproteins and cells after in vitro incubations. *Journal of Chromatography. B, Biomedical Sciences and Applications*, **702**, 103–10.

Bode, L.M., Bunzel, D., Huch, M., *et al.* (2013) In vivo and in vitro metabolism of trans-resveratrol by human gut microbiota. *American Journal of Clinical Nutrition*, **97**, 295–309.

Böhmdorfer, M., Szakmary, A., Schiestl, R.H., *et al.* (2017) Involvement of UDP-glucuronosyltransferases and sulfotransferases in the excretion and tissue distribution of resveratrol in mice. *Nutrients*, **9**, 1347.

Boocock, D.J., Faust, G.E., Patel, K.R., *et al.* (2007) Phase I dose escalation pharmacokinetic study in healthy volunteers of resveratrol, a potential cancer chemopreventive agent. *Cancer Epidemiology, Biomarkers & Prevention*, **16**, 1246–52.

Borra, M.T., Smith, B.C., and Denu, J.M. (2005) Mechanism of human SIRT1 activation by resveratrol. *Journal of Biological Chemistry*, **280**, 17187–95.

Burns, J., Yokota, T., Ashihara, H., Lean, M.E., and Crozier, A. (2002) Plant foods and herbal sources of resveratrol. *Journal of Agricultural and Food Chemistry*, **50**, 3337–40.

Calleri, E., Pochetti, G., Dossou, K.S., *et al.* (2014) Resveratrol and its metabolites bind to PPARs. *Chembiochem*, **15**, 1154–60.

Chaplin, A., Carpéné, C., and Mercader, J. (2018) Resveratrol, metabolic syndrome, and gut microbiota. *Nutrients*, **10**, 1651.

Chedea, V.S., Vicaş, S.I., Sticozzi, C., *et al.* (2017) Resveratrol: from diet to topical usage. *Food & Function*, **8**, 3879–92.

Chen, W., Guo, J.X., and Chang, P. (2012) The effect of taurine on cholesterol metabolism. *Molecular Nutrition & Food Research*, **56**, 681–90.

Cho, S., Namkoong, K., Shin, M., *et al.* (2017) Cardiovascular protective effects and clinical applications of resveratrol. *Journal of Medicinal Food*, **20**, 323–34.

Chow, H.H., Garland, L.L., Hsu, C.H., *et al.* (2010) Resveratrol modulates drug- and carcinogen-metabolizing enzymes in a healthy volunteer study. *Cancer Prevention Research*, **3**, 1168–75.

Das, M. and Das, D.K. (2010) Resveratrol and cardiovascular health. *Molecular Aspects of Medicine*, **31**, 503–12.

Delmas, D. and Lin, H.Y. (2011) Role of membrane dynamics processes and exogenous molecules in cellular resveratrol uptake: consequences in bioavailability and activities. *Molecular Nutrition & Food Research*, **55**, 1142–53.

Delmas, D., Jannin, B., and Latruffe, N. (2005) Resveratrol: preventing properties against vascular alterations and ageing. *Molecular Nutrition & Food Research*, **49**, 377–95.

De Santi, C., Pietrabissa, A., Mosca, F., and Pacifici, G.M. (2000a) Glucuronidation of resveratrol, a natural product present in grape and wine, in the human liver. *Xenobiotica*, **30**, 1047–54.

De Santi, C., Pietrabissa, A., Spisni, R., Mosca, F., and Pacifici, G.M. (2000b) Sulphation of resveratrol, a natural compound present in wine, and its inhibition by natural flavonoids. *Xenobiotica*, **30**, 857–66.

Di Leva, G., Garofalo, M., and Croce, C.M. (2014) MicroRNAs in cancer. *Annuual Review of Pathology*, **9**, 287–314.

Duplain, H., Burcelin, R., Sartori, C., *et al.* (2001) Insulin resistance, hyperlipidemia, and hypertension in mice lacking endothelial nitric oxide synthase. *Circulation*, **104**, 342–5.

Dweep, H. and Gretz, N. (2015) miRWalk2.0: a comprehensive atlas of microRNA-target interactions. http://mirwalk.umm .uni-heidelberg.de/

Eseberri, I., Lasa, A., Churruca, I., and Portillo, M.P. (2013) Resveratrol metabolites modify adipokine expression and secretion in 3T3-L1 pre-adipocytes and mature adipocytes. *PLoS One*, **8**, e63918.

Eseberri, I., Lasa, A., Miranda, J., Gracia, A., Portillo, M.P. (2017) Potential miRNA involvement in the anti-adipogenic effect of resveratrol and its metabolites. *PLoS One*, **12**, e0184875.

Falany, C.N. (1997) Enzymology of human cytosolic sulfotransferases. *FASEB Journal*, **11**, 206–16.

Fischer-Posovszky, P., Kukulus, V., Tews, D., *et al.* (2010) Resveratrol regulates human adipocyte number and function in a Sirt1-dependent manner. *American Journal of Clinical Nutrition*, **92**, 5–15.

Gambini, J., Inglés, M., Olaso, G., *et al.* (2015) Properties of resveratrol: in vitro and in vivo studies about metabolism, bioavailability, and biological effects in animal models and humans. *Oxidative Medicine and Cellular Longevity*, **2015**, e837042.

Gescher, A., Steward, W.P., and Brown, K. (2013) Resveratrol in the management of human cancer: how strong is the clinical evidence? *Annals of the New York Academy of Sciences*, **1290**, 12–20.

Honari, M., Shafabakhsh, R., Reiter, R.J., Mirzaei, H., and Asemi, Z. (2019) Resveratrol is a promising agent for colorectal cancer prevention and treatment: focus on molecular mechanisms. *Cancer Cell International*, **19**, 180.

Hosogai, N., Fukuhara, A., Oshima, K., *et al.* (2007) Adipose tissue hypoxia in obesity and its impact on adipocytokine dysregulation. *Diabetes*, **56**, 901–11.

Houshmand-Oeregaard, A., Hansen, N.S., Hjort, L., *et al.* (2017) Differential adipokine DNA methylation and gene expression in subcutaneous adipose tissue from adult offspring of women with diabetes in pregnancy. *Clinical Epigenetics*, **9**, 37.

Huang, X.T., Li, X., Xie, M.L., *et al.* (2019) Resveratrol: review on its discovery, anti-leukemia effects and pharmacokinetics. *Chemico-Biological Interactions*, **306**, 29–38.

Jannin, B., Menzel, M., Berlot, J.P., Delmas, D., Lançon, A., and Latruffe, N. (2004) Transport of resveratrol, a cancer chemopreventive agent, to cellular targets: plasmatic protein binding and cell uptake. *Biochemical Pharmacology*, **68**, 1113–18.

Juan, M.E., Maijó, M., and Planas, J.M. (2010) Quantification of trans-resveratrol and its metabolites in rat plasma and tissues by HPLC. *Journal of Pharmaceutical and Biomedical Analysis*, **51**, 391–8.

Kamal-Eldin, A., Frank, J., Razdan, A., Tengblad, S., Basu, S., and Vessby, B. (2000) Effects of dietary phenolic compounds on tocopherol, cholesterol, and fatty acids in rats. *Lipids*, **35**, 427–35.

Kapetanovic, I.M., Muzzio, M., Huang, Z., Thompson, T.N., and McCormick, D.L. (2011) Pharmacokinetics, oral bioavailability, and metabolic profile of resveratrol and its dimethylether analog, pterostilbene, in rats. *Cancer Chemotherapy and Pharmacology*, **68**, 593–601.

Kennedy, D.O., Wightman, E.L., Reay, J.L., *et al.* (2010) Effects of resveratrol on cerebral blood flow variables and cognitive performance in humans: a double-blind, placebo-controlled, crossover investigation. *American Journal of Clinical Nutrition*, **91**, 1590–7.

Kim, S., Jin, Y., Choi, Y., and Park, T. (2011) Resveratrol exerts anti-obesity effects via mechanisms involving down-regulation of adipogenic and inflammatory processes in mice. *Biochemical Pharmacology*, **81**, 1343–51.

Ko, J.H., Sethi, G., Um, J.Y., *et al.* (2017) The role of resveratrol in cancer therapy. *International Journal of Molecular Sciences*, **18**, 2589.

Kulkarni, S.S. and Cantó, C. (2015) The molecular targets of resveratrol. *Biochimica et Biophysica Acta*, **1852**, 1114–23.

Ladurner, A., Schachner, D., Schueller, K., *et al.* (2014) Impact of trans-resveratrol-sulfates and -glucuronides on endothelial nitric oxide synthase activity, nitric oxide release and intracellular reactive oxygen species. *Molecules*, **19**, 16724–36.

Langcake, P. and Price, R.J. (1976) The production of resveratrol by Vitis vinifera and other members of the Vitaceae as a response to infection or injury. *Physiological Plant Pathology*, **9**, 77–86.

Lasa, A., Churruca, I., Eseberri, I., Andrés-Lacueva, C., and Portillo, M.P. (2012) Delipidating effect of resveratrol metabolites in 3T3-L1 adipocytes. *Molecular Nutrition & Food Research*, **56**, 1559–68.

Le Gouill, E., Jimenez, M., Binnert, C., *et al.* (2007) Endothelial nitric oxide synthase (eNOS) knockout mice have defective mitochondrial beta-oxidation. *Diabetes*, **56**, 2690–6.

Lekakis, J., Rallidis, L.S., Andreadou, I., *et al.* (2005) Polyphenolic compounds from red grapes acutely improve endothelial function in patients with coronary heart disease. *European Journal of Cardiovascular Prevention and Rehabilitation*, **12**, 596–600.

Li, X. (2013) SIRT1 and energy metabolism. *Acta Biochimica et Biophysica Sinica*, **45**, 51–60.

Lin, H.Y., Lansing, L., Merillon, J.M., *et al.* (2006) Integrin alphaVbeta3 contains a receptor site for resveratrol. *FASEB Journal*, **20**, 1742–4.

Lin, S.P., Chu, P.M., Tsai, S.Y., Wu, M.H., and Hou, Y.C. (2012) Pharmacokinetics and tissue distribution of resveratrol, emodin and their metabolites after intake of Polygonum cuspidatum in rats. *Journal of Ethnopharmacology*, **144**, 671–6.

Liu, Z. and Hu, M. (2007) Natural polyphenol disposition via coupled metabolic pathways. *Expert Opinion on Drug Metabolism & Toxicology*, **3**, 389–406.

Manson, J.E. and Bassuk, S.S. (2015) Biomarkers of cardiovascular disease risk in women. *Metabolism*, **64**(Suppl 1), S33–S39.

Marchal, J., Pifferi, F., and Aujard, F. (2013) Resveratrol in mammals: effects on aging biomarkers, age-related diseases, and life span. *Annals of the New York Academy of Sciences*, **1290**, 67–73.

Marier, J.F., Vachon, P., Gritsas, A., Zhang, J., Moreau, J.P., and Ducharme, M.P. (2002) Metabolism and disposition of resveratrol in rats: extent of absorption, glucuronidation, and enterohepatic recirculation evidenced by a linked-rat model. *Journal of Pharmacology and Experimental Therapeutics*, **302**, 369–73.

Marín, L., Miguélez, E.M., Villar, C.J., and Lombó, F. (2015) Bioavailability of dietary polyphenols and gut microbiota metabolism: antimicrobial properties. *BioMed Research International*, **2015**, e905215.

Miksits, M., Wlcek, K., Svoboda, M., *et al.* (2010) Expression of sulfotransferases and sulfatases in human breast cancer: impact on resveratrol metabolism. *Cancer Letters*, **289**, 237–45.

Milton-Laskibar, I., Aguirre, L., Etxeberria, U., Milagro, F.I., Martínez, J.A., and Portillo, M.P. (2018) Do the effects of resveratrol on thermogenic and oxidative capacities in ibat and skeletal muscle depend on feeding conditions? *Nutrients*, **10**, 1446.

Pallauf, K., Chin, D., Günther, I., *et al.* (2019) Resveratrol, lunularin and dihydroresveratrol do not act as caloric restriction mimetics when administered intraperitoneally in mice. *Scientific Reports*, **9**, 4445.

Patel, K.R., Andreadi, C., Britton, R.G., *et al.* (2013) Sulfate metabolites provide an intracellular pool for resveratrol generation and induce autophagy with senescence. *Science Translational Medicine*, **5**, 205ra133.

Pervaiz, S. (2003) Resveratrol: from grapevines to mammalian biology. *FASEB Journal*, **17**, 1975–85.

Poulsen, M.M., Vestergaard, P.F., Clasen, B.F., *et al.* (2013) High-dose resveratrol supplementation in obese men: an investigator-initiated, randomized, placebo-controlled clinical trial of substrate metabolism, insulin sensitivity, and body composition. *Diabetes*, **62**, 1186–95.

Rayalam, S., Yang, J.Y., Ambati, S., Della-Fera, M.A., and Baile, C.A. (2008) Resveratrol induces apoptosis and inhibits adipogenesis in 3T3-L1 adipocytes. *Phytotherapy Research*, **22**, 1367–71.

Rotches-Ribalta, M., Urpi-Sarda, M., Llorach, R., *et al.* (2012) Gut and microbial resveratrol metabolite profiling after moderate long-term consumption of red wine versus dealcoholized red wine in humans by an optimized ultra-high-pressure liquid chromatography tandem mass spectrometry method. *Journal of Chromatography. A*, **1265**, 105–13.

Ryan, V.H., German, A.J., Wood, I.S., Hunter, L., Morris, P., and Trayhurn, P. (2010) Adipokine expression and secretion by canine adipocytes: stimulation of inflammatory adipokine production by LPS and TNFalpha. *Pflugers Archiv*, **460**, 603–16.

Santos, J.A., de Carvaho, G.S., Oliveira, V., Raposo, N.R., and da Silva, A.D. (2013) Resveratrol and analogues: a review of antioxidant activity and applications to human health. *Recent Patents on Food, Nutrition & Agriculture*, **5**, 144–53.

Schueller, K., Pignitter, M., and Somoza, V. (2015) Sulfated and glucuronated trans-resveratrol metabolites regulate chemokines and sirtuin-1 expression in U-937 macrophages. *Journal of Agricultural and Food Chemistry*, **63**, 6535–45.

Shankar, R.R., Wu, Y., Shen, H.Q., Zhu, J.S., and Baron, A.D. (2000) Mice with gene disruption of both endothelial and neuronal nitric oxide synthase exhibit insulin resistance. *Diabetes*, **49**, 684–7.

Shao, D., Wang, Y., Huang, Q., *et al.* (2016) Cholesterol-lowering effects and mechanisms in view of bile acid pathway of resveratrol and resveratrol glucuronides. *Journal of Food Science*, **81**, H2841–H2848.

Simini, B. (2000) Serge Renaud: from French paradox to Cretan miracle. *Lancet*, **355**, 48.

Soleas, G.J., Angelini, M., Grass, L., Diamandis, E.P., and Goldberg, D.M. (2001) Absorption of trans-resveratrol in rats. *Methods in Enzymology*, **335**, 145–54.

Steinberg, H.O., Chaker, H., Leaming, R., Johnson, A., Brechtel, G., and Baron, A.D. (1996) Obesity/insulin resistance is associated with endothelial dysfunction. Implications for the syndrome of insulin resistance. *Journal of Clinical Investigation*, **97**, 2601–10.

Švajger, U. and Jeras, M. (2012) Anti-inflammatory effects of resveratrol and its potential use in therapy of immune-mediated diseases. *International Reviews of Immunology*, **3**, 202–22.

Švajger, U., Obermajer, N., and Jeras, M. (2010) Dendritic cells treated with resveratrol during differentiation from monocytes gain substantial tolerogenic properties upon activation. *Immunology*, **129**, 525–35.

Takaoka, M. (1939) Resveratrol, a new phenolic compound, from Veratrum grandiflorum. *Journal of Chemistry Society of Japan*, **60**, 1090–100.

Van Gaal, L.F., Mertens, I.L., and de Block, C.E. (2006) Mechanisms linking obesity with cardiovascular disease. *Nature*, **444**, 875–80.

Walker, J., Schueller, K., Schaefer, L.M., Pignitter, M., Esefelder, L., and Somoza, V. (2014) Resveratrol and its metabolites inhibit pro-inflammatory effects of lipopolysaccharides in U-937 macrophages in plasma-representative concentrations. *Food & Function*, **5**, 74–84.

Walle, T. (2011) Bioavailability of resveratrol. *Annals of the New York Academy of Sciences*, **1215**, 9–15.

Walle, T., Hsieh, F., DeLegge, M.H., Oatis, J.E., and Walle, U.K. (2004) High absorption but very low bioavailability of oral resveratrol in humans. *Drug Metabolism and Disposition*, **32**, 1377–82.

Wang, D., Xu, Y., and Liu, W. (2008) Tissue distribution and excretion of resveratrol in rat after oral administration of Polygonum cuspidatum extract (PCE). *Phytomedicine*, **15**, 859–66.

Wang, P., Keijer, J., Bunschoten, A., Bouwman, F., Renes, J., and Mariman, E. (2006) Insulin modulates the secretion of proteins from mature 3T3-L1 adipocytes: a role for transcriptional regulation of processing. *Diabetologia*, **49**, 2453–62.

Wenzel, E. and Somoza, V. (2005) Metabolism and bioavailability of trans-resveratrol. *Molecular Nutrition & Food Research*, **49**, 472–81.

Wu, J.M., Wang, Z.R., Hsieh, T.C., Bruder, J.L., Zou, J.G., and Huang, Y.Z. (2001) Mechanism of cardioprotection by resveratrol, a phenolic antioxidant present in red wine (review). *International Journal of Molecular Medicine*, **8**, 3–17.

Xu, Q. and Si, L.Y. (2012) Resveratrol role in cardiovascular and metabolic health and potential mechanisms of action. *Nutrition Research*, **32**, 648–58.

Yu, W., Fu, Y.C., and Wang, W. (2012) Cellular and molecular effects of resveratrol in health and disease. *Journal of Cellular Biochemistry*, **113**, 752–9.

Zamora-Ros, R., Knaze, V., Rothwell, J.A., *et al.* (2016) Dietary polyphenol intake in Europe: the European Prospective

Investigation into Cancer and Nutrition (EPIC) study. *European Journal of Nutrition*, **55**, 1359–75.
Zeigerer, A., Rodeheffer, M.S., McGraw, T.E., and Friedman, J.M. (2008) Insulin regulates leptin secretion from 3T3-L1 adipocytes by a PI 3 kinase independent mechanism. *Experimental Cell Research*, **314**, 2249–56.
Zhu, W., Qin, W., Zhang, K., *et al.* (2012) Trans-resveratrol alters mammary promoter hypermethylation in women at increased risk for breast cancer. *Nutrition and Cancer*, **64**, 393–400.

12

Flavanones

*Gema Pereira-Caro[1], Colin D. Kay[2], Michael N. Clifford[3], and Alan Crozier[4,5]**

[1] *Department of Food Science and Health, Andalusian Institute of Agricultural and Fishery Research and Training, Alameda del Obispo, Córdoba, Spain*
[2] *Food Bioprocessing and Nutritional Sciences, Plants for Human Health Institute, North Carolina State University, Kannapolis, North Carolina, USA*
[3] *School of Biosciences and Medicine, University of Surrey, Guildford, UK*
[4] *Department of Nutrition, University of California, Davis, California, USA*
[5] *School of Medicine, Dentistry and Nursing, University of Glasgow, Glasgow, UK*

12.1 Introduction

The absorption, distribution, metabolism, and excretion (ADME), synonymous with bioavailability in nutrition research, of flavonoid glycosides and related phenolic compounds have been a topic of increasing investigation in recent years. Following ingestion, absorption into the circulatory system initially occurs in the small intestine (Del Rio *et al.*, 2013). Typically, absorption is associated with hydrolysis releasing aglycones as a result of the action of lactase-phlorizin hydrolase (LPH) in the brush-border of the small intestine epithelial cells. LPH exhibits broad substrate specificity for flavonoid-*O*-β-D-glucosides and the released aglycone may then enter the epithelial cells by passive diffusion as a result of its increased lipophilicity and its proximity to the enterocyte membrane (Day *et al.*, 2000). An alternative hydrolytic step mediated by a cytosolic β-glucosidase (CBG) within the epithelial cells has been proposed. In order for CBG-catalyzed hydrolysis to occur, the polar glucosides

*Corresponding author.

Dietary Polyphenols: Metabolism and Health Effects, First Edition.
Edited by Francisco A. Tomás-Barberán, Antonio González-Sarrías, and Rocío García-Villalba.

must be transported into the epithelial cells, possibly with the involvement of the active sodium-dependent glucose transporter, SGLT1 (Gee *et al.*, 2000; Williamson *et al.*, 2018). Thus, there are two possible routes by which the glucoside conjugates are hydrolyzed and the resultant aglycones appear in the epithelial cells, namely LPH/diffusion and CBG/transport.

Prior to passage into the bloodstream, the aglycones undergo phase II metabolism in the enterocytes forming sulfate, glucuronide, and/or methylated metabolites through the respective action of sulfotransferases, uridine-5′-diphosphate glucuronosyltransferases, and catechol-*O*-methyltransferases. There is efflux of at least some of the metabolites back into the lumen of the small intestine and this is thought to involve members of the ATP-binding cassette family of transporters, including multidrug resistance protein and P-glycoprotein. Once in the portal bloodstream, metabolites pass to the liver, where they may be subjected to further phase II metabolism (Williamson *et al.*, 2018). Although there is much speculation, as yet there is no compelling evidence that enterohepatic recirculation in humans results in substantial recycling of (poly)phenol metabolites back into the small intestine via biliary excretion (Crozier, 2013).

Analysis of ileal fluid collected from ileostomists after the ingestion of various foodstuffs indicates that even when dietary (poly)phenols are absorbed in the small intestine, substantial quantities can nonetheless pass to the large intestine (Borges *et al.*, 2013; Hollman *et al.*, 1995; Stalmach *et al.*, 2010a,b) where the intestinal microbiota cleave conjugating moieties and the resultant aglycones may be absorbed as phase II metabolites and/or undergo C-ring fission leading to the production of phenolic catabolites and aromatic compounds. These phenolic and aromatic catabolites can be absorbed with a portion being subjected to phase II metabolism before being excreted in urine in substantial quantities, which in most instances appear in excess of the metabolites that enter the circulatory system via the upper small intestine (Del Rio *et al.*, 2013; Williamson *et al.*, 2018).

This chapter will focus on research on the ADME in humans of flavanones which, because of their occurrence in oranges (*Citrus sinensis*), has been carried out principally with orange juice (OJ). The evidence for protective effects of dietary OJ flavanones will also be discussed along with the potential mechanisms underlying such effects.

12.2 Flavanones and Their Occurrence

Flavanones are characterized by a C_6-C_3-C_6 structure with the absence of a $\Delta^{2,3}$ double bond and the presence of a chiral centre at C2 (structure **1**). Citrus fruit contains relatively high amounts of flavanones which are present in both the juice and the tissues that are ingested when fruit segments are consumed. The tissues are a particularly rich source but are only consumed as an adjunct to the consumption of OJ "with pulp" or "with bits" which are now common on supermarket shelves. The main flavanones in oranges and OJ are hesperetin-7-*O*-rutinoside (**2**, hesperidin), and naringenin-7-*O*-rutinoside (**3**, narirutin). OJ also contains smaller quantities of hesperetin-7-*O*-rutinoside-3′-*O*-glucoside (**4**), 4′-*O*-methyl-naringenin-7-*O*-rutinoside (**5**, didymin), and eriodictyol-7-*O*-rutinoside (**6**, eriocitrin). Additional (poly) phenolic compounds that can occur in OJ in trace amounts include the flavone apigenin-6,8-C-diglucoside (7, vicenin-2), ferulic acid-4′-*O*-glucoside (**8**) and coumaric acid-4′-*O*-glucoside (**9**). The alkaloid *p*-sympatol (**10**) is found in OJ (Pereira-Caro *et al.*, 2016) which also contains a series of polymethoxylated flavones (Crozier *et al.*, 2006). Sources of other flavanones are covered in a recent review by Najmanová *et al.* (2019).

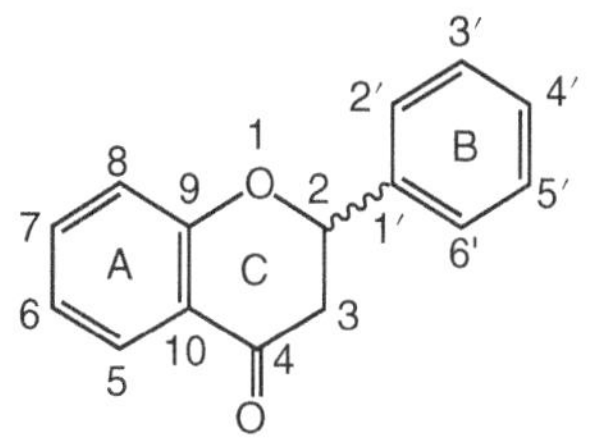

Flavanone skeleton (**1**)

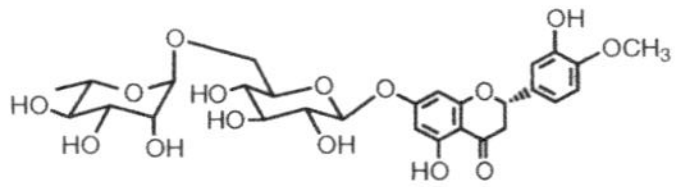

Hesperetin-7-*O*-rutinoside (**2**)

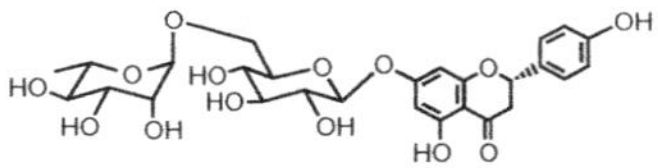

Naringenin-7-*O*-rutinoside (**3**)

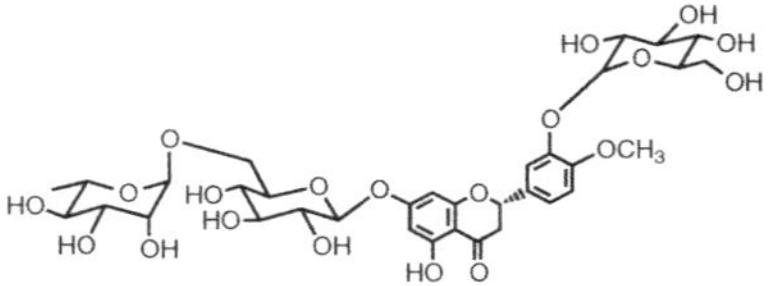

Hesperetin-7-*O*-rutinoside-3'-*O*-glucoside (**4**)

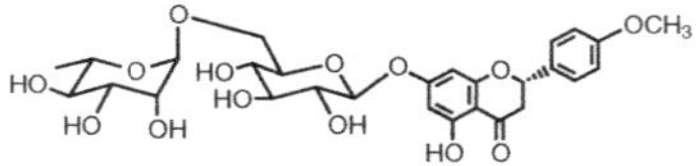

4'-*O*-Methyl-naringenin-7-*O*-rutinoside (**5**)

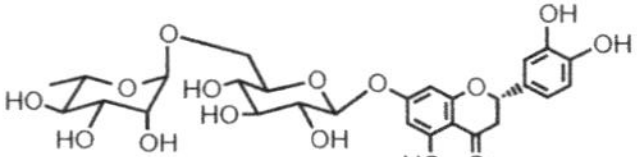

Eriodictyol-7-*O*-rutinoside (**6**)

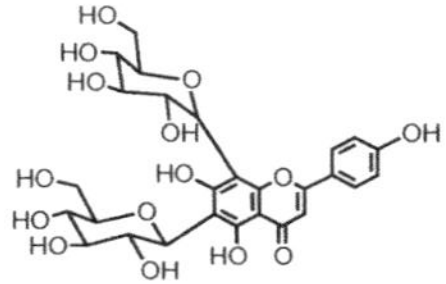

Apigenin-6,8-*C*-diglucoside (**7**)

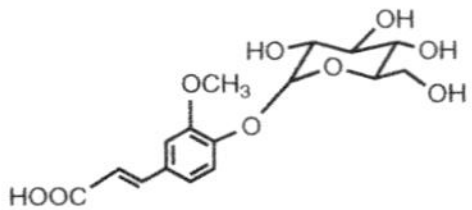

Ferulic acid-4'-*O*-glucoside (**8**)

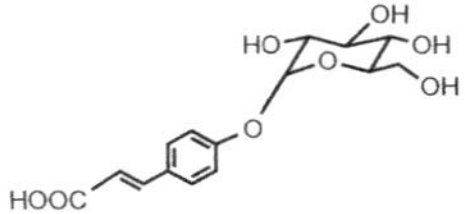

Coumaric acid-4'-*O*-glucoside (**9**)

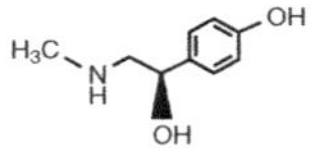

p-Sympatol (**10**)

Both hesperetin-7-*O*-rutinoside and naringenin-7-*O*-rutinoside are tasteless. In contrast, flavanone neohesperidoside conjugates, such as hesperetin-7-*O*-neohesperidoside (**11**, neohesperidin) from bitter orange (*Citrus aurantium*) and naringenin-7-*O*-neohesperidoside (**12**, naringin) from grapefruit peel (*Citrus paradise*), have an intense bitter taste. The two main flavanones in lemons (*Citrus limon*) are hesperetin-7-*O*-rutinoside and eriodictyol-7-*O*-rutinoside while hesperetin-7-*O*-rutinoside predominates in limes (*Citrus aurantifolia*) (Crozier *et al.*, 2006; Neveu *et al.*, 2010).

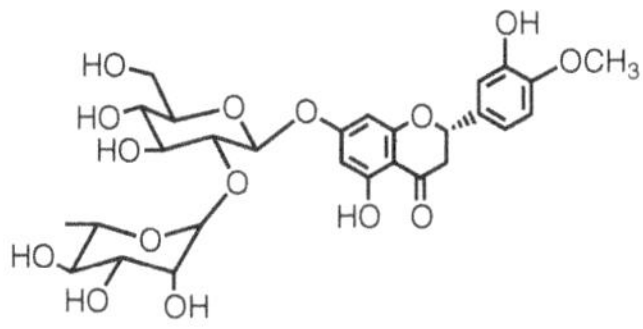

Hesperetin-7-*O*-neohesperidoside (**11**)

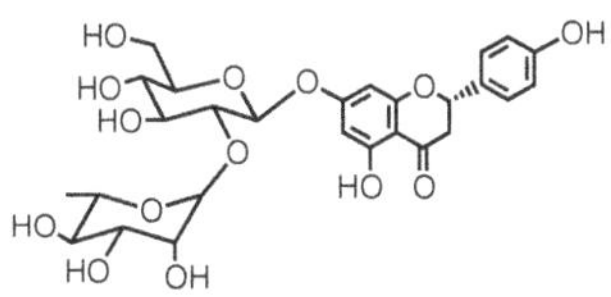

Naringenin-7-*O*-neohesperidoside (**12**)

12.3 Absorption of Flavanone Metabolites in the Proximal and Distal Gastrointestinal Tract

There was, and indeed still is, a general belief that flavanones, and related compounds, are poorly bioavailable with only relatively small amounts of the ingested dose entering the systemic circulation, in the form of metabolites, prior to being excreted in urine. However, the absorbed C_6-C_3-C_6 metabolites are treated as xenobiotics so their presence in the circulatory system is transient as they are rapidly removed from the bloodstream via renal excretion. Thus, plasma pharmacokinetic measurements, while useful, provide only a "snapshot" of the absorbed circulating metabolites. When 0–24-hour urinary excretion of flavonoid phase II metabolites, as well as their microbiota-derived phenolic catabolites, are included in the overall estimation of ADME, it is evident that flavanones are highly bioavailable (Del Rio *et al.*, 2013; Najmanová *et al.*, 2019; Pereira-Caro *et al.*, 2014; Rodriguez-Mateos *et al.*, 2014; Williamson *et al.*, 2018).

Borges *et al.* (2010) described an investigation in which volunteers ingested a (poly)phenol-rich drink containing a diverse spectrum of phenolic and flavonoid compounds including 45 μmol of hesperetin-7-*O*-rutinoside.

Hesperetin-7-*O*-glucuronide (**13**) and a second glucuronide, probably hesperetin-3′-*O*-glucuronide (**14**), were detected in plasma with a peak plasma concentration (C_{max}) of 168 ± 45 nmol/L which was attained (T_{max}) 3.7 ± 0.2 h after intake. This is in keeping with the findings of earlier feeding studies with OJ in which hesperetin metabolite T_{max} values ranged from 3.7 ± 0.5 h to 7.4 ± 2.0 h (Urpi-Sarda *et al.*, 2012), suggesting absorption in the lower bowel rather than the upper small intestine. The same two glucuronides, together with a sulfated-glucuronide, were excreted in urine collected 0–24 hours after drink intake in amounts corresponding to 12.0 ± 1.6% of the ingested hesperetin-7-*O*-rutinoside (Borges *et al.*, 2010). In a subsequent study the (poly)phenol-rich beverage was consumed by ileostomists and the quantity of hesperetin metabolites detected in urine was 3.5 ± 0.4% of intake (Borges *et al.*, 2013). The differing levels of hesperetin metabolites excreted in urine by the ileostomists and volunteers with functioning colon indicate that absorption of hesperetin is not exclusive to the large intestine with one-third being absorbed in the upper gastrointestinal tract (GIT) and two-thirds in the distal GIT.

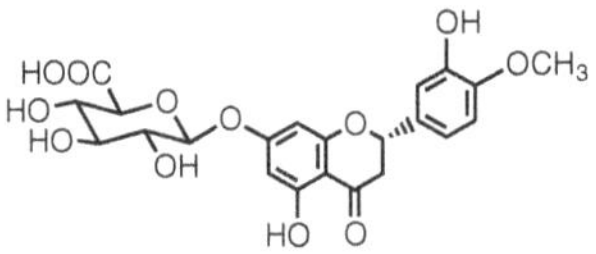

Hesperetin-7-*O*-glucuronide (**13**)

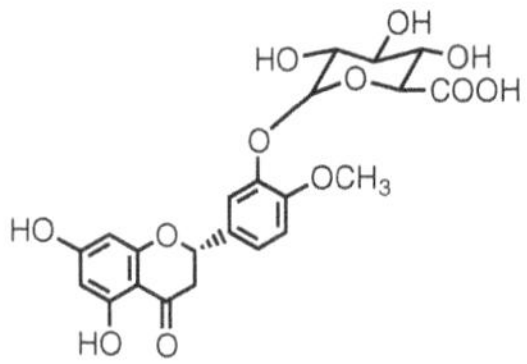

Hesperetin-3′-*O*-glucuronide (**14**)

Nielson *et al.* (2006) carried out studies in which the bioavailability of hesperetin-7-*O*-rutinoside was compared with that of OJ treated with α-rhamnosidase to convert the rutinoside to hesperetin-7-*O*-glucoside (**15**). They demonstrated that the conjugating sugar moiety has a major impact on flavanone absorption. The glucoside was much more bioavailable with its metabolites having substantially higher C_{max}, area-under-the-curve values and urinary excretion, with a T_{max} 0.6 ± 0.1 h compared to that of 7.4 ± 2.0 h for metabolites of hesperetin-7-*O*-rutinoside. In a similar manner, naringenin-7-*O*-glucoside (**16**) was absorbed more rapidly with metabolites excreted in urine in ~7-fold higher amounts than those derived from naringenin-7-*O*-rutinoside (Bredsdorff *et al.*, 2010).

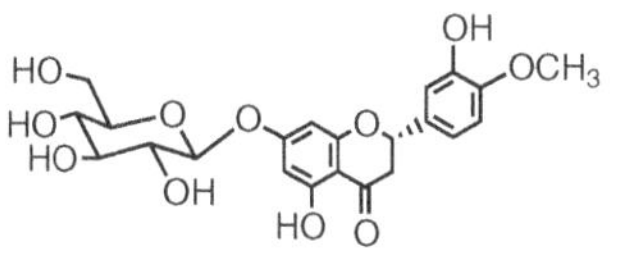

Hesperetin-7-*O*-glucoside (**15**)

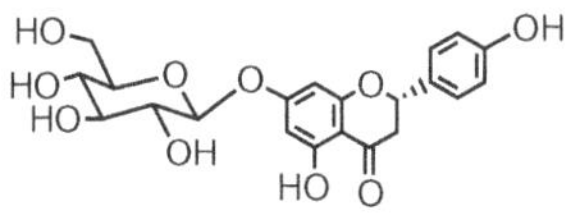

Naringenin-7-*O*-glucoside (**16**)

These findings indicate that CBG/LPH enzymes in the small intestine are able to hydrolyze the flavanone-*O*-monoglucoside, releasing the aglycone which in epithelial cells is subject to phase II metabolism before entering the circulatory system. In contrast, only limited hydrolysis of flavanone-*O*-rutinosides occurs in the small intestine and, as a consequence, they pass to the distal GIT where the colonic microbiota cleave the sugar moiety, releasing the aglycone, some of which enters the bloodstream after being subjected to phase II metabolism in the colonocytes and/or hepatocytes. The urinary excretion of glucuronide and sulfate metabolites after OJ consumption varies somewhat from study to study with urinary excretion, which is typically higher for naringenin than hesperetin, ranging from 0.7% to 17.7% of intake (Silveira *et al.*, 2014). This variability is probably a consequence of dose, the manner in which the juice was processed, the level of soluble and precipitated flavanones (Vallejo *et al.*, 2010), person-to-person variations in the colonic microbiota and the analytical methodologies utilized (see section 12.5.5).

A more detailed quantitative estimate of flavanone bioavailability was obtained in an OJ feeding study where urinary excretion of both hesperetin and naringenin metabolites and phenolic and aromatic acids was determined (Pereira-Caro *et al.*, 2014). A diversity of flavanone metabolites were detected in 0–24 hour urine after OJ consumption that were not present in the urine of the volunteers collected after drinking a flavanone-free placebo drink. This study reported an 83 μmol excretion of metabolites, corresponding to 16% of the flavanone intake. The principal metabolites were the 3′-*O*- and 7-*O*-glucuronides of hesperetin and hesperetin-3′-sulfate (**17**), together with lower quantities of naringenin-4′-*O*-glucuronide (**18**), naringenin-7-*O*-glucuronide (**19**), naringenin-4′-sulfate (**20**), hesperetin-5-*O*-glucuronide (**21**), and as series of hesperetin and naringenin diglucuronides (**22–27**).

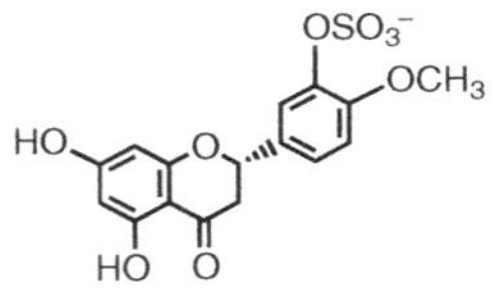

Hesperetin-3′-sulfate (**17**)

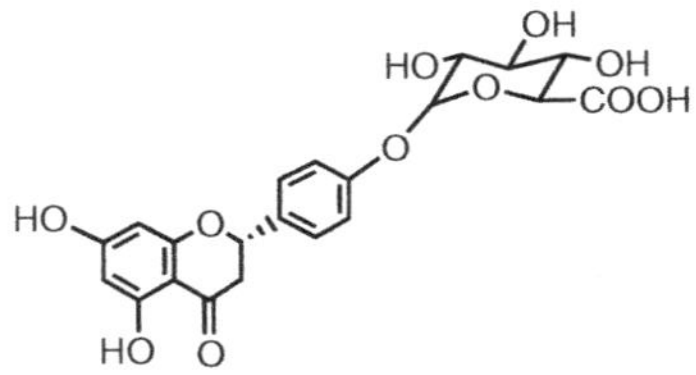

Naringenin-4′-*O*-glucuronide (**18**)

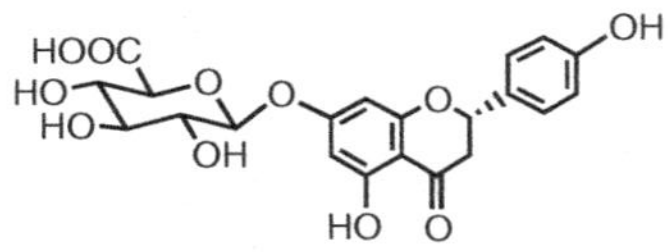

Naringenin-7-*O*-glucuronide (**19**)

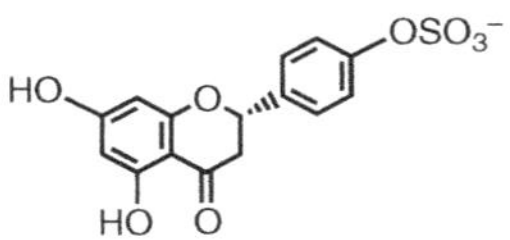

Naringenin-4′-sulfate (**20**)

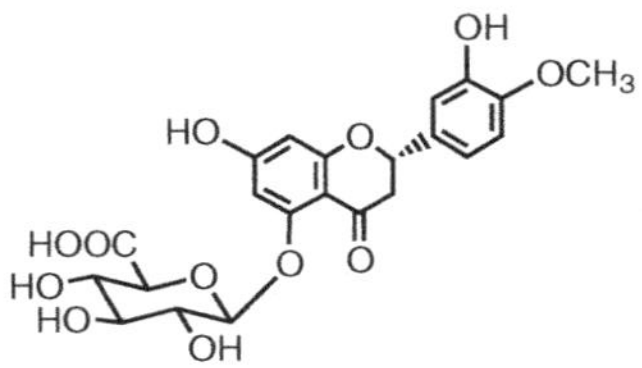

Hesperetin-5-*O*-glucuronide (**21**)

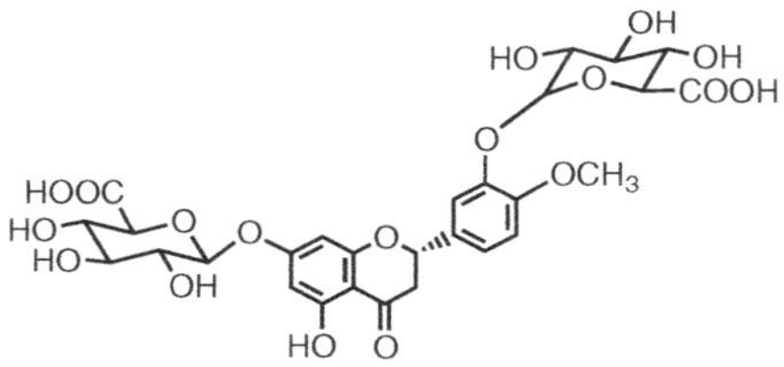

Hesperetin-3′,7-*O*-diglucuronide (**22**)

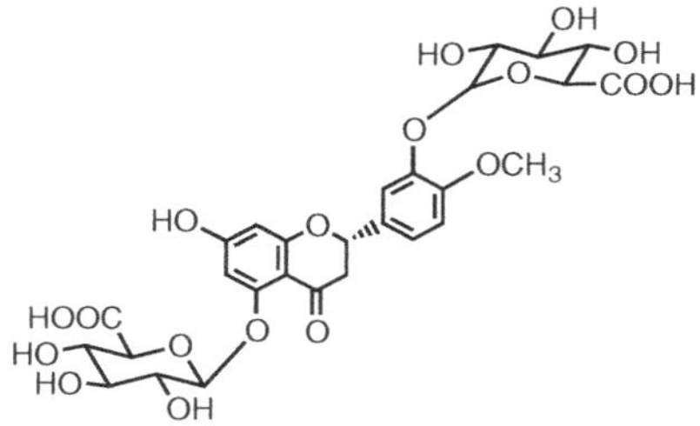

Hesperetin-3′,5-*O*-diglucuronide (**23**)

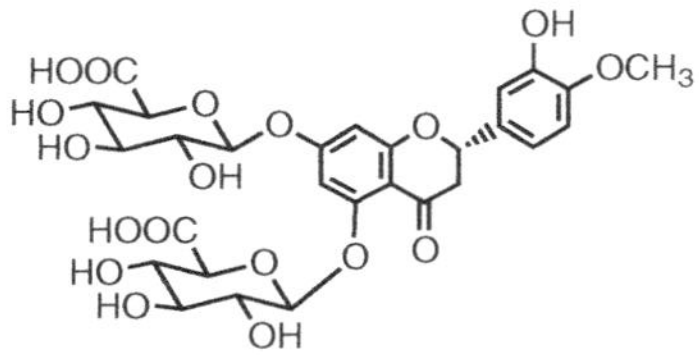

Hesperetin-5,7-*O*-diglucuronide (**24**)

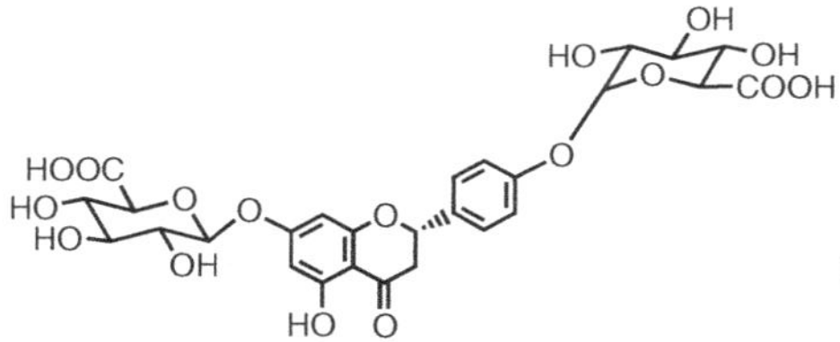

Naringenin-4′,7-*O*-diglucuronide (**25**)

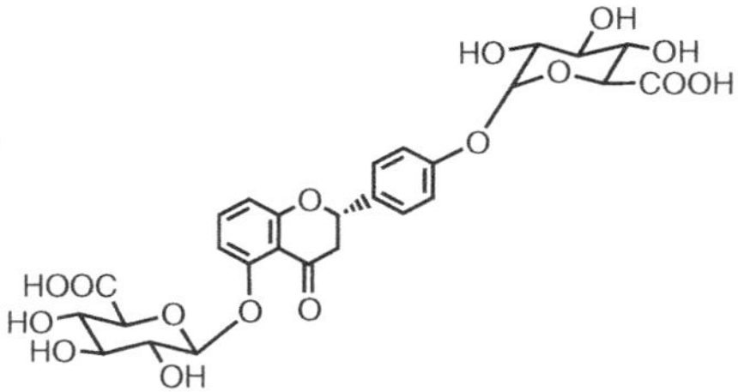

Naringenin-4′,5-*O*-diglucuronide (**26**)

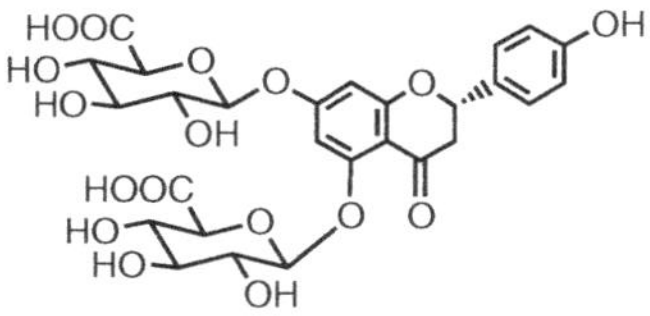

Naringenin-5,7-*O*-diglucuronide (**27**)

Unlike the flavanone metabolites, the baseline urine does contain phenolic and aromatic compounds as they are produced by endogenous pathways as well as being metabolic products of dietary flavanones and other (poly)phenolic compounds. However, eight phenolic catabolites were excreted in significantly higher amounts in 0–24 hour urine after OJ consumption compared with intake of the flavanone-free placebo drink (Table 12.1) (Pereira-Caro *et al.*, 2014). 3-(3′-Hydroxy-4′-methoxyphenyl)hydracrylic acid (**28**) may be a key indicator of OJ intake as it was excreted in substantial amounts and was absent in placebo urine. 3-(3′-Hydroxyphenyl)hydracrylic acid (**29**), although present in

Table 12.1 Phenolic catabolites excreted in urine in significantly increased amounts ($P < 0.05$) 0–24 h after the ingestion of 250 mL of orange juice containing 584 μmol of (poly)phenolic compounds compared to a (poly)phenol-free placebo drink (based on Pereira-Caro *et al.*, 2014)[a]

Phenolic catabolites	Placebo drink	Orange juice	Increase with orange juice
3-(3′-Hydroxy-4′-methoxyphenyl)hydracrylic acid	n.d	72 ± 10	72
3-(3′-Hydroxyphenyl) hydracrylic acid	2.5 ± 0.8	31 ± 4	29
3-(3′-Hydroxy-4′-methoxyphenyl)propionic acid (dihydro-isoferulic acid)	n.d.	4.7 ± 0.8	4.7
3-(3′-Methoxy-4′-hydroxyphenyl)propionic acid (dihydroferulic acid)	n.d.	6.3 ± 1.0	6.3
3′-Methoxy-4′-hydroxyphenylacetic acid	4.7 ± 0.7	7.3 ± 0.8	2.6
3′-Hydroxyhippuric acid	n.d.	0.8 ± 0.3	0.8
4′-Hydroxyhippuric acid	0.2 ± 0.1	19 ± 4	19
Hippuric acid	232 ± 21	610 ± 48	378
Total minus hippuric acid	*7*	*141*	*134 (23%)*
Total with hippuric acid	*239*	*751*	*512 (88%)*

a) Data expressed in μmol as mean values ± S.E. Figures in italics are excretion as a percentage of (poly)phenol intake.

n.d., not detected.

trace amounts in the control urine, increased >10-fold after OJ consumption, while excretion of 4′-hydroxhippuric acid (**30**) increased ~100-fold from 0.2 to 19 μmol. The overall increase in phenolic catabolites, excluding hippuric acid (**31**), after OJ consumption was 134 μmol, which is 23% of the 584 μmol intake of (poly)phenolic compounds (Table 12.2) (see structures 28–31). Addition of 134 μmol to the 83 μmol excretion of flavanone glucuronide and sulfate metabolites equates to a recovery of 37% of intake (see Table 12.1).

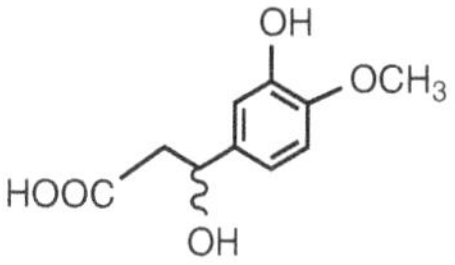

3-(3'-Hydroxy-4'-methoxyphenyl) hydracrylic acid (**28**)

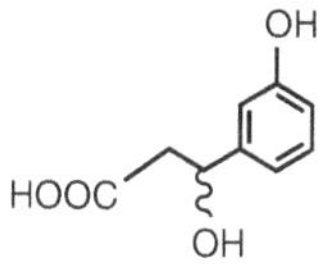

3-(3'-Hydroxyphenyl) hydracrylic acid (**29**)

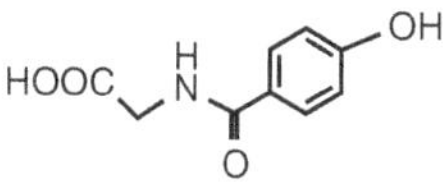

4'-Hydroxyhippuric acid (**30**)

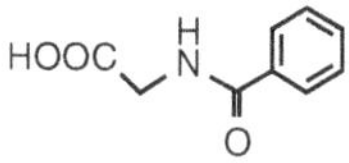

Hippuric acid (**31**)

Excretion of hippuric acid was not included in the reported 37% recovery as the placebo 0–24 hour urine contained 232 ± 21 μmol of the glycinated benzoic acid derivative (Pereira-Caro *et al.*, 2014). Although postintervention levels increased significantly by 378 μmol to 610 ± 48 μmol following OJ intake, exactly how much of this increase in hippuric acid is related to OJ (poly)phenol metabolism intake is difficult to ascertain because of the contribution of endogenous sources of hippuric acid and the person-to-person variation in the amounts excreted. If increased hippuric acid was included in the calculation, the overall (poly)phenol recovery would equate to ~100%. Nonetheless, it is apparent that when metabolites and catabolites are taken into consideration, the OJ (poly)phenols, principally hesperetin- and naringenin-7-*O*-rutinosides, are highly bioavailable (Pereira-Caro *et al.*, 2014).

In order to further elucidate the ADME of OJ flavanones and, in particular, the pathways by which the various phenolic catabolites are formed, hesperetin, naringenin, and ferulic acid

Table 12.2 Intervolunteer variations in the urinary excretion of hesperetin and naringenin phase II metabolites over a 24-h period after acute ingestion of an orange juice containing 167 μmol of hesperetin and naringenin glycosides. Data from a study by Pereira-Caro *et al*. (2015b)

	Volunteers														
Total metabolites	1	2	3	4	5	6	7	8	9	10	11	12	13	14	Mean ± SE
μmole	86	60	57	54	51	48	46	45	37	31	18	17	16	10	41 ± 6
Percentage of intake	*52%*	*36%*	*34%*	*32%*	*31%*	*29%*	*28%*	*27%*	*22%*	*18%*	*11%*	*10%*	*10%*	*6%*	*25 ± 3*

SE, standard error.

were incubated *in vitro* with human fecal material cultured under anaerobic conditions (Pereira-Caro *et al.*, 2015a). The data obtained were compared with the *in vivo* urinary excretion of phenolic catabolites after OJ intake, and taken together with the potential involvement of microbial and mammalian enzymes as outlined in the previous paragraphs, to produce the proposed catabolic routes that are illustrated in Figures 12.1 and 12.2.

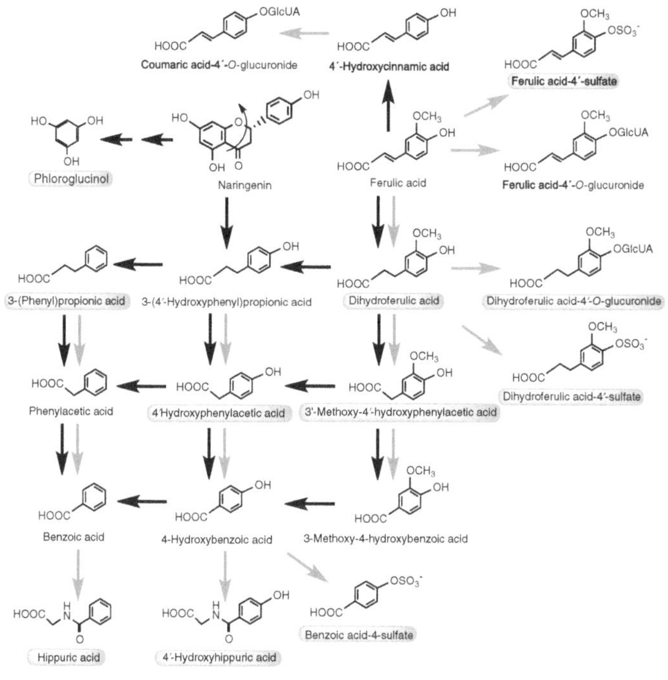

Figure 12.1 Proposed pathways for the metabolism of naringenin and ferulic acid released, respectively, by cleavage of naringenin-7-*O*-rutinoside and ferulic acid-4′-*O*-glucoside, after the consumption of orange juice by humans. Based on urinary excretion in feeding studies and *in vitro* fecal incubations. Compounds with boxed names indicate major components in 0–24-h urine. The curved arrow indicates the site of ring fission. Black arrows are potential microbiota-mediated conversions and gray arrows indicate steps potentially catalyzed by mammalian enzymes. *, potential intermediates that did not accumulate in detectable amounts; GlcUA, glucuronide. Source: Based on data from Pereira-Caro *et al.* (2014, 2015a,b, 2016, 2018).

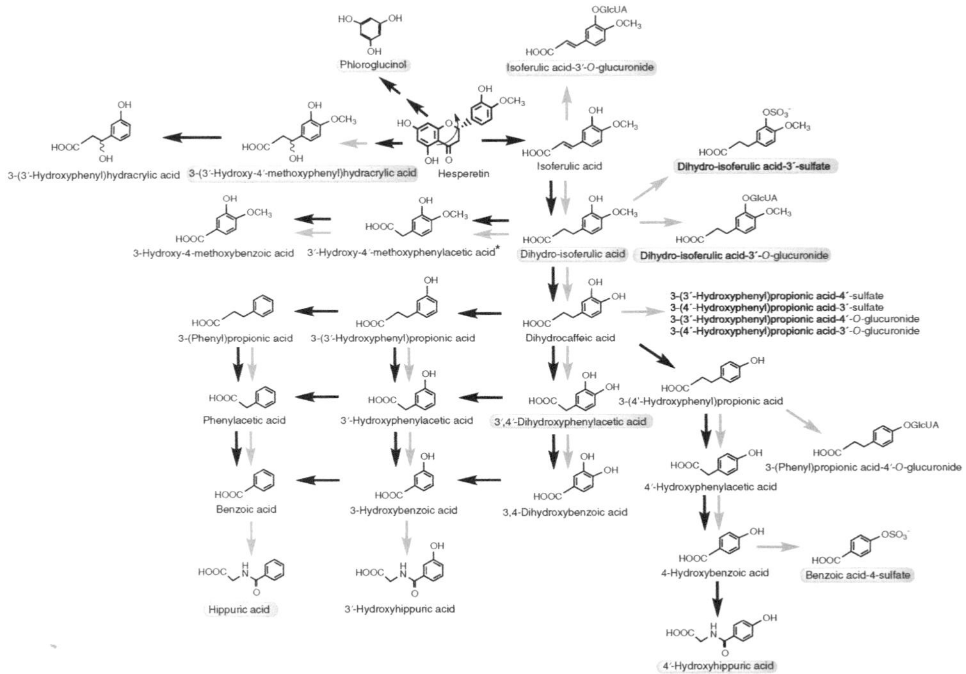

Figure 12.2 Proposed pathways for the colonic metabolism of hesperetin released by cleavage of hesperetin-7-*O*-rutinoside after the consumption of orange juice by humans. Based on urinary excretion in feeding studies and *in vitro* fecal incubations. Compounds with boxed names indicate major components in 0–24-h urine. Curved arrow indicates ring fission. Black arrows are potential microbiota-mediated conversions and gray arrows are steps potentially catalyzed by mammalian enzymes. *, intermediate that did not accumulate in detectable amounts; GlcUA, glucuronide. Source: Based on data from Pereira-Caro *et al*. (2014, 2015a,b, 2016, 2018).

In vitro naringenin is likely to undergo ring fission producing 3-(4′-hydroxyphenyl)propionic acid which is dehydroxylated and converted mainly to 3-(phenyl)propionic acid, which *in vivo* could potentially contribute to the urinary pool of hippuric acid (see Figure 12.1). Intestinal microbial-mediated ring fission of hesperetin produces dihydro-isoferulic acid (aka 3-(3′-hydroxy-4′-methoxyphenyl)propionic acid) which is demethylated producing dihydrocaffeic acid (aka 3-(3′,4′-dihydroxyphenyl)propionic acid). This, in turn, via successive dehydroxylations is converted to 3-(3′-hydroxyphenyl)propionic acid and 3-(phenyl)propionic acid, which could be further converted to the respective urinary pools of 3′-hydroxyhippuric acid and hippuric acid (see Figure 12.1). *In vitro* degradation of hesperetin also yielded small quantities of 3-(4′-hydroxyphenyl)propionic acid and 4′-hydroxyphenylacetic acid which, via the pathway illustrated in Figure 12.2, and like the naringenin catabolites, could be further converted to 4′-hydroxyhippuric acid. Potential intestinal bacteria involved in some of the proposed conversions illustrated in Figures 12.1 and 12.2 are *Eubacterium limosum* and members of the *Enterobacter* and *Escherichia* genera which possess *O*-demethylase and hydrolase activity (de Weerd *et al.*, 1988; Grbić-Galić, 1986; Hur and Rafii, 2000).

Phenolic acids of interest that were excreted in urine in low but increased amounts after OJ consumption (see Table 12.1) are 3-(3′-methoxy-4′-hydroxyphenyl)propionic acid (aka dihydroferulic acid) and 3′-methoxy-4′-hydroxyphenylacetic acid which because of the 3′-methoxy group are unlikely to be catabolites of hesperetin. They could, however, be derived from ferulic acid-4′-*O*-glucoside which has been tentatively identified in OJ (Pereira-Caro *et al.*, 2016). *In vitro* fecal incubations with ferulic acid indicate this is feasible as the hydroxycinnamate was converted principally to 3′-methoxy-4′-hydroxyphenyl)propionic acid and its demethoxy-derivative 3-(4′-hydroxyphenyl)propionic acid, along with smaller amounts of their phenylacetic acid counterparts (Pereira-Caro *et al.*, 2015a). Potential routes for these transformations are illustrated in Figure 12.1 which again shows the possible conversion of 4′-hydroxyphenylacetic acid to 4′-hydroxyhippuric acid.

Orange juice consumption results in elevated levels of urinary 4′-hydroxymandelic acid (**32**), the likely source of which is *p*-sympatol (see structure 10) which was present in the ingested OJ (Pereira-Caro *et al.*, 2016). Following ingestion by humans, ^{3}H-labeled *p*-sympatol is metabolized to 4′-hydroxymandelic acid, possibly via a three-step pathway that is enterocyte/hepatocyte in origin (Hengstmann and Aulepp, 1978).

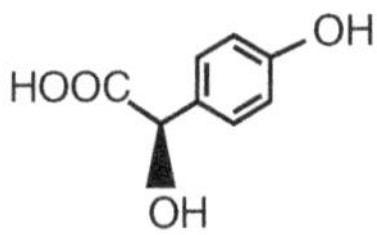

4′-Hydroxymandelic acid (**32**)

12.4 Formation of 3-(3′-Hydroxy-4′-Methoxyphenyl)Hydracrylic Acid

Orange juice intake is characterized by the subsequent urinary excretion of 3-(3′-hydroxy-4′-methoxyphenyl)hydracrylic acid (see structure 28) and 3-(3′-hydroxyphenyl)hydracrylic acid (see structure 29) (Pereira-Caro *et al.*, 2014). The formation of these phenylhydracrylic acids is of interest. They are not produced from flavanones *in vitro* by the gut microflora (Pereira-Caro *et al.*, 2015a) and their metabolic origin is uncertain. Because of their compatible B-ring substituent pattern, they would appear to be catabolites of hesperetin.

The most plausible route for the production of 3-(3′-hydroxy-4′-methoxyphenyl)hydracrylic acid is from hesperetin and would appear to be a postabsorption 3-hydroxylation of the side chain of 3-(3′-hydroxy-4-methoxyphenyl)propionic acid by a P450 monooxygenase or its oxidation to a cinnamic acid and the addition of water across the double bond by a hydratase enzyme. These routes, however, are improbable, because while such phenylpropionic acids are quantitatively significant catabolites of caffeoylquinic acids, in HPLC-MS-based studies phenylhydracrylic acids have not been detected in urine after the consumption of either coffee or other sources of chlorogenic acids (Borges *et al.*, 2010;

Stalmach *et al.*, 2009). However, formation from phenylpropionic acids cannot be eliminated unequivocally because a study, based on analysis by paper chromatography, reported the presence of 3-(3′-hydroxyphenyl)hydracrylic acid and 3-(3′-methoxy-4′-hydroxyphenyl)hydracrylic acid in urine from a single volunteer after the consumption of 1 L of beverage produced by infusing 78 g of roasted, ground coffee (Shaw and Trevarthen, 1958).

An alternative route is the direct formation of 3-(hydroxyphenyl)hydracrylic acids via an unusual ring fission of hesperetin in which the pyran oxygen is retained on the side chain of the B-ring fragment either as the required hydroxyl or as a ketone. Aromatic ketones such as acetophenone and propiophenone are metabolized, respectively, to mandelic and phenylhydracrylic acid, by several animals (Adams *et al.*, 2007). Such a gut microflora transformation would utilize the C–C hydrolase enzyme produced by several gut microflora anaerobes (Braune *et al.*, 2005) and simultaneously generate an A-ring derivative with a 2,4-dihydroxy substitution, such as 2,4-dihydroxybenzoic acid (Stolz, 2005) or resorcinol which has been found in human urine as a resorcinol sulfate (Curzon and Pratt, 1964). Stoupi *et al.* (2010) detected an A-ring-derived 2,4-dihydroxyphenyl metabolite when procyanidin B2 was incubated with human gut microbiota, and there are reports of volunteers excreting 3-(3′-hydroxyphenyl)hydracrylic acid after consuming (+)-catechin (Das, 1971) and green tea (Roowi *et al.*, 2010) and mixed wine and grape juice (van Dorsten *et al.*, 2010). While definitive human data do not seem to be available, a combination of such a gut flora transformation followed by absorption, β-oxidation and, if necessary, ketone reduction might account for the formation of the phenylhydracrylic acids from the flavanones, and also flavan-3-ols.

In the case of hesperetin, this could involve an unusual ring fission in which the pyran oxygen is retained on the side chain of the B-ring fragment either as the required hydroxyl or as a ketone, and this could proceed via the pathway illustrated in Figure 12.3. The initial conversion (I) would involve A-ring hydroxylation at C-8 catalyzed by a pyrogallol-phloroglucinol isomerase (Jeffrey *et al.*, 1972a; Krumholz and Bryan, 1988; Krumholz *et al.*, 1987; Messerschmidt *et al.*, 2004) and result

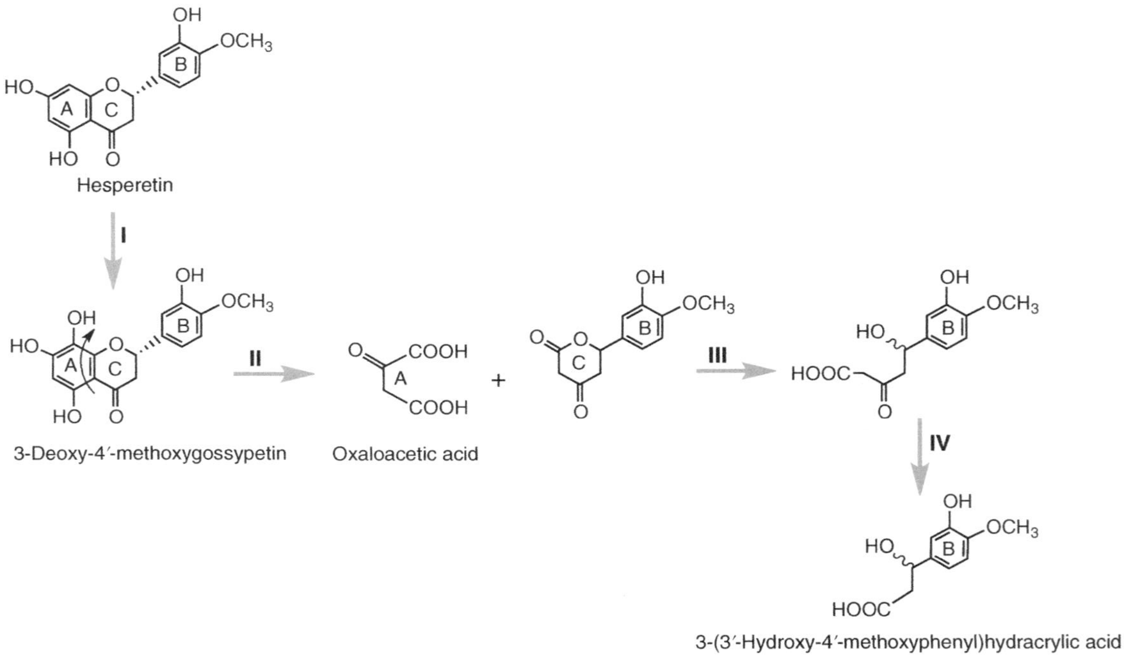

Figure 12.3 Potential route for the conversion of hesperetin to 3-(3′-hydroxy-4′-methoxyphenyl)hydracrylic acid. The curved arrow indicates the site of ring fission.

in the formation of 3-deoxy-4′-methyoxygossypetin. This is followed by a reductive A-ring cleavage (II), as observed with taxifolin (Jeffrey *et al.*, 1972b), resulting in the elimination of oxaloacetic acid, as observed with many flavonoids, and the subsequent opening of the δ-lactone of the C-B-ring fragment (III), possibly by acid hydrolysis, followed by side chain β-oxidation (IV) to produce 3-(3′-hydroxy-4′-methoxyphenyl)hydracrylic acid.

The involvement of the intestinal microbiome in the formation of the phenylhydracrylic acids is supported by a study which showed urinary excretion of 3-(3′-hydroxy-4′-methoxyphenyl) hydracrylic acid following oral intake of hesperetin-7-*O*-rutinoside, but only trace amounts when absorption of the rutinoside was confined to a 20 cm segment of the proximal jejunum by using a multilumen intestinal perfusion catheter (Actis-Goretta *et al.*, 2015).

12.5 Factors Affecting the Bioavailability of Flavanones

12.5.1 Impact of Physical Activity

Two investigations on the impact of exercise on flavanone bioavailability have produced conflicting findings. Medina *et al.* (2012) reported that after acute consumption of an aronia-citrus juice, the quantity of flavanone metabolites excreted in urine by triathletes was >5-fold higher than that of control, more sedentary volunteers. The two groups of volunteers consumed a juice containing 80 mg flavanones, which, assuming they were mainly hesperetin- and naringenin-*O*-rutinosides, equates to a dose of ~130 μmol. The aglycones hesperetin and naringenin, released by enzyme hydrolysis of urine before analysis, were quantified by HPLC-mass spectrometry. The 0–24-hour postconsumption excretion was estimated to be ~0.3% of intake by the control group and ~1.3% by the triathletes.

Hydrolysis of flavonoid glucuronide and sulfate metabolites by mollusc enzymes, which have inconsistent titers, does not result in complete cleavage, especially of sulfates, and, as a consequence, underestimates metabolite levels (Donovan *et al.*, 2006; Saha *et al.*, 2012). However, the inefficiency of this step is unlikely to have accounted for estimates of urinary excretion obtained by Medina *et al.* (2012) being ~50-fold lower than those detected by Pereira-Caro *et al.* (2014, 2015b), as well as being substantially below flavanone excretion levels reported by other investigators (Erlund *et al.*, 2001; Manach *et al.*, 2003; Mullen *et al.*, 2008; Tomás-Navarro *et al.*, 2014; Vallejo *et al.*, 2010).

A later study investigated the impact of a seven-day cessation of training by male endurance athletes on the bioavailability of OJ flavanones (Pereira-Caro *et al.*, 2017). Whereas stopping training significantly increased 0–24-hour urinary excretion of flavanones metabolites from 4.2% to 5.2% of intake, a longer period without training would appear to be required to attain the 16% excretion of flavanones metabolites observed when OJ was consumed by volunteers who were not endurance athletes (Pereira-Caro *et al.*, 2016). The data indicate, in contrast to the findings of Medina *et al.* (2012), that exercise reduces flavanone bioavailability. The underlying mechanism(s) responsible for this effect require further investigation. It is perhaps of note that endurance training brings about changes in intestinal microbiota (Clarke *et al.*, 2014) which arguably could limit microbiota-mediated cleavage of flavanone rutinosides and so reduce the amount of hesperetin and naringenin released for absorption in the distal GI tract. Furthermore, training results in adverse physiological adaptations in the gut (Casey *et al.*, 2005) and enhances muscle blood flow $\leq$12–16 hour after the last exercise session (Malkova *et al.*, 2000). These events could affect gut function and further reduce the absorption of flavanone aglycones.

12.5.2 Matrix Effects

Most investigations on the bioavailability of (poly)phenols, with few exceptions, have focused on the acute intake of a single product, such as a juice, fruit or vegetable, after volunteers had been on low-(poly)phenol diets for 1–3 days. Information on matrix effects and the impact of other dietary constituents of (poly)phenol ADME is, therefore, limited. Very little is currently known about the bioavailability of dietary (poly)phenols in the context of a typical multicomponent meal/diet.

In one study, the bioavailability of flavanones was investigated when volunteers drank an OJ with and without 150 mL of yogurt containing 3.8% fat (Mullen *et al.*, 2008). The yogurt did not impact significantly on plasma C_{max} and T_{max} of hesperetin-7-*O*-glucuronide and a second metabolite, most probably hesperetin-3′-*O*-glucuronide. The 0–5-hour urinary excretion of the glucuronides and other minor metabolites was reduced significantly by the yogurt but the quantity of metabolites in the 0–24-hour urine was not significantly different. Perhaps because of its low fat content, the yogurt had no effect on either gastric emptying or the mouth-to-cecum transit time of the "head" of the meal (Mullen *et al.*, 2008). The initial significant delay in urinary excretion of flavanone metabolites when OJ was consumed with the yogurt may, therefore, reflect differences in gastric emptying and intestinal transit (i.e., the slower delivery of the bulk of the meal).

Analysis of microbial-derived urinary phenolic catabolites in the OJ-yogurt study was reported by Roowi *et al.* (2009). When OJ was consumed on its own, 0–24-hour urine contained a total of 62 ± 18 μmol of hydroxy- and methoxyphenolic catabolites, equivalent to 37% of the ingested flavanones. When the OJ was ingested with the yogurt, excretion of the catabolites fell to 9.3 ± 4.4 μmol. The reasons for this marked decline are as yet unclear, but potentially the yogurt could have slowed the bulk of the meal reaching the large intestine which may have altered the microbiota-mediated breakdown of the flavanones.

12.5.3 Probiotics

Pereira-Caro *et al.*, (2015b) investigated the impact of an orally ingested microencapsulated probiotic (*Bifidobacterium longum* R0175) on the bioavailability of OJ flavanones. In an acute study, volunteers consumed a single bolus of OJ with and without the probiotic. In the second part of the study, the same OJ was consumed acutely after daily supplementation with the probiotic for a period of four weeks. Bioavailability was assessed by 0–24-hour urinary excretion and was similar when the OJ was consumed with and without acute probiotic intake. Excretion of hesperetin and naringenin metabolites corresponded to 22% of intake, while excretion of key phenolic and aromatic acids was equivalent to 21% of the ingested OJ (poly)phenols. Acute OJ consumption after four weeks of probiotic intake significantly increased excretion of flavanone metabolites to 27% of intake while excretion of the phenolic catabolites also increased significantly to 43% of (poly)phenol intake, giving an overall bioavailability of 70%. It is unclear how this increased bioavailability is mediated as neither the probiotic bacterial profiles of stools, nor stool moisture, weight, pH or levels of short-chain fatty acids and phenols differed significantly between treatments.

In vitro, the *B. longum* R0175 possessed rhamnosidase activity when assayed by the method of Ávila *et al.* (2009) using *p*-nitrophenyl-α-L-rhamnopyranoside as a substrate. However, *in vitro* anaerobic incubations of the bacterium with hesperetin-7-*O*-rutinoside and naringenin-7-*O*-rutinoside did not result in the conversion of either rutinoside to a glucoside or glycoside to aglycone (Pereira-Caro *et al.*, 2018). Assuming the gut environment does not induce rhamnosidase and/or glucosidase expression *in vivo*, it is unlikely that cleavage of sugar residues represents a key mechanism by which chronic intake of the probiotic enhances the bioavailability of the OJ flavanones. The data do, however, raise the question of whether long-term supplementation with a probiotic could provide health-promoting benefits by altering the intestinal microbiota in a manner that improves the bioavailability of flavanones (Pereira-Caro *et al.*, 2015b).

12.5.4 Inter- and Intraindividual Variability

The interindividual variation in the absorption of OJ flavanones (Tomás-Navarro *et al.*, 2014; Vallejo *et al.*, 2010), and the bioavailability of dietary (poly)phenols in general, is an area of increasing interest because information on variations in the capacity to metabolize these compounds may lead to a better understanding of the beneficial effects of plant bioactive compounds against diseases, particularly their role in healthy aging and cardiometabolic risk reduction (Manach *et al.*, 2017).

The data in Table 12.2 were obtained in a study by Pereira-Caro *et al.*, (2015b) in which 14 volunteers ingested an OJ containing

167 μmol of hesperetin and naringenin glycosides after which 0–24-hour urine was collected and flavanone phase II metabolites was analyzed by HPLC-MS. As noted in an earlier study by Nielsen *et al.* (2006), there were substantial differences in the quantity of metabolites excreted by the individual volunteers. For instance, excretion of flavanone metabolites ranged from 10 to 86 μmol which is equivalent to 6–52% of flavanone intake (see Table 12.2). Volunteers V1–V5 all excreted ≥30% of the flavanone intake while volunteers V11–V14 excreted ≤11%. There was an 8.6-fold difference in the excretion level in volunteers V1 and V14 so it is evident that there were substantial differences between high and low excretors. The average excretion level was 25% of intake and there was a percentage coefficient of variance (%CV) of 51%. This is substantially higher than a %CV of 30% obtained with 0–24-hour urinary (−)-epicatechin metabolites in a study in which volunteers consumed [^{14}C](−)-epicatechin (Borges *et al.*, 2018). This could be due to the (−)-epicatechin being ingested in an aqueous solution while the flavanones were minor ingredients in a multicomponent juice.

However, another factor could be that (−)-epicatechin is an aglycone, and is absorbed in the small intestine without cleavage of a sugar unit being a prerequisite step (Ottaviani *et al.*, 2016) while, as noted in section 12.3, absorption of flavanones occurs principally in the lower intestine and is dependent upon cleavage of the rutinoside moiety by the resident microbiota, which can vary substantially between individuals. This is in keeping with the observations of Almeida *et al.* (2018) who, in a review on the bioavailability of quercetin glycosides, noted that that there was less interindividual variation in metabolite absorption in the upper small intestine compared to those derived from the action of microbiota in the lower bowel.

Intraindividual differences were assessed in the study in which OJ was consumed by endurance athletes before and after cessation of training for seven days that was discussed in section 12.5.1. In this instance, the differences were much smaller than the interindividual variations, with high excretors producing similar amounts of metabolites in both the trained state and after cessation of training, while low excretors maintained this condition after cessation of training (Pereira-Caro *et al.*, 2017).

12.5.5 Other Effects

A study with orange fruit and OJ found no impact of body mass index or contraceptive pill use on the bioavailability of flavanones but did reveal a slight but significant decline in urinary excretion of hesperetin, but not naringenin metabolites, with increasing volunteer age (Brett *et al.*, 2009).

12.6 Analysis of Flavanone Metabolites and Catabolites

The first papers on the ADME of flavanones involved analysis of hesperetin and narigenin, released from phase II metabolites by enzyme hydrolysis, by HPLC with electrochemical detection (Erlund *et al.*, 2001; Manach *et al.*, 2003). The shortcomings of enzyme hydrolysis, as noted in section 12.5, were revealed in later studies. With the advent of mass spectrometric detection for HPLC, it became possible to analyze metabolites without recourse to enzyme hydrolysis.

The first study on flavanone bioavailability using HPLC-MS, and which was able to make use of an extensive set of "in-house" synthesized authentic metabolites as reference compounds, was that of Nielsen *et al.* (2006). Not all investigators had access to authentic reference compounds of the main flavanone metabolites and, as a consequence, MS-based quantifications have been made by reference to hesperetin or naringenin. This can result in substantial inaccuracies in estimates of metabolite levels with, for instance, the analytical response (i.e., mass detector intensity) of the hesperetin being ~50 times greater than that of hesperetin-3′-*O*-glucuronide (Ottaviani *et al.*, 2018), leading to significant overestimation or underestimation when calculating concentration using response curve analysis. A large difference in the MS response to hesperetin and hesperetin-7-*O*-glucuronide is also evident in Figure 12.4, whereas with the conditions used by Pereira-Caro *et al.* (2016), there are similar response curves for the aglycone and hesperetin-3′-sulfate. This source of inaccuracy/uncertainty can now be avoided. Details of procedures for the syntheses of many (poly)phenol methyl, glucuronide, and sulfate metabolites, including those of

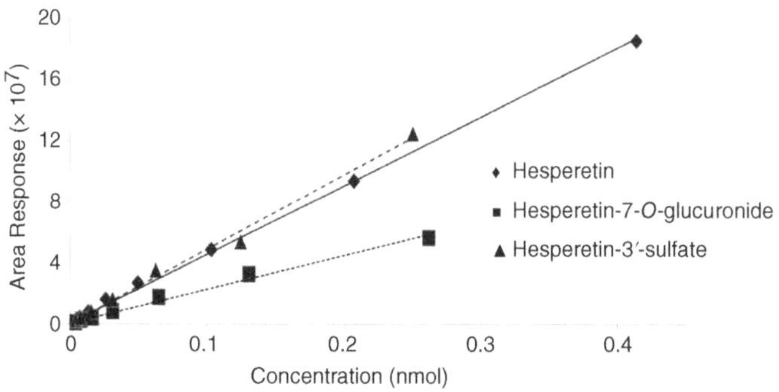

Figure 12.4 Correlation between tested concentration of standards of hesperetin, hesperetin-7-*O*-glucuronide, and hesperetin-3′–sulfate and the response of the MS detector. The HPLC-HRMS conditions using an Exactive Orbitrap mass spectrometer were those described by Pereira-Caro *et al.* (2016).

naringenin and hesperetin, have been published (Boumendjel *et al.*, 2009). Several companies offer custom synthesis of metabolites and have catalogues of off-the-shelf compounds that contain a number of the major flavanone metabolites (Table 12.3).

Gas chromatography-mass spectrometry was originally used to analyze flavanone-derived phenolic catabolites in biofluids after OJ intake (Pereira-Caro *et al.*, 20014, 2015; Roowi *et al.*, 2009). This involved a mandatory derivatization of samples with *N*-trimethylsilyl-*N*-methyl trifluoroacetamide (MSTFA) and while the method provided an accurate assessment of free phenolics, it is not suitable for the analysis of sulfate and glucuronide conjugates of phenolics because even as MSTFA derivatives they lack the volatility required for GC. The method of choice for the analysis of these compounds is now HPLC-MS with electrospray ionization in negative ionization mode. The use of UHPLC for enhanced chromatographic resolution in combination with a high-resolution Orbitrap mass spectrometer represents a versatile and very sensitive tool for the identification and quantification of a wide range of metabolites, without extensive preanalysis sample treatment. This methodology has been applied effectively in recent studies for the analysis of flavanone phase II metabolites and their

Table 12.3 List of flavanone metabolites and catabolites available from commercial sources

Standards	Suppliers	Quantity	Price
Flavanone metabolites			
Hesperetin-7-*O*-glucuronide	Synthose	5 mg	\$100
Hesperetin-3′-*O*-glucuronide	Synthose	5 mg	\$100
Hesperetin-3′-*O*-sulfate	Toronto Research Chemicals	5 mg	\$150
Naringenin-7-*O*-glucuronide	Toronto Research Chemicals	1 mg	\$140
Flavanone-derived phenolic acid catabolites			
4-Hydroxybenzoic acid	Sigma-Aldrich	50 mg	\$80
3-Hydroxybenzoic acid	Sigma-Aldrich	100 mg	\$54
Ferulic acid	Sigma-Aldrich	25 mg	\$340
Isoferulic acid	Sigma-Aldrich	10 mg	\$302
4′-Hydroxyphenylacetic acid	Sigma-Aldrich	10 mg	\$49
3′-Hydroxyphenylacetic acid	Sigma-Aldrich	5 g	\$64
3′,4′-Dimethoxyphenylacetic acid	Sigma-Aldrich	25 mg	\$31
3′-Methoxy-4′-hydroxy-phenylacetic acid	Sigma-Aldrich	25 mg	\$52
Hippuric acid	Sigma-Aldrich	10 mg	\$69
3,4-Dihydroxybenzoic acid	Sigma-Aldrich	50 mg	\$99
3-Methoxy-4-hydroxybenzoic acid	Sigma-Aldrich	50 mg	\$94
3-Hydroxy-4-methoxybenzoic acid	Sigma-Aldrich	25 mg	\$86
3-(3′-Methoxy-4′-hydroxy-phenyl)propionic acid	Sigma-Aldrich	5 g	\$303
3-(3′-Hydroxy-4′-metho-xyphenyl)propionic acid	Sigma-Aldrich	50 g	\$53
3-(3′,4′-Dihydroxyphenyl) acetic acid	Sigma-Aldrich	25 g	\$52
3-(4′-Hydroxyphenyl)propionic acid	Sigma-Aldrich	5 g	\$23
3-(3′-Hydroxyphenyl) propionic acid	Sigma-Aldrich	10 mg	\$27

Table 12.3 (Continued)

Standards	Suppliers	Quantity	Price
3-(Phenyl)propionic acid	Sigma-Aldrich	100 g	$61
3′-Hydroxyhippuric acid	Toronto Research Chemicals	250 mg	$125
3-(3′-Hydroxyphenyl) hydracrylic acid	Toronto Research Chemicals	25 mg	$110
Coumaric acid-4′-*O*-glucuronide	Toronto Research Chemicals	10 mg	$150
Caffeic acid-3′-*O*-glucuronide	Toronto Research Chemicals	1 mg	$110
Caffeic acid-4′-*O*-glucuronide	Toronto Research Chemicals	2.5 mg	$140
Ferulic acid-4′-*O*-glucuronide	Toronto Research Chemicals	5 mg	$150
Ferulic acid-4′-sulfate	Toronto Research Chemicals	100 mg	$150
Isoferulic acid-3′-*O*-glucuronide	Toronto Research Chemicals	5 mg	$175
4′-Hydroxyhippuric acid	Toronto Research Chemicals	25 mg	$45

catabolites in biological samples collected after OJ intake by humans (Ordoñez *et al.*, 2018; Pereira-Caro *et al.*, 2016, 2017).

12.7 Biomarkers and Metabolomics

A focal point in human nutrition metabolomics studies is to discover markers related to diet, food intake or exposure, including those derived from human metabolism and microbiota-mediated transformations. The metabolites derived from bioavailable dietary components comprise many different compounds which could be included in what is termed the "food metabolome" (González-Dominguez *et al.*, 2019; Scalbert *et al.*, 2014). In this sense, diet plays a pivotal role in shaping the "human metabolome" as the metabolites of exogenous phytochemicals that are part of the food metabolome are likely to alter human metabolic processes (Manach *et al.*, 2009).

As discussed earlier in section 12.3, targeted HPLC-MS, and increasingly UHPLC-HR-MS, have been used to quantify flavanone metabolites in biofluids after OJ consumption. From this work it is evident that hesperetin-3′-*O*-glucuronide (see structure 14), hesperetin-3′-sulfate (see structure 17) and 3-(3′-hydroxy-4′-methoxyphenyl)hydracrylic acid (see structure 28) are candidate biomarkers of hesperetin intake derived principally for OJ consumption.

The identification of other potentially bioactive metabolites requires untargeted analysis which presents major challenges for metabolite identification as reference compounds are often not available and existing databases are not complete and do not contain many human and animal phytochemical metabolites. Information on the phytochemical composition of the food of interest, associated with knowledge of the human metabolism of these components, greatly aids identification. Moreover, samples are complex and can contain hundreds or thousands of chemical species, depending on the biological system and sample type being studied. Human biofluids, for instance, contain endogenous metabolites as well as exogenous metabolites derived from diet (Lloyd *et al.*, 2011), lifestyle and physical activity (Pechlivanis *et al.*, 2010) and the gut microbiota (Wikoff *et al.*, 2009), many at sub-μmol/L concentrations or lower.

The use of metabolomics has facilitated the identification of a large number of putative novel biomarkers of food, as opposed to flavonoid intake biomarkers. To date, metabolomic profiling strategies and subsequent multivariate data analysis allowed the identification of proline betaine (**33**, aka stachydrine) (Heinzmann *et al.*, 2010; Lang *et al.*, 2017; Llorach *et al.*, 2014; Lloyd *et al.*, 2011), hydroxyproline betaine (**34**, aka betonicine), and hesperetin and naringenin glucuronides (Pujos-Guillot *et al.*, 2013) as urinary biomarkers of citrus fruit intake. Proline betaine is a major component in OJ where it is found in high concentrations ranging from 1 to 7.9 mmol/L which is well in excess of levels found in other dietary components (Lang *et al.*, 2017). Following a 12-week intake of a high-polyphenol OJ by volunteers, Rangel-Huertas *et al.* (2017) proposed proline betaine, 4-hydroxyproline betaine, 1-*O*-methylglucopyranoside

(**35**), dihydroferulic acid (**36**), and galactonic acid (**37**) as metabolic biomarkers of OJ intake, which they associated with improved oxidative stress and inflammation biomarkers by decreasing serum concentrations of hydroxyoctadecadienoic and dihydroxyocctadecanoic acid and increasing levels of 12-hydroxyeicosatetraenoic acid.

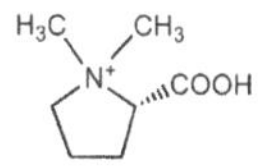

Proline betaine (**33**)

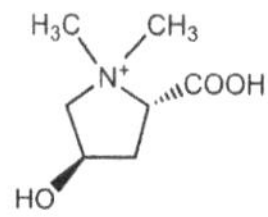

4-Hydroxyproline betaine (**34**)

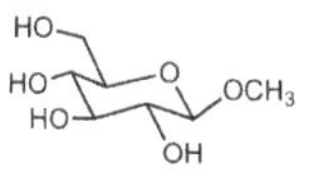

1-*O*-Methylglucopyranoside (**35**)

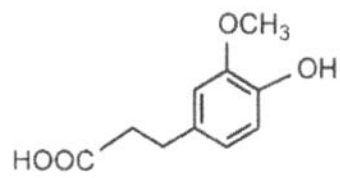

Dihydroferulic acid (**36**)

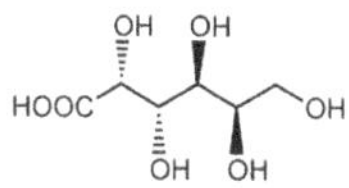

Galactonic acid (**37**)

A metabolomics approach has been used to assess the effect of a two-week OJ supplementation on lipid metabolism in healthy volunteers (Moreira *et al.*, 2018). It was shown that OJ stimulated and regulated mitochondrial and peroxisomal fatty acid β-oxidation in the liver and decreased medium- and long-chain and increased short-chain acylcarnitine blood levels. On the basis of these findings, it was proposed that OJ may have a positive impact on metabolic diseases, such as obesity and liver steatosis. The mechanisms associated with such biological effects remain to be clarified and the extent to which they involve flavanones and/or other OJ constituents has not been determined. On the basis of their findings, Moreira *et al.* (2018) proposed that OJ may have a positive impact on metabolic diseases, such as obesity and liver steatosis.

12.8 Protective Effects

The beneficial health effects of citrus have been reported since the 1930s, when researchers identified a synergistic activity between citrus extract and ascorbic acid on hemorrhagic symptoms associated with capillary fragility in the treatment of scurvy. The isolated citrus bioactive was originally termed "citrin" and reported to have "vitamin-like" qualities. Citrin was eventually established as comprising two flavanones, hesperetin-7-*O*-rutinoside (see structure 2) and an eriodictyol-glucoside (Roger, 1988; Zilva, 1937). Despite this early interest in the beneficial effects of citrus, flavanone research has been slow to progress relative to other subclasses of dietary flavonoids.

12.8.1 Cardiovascular Disease

A metaanalysis of 14 prospective studies found a reduced cardiovascular disease risk associated with the consumption of flavanones (Wang *et al.*, 2014). A prospective study of 1800 incident strokes over a 14-year period revealed that an increased intake of flavanones in women was associated with a 19% reduction in risk of ischemic stroke (Cassidy *et al.*, 2012). In addition, human intervention studies have reported positive impacts of flavanone intake on CVD risk factors, most notably hypertension. A naringenin glycoside-rich juice consumed for five weeks was associated with lowered diastolic blood pressure in 12 individuals with hypertension (Reshef *et al.*, 2005), and vascular effects in subsequent studies were attributed to improvements in flow-mediated dilation and postprandial microvascular endothelial reactivity.

In a later intervention feeding either one or three cups of OJ to 16 healthy participants over four weeks, the three cups per day (750 mL) treatment was associated with a 21% increase in HDL cholesterol (Kurowska *et al.*, 2000). A cross-over study in 24 healthy adults consuming 500 mL OJ daily for four weeks showed reductions in diastolic blood pressure relative to controls (Morand *et al.*, 2011). Another study feeding 400 mg/day of encapsulated naringenin-7-*O*-neohesperidoside (see structure 12) to 60 volunteers for eight weeks found the flavanone lowered plasma total cholesterol by 14% and LDL-cholesterol by 17% in hypercholesterolemic participants relative to healthy controls (Jung *et al.*, 2003). Hesperetin-7-*O*-rutinoside has also been shown to improve endothelial dysfunction and reduce inflammatory markers in individuals with metabolic syndrome supplemented with 500 mg of the rutinoside for three weeks (Rizza *et al.*, 2011).

Finally, hypercholesterolemic individuals treated daily for four weeks with 800 mg hesperetin-7-*O*-rutinoside without the citrus matrix experienced no cholesterol-lowering effects (Demonty *et al.*, 2010), suggesting the food matrix impacts on the vascular effect of flavanones. Based on the sparse and conflicting human data, there is not enough evidence to disentangle the impact of matrix on the cardiovascular bioactivity of flavanones in humans.

There are far greater numbers of rodent studies on flavanone cardiovascular activity, where hypotensive and hypolipidemic effects are most often observed, than investigations with human. Several rodent studies, using either high-fat, high-carbohydrate or drug-induced hypertensive models, and feeding daily doses between 10 and 1000 mg/kg of naringenin (**38**) or naringenin-7-*O*-neohesperidoside (see structure 4), have reported improvement in hypertensive outcomes. In these studies, mechanisms including lowering of blood pressure, blood lipids and lipoproteins, improving endothelium-dependent vasorelaxation, reducing inflammation and lipid peroxidation, and/or improving mitochondrial function have been reported (Alam *et al.*, 2013). A study using a diabetic rat model feeding hesperetin-7-*O*-rutinoside (10 g/kg) for four weeks reported a lowering of hypertensive activities of the treatment relative to controls (Akiyama *et al.*, 2010). Wistar rats fed a high-cholesterol diet for 12 weeks showed improved hypercholesterolemia following hesperetin-7-*O*-rutinoside supplementation, which was associated with inhibition of both cholesterol synthesis and absorption (Wang *et al.*, 2011).

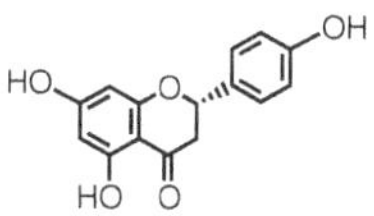

Naringenin (**38**)

Naringenin-7-*O*-neohesperidoside was shown to attenuate bisphenol A-induced cardiotoxicity in male Wistar rats fed 80 or 160 mg/kg orally for 30 days. Here lipid-lowering and antioxidative properties were observed, including amelioration of increased triglyceride and lipid peroxidation and decreased levels of glutathione, superoxide dismutase, glutathione peroxidase, and catalase triggered by bisphenol A (Khodayar *et al.*, 2018). Hesperetin-7-*O*-rutinoside and its aglycone hesperetin (**39**) were shown to attenuate atherosclerotic plaque progression and lesion macrophage infiltration in apolipoprotein E knockout (apoE−/−) mice fed a high-fat diet, where the flavanone-treated mice exhibited reduced total cholesterol, improved endothelial function, and reduced inflammation (Sugasawa *et al.*, 2019). A high-fat fed apoE−/− mouse model and *in vitro* macrophage culture studies indicated enhanced cholesterol efflux by naringenin via mediation of the PI3K/AKT pathway and endoplasmic reticulum stress (Xu *et al.*, 2019). Burke *et al.* (2019) reported that an eight-week naringenin-7-*O*-neohesperidoside treatment in lean chow-fed mice (Pair-Fed Ldlr(−/−) mice) reduced lipids and enhanced insulin sensitivity and energy expenditure (increased respiratory exchange ratio), indicating that the metabolic properties of the flavanone appear to be present even in the absence of high-fat diets (Burke *et al.*, 2019) (see structures **38-40**).

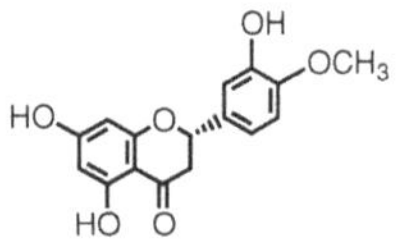

Hesperetin (**39**)

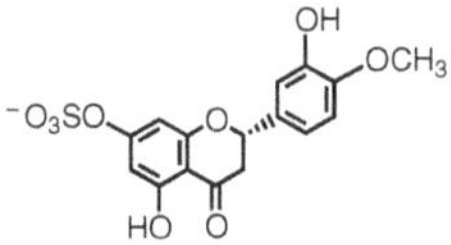

Hesperetin-7-sulfate (**40**)

Gavage-fed 25, 50, or 100 mg/kg of naringenin attenuated balloon injury-induced carotid restenosis in rats, where antioxidative and antiinflammatory mechanisms were observed, including decreased expression of cluster of differentiation 163 (CD163), proliferating cell nuclear antigen and reduced TNF-α and IL-1β (Huang *et al.*, 2019). Hesperetin-7-*O*-rutinoside was also shown to regress cardiac hypertrophy in male, albino Wistar rats through suppressing oxidative stress and apoptosis in addition to having agonistic activity on proliferator-activated receptor (PPAR)-γ, an inhibitor of cardiac hypertrophy signaling pathways (Bhargava *et al.*, 2019).

When physiologically relevant forms and concentrations of pure flavanones are applied to cultured vascular cells (monocytes and vascular endothelial cells), antiinflammatory and antiadhesion mechanisms of action are frequently reported. *In vitro* 0.5–10 μmol/L concentrations of the flavanone metabolites hesperetin-7-*O*-glucuronide (see structure 13), hesperetin-3′-*O*-glucuronide (see structure 14), hesperetin-3-sulfate (see structure 17) and naringenin-7-*O*-glucuronide (see structure 19), exerted antiatherogenic effects in proinflammation models, via amelioration of monocyte adhesion to endothelial cells and modulation of the expression of proteins associated with inflammation (Chanet *et al.*, 2013). A subsequent study by Gimenez-Bastida *et al.* (2016) demonstrated that hesperetin-3′-*O*-glucuronide, hesperetin-7-*O*-glucuronide, hesperetin-3′-sulfate, and hesperetin-7-sulfate (see structure 40), at similar physiological concentrations, efficiently reduced TNF-α-induced migration of human aortic endothelial cells. This was mediated by significant decreases in plasminogen activator inhibitor levels, a thrombogenic protein involved in a wide range of cardiovascular diseases mechanisms, including cell migration.

12.8.2 Diabetic and Metabolic Syndrome

A limited number of human intervention studies have directly investigated the impact of flavanones on diabetics or participants with metabolic syndrome. In a study feeding 10 healthy women OJ for 10 months, improved insulin sensitivity, blood glucose, and LDL-cholesterol were observed relative to controls (Lima *et al.*, 2019). These are all indicators of improved diabetic or metabolic symptoms. However, transfer to human patients requires confirmation in appropriate at-risk populations.

There are a number of rodent and *in vitro* studies describing the activity of flavanones on risk factors and mechanisms of diabetes and metabolic syndrome, including improving hyperglycemia and hyperlipidemia. After consuming a high-fat diet, Wistar rats with streptozotocin-induced diabetes showed improved glycemic and lipidemic effects following treatment with hesperetin-7-*O*-rutinoside (10 mg/kg diet) for four weeks relative to control fed rodents (Akiyama *et al.*, 2010). A study feeding 50 mg/kg of hesperetin-7-*O*-rutinoside or naringenin-7-*O*-neohesperidoside for four weeks also revealed attenuated hyperglycemia-induced inflammation and oxidative stress (Mahmoud *et al.*, 2012). Furthermore, 50 mg/kg naringenin fed for 45 days to streptozotocin-induced diabetic mice showed significant improvements in lipid profile and blood glucose levels (Rajappa *et al.*, 2018).

Mechanisms of action which might explain the reported animal and human activities observed in cell culture models often relate to regulators of lipid metabolism and redox cycling. Naringenin-7-*O*-neohesperidoside was shown to activate some PPARs while suppressing the LXR-α response element in human hepatocytes, thereby inducing fatty acid oxidation while inhibiting lipogenesis (Goldwasser *et al.*, 2010). In addition, naringenin has been shown to inhibit cellular aptosis in MIN6 cells through elevation of Nrf2 transcription target genes *NQO1* and *GSH* (Rajappa *et al.*, 2018).

12.8.3 Cancer

A significant decrease in colorectal cancer incidence was identified across eight flavanone intervention studies involving 390 769 participants spanning 10 years in a 2013 Cochrane review (Andrews 2013). However, an updated 2018 metaanalysis revealed no significant effect (Chang *et al.*, 2018). There are, however, some rodent and cell culture data suggesting that flavanones have anticancer activity.

In a Sprague–Dawley rat model of dimethylbenz[α]anthracene-induced breast carcinogenesis, feeding 30 mg/kg hesperetin-7-*O*-rutinoside (Nandakumar *et al.*, 2014) and in a Swiss albino mouse benzo(a)pyrene-induced lung carcinogenesis model, feeding 25 mg/kg hesperetin-7-*O*-rutinoside (Kamaraj *et al.*, 2011), alleviations in carbohydrate and lipid metabolism and mitochondrial dysfunction were observed. The flavanone also significantly attenuated the elevation in liver function enzymes and oxidative stress markers in an adult Wistar rat model of diethylnitrosamine-induced hepatocellular carcinoma. Here, a 200 mg/kg daily dose administered orally for 16 weeks was protective through downregulation of the P13K/Akt pathway (Mo′men *et al.*, 2019). In a female Sprague–Dawley rat model of letrozole-induced polycystic ovary syndrome, naringenin treatment at 20 mg/kg also normalized estradiol and testosterone levels and increased antioxidant defense enzymes GPX, SOD, and CAT (Hong *et al.*,

2019). Furthermore, hesperetin-7-*O*-rutinoside attenuated inflammation, oxidation, and accelerated proliferation in a colon cancer model using male Swiss albino mice fed 25 mg/kg daily for one week. The reported mechanisms of action included downregulation of NF-κB, iNOS, and COX-2 (Saiprasad *et al.*, 2013).

In vitro studies have found that hesperetin-7-*O*-rutinoside inhibits testosterone-induced cell proliferation in LNCaP prostate cancer and MCF-7 human breast cancer cells (Lee *et al.*, 2010). With *in vitro* zymography and transwell culture assays, naringenin has been shown to suppress prostate cancer cell invasion (Han *et al.*, 2018). Chemopreventive mechanisms identified in MDA-MB-231 breast cancer cells were linked to the induction of apoptosis as a result of caspase-3 and -9 inhibition (Wang *et al.*, 2019), while others report inhibition of cell migration via antiinflammatory processes and apoptotic cell signaling pathways (Zhao *et al.*, 2019).

12.8.4 Cognition and Neuroprotection

A few studies have been conducted with humans on cognition and the impact of flavanone-rich foods. A study feeding 240 mL polyphenol-rich OJ positively impacted cognitive performance in 24 healthy middle-aged men over a six-hour period post consumption (Alharbi *et al.*, 2016). Increased human cerebral blood flow post OJ consumption was suggested as a possible cognitive or neuroprotective mechanism of action (Lamport *et al.*, 2016).

Hesperetin has been shown to attenuate the progression of diabetic nephropathy in streptozotocin-induced diabetic rats. Rats treated orally with 50 and 150 mg/kg hesperetin showed improved renal function, upregulated glyoxalase (Glo-1), inhibition of the advanced glycation endproducts/receptor for advanced glycation endproducts (AGEs/RAGE)-axis, reduced inflammation, and increased Nrf2 expression (Chen *et al.*, 2019). Hesperetin-7-*O*-rutinoside pretreatment at 100 mg/kg was also shown to attenuate mitochondrial enzyme complex dysfunction and oxidative stress in an earlier study of neurotoxicity with male Wistar rats (Menze *et al.*, 2012). Hesperetin and its 7-*O*-rutinoside have been reported to have activity against amyloid β-induced neurotoxicity in Neuro A2 cells (Huang *et al.*, 2012), and to protect against neurotoxicity through a voltage-dependent anion channel 1 (VDAC1)-regulated mitochondrial apoptotic pathway in a pheochromocytoma cell culture model of Alzheimer's disease (Wang *et al.*, 2013). Furthermore, mitochondrial function was maintained and apoptosis prevented though antioxidant mechanisms in a neuroblastoma cell model of Parkinson's disease (Tamilselvam *et al.*, 2013). In addition, naringenin has been reported to decrease neuronal apoptosis in NSC34 motor neuron-like cells, through activation of antioxidant defense proteins GSH and SOD and increasing Nrf2 expression (Tseng *et al.*, 2019).

12.8.5 Bones

Hesperetin-7-*O*-rutinoside inhibited bone resorption in male, ddY mice fed 590 mg/kg (Chiba *et al.*, 2014). In a rat model of LPS-induced osteoporosis, the rutinoside suppressed receptor activator of NF-κB-induced osteoclastogenesis and osteoclastic bone resorption. Subsequent *in vitro* culture studies using RAW 264.7 cells linked mechanistic activity to inhibition of NF-κB and MAPK signaling activation and activation of antioxidant defense, primarily Nrf2/HO-1 (Liu *et al.*, 2019).

12.8.6 Liver

N-acetyl-p-aminophenol-induced liver injury (hepatotoxicity) in male Wistar rats treated with an orange peel extract administered for four week by oral gavage (50 mg/kg) showed increased antioxidant defense enzymes/thiols GSH, SOD, GST, and GPX, and suppression of inflammation and apoptosis relative to control rats (Ahmed *et al.*, 2019).

12.8.7 Immunomodulation and Antiinflammatory Activity

Hesperetin-7-*O*-rutinoside has potential therapeutic implications as observed in multiple rodent models of proinflammatory conditions including arthritis (Li and Schluesener 2017; Umar *et al.*, 2013), dextran sulfate sodium-induced colitis (Xu *et al.*, 2009), UV-induced skin damage (Petrova *et al.*, 2011), arthritis (Li *et al.*, 2010; Umar *et al.*, 2013), and asthma (Kim *et al.*, 2011). Moreover, naringenin-4′-*O*-glucuronide (see structure 18) and naringenin-7-*O*-glucuronide (see structure 19) were able to affect gene expression of M1 and M2a macrophages in culture at physiological concentrations (Dall'Asta *et al.*, 2013).

12.8.8 Gastric Function and the Microbiome

Gastrointestinal health problems are often associated with elevated intestinal permeability and associated inflammation (Fukui, 2016) and a significant focus in modern polyphenol research has been centered around gut health, including the microbiome and implications for gastric integrity. Evidence suggests one plausible mechanism for the activity of flavanones and their microbial catabolites could be through improved gastric barrier function, reduced mucosal cell damage and proinflammatory cytokine reduction, as reported in numerous animal and *in vitro* cell models of colitis and Caco-2 cell permeability (Azuma *et al.*, 2013; Noda *et al.*, 2012, 2013; Xu *et al.*, 2009). In an ulcerative colitis mouse model, oral hesperetin- and naringenin-treatment protected the integrity of the colonic mucus layer, and a subsequent study using an *in vitro* Caco-2 and a RAW264.7 co-culture model of LPS-induced epithelial monolayer damage revealed a mechanism associated with increased epithelial monolayer integrity (He *et al.*, 2019). More recent studies, primarily in animal and *in vitro* models, have also shown impact of flavanones on gastric healing (da Silva *et al.*, 2019) and antiallergic activity (Escribano-Ferrer *et al.*, 2019), but further human research is required to confirm these findings.

12.8.9 Modulation of the Microbiota and Biological Activity of Microbial Metabolites

Many flavonoid interventions have reported alterations in the microbiome and associated microbial metabolic activity (Williamson *et al.*, 2018). Consumption of OJ in humans resulted in significant increases in *Lactobacillus* spp. (Lima *et al.*, 2019) and abundance of clostridia (Brasili *et al.*, 2019), while daily supplementation with a 500 mg citrus extract for 12 weeks resulted in an increased butyrate to short chain fatty acid (SCFA) ratio, but not total fecal SCFA levels (Stevens *et al.*, 2019). Furthermore, the daily consumption of OJ for two months in a small clinical trial of 10 healthy women showed increasing metabolic activity of the microbiota, particularly *Bifidobacterium* spp. and *Lactobacillus* spp., increase in SCFAs, and a reduction in ammonia (Lima *et al.*, 2019).

An *in vitro* study demonstrated that naringenin exerted dose-dependent inhibitory effects on intestinal bacterial growth of *Bifidobactrium catenulatum*, *Bacteroides galacturonicus*, *Enterococcus caccae*, *Escherichia coli*, *Lactobacillus* spp., and *Ruminococcus gauvreauii* while its 7-*O*-rutinoside was inactive. Here inhibitory effects were observed at doses exceeding 200 μg/mL (Duda-Chodak 2012). Furthermore, *in vitro* studies involving hesperetin-7-*O*-rutinoside in simulated digestion and Caco-2 cell transport models (van Rymenant *et al.*, 2018) corroborate the changes in microbiota and increased levels of SCFAs reported in human and animal studies. Interestingly, van Rymenant *et al.* (2018) report the accumulation of the flavanone catabolites dihydrocaffeic acid (**41**, 3-(3′,4′-dihydroxyphenyl)propionic acid) and dihydroisoferulic acid (**42**, 3-(3′-hydroxy-4′-methoxyphenyl)propionic acid) The antimicrobial effects of microbial metabolites of flavanones, such as caffeic acid (**43**), ferulic acid (**44**), isoferulic acid (**45**), 4′-hydroxyphenylacetic acid (**46**) and 4-hydroxybenzoic acid (**47**) and 3-methoxy-4-hydroxybenzoic acid (**48**, vanillic acid), have previously been reported by Parkar *et al.* (2008) and Cueva *et al.* (2010) (see structures 41–50).

Numerous catabolites of flavanones have been reported in human, animal, and *in vitro* fermentation studies (see section 12.3), including ferulic acid and isoferulic acid, hydroxycinnamic acids, phenylpropanoic acids, phenylacetic acids, and hydroxybenzoic acids (see Figures 12.1 and 12.2) (Gao *et al.*, 2006; Pereira-Caro *et al.*, 2015a, 2018; van Rymenant *et al.*, 2018). Many of these catabolites, and their phase II conjugates formed during absorption and first-past metabolism, are also reported to have biological activity in cell culture modes (Williamson *et al.*, 2018). As the A- and B-ring configuration patterns of many flavanones are analogous to other polyphenols, they will share common catabolites such as hydroxybenzoic and hydroxycinnamic acids (Williamson *et al.*, 2018), allowing for comparison across bioactivity studies. Such comparisons

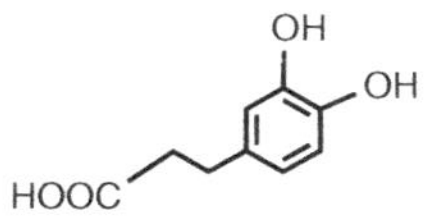

Dihydrocaffeic acid
(**41**, 3′,4′-dihydroxyphenyl)propionic acid)

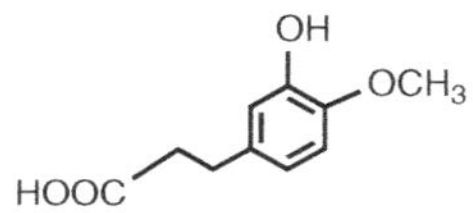

Dihydrisooferulic acid
(**42**, 3′-hydroxy-4′-methoxyphenyl)propionic acid)

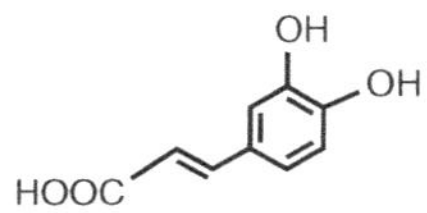

Caffeic acid (**43**)

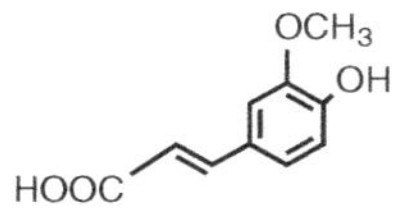

Ferulic acid (**44**)

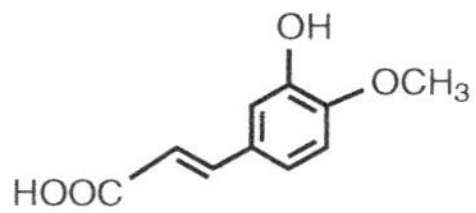

Isoferulic acid (**45**)

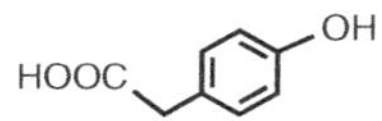

4′-Hydroxyphenylacetic acid (**46**)

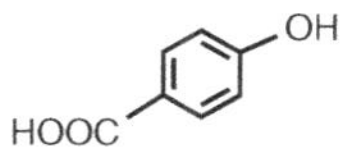

4-Hydroxybenzoic acid (**47**)

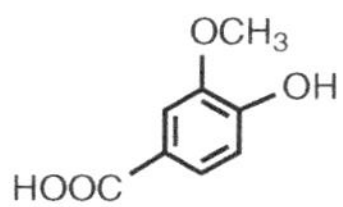

Vanillic acid
(**48** 3-Methoxy-4-hydroxybenzoic acid)

are useful as at present, studies directly exploring the activity of flavanone catabolites or comparing precursor flavanones to their microbial catabolites are limited.

Most of the evidence reported for phenolic catabolites of flavonoids stems from studies with ferulic acid (Ou and Kwokm 2004) and 3,4-dihydroxybenzoic acid (**49**, protocatechuic acid) (Kakkar and Bais, 2014), which have revealed a number of activities, including antiatherosclerotic, antithrombosis, antidiabetic, anticancer, and antiinflammatory properties. A growing number of *in vitro* studies report the activity of other microbial catabolites of flavonones and other flavonoids, such as caffeic acid (see structure 43), vanillic acid (see structure 48), and gallic acid (see structure 50) (Williamson *et al.*, 2018).

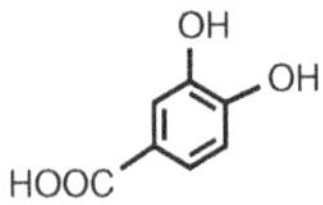

Protocatechuic acid
(**49**, 3,4-dihydroxybenzoic acid)

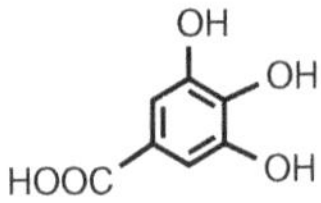

Gallic acid
(**50**, 3,4,5-trihydroxybenzoic acid)

Dihydroferulic acid (see structure 41), a ferulic acid intestinal microbial-derived OJ catabolite (see Figure 12.1), has been reported to protect neurons in a model of 2,3-dimethoxy-1,4-naphthoquinone-induced toxicity in human neuroblastoma SK-N-MC clonal cells (Verzelloni *et al.*, 2011). In screens involving vascular inflammation and adhesion, catabolites including ferulic acid and benzoic derivatives, along with their sulfate and glucuronide conjugates, were observed to reduce levels of proinflammatory chemokines and adhesion molecules such as IL-6, TNF-α, and sVCAM-1 (Amin *et al.*, 2015; di Gesso *et al.*, 2015; Warner *et al.*, 2016). Evidence suggests mixtures of these catabolites, as found in the circulation, may have additive or synergistic effects (di Gesso *et al.*, 2015; Warner *et al.*, 2017). Furthermore, the activity of some flavonone catabolites, including 4-hydroxybenzoic acid (see structure 47), its glucuronide and sulfate conjugates, and ferulic acid, has been reported to have greater biological activity than many precursor flavonoids (Amin *et al.*, 2015; Warner *et al.*, 2016). Caffeic acid (see structure 43) (Xu *et al.*, 2005) and gallic acid (see structure 50) (Ho *et al.*, 2010) have also been reported to downregulate Ras-related C3 botulinum toxin substrate 1 (RAC1) in cultured cells, which is believed to be involved in the coordination of inflammation, oxidative and metabolic pathways through cross-talk between the N-Kβ and Nrf2 pathways (Cuadrado *et al.*, 2014; Wardyn *et al.*, 2015). Taken together, this and previously reported activity of flavanones on PPARs, described earlier, which are also influenced by Nrf2 (Hayes and Dinkova-Kostova, 2014), provide further support for a mechanism of action of flavanones involving redox signaling and metabolism.

References

Actis-Goretta, L., Dew, T.P., Lévèsques, A., *et al.* (2015) Gastrointestinal absorption and metabolism of hesperetin-7-*O*-rutinoside and hesperetin-7-*O*-glucoside in healthy humans. *Molecular Nutrition & Food Research*, **59**, 1651–62.

Adams, T.B., McGowen, M.M., Williams, M.C., *et al.* (2007) The FEMA GRAS assessment of aromatic substituted secondary alcohols, ketones, and related esters used as flavor ingredients. *Food and Chemical Toxicology*, **45**, 171–201.

Ahmed, O.M., Fahim, H.I., Ahmed, H.Y., *et al.* (2019) The preventive effects and the mechanisms of action of navel orange peel hydroethanolic extract, naringin, and naringenin in N-acetyl-*p*-aminophenol-induced liver injury in Wistar rats. *Oxidative Medicine and Cellular Longevity*, **2019**, e2745352.

Akiyama, S., Katsumata, S., Suzuki, K., Ishimi, Y., Wu, J., and Uehara, M. (2010) Dietary hesperidin exerts hypoglycemic and hypolipidemic effects in streptozotocin-induced marginal type 1 diabetic rats. *Journal of Clinical Biochemistry and Nutrition*, **46**, 87–92.

Alam, M.A., Kauter, K., and Brown, L. (2013) Naringin improves diet-induced cardiovascular dysfunction and obesity in high carbohydrate, high fat diet-fed rats. *Nutrients*, **5**, 637–50.

Alharbi, M.H., Lamport, D.J., Dodd, G.F., *et al.* (2016) Flavonoid-rich orange juice is associated with acute improvements in cognitive function in healthy middle-aged males. *European Journal of Nutrition*, **55**, 2021–9.

Almeida, A.F., Borge, G.I.A., Piskula, M., *et al.* (2018) Bioavailability of quercetin in humans with a focus on interindividual variation. *Comprehensive Reviews in Food Science and Food Safety*. Available at: https://doi.org/10.1111/1541-4337.12342.

Amin, H.P., Czank, C., Raheem, S., *et al.* (2015) Anthocyanins and their physiologically relevant metabolites alter the expression of IL-6 and VCAM-1 in CD40L and oxidized LDL challenged vascular endothelial cells. *Molecular Nutrition & Food Research*, **59**, 1095–106.

Andrews, L. (2013) Dietary flavonoids for the prevention of colorectal cancer. *Clinical Journal of Oncology Nursing*, **17**, 671–2.

Ávila, M., Jaquet, M., Moine, D., *et al.* (2009) Physiological and biochemical characterization of two α-L-rhamnosidases in *Lactobacillus plantarum* NCC245. *Microbiology*, **155**, 2739–49.

Azuma, T., Shigeshiro, M., Kodama, M., Tanabe, S., and Suzuki, T. (2013) Supplemental naringenin prevents intestinal barrier defects and inflammation in colitic mice. *Journal of Nutrition*, **143**, 827–34.

Bhargava, P., Verma, V.K., Malik, S., Khan, S.I., Bhatia, J., and Arya, D.S. (2019) Hesperidin regresses cardiac hypertrophy by virtue of PPAR-γ agonistic, anti-inflammatory, antiapoptotic, and antioxidant properties. *Journal of Biochemical and Molecular Toxicology*, **33**, e22283.

Boumendjel, A., Blanc, M., Williamson, G., and Barron, D. (2009) Efficient synthesis of flavanone glucuronides. *Journal of Agricultural and Food Chemistry*, **57**(16), 7264–7.

Borges, G., Mullen W., Mullan, A., Lean, M.E.J., Roberts, S.A., and Crozier, A. (2010) Bioavailability of multiple components following acute ingestion of a polyphenol-rich drink. *Molecular Nutrition & Food Research*, **54** (Suppl 2), S268–S277.

Borges, G., Lean, M.E.J., Roberts, S.A., and Crozier, A. (2013) Bioavailability of dietary (poly)phenols: a study with ileostomists to discriminate between absorption in the small and large intestine. *Food & Function*, **4**, 754–62.

Borges, G., Ottaviani, J.I., van der Hooft, J.J.J., Schroeter, H., and Crozier, A. (2018) Absorption, metabolism, distribution and excretion of (–)-epicatechin: a review of recent findings. *Molecular Aspects of Medicine*, **61**, 18–30.

Brasili, E., Hassimotto, N.M.A., del Chierico, F., *et al.* (2019) Daily consumption of orange juice from *Citrus sinensis* L. Osbeck cv. Cara Cara and cv. Bahia differently affects gut microbiota profiling as unveiled by an integrated meta-omics approach. *Journal of Agricultural and Food Chemistry*, **67**, 1381–91.

Braune, A., Engst, W., and Blaut, M. (2005) Degradation of neohesperidin dihydrochalcone by human intestial bacteria. *Journal of Agricultural and Food Chemistry*, **53**, 1782–90.

Bredsdorff, L., Nielsen, I.L.F., Rasmussen, S.E., *et al.* (2010) Absorption, conjugation and excretion of the flavones naringenin and hesperetin from α-rhamnosidase-treated orange juice in human subjects. *British Journal of Nutrition*, **103**, 1602–9.

Brett, G.M., Hollands, W., Needs, P.W., *et al.* (2009) Absorption, metabolism and excretion of flavanones from single portions of orange fruit and juice and effects of anthropometric variables and contraceptive pill use on flavanone excretion. *British Journal of Nutrition*, **101**, 664–75.

Burke, A.C., Telford, D.E., Edwards, J.Y., Sutherland, B.G., Sawyez, C.G., and Huff, M.W. (2019) Naringenin supplementation to a chow diet enhances energy expenditure and fatty acid oxidation, and reduces adiposity in Lean, Pair-Fed Ldlr(−/−) mice. *Molecular Nutrition & Food Research*, **63**, e1800833.

Casey, E., Mistry, D.L., and MacKnight, J.M. (2005) Training room management of medical conditions: sports gastroenterology. *Clinics in Sports Medicine*, **24**, 525–40.

Cassidy, A., Rimm, E.B., O'Reilly, E.J., *et al.* (2012) Dietary flavonoids and risk of stroke in women. *Stroke*, **43**, 946–51.

Chanet, A., Milenkovic, D., Claude, S., *et al.* (2013) Flavanone metabolites decrease monocyte adhesion to TNF-α-activated endothelial cells by modulating expression of atherosclerosis-related genes. *British Journal of Nutrition*, **110**, 587–98.

Chang, H., Lei, L., Zhou, Y., Ye, F., and Zhao, G. (2018) Dietary flavonoids and the risk of colorectal cancer: an updated meta-analysis of epidemiological studies. *Nutrients*, **10**, e950.

Chen, Y.J., Kong, L., Tang, Z.Z., *et al.* (2019) Hesperetin ameliorates diabetic nephropathy in rats by activating Nrf2/ARE/glyoxalase 1 pathway. *Biomedicine & Pharmacotherapy*, **111**, 1166–175.

Chiba, H., Kim, H., Matsumoto, A., *et al.* (2014) Hesperidin prevents androgen deficiency-induced bone loss in male mice. *Phytotherapy Research*, **28**, 289–95.

Clarke, S.F., Murphy, E.F., O'Sullivan, O., *et al.* (2014) Exercise and associated dietary extremes impact on gut microbial diversity. *Gut*, **63**, 1913–20.

Crozier, A. (2013) Absorption, metabolism and excretion of (−)-epicatechin in humans: an evaluation of recent findings. *American Journal of Clinical Nutrition*, **98**, 861–2.

Crozier, A., Yokota, T., Jaganath, I.B., Marks, S., Saltmarsh, M., and Clifford, M.N. (2006) Secondary metabolites as dietary components in plant-based foods and beverages. In A. Crozier, M.N. Clifford, and H. Ashihara (Eds.), *Plant Secondary Metabolites: Occurrence, Structure and Role in the Human Diet*. Oxford: Blackwell Publishing.

Cuadrado, A., Martin-Moldes, Z., Ye, J., and Lastres-Becker, I. (2014) Transcription factors NRF2 and NF-κB are coordinated effectors of the Rho family, GTP-binding protein RAC1 during inflammation. *Journal of Biological Chemistry*, **289**, 15244–58.

Cueva, C,. Moreno-Arribas, M.V., Martin-Alvarez, P.J., *et al.* (2010) Antimicrobial activity of phenolic acids against commensal, probiotic and pathogenic bacteria. *Research in Microbiology*, **161**, 372–82.

Curzon, G. and Pratt, R.T. (1964) Origin of urinary resorcinol sulphate. *Nature*, **204**, 383–4.

Dall'Asta, M., Derlindati, E., Curella, V., *et al.* (2013) Effects of naringenin and its phase II metabolites on *in vitro* human macrophage gene expression. *International Journal of Food Sciences and Nutrition*, **64**, 843–9.

Das, N.P. (1971) Studies on flavonoid metabolism. Absorption and metabolism of (+)-catechin in man. *Biochemical Pharmacology*, **20**, 3435–45.

da Silva, L.M., Pezzini, B.C., Somensi, L.B., *et al.* (2019) Hesperidin, a citrus flavanone glycoside, accelerates the gastric healing process of acetic acid-induced ulcer in rats. *Chemico-Biological Interactions*, **308**, 45–50.

Day, A.J., Canada, F.J., Diaz, J.C., *et al.* (2000) Dietary flavonoid and isoflavone glycosides are hydrolysed by the lactase site of lactase phlorizin hydrolase. *FEBS Letters*, **468**, 166–70.

Del Rio, D., Rodriguez-Mateos, A.M., Spencer, J.P.E., Tognolini, M., Borges, G., and Crozier, A. (2013) Dietary (poly)phenolics in human health and disease: structures, bioavailability, evidence of protective effects and potential mechanisms. *Antioxidants & Redox Signaling*, **18**, 1818–92.

Demonty, I., Lin, Y., Zebregs, Y.E., Vermeer, M.A., van der Knaap, H.C., and Trautwein, E.A. (2010) The citrus flavonoids hesperidin and naringin do not affect serum cholesterol in moderately hypercholesterolemic men and women. *Journal of Nutrition*, **140**, 1615–20.

De Weerd, K.A., Saxena, A., Nagle, D.P. Jr.,, and Suflita, J.M. (1988) Metabolism of the ^{18}O-methoxy substituent of 3-methoxybenzoic acid and other unlabeled methoxybenzoic acids by anaerobic bacteria. *Applied and Environmental Microbiology*, **54**, 1237–42.

di Gesso, J.L., Kerr, J.S., Zhang, Q., *et al.* (2015) Flavonoid metabolites reduce tumor necrosis factor-α secretion to a greater extent than their precursor compounds in human THP-1 monocytes. *Molecular Nutrition & Food Research*, **59**, 1143–54.

Donovan, J.L., Manach, C., Faulks, R.M., and Kroon, P. (2006) Absorption and metabolism of dietary secondary metabolites. In A. Crozier, M.N. Clifford, and H. Ashihara (Eds.), *Plant Secondary Metabolites: Occurrence, Structure and Role in the Human Diet*. Oxford: Blackwell Publishing.

Duda-Chodak, A. (2012) The inhibitory effect of polyphenols on human gut microbiota. *Journal of Physiology and Pharmacology*, **63**, 497–503.

Erlund, I., Merirrinne, E., Alfthan, G., and Aro, A. (2001) Plasma kinetics and urinary excretion of the flavanones naringenin and hesperetin in humans after ingestion of orange juice and grapefruit juice. *Journal of Nutrition*, **131**, 235–41.

Escribano-Ferrer, E., Queralt Regue, J., Garcia-Sala, X., Boix Montanes, A., and Lamuela-Raventos, R.M. (2019) *In vivo* anti-inflammatory and antiallergic activity of pure naringenin, naringenin chalcone, and quercetin in mice. *Journal of Natural Products*, **82**, 177–82.

Fukui, H. (2016) Increased intestinal permeability and decreased barrier function: does it really influence the risk of inflammation? *Inflammatory Intestinal Diseases*, **1**, 135–45.

Gao, K., Xu, A., Krul, C., *et al.* (2006) Of the major phenolic acids formed during human microbial fermentation of tea, citrus, and soy flavonoid supplements, only 3,4-dihydroxyphenylacetic acid has antiproliferative activity. *Journal of Nutrition*, **136**, 52–7.

Gee, J.M., DuPont, S.M., Day, A.J., Plumb, G.W., Williamson, G., and Johnson, I.T. (2000) Intestinal transport of quercetin glycosides in rats involves both deglycosylation and interaction with the hexose transport pathway. *Journal of Nutrition*, **130**, 2765–71.

Gimenez-Bastida, J.A., Gonzalez-Sarrias, A., Vallejo, F., Espín, J.C., and Tomás-Barberán, F.A. (2016) Hesperetin and its sulphate and glucuronide metabolites inhibit TNF-α-induced human aortic endothelial cell migration and decrease plasminogen activator inhibitor-1 (PAI-1) levels. *Food & Function*, **7**, 118–26.

Goldwasser, J., Cohen, P.Y., Yang, E., Balaguer, P., Yarmush, M.L., and Nahmias, Y. (2010) Transcriptional regulation of human and rat hepatic lipid metabolism by the grapefruit flavonoid naringenin: role of PPARα, PPARγ and LXRα. *PloS One*, **5**, e12399.

González-Domínguez, R., Urpi-Sarda, M., Jauregui, O., Needs, P., Kroon, P.A., and Andrés-Lacueva, C. (2019) Quantitative dietary fingerprinting (QDF) – a novel tool for comprehensive dietary assessment based on urinary nutrimetabolomics. *Journal of Agricultural and Food Chemistry*, **67**, 1851–61.

Grbić-Galić, D. (1986) *O*-Demethylation, dehydroxylation, ring-reduction and cleavage of aromatic substrates by Enterobacteriaceae under anaerobic conditions. *Journal of Applied Bacteriology*, **61**, 491–7.

Han, K.Y., Chen, P.N., Hong, M.C., *et al.* (2018) Naringenin attenuated prostate cancer invasion via reversal of epithelial-to-mesenchymal transition and inhibited uPA activity. *Anticancer Research*, **38**, 6753–8.

Hayes, J.D. and Dinkova-Kostova, A.T. (2014) The Nrf2 regulatory network provides an interface between redox and intermediary metabolism. *Trends in Biochemical Sciences*, **39**, 199–218.

He, W., Liu, M., Li, Y., *et al.* (2019) Flavonoids from Citrus aurantium ameliorate TNBS-induced ulcerative colitis through protecting colonic mucus layer integrity. *European Journal of Pharmacology*, **857**, e172456.

Heinzmann, S.S., Brown, I.J., Chan, Q., *et al.* (2010) Metabolic profiling strategy for discovery of nutritional biomarkers: proline betaine as a marker of citrus consumption. *American Journal of Clinical Nutrition*, **92**, 436-43.

Hengstmann, J.H. and Aulepp, H. (1978) Pharmacokinetics and metabolism of synephrine-H^3. *Arzneimittel-Forschung*, **28**, 2326–31.

Ho, H.H., Chang, C.S., Ho, W.C., Liao, S.Y., Wu, C.H., and Wang, C.J. (2010) Anti-metastasis effects of gallic acid on gastric cancer cells involves inhibition of NF-κB activity and downregulation of PI3K/AKT/small GTPase signals. *Food and Chemical Toxicology*, **48**, 2508–16.

Hollman., P.C.H., de Vries, J.H.M., van Leeuwen, S.D., Mengelers, M.J.B., and Katan, M.B. (1995) Absorption of dietary quercetin glycosides and quercetin in healthy ileostomy volunteers. *American Journal of Clinical Nutrition*, **62**, 1276–82.

Hong, Y., Yin, Y., Tan, Y., Hong, K., and Zhou, H. (2019) The flavanone, naringenin, modifies antioxidant and steroidogenic enzyme activity in a rat model of letrozole-induced polycystic ovary syndrome. *Medical Science Monitor*, **25**, 395–401.

Huang, B., Hu, P., Hu, A., *et al.* (2019) Naringenin attenuates carotid restenosis in rats after balloon injury through its anti-inflammation and anti-oxidative effects via the RIP1-RIP3-MLKL signaling pathway. *European Journal of Pharmacology*, **855**, 167–74.

Huang, S.M., Tsai, S.Y., Lin, J.A., Wu, C.H., and Yen, G.C. (2012) Cytoprotective effects of hesperetin and hesperidin against amyloid beta-induced impairment of glucose transport through downregulation of neuronal autophagy. *Molecular Nutrition & Food Research*, **56**, 601–9.

Hur, H. and Rafii, F. (2000) Biotransformation of the isoflavonoids biochanin A, formononetin, and glycitein by *Eubacterium limosum*. *FEMS Microbiology Letters*, **192**, 21–5.

Jeffrey, A.M., Knight, M., and Evans, W.C. (1972a) The bacterial degradation of flavonoids. Hydroxylation of the A-ring of taxifolin by a soil pseudomonad. *Biochemical Journal*, **130**, 373–81.

Jeffrey, A.M., Jerina, D.M., Self, R., and Evans, W.C. (1972b) The bacterial degradation of flavonoids. Oxidative fission of the

A-ring of dihydrogossypetin by a *Pseudomonas* sp. *Biochemical Journal*, **130**, 383–90.

Jung, U.J., Kim, H.J., Lee, J.S., *et al.* (2003) Naringin supplementation lowers plasma lipids and enhances erythrocyte antioxidant enzyme activities in hypercholesterolemic subjects. *Clinical Nutrition*, **22**, 561–8.

Kakkar, S. and Bais, S. (2014) A review on protocatechuic acid and its pharmacological potential. *ISRN Pharmacology*, **2014**, e952943.

Kamaraj, S., Anandakumar, P., Jagan, S., Ramakrishnan, G., and Devaki, T. (2011) Hesperidin attenuates mitochondrial dysfunction during benzo(a)pyrene-induced lung carcinogenesis in mice. *Fundamental & Clinical Pharmacology*, **25**, 91–8.

Khodayar, M.J., Kalantari, H., Mahdavinia, M., *et al.* (2018) Protective effect of naringin against BPA-induced cardiotoxicity through prevention of oxidative stress in male Wistar rats. *Drug and Chemical Toxicology*, **43**, 85–95.

Kim, S.H., Kim, B.K., and Lee, Y.C. (2011) Antiasthmatic effects of hesperidin, a potential Th2 cytokine antagonist, in a mouse model of allergic asthma. *Mediators of Inflammation*, **2011**, e485402.??

Krumholz, L.R. and Bryan, M.P. (1988) Characterization of the pyrogallol-phloroglucinol isomerase of *Eubacterium oxidaoreducens*. *Journal of Bacteriology*, **170**, 2472–9.

Krumholz, L.R., Crawford, R.L., Hemling, M.E., and Bryant, M.P. (1987) Metabolism of gallate and phloroglucinol in *Eubacteruim oxidoreductens* via 3-hydroxy-5-oxohexanoate. *Journal of Bacteriology*, **169**, 1886–90.

Kurowska, E.M., Spence, J.D., Jordan, J., *et al.* (2000) HDL-cholesterol-raising effect of orange juice in subjects with hypercholesterolemia. *American Journal of Clinical Nutrition*, **72**, 1095–100.

Lamport, D.J., Pal, D., Macready, A.L., *et al.* (2016) The effects of flavanone-rich citrus juice on cognitive function and cerebral blood flow: an acute, randomised, placebo-controlled cross-over trial in healthy, young adults. *British Journal of Nutrition*, **116**, 2160–8.

Lang, R., Lang, T., Bader, M., Beusch, A., Schlagbauer, V., and Hofmann, T. (2017) High-Throughput quantitation of proline betaine in foods and suitability as valid biomarker for citrus consumption. *Journal of Agricultural and Food Chemistry*, **65**, 1613–19.

Lee, C.J., Wilson, L., Jordan, M.A., Nguyen, V., Tang, J., and Smiyun, G. (2010) Hesperidin suppressed proliferations of both human breast cancer and androgen-dependent prostate cancer cells. *Phytotherapy Research*, **24** (Suppl. 1), S15–S19.

Li, C. and Schluesener, H. (2017) Health-promoting effects of the citrus flavanone hesperidin. *Critical Reviews in Food Science and Nutrition*, **57**, 613–31.

Li, R., Cai, L., Xie, X. F., Yang, F., and Li, J. (2010) Hesperidin suppresses adjuvant arthritis in rats by inhibiting synoviocyte activity. *Phytotherapy Research*, **24** (Suppl. 1), S71–S76.

Lima, A.C.D., Cecatti, C., Fidelix, M.P., *et al.* (2019) Effect of daily consumption of orange juice on the levels of blood glucose, lipids, and gut microbiota metabolites: controlled clinical trials. *Journal of Medicinal Food*, **22**, 202–10.

Liu, H., Dong, Y., Gao, Y., *et al.* (2019) Hesperetin suppresses RANKL-induced osteoclastogenesis and ameliorates lipopolysaccharide-induced bone loss. *Journal of Cell Physiology*, **234**, 11009–22.

Llorach, R., Medina, S., García-Viguera, C., *et al.* (2014) Discovery of human urinary biomarkers of aronia-citrus juice intake by HPLC-q-TOF-based metabolomic approach. *Electrophoresis*, **35**, 1599–606.

Lloyd, A.J., Beckman, M., Fave, G., Mathers, J.C., and Draper, J. (2011) Proline betaine and its biotransformation products in fasting urine samples are potential biomarkers of habitual citrus fruit consumption. *British Journal of Nutrition*, **106**, 812–24.

Mahmoud, A.M., Ashour, M.B., Abdel-Moneim, A., and Ahmed, O.M. (2012) Hesperidin and naringin attenuate hyperglycemia-mediated oxidative stress and proinflammatory cytokine production in high fat fed/streptozotocin-induced type 2 diabetic rats. *Journal of Diabetes and its Complications*, **26**, 483–90.

Malkova, D., Evans, R., Frayn, K., Humphreys, S., Jones, P., and Hardman, A. (2000) Prior exercise and postprandial substrate extraction across the human leg. *American Journal of Physiology. Endocrinology and Metabolism*, **279**, E1020–E1028.

Manach, C., Morand, C., Gil-Izquierdo, A., Bouteloup-Demange, C., and Rémésy. C. (2003) Bioavailability in humans of the flavanones hesperidin and narirutin after the ingestion of two doses of orange juice. *European Journal of Clinical Nutrition*, **57**, 235–42.

Manach, C., Hubert, J., Llorach, R., and Scalbert, A. (2009) The complex links between dietary phytochemicals and human health deciphered by metabolomics. *Molecular Nutrition & Food Research*, **53**, 1303–15.

Manach, C., Milenkovic. D., van de Wiele, T., *et al.* (2017) Addressing the inter-individual variation in response to consumption of plant food bioactives - towards a better understanding of their role in healthy ageing and cardiometabolic risk reduction. *Molecular Nutrition & Food Research*, **61**, e1600557.

Medina, S., Domíngues-Peries, R., García-Viguera, C., *et al.* (2012) Physical activity increases the bioavailability of flavanones after dietary aronia-citrus intake in triathletes. *Food Chemistry*, **135**, 2133–7.

Menze, E.T., Tadros, M.G., Abdel-Tawab, A.M., and Khalifa, A.E. (2012) Potential neuroprotective effects of hesperidin on 3-nitropropionic acid-induced neurotoxicity in rats. *Neurotoxicology*, **33**, 1265–75.

Messerschmidt, A., Niessen, H., Abt, D., Einnsle, O., Schink, B., and Kronek, P.M. (2004) Crystal structure of pyrogallol-phloroglucinol transhydroxylase, a Mo enzyme capable of intermolecular hydroxyl transfer between phenols. *Proceedings of the National Academy of Sciences of the U.S.A.*, **1010**, 11571–6.

Mo'men, Y.S., Hussein, R.M., and Kandeil, M.A. (2019) Involvement of PI3K/Akt pathway in the protective effect of hesperidin against a chemically induced liver cancer in rats. *Journal of Biochemical and Molecular Toxicology*, **2019**, e22305.

Morand, C., Dubray, C., Milenkovic, D., *et al.* (2011) Hesperidin contributes to the vascular protective effects of orange juice: a randomized crossover study in healthy volunteers. *American Journal of Clinical Nutrition*, **93**, 73–80.

Moreira, V., Brasili, E., Fiamoncini, J., *et al.* (2018) Orange juice affects acylcarnitine metabolism in healthy volunteers as revealed by a mass-spectrometry based metabolomics approach. *Food Research International*, **107**, 346–52.

Mullen, W., Archeveque, M.-A., Edwards, C. A., and Crozier, A. (2008) Bioavailability and metabolism of orange juice flavanones in humans: impact of a full fat yogurt. *Journal of Agricultural and Food Chemistry*, **56**, 11157–64.

Najmanová, I., Vopršalová, M., Saso, L., and Mladěnka, P. (2019) The pharmacokinetics of flavanones. *Critical Reviews in Food Science and Nutrition*. Available at: https://doi.org/10.1080/10408398.2019.1679085.

Nandakumar, N., Rengarajan, T., Balamurugan, A., and Balasubramanian, M.P. (2014) Modulating effects of hesperidin on key carbohydrate-metabolizing enzymes, lipid profile, and membrane-bound adenosine triphosphatases against 7,12-dimethylbenz(a)anthracene-induced breast carcinogenesis. *Human & Experimental Toxicology*, **33**, 504–16.

Neveu, V., Perez-Jimenez, J., Vos, F., *et al.* (2010) Phenol-Explorer: an online comprehensive database on polyphenol contents in foods. *Database*, **2010**, bap024.

Nielsen, I.L.F., Chee, W.S.S., Poulsen, L., *et al.* (2006) Bioavailability is improved by enzymatic modification of the citrus flavonoid hesperidin in humans: a randomised, double-blind, crossover trial. *Journal of Nutrition*, **136**, 404–8.

Noda, S., Tanabe, S., and Suzuki, T. (2012) Differential effects of flavonoids on barrier integrity in human intestinal Caco-2 cells. *Journal of Agricultural and Food Chemistry*, **60**, 4628–33.

Noda, S., Tanabe, S., and Suzuki, T. (2013) Naringenin enhances intestinal barrier function through the expression and cytoskeletal association of tight junction proteins in Caco-2 cells. *Molecular Nutrition & Food Research*, **57**, 2019–28.

Ordóñez, J.L., Pereira-Caro, G., Ludwig, I., *et al.* (2018) An evaluation of GC-MS and HPLC-HRMS analysis of microbial metabolites in human urine after consumption of orange juice. *Journal of Chromatography. A*, **1575**, 100–12.

Ottaviani, J.I., Borges, G., Momma, T., *et al.* (2016) The metabolome of [2-^{14}C](–)-epicatechin in humans: implications for the assessment of efficacy safety, and mechanisms of action of polyphenolic bioactives. *Scientific Reports*, **6**, 29034.

Ottaviani, J.I., Fong. R.Y., Borges, G., Schroeter, H., and Crozier, A. (2018) Use of LC-MS for the quantitative analysis of (poly)phenol metabolites does not necessarily yield accurate results: implications for assessing existing data and conducting future research. *Free Radical Biology & Medicine*, **124**, 97–103.

Ou, S. and Kwok, K.-C. (2004) Ferulic acid: pharmaceutical functions, preparation and applications in foods. *Journal of the Science of Food and Agriculture*, **84**, 1261–9.

Parkar, S.G., Stevenson, D.E., and Skinner, M.A. (2008) The potential influence of fruit polyphenols on colonic microflora and human gut health. *International Journal of Food Microbiology*, **124**, 295–8.

Pechlivanis, A., Kostidis, S., Saraslanidis, P., *et al.* (2010) ^{1}H NMR-based metabonomic investigation of the effect of two different exercise sessions on the metabolic fingerprint of human urine. *Journal of Proteome Research*, **9**, 6405–16.

Pereira-Caro, G., Borges, G., van der Hooft, J., *et al.* (2014) Orange juice (poly)phenols are highly bioavailable in humans. *American Journal of Clinical Nutrition*, **100**, 1378–84.

Pereira-Caro, G., Borges, G., Ky, I., *et al.* (2015a) *In vitro* colonic catabolism of orange juice (poly)phenols. *Molecular Nutrition & Food Research*, **59**, 465–75.

Pereira-Caro, G., Oliver, C.M., Weerakkody, R., *et al.* (2015b) Chronic administration of a microencapsulated probiotic enhances the bioavailability of orange juice flavanaones in humans. *Free Radical Biology & Medicine*, **84**, 206–14.

Pereira-Caro, G., Ludwig, I.A., Polyviou, T., *et al.* (2016) Plasma and urinary metabolites and catabolites derived from orange juice (poly)phenols: analysis by high performance liquid chromatography-high resolution mass spectrometry. *Journal of Agricultural and Food Chemistry*, **64**, 5724–36.

Pereira-Caro, G., Polyviou, T., Ludwig, I., *et al.* (2017) Bioavailability of orange juice (poly)phenols: the impact of short-term cessation of training by male endurance athletes. *American Journal of Clinical Nutrition*, **106**, 791–800.

Pereira-Caro, G., Fernández-Quirós, B., Ludwig, I., Pradas, I., Crozier. A., and Moreno-Rojas, J.M. (2018) Catabolism of citrus flavanones by the probiotics *Bifidobacterium longum* and *Lactobacillus rhamnosus*. *European Journal of Nutrition*, **57**, 231–42.

Petrova, A., Davids, L.M., Rautenbach, F., and Marnewick, J.L. (2011) Photoprotection by honeybush extracts, hesperidin and mangiferin against UVB-induced skin damage in SKH-1 mice.

Journal of Photochemistry and Photobiology. B, Biology, **103**, 126–39.

Pujos-Guillot, E., Hubert, J., Martin, J. F., *et al.* (2013) Mass spectrometry-based metabolomics for the discovery of biomarkers of fruit and vegetable intake: citrus fruit as a case study. *Journal of Proteome Research*, **12**, 1645–59.

Rajappa, R., Sireesh, D., Salai, M.B., Ramkumar, K. M., Sarvajayakesavulu, S., and Madhunapantula, S.V. (2018) Treatment with naringenin elevates the activity of transcription factor Nrf2 to protect pancreatic β-cells from streptozotocin-induced diabetes *in vitro* and *in vivo*. *Frontiers in Pharmacology*, **9**, 1562.

Rangel-Huerta, O.D., Aguilera, C.M., Perez de la Cruz, A., *et al.* (2017) A serum metabolomics-driven approach predicts orange juice consumption and its impact on oxidative stress and inflammation in subjects from the BIONAOS study. *Molecular Nutrition & Food Research*, **61**, e.201600120.

Reshef, N., Hayari, Y., Goren, C., Boaz, M., Madar, Z., and Knobler, H. (2005) Antihypertensive effect of sweetie fruit in patients with stage I hypertension. *American Journal of Hypertension*, **18**, 1360–3.

Rizza, S., Muniyappa, R., Iantorno, M., *et al.* (2011) Citrus polyphenol hesperidin stimulates production of nitric oxide in endothelial cells while improving endothelial function and reducing inflammatory markers in patients with metabolic syndrome. *Journal of Clinical Endocrinology and Metabolism*, **96**, E782–E792.

Rodriguez-Mateos, A.M., Vauzour, D., Kreuger, C.G., *et al.* (2014) Flavonoids and related compounds, bioavailability, bioactivity and impact on human health: an update. *Archives of Toxicology*, **88**, 1803–53.

Roger, C.R. (1988) The nutritional incidence of flavonoids: some physiological and metabolic considerations. *Experientia*, **44**, 725–33.

Roowi, S., Mullen, W., Edwards, C. A., and Crozier, A. (2009) Yoghurt impacts on the excretion of phenolic acids derived from colonic breakdown of orange juice flavanones in humans. *Molecular Nutrition & Food Research*, **53**, S68–S75.

Roowi, S., Stalmach, A., Mullen, W., Lean, M.E.J., Edwards, C.A., and Crozier, A. (2010) Green tea flavan-3-ols: colonic degradation and urinary excretion of catabolites by humans. *Journal of Agricultural and Food Chemistry*, **58**, 1296–304.

Saha, S., Holland, W., Needs, P.W., *et al.* (2012) Human *O*-sulphated metabolites of (−)-epicatechin and methylated-(−)-epicatechin are poor substrates for commercial arylsulatases: implications for studies concerned with quantifying epicatechin bioavailability. *Pharmacological Research*, **65**, 592–602.

Saiprasad, G., Chitra, P., Manikandan, R., and Sudhandiran, G. (2013) Hesperidin alleviates oxidative stress and downregulates the expressions of proliferative and inflammatory markers in azoxymethane-induced experimental colon carcinogenesis in mice. *Inflammation Research*, **62**, 425–40.

Scalbert, A., Brennan, L., Manach, C., *et al.* (2014) The food metabolome: a window over dietary exposure. *American Journal of Clinical Nutrition*, **99**, 1286–308.

Shaw, K.N. and Trevarthen, J. (1958) Exogenous sources of urinary phenol and indole acids. *Nature*, **182**, 797–8.

Silveira, J.Q., Cesar, T.B., Manthey, J.A., Baldwin, E.A., Bai, J., and Raithore, S. (2014) Pharmacokinetics of flavanone glycosides after ingestion of single doses of freshly squeezed orange juice versus commercially processed orange juice in healthy humans. *Journal of Agricultural and Food Chemistry*, **62**, 12576–84.

Stalmach, A., Mullen, W., Barron, D., *et al.* (2009) Metabolite profiling of hydroxycinnamate derivatives in plasma and urine after ingestion of coffee by humans: identification of biomarkers of coffee consumption. *Drug Metabolism and Disposition*, **37**, 1749–58.

Stalmach, A., Mullen, W., Steiling, H., Williamson, G., Lean, M.E.J., and Crozier, A. (2010a) Absorption, metabolism, and excretion of green tea flavan-3-ols in humans with an ileostomy. *Molecular Nutrition & Food Research*, **54**, 323–34.

Stalmach, A., Steiling, H., Williamson, G., and Crozier, A. (2010b) Bioavailability of chlorogenic acids following acute ingestion of coffee by humans with an ileostomy. *Archives of Biochemistry and Biophysics*, **501**, 98–105.

Stevens, Y., van Rymenant, E., Grootaert, C., *et al.* (2019) The intestinal fate of citrus flavanones and their effects on gastrointestinal health. *Nutrients*, **11**, e1464.

Stolz, N. (2005) The presence of β-m-hydroxyphenylhydracrylic acid in the urine of patients with ADHD and other neurodegenerative metabolic disorders. MSc thesis, South Western University, South Africa.

Stoupi, S., Williamson, G., Drynan, J.W., Barron, D., and Clifford, M.N. (2010) A comparison of the in vitro biotransformation of (–)-epicatechin and procyanidin B2 by human faecal microbiota. *Molecular Nutrition & Food Research*, **54**, 747–59.

Sugasawa, N., Katagi, A., Kurobe, H., *et al.* (2019) Inhibition of atherosclerotic plaque development by oral administration of α-glucosyl hesperidin and water-dispersible hesperetin in apolipoprotein e knockout mice. *Journal of the American College of Nutrition*, **38**, 15–22.

Tamilselvam, K., Braidy, N., Manivasagam, T., *et al.* (2013) Neuroprotective effects of hesperidin, a plant flavanone, on rotenone-induced oxidative stress and apoptosis in a cellular model for Parkinson's disease. *Oxidative Medicine and Cellular Longevity*, **2013**, 102741.

Tomás-Navarro, M., Vallejo, F., Sentandeu, E., Navarro, J.L., and Tomás-Barberán, F.A. (2014) Volunteer stratification is more relevant than technological treatment in orange juice flavanone bioavailability. *Journal of Agricultural and Food Chemistry*, **62**, 24–7.

Tseng, Y.T., Hsu, H.T., Lee, T.Y., Chang, W.H., and Lo, Y.C. (2019) Naringenin, a dietary flavanone, enhances insulin-like growth factor 1 receptor-mediated antioxidant defense and attenuates methylglyoxal-induced neurite damage and apoptotic death. *Nutritional Neuroscience*, **2019**, 1594554.

Umar, S., Kumar, A., Sajad, M., *et al.* (2013) Hesperidin inhibits collagen-induced arthritis possibly through suppression of free radical load and reduction in neutrophil activation and infiltration. *Rheumatology International*, **33**, 657–63.

Urpi-Sarda, M., Rothwell, J., Morand, C., and Manach, C. (2012) Bioavailability of flavanones. In J.P.E. Spencer and A. Crozier (Eds.), *Bioavailability and Function of Flavonoids*. Boca Raton: CRC Press.

Vallejo, F., Larrosa, M., Escudero, E., *et al.* (2010) Concentration and solubility of flavanones in orange juice beverages affect their bioavailability in humans. *Journal of Agricultural and Food Chemistry*, **58**, 6516–24.

Van Dorsten, F.A., Grün, C.H., van Velzen, E.J., Jacobs, D.M., Daijer, R., and van Duynhoven, J.P. (2010) The metabolic fate of red wine and grape juice polyphenols in humans assessed by metabolomics. *Molecular Nutrition & Food Research*, **54**, 897–908.

Van Rymenant, E., Salden, B., Voorspoels, S., *et al.* (2018) A critical evaluation of *in vitro* hesperidin 2S bioavailability in a model combining luminal (microbial) digestion and Caco-2 cell absorption in comparison to a randomized controlled human trial. *Molecular Nutrition & Food Research*, **62**, e1700881.

Verzelloni, E., Pellacani, C., Tagliazucchi, D., *et al.* (2011) Antiglycative and neuroprotective activity of colon-derived polyphenol catabolites. *Molecular Nutrition & Food Research*, **55** (Suppl. 1), S35–S43.

Wang, D.M., Li, S. Q., Zhu, X.Y., Wang, Y., Wu, W.L., and Zhang, X.J. (2013) Protective effects of hesperidin against amyloid-β induced neurotoxicity through the voltage dependent anion channel 1 (VDAC1)-mediated mitochondrial apoptotic pathway in PC12 cells. *Neurochemical Research*, **38**, 1034–44.

Wang, R., Wang, J., Dong, T., Shen, J., Gao, X., and Zhou, J. (2019) Naringenin has a chemoprotective effect in MDA-MB-231 breast cancer cells via inhibition of caspase-3 and -9 activities. *Oncology Letters*, **17**, 1217–22.

Wang, X., Hasegawa, J., Kitamura, Y., *et al.* (2011) Effects of hesperidin on the progression of hypercholesterolemia and fatty liver induced by high-cholesterol diet in rats. *Journal of Pharmacological Sciences*, **117**, 129–38.

Wang, X., Ouyang, Y.Y., Liu, J., and Zhao, G. (2014) Flavonoid intake and risk of CVD: a systematic review and meta-analysis of prospective cohort studies. *British Journal of Nutrition*, **111**, 1–11.

Wardyn, J.D., Ponsford, A.H., and Sanderson, C.M. (2015) Dissecting molecular cross-talk between Nrf2 and NF-κB response pathways. *Biochemical Society Transactions*, **43**, 621–6.

Warner, E.F., Zhang, Q., Raheem, K.S., O'Hagan, D., O'Connell, M.A., and Kay, C.D. (2016) Common phenolic metabolites of flavonoids, but not their unmetabolized precursors, reduce the secretion of vascular cellular adhesion molecules by human endothelial cells. *Journal of Nutrition*, **146**, 465–73.

Warner, E.F., Smith, M.J., Zhang, Q., *et al.* (2017) Signatures of anthocyanin metabolites identified in humans inhibit biomarkers of vascular inflammation in human endothelial cells. *Molecular Nutrition & Food Research*, **61**, e00053.

Wikoff, W.R., Anfora, A.T., Liu, J., *et al.* (2009) Metabolomics analysis reveals large effects of gut microflora on mammalian blood metabolites. *Proceedings of the National Academy of Sciences of the USA*, **106**, 3698–703.

Williamson, G., Kay, C.D., and Crozier, A. (2018) The bioavailability, transport and bioactivity of dietary flavonoids: a review from a historical perspective. *Comprehensive Reviews in Food Science and Food Safety*, **17**, 1054–112.

Xu, J.W., Ikeda, K., Kobayakawa, A., *et al.* (2005) Downregulation of Rac1 activation by caffeic acid in aortic smooth muscle cells. *Life Sciences*, **76**, 2861–72.

Xu, L., Yang, Z.L., Li, P., and Zhou, Y.Q. (2009) Modulating effect of Hesperidin on experimental murine colitis induced by dextran sulphate sodium. *Phytomedicine*, **16**, 989–95.

Xu, X., Lei, T., Li, W., and Ou, H. (2019) Enhanced cellular cholesterol efflux by naringenin is mediated through inhibiting endoplasmic reticulum stress – ATF6 activity in macrophages. *Biochimica et Biophysica Acta. Molecular and Cell Biology of Lipids*, **1864**, 1472–82.

Zhao, Z., Jin, G., Ge, Y., and Guo, Z. (2019) Naringenin inhibits migration of breast cancer cells via inflammatory and apoptosis cell signaling pathways. *Immunopharmacology*, **27**, 1020–36.

Zilva, S.S. (1937) Vitamin P. *Biochemical Journal*, **31**, 915–19.

13

Understanding Polyphenols' Health Effects Through the Gut Microbiota

Maria V. Selma, Francisco A. Tomás-Barberán, Maria Romo-Vaquero, Adrian Cortés-Martín, and Juan C. Espín*

Laboratory of Food and Health, Research Group on Quality, Safety, and Bioactivity of Plant Foods, Dept. Food Science and Technology, CEBAS-CSIC, Murcia, Spain

13.1 Microbial Metabolism of Dietary Polyphenols

Phenolics are usually found conjugated to organic acids and sugars. Gut microbiota can hydrolyze glycosides, glucuronides, and organic acids, resulting in the release of the corresponding aglycones (Selma *et al.*, 2009). These aglycones are potentially more bioactive than their precursors. Deconjugated metabolites are absorbed from the colon and transformed into phase II conjugates in the intestinal tissues and in the liver, resulting in their sulfated and glucuronidated conjugates (Selma *et al.*, 2009; Yeh and Yen, 2003; Zhao *et al.*, 2004). Through the entero-hepatic recirculation, the liver excretes conjugated compounds into the gut as biliary constituents, and these compounds are deconjugated by microbial enzymes before being reabsorbed or further transformed. Deconjugation in the gut is catalyzed by fecal microbial enzymes (*β*-glucosidase, *α*-rhamnosidase, esterase, and *β*-glucuronidase) (Aura, 2008). The type of glycosylation (*O*- or *C*-glycoside) of polyphenols has a considerable impact on the microbial degradation rate. Gut microbiota also performs ring and lactone fission, dehydroxylation, reduction,

*Corresponding author.

Dietary Polyphenols: Metabolism and Health Effects, First Edition.
Edited by Francisco A. Tomás-Barberán, Antonio González-Sarrías, and Rocío García-Villalba.

decarboxylation, demethylation and isomerization reactions, among others (Aura, 2008; Braune and Blaut, 2016; Ozdal *et al.*, 2016; Selma *et al.*, 2009). Microbial transformations are different depending on the phenolic structure (flavonoid or nonflavonoid phenolics), polymerization degree, and spatial configuration.

Flavonoid type phenolics, including flavonols, flavanones, flavan-3-ols, isoflavones, and anthocyanins, contain a common structure consisting of two benzene rings (A and B) linked through a heterocyclic pyrone *C*-ring (Braune and Blaut, 2016; Selma *et al.*, 2009) (Figure 13.1). Some flavonoids are planar (anthocyanins, flavonols, flavones, and isoflavones) while others are not (flavan-3-ols and flavanones). Most of the flavonoids in foods occur as glycosides, mainly *O*-glycosides but also *C*-glycosides. Exceptions to this are flavan-3-ols, which are not conjugated but can form oligomeric and polymeric structures called proanthocyanidins or condensed tannins as a whole or procyanidins, prodelphinidins or propelargonidins when they are only composed of (epi)catechin, (epi)gallocatechin or (epi)azfelechin, respectively.

Gut microbiota is able to break down the flavonoid *C*-ring in different positions, releasing a higher number of simple phenolics derived from A and B rings (see Figure 13.1). The hydroxylation pattern of the B-ring, as well as its position, affect the type of phenolic compounds produced. Consequently, *C*-ring cleavages at 1-2 and 4-10 bonds or 1-2 and 3-4 bonds are produced in flavonols, flavan-3-ols, and proanthocyanidins, rendering hydroxyphenyl-propionic acids or hydroxyphenyl-acetic acids. In contrast, metabolites derived from flavanones and isoflavones suggest that *C*-ring cleavage is produced in these cases breaking the bonds between 1 and 2 or 4 and 10 positions but not between the 3 and 4 positions (Selma *et al.*, 2009). Particularly in isoflavones, the benzenoid B ring is in the 3-position. Consequently, *C*-ring cleavage occurs to break the bond between the oxygen in the pyrone ring and the *C*-2, only if a hydroxyl at 5 position is missing, as is the case of daidzein, and both the 1 and 2 positions and the 4 and 10 positions only if a hydroxyl at 5 position is present, as is the case of genistein, and this breakdown is necessary to produce *O*-demethylangolensin (*O*-DMA).

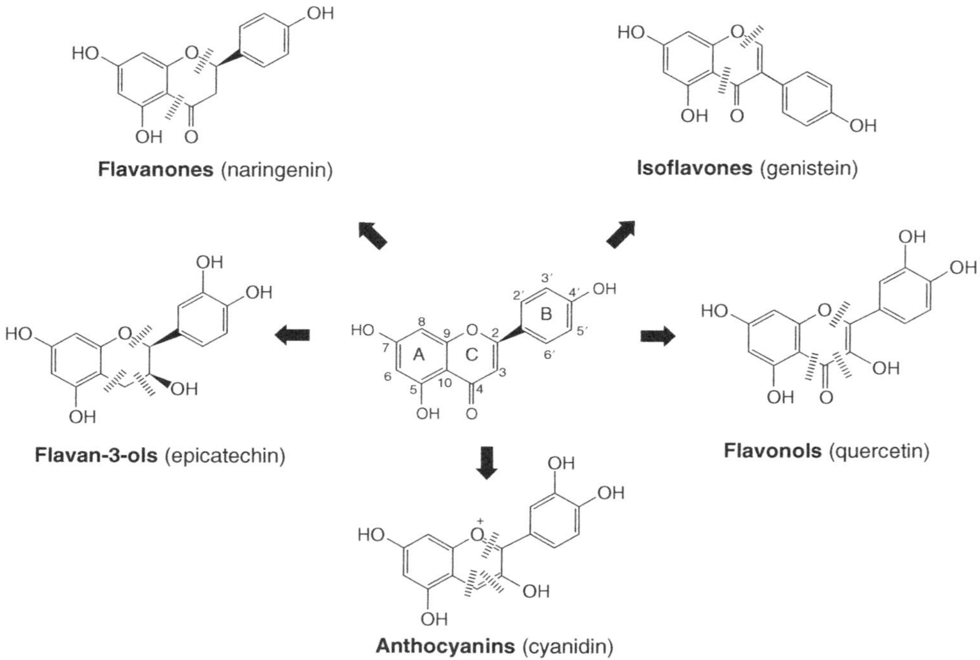

Figure 13.1 Potential heterocyclic *C*-ring cleavage of flavonoids by the gut microbiota. (IIIIII) Positions of the potential *C*-ring cleavages.

After C-ring-fission, flavonoids can be further metabolized by gut microbiota via demethylation and dehydroxylation reactions (Braune and Blaut, 2016). Simpler phenolic compounds derived from A and B rings are released as a product of this microbial degradation of flavonoids in the gut. The majority of these metabolites consist of acid or aldehyde phenolics with one, two or three hydroxyl and methyl ester radicals. Some examples are 3-(3,4-dihydroxyphenyl)-propionic acid from the flavonol quercetin and the flavanone naringenin, 5-(3′,4′-dihydroxyphenyl)-γ-valerolactone from the flavan-3-ol catechin, 1,3,5-trihydroxybenzene (phloroglucinol) from the anthocyanin kuromanin as well as O-DMA and equol from the isoflavone daidzein (Figure 13.2). Apart from the C-ring cleavage and the dehydroxylation, some isoflavones such as glycitein are demethoxylated and demethylated (Simons *et al.*, 2005).

Nonflavonoid phenolics are a phenolic group with higher heterogeneity than flavonoids in both structure and polymerization level (Figure 13.3). Among them, benzoic acids, hydroxycinnamates, and stilbenes are simpler groups than lignans which are in

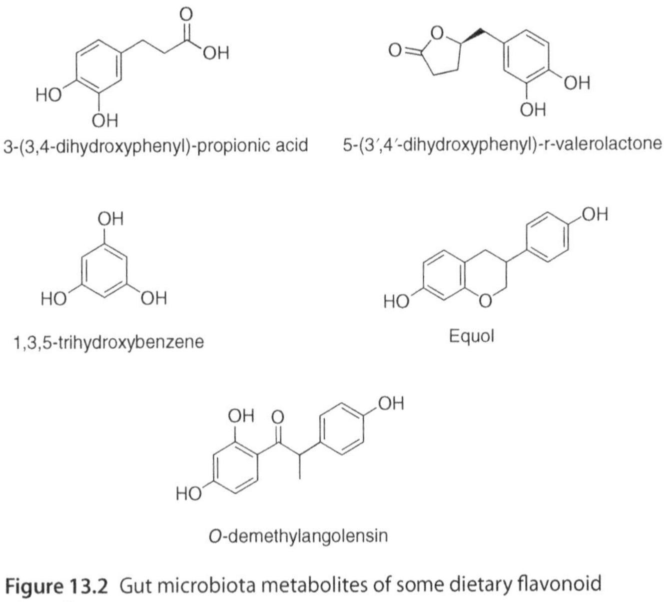

Figure 13.2 Gut microbiota metabolites of some dietary flavonoid phenolics.

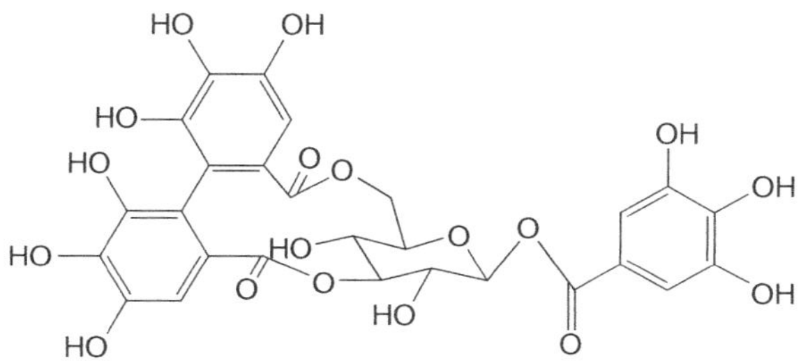

Hydrolyzable tannins (corilagin)

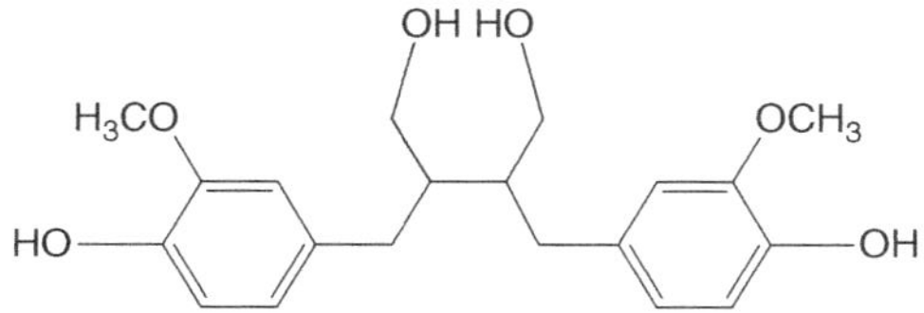

Lignans (secoisolariciresinol)

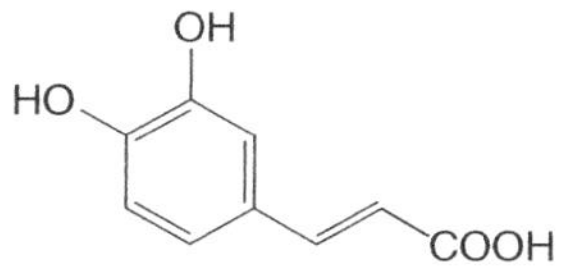

Hydroxycinnamates (caffeic acid)

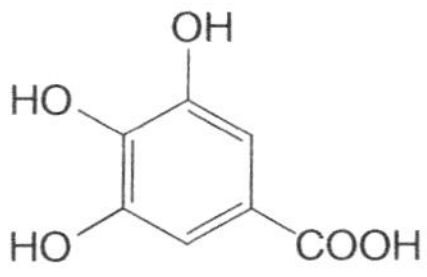

Benzoic acids (gallic acid)

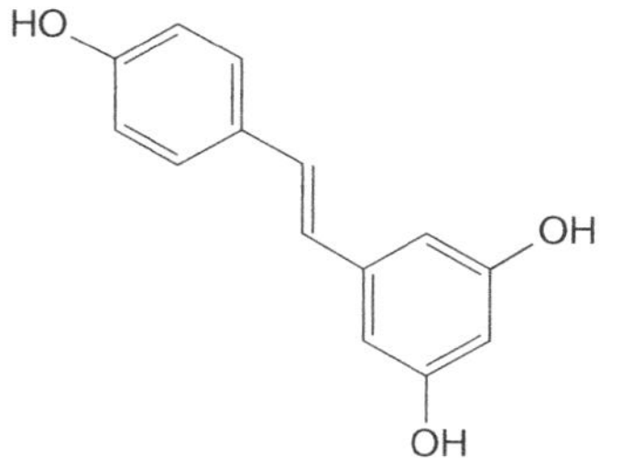

Stilbenes (resveratrol)

Figure 13.3 Dietary nonflavonoid phenolics that are metabolized by the gut microbiota.

turn simpler than hydrolyzable tannins. Nonflavonoid phenolics are an example of the enormous diversity of phenolic structures commonly found in plant foods. Depending on their chemical complexity, they are more or less absorbed in the small intestine, and consequently, the gut microbiota has less or more opportunity to metabolize them.

Hydrolyzable tannins are the most complex phenolics, and they include gallotannins and ellagitannins. Gallotannins are composed of monomeric units of gallic acid, whereas ellagitannins (ETs) are composed of ellagic acid (EA) monomers. Deconjugation of some ETs (*C*-glycosidic ETs such as castalagin and vescalagin) has not been described yet, but they are nevertheless, for historical reasons, classified as hydrolyzable tannins (Mingshu *et al.*, 2006). It is known, however, that flavone *C*-glycosides are degraded by gut bacteria to release the flavone aglycone (Braune and Blaut, 2016). Something similar could happen in ellagitannin-*C*-glycosides, but further research is needed to demonstrate this. In contrast, tannin-*O*-glycosides such as punicalagin and pedunculagin are largely metabolized by the gut microbiota of humans and other mammals.

Gut microbial enzymes involved in tannin hydrolysis are known as tannases (tannin acyl hydrolase) which hydrolyze the ester bonds present in gallotannins and ETs although they do not affect the *C-C* bonds (Aguilar-Zarate *et al.*, 2014; Jimenez *et al.*, 2014; Tomás-Barberán *et al.*, 2017) (see Figure 13.3). After hydrolysis of ellagitannin-O-glycosides by tannase activity, the released hexa-hydroxydiphenic acid suffers a double lactonization to form EA. The gut microbiota further transforms gallic acid from gallotannins and EA from ETs. Gallic acid can undergo decarboxylation and dehydroxylations. Similarly, EA can dehydroxylate to render EA metabolites called nasutins in which two hydroxyls have been removed (Tomás-Barberán *et al.*, 2017). However, EA is also further transformed by lactone ring cleavage, decarboxylation, and dehydroxylation reactions to render the EA metabolites called urolithins (Figure 13.4) (Tomás-Barberán *et al.*, 2017).

In the conversion pathway from EA to urolithins, it seems that the first step is the opening of one of the lactone rings by lactonases to yield luteic acid that is further transformed by decarboxylases from the gut microbiota to produce

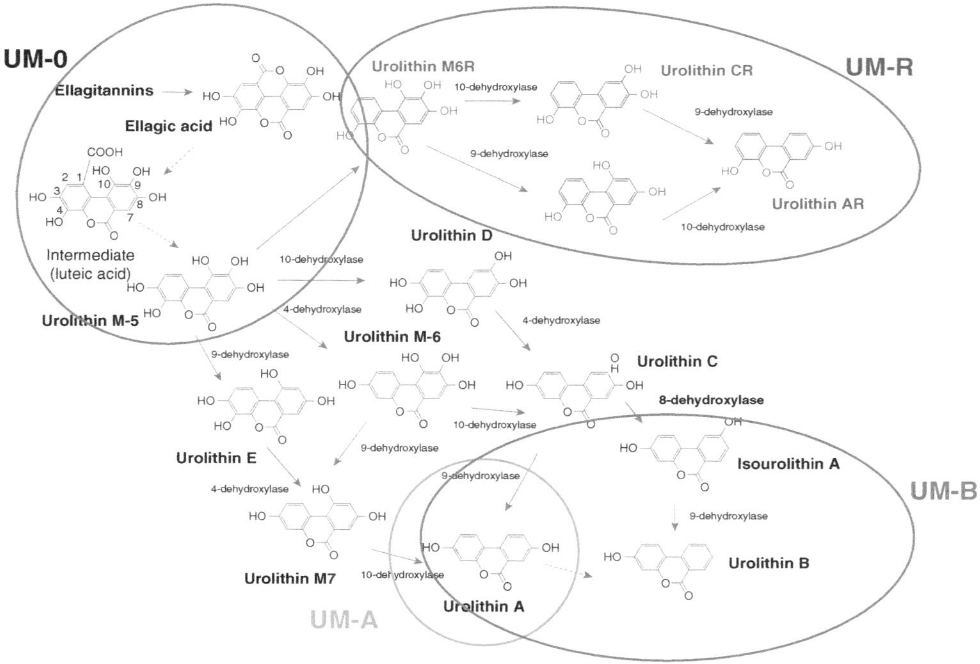

Figure 13.4 Microbial conversion of ellagic acid and urolithins produced by each metabotype (UM-A, UM-B, and UM-0).

pentahydroxy-urolithin (Uro-M5), which is a key intermediate in the production of the different urolithins. From Uro-M5, consecutive dehydroxylations render tetrahydroxy-urolithins (Uro-D, Uro-M6, and Uro-E) and trihydroxy-urolithins (Uro-C and M7) that finally lead to the main metabolites detected *in vivo* which are the dihydroxy-urolithins urolithin-A (Uro-A) and Isourolithin-A (IsoUro-A) and the hydroxy-urolithin, which is known as urolithin-B (Uro-B) (Tomás-Barberán *et al.*, 2017) (see Figure 13.4).

Lignans with a structure somewhat similar to urolithins include a number of diphenolic compounds with a 1,4-diarylbutane structure such as pinoresinol, secoisolariciresinol, matairesinol, lariciresinol, isolariciresinol, and syringaresinol (Possemiers *et al.*, 2007) (see Figure 13.3). Dietary lignans are transformed by gut microbes to render enterolignans (enterodiol and enterolactone) (Figure 13.5), which are mammalian phytoestrogens with agonist and antagonist properties (Clavel *et al.*, 2005). Microbial activities on lignans include reductions,

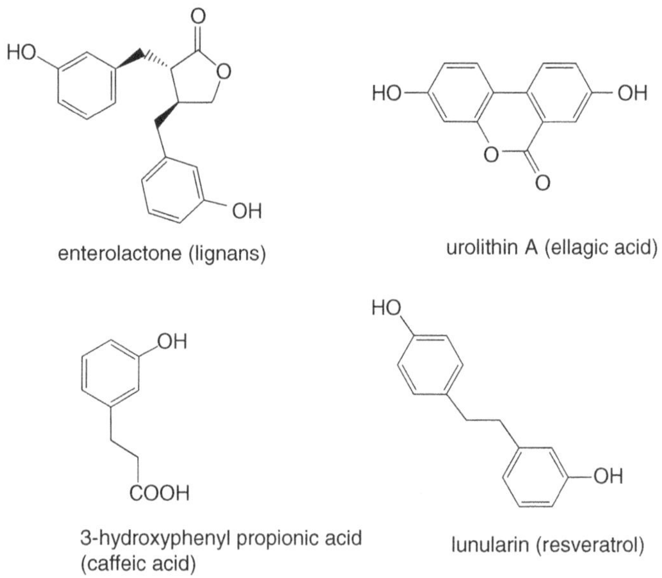

Figure 13.5 Gut microbiota metabolites of some dietary nonflavonoid phenolics.

demethylation followed by dehydroxylation and lactonization reactions (Bess *et al.*, 2020). For example, pinoresinol and lariciresinol are transformed by the gut microbiota via four distinct types of chemical reactions: benzyl ether reduction, guaiacol demethylation, catechol dehydroxylation, and diol lactonization (Bess *et al.*, 2020). Therefore, a complex metabolic pathway, involving several precursors, different intermediary metabolites, diverse conjugation patterns and different bacterial species, is needed to yield the final product enterolactone from dietary lignans (see Figure 13.5).

Stilbenes, based on a C6-C2-C6 polyphenolic structure, are nonflavonoid phenolics also transformed by the gut microbiota (see Figure 13.3). Among stilbenes, *trans*-resveratrol is the main compound studied due to its acknowledged health benefits (Chaplin *et al.*, 2018; Larrosa *et al.*, 2009; Tomé-Carneiro *et al.*, 2013). Gut microbiota has deglycosylation activity able to transform resveratrol *O*-glucosides such as *trans*-piceid (trans-resveratrol-3-*O*-β-D-glucopyranoside) into the resveratrol aglycone (Bode *et al.*, 2013). Conversely, resveratrol can be glycosylated in the gut to produce piceid again. Piceid and resveratrol are absorbed and extensively conjugated, being the sulfate and glucuronide derivatives of the primary circulating metabolites. On the other hand, gut bacteria metabolize resveratrol and its precursors, including piceid, resulting in certain resveratrol derivatives by means of reduction activities. Dihydroresveratrol was the first derivative identified followed by two other bacterial *trans*-resveratrol metabolites, 3,4'-dihydroxy-*trans*-stilbene and 3,4'-dihydroxydihydro-stilbene (lunularin) (see Figure 13.5) (Bode *et al.*, 2013). Piceid is also metabolized by the gut microbiota to produce dihydropiceid and dihydroresveratrol (Wang *et al.*, 2012). As resveratrol, its microbial metabolites are also conjugated to its sulfate and glucuronide forms.

Hydroxycinnamates are one of the simplest group of nonflavonoid compounds (see Figure 13.3). The most common dietary C6-C3 hydroxycinnamates include *p*-coumaric acid (4-hydroxycinnamic acid), caffeic acid (3,4-dihydroxycinnamic acid), ferulic acid (4-hydroxy 3-methoxycinnamic acid), and sinapic acid (4-hydroxy 3,5-dimethoxycinnamic acid), and their esters with quinic, malic, and tartaric acids (chlorogenic, caftaric, etc.) among others. Hydroxycinnamates, esterified or not with sugars and hydroxylated acids (quinic, tartaric, malic), are substrates of the gut microbiota. Gonthier and colleagues (2006) showed that caffeic acid and its esters, caftaric acid and chlorogenic acid, were transformed into 3-(3′-hydroxyphenyl)propionic acid and minor quantities of benzoic acid. Generally, the free acid is released from esters by bacterial esterases (e.g., cinnamoyl esterase) (Andreasen *et al.*, 2001), and then metabolized to reduce the double bond to give rise to phenylpropionic acids (see Figure 13.5). The reduction of the double bond or the opening and decarboxylation of the quinic acids can also occur before the action of the esterases (Tomás-Barberan *et al.*, 2014a). The side-chain shortening of phenylpropionic acid takes place through β-oxidation (Gonthier *et al.*, 2006; Ozdal *et al.*, 2016) at different levels of the degradation pathway. Finally, hydroxycinnamate metabolites are decarboxylated to produce hydroxybenzenes such as catechol (1,2-dihydroxyphenol). Dehydroxylation usually occurs before reduction and decarboxylation to remove the hydroxyl at C4-position of the caffeic acid residue.

Although there are potentially thousands of different phenolic compounds in the diet, they are typically transformed into a much smaller number of final metabolites. Hydroxy-benzoic acids are the common microbial degradation metabolites obtained in the gut from flavonoid and non-flavonoid phenolics. These compounds are common microbial metabolites produced from different phenolic groups but also are commonly found in most fruits. These phenolics are further transformed by microbial decarboxylase enzymes when a free hydroxyl group is present in the 4-position. This is the case for gallic acid (3,4,5-trihydroxybenzoic acid), protocatechuic acid (3,4-dihydroxybenzoic), and vanillic acid (3-methoxy-4-hydroxybenzoic acid) which are transformed to pyrogallol (1,2,3-trihydroxyphenol), catechol (1,2-dihydroxyphenol), and *O*-methylcatechol, respectively (Selma *et al.*, 2009). These metabolites are rapidly absorbed in the GI tract but can also be conjugated with glycine, increasing urinary hippuric acid concentration (Torrallardona *et al.*, 2007).

13.2 Bacteria Responsible for Dietary Polyphenols Transformations and Health Implications

The intestinal bacteria play a crucial role in the conversion of dietary polyphenols and thereby affect their health-promoting effects in the human host. This was demonstrated by the fact that germ-free or antibiotic-treated animals did not produce the polyphenolic metabolites (Griffiths and Barrow, 1972). The identification of bacteria involved in intestinal polyphenol conversion has gained increasing interest. Several studies have been performed to identify the microorganisms implicated in phenolic transformations (see references in Table 13.1 and 13.2). However, bacterial species implicated in the transformation of many phenolic compounds are still unknown.

Organic acids and sugars of conjugated flavonoids and nonflavonoids are hydrolyzed by the gut microbiota to render phenolic aglycones. Bacterial species reported to catalyze deconjugation reactions are listed in Table 13.1. Species from different phyla (Firmicutes, Bacteroidetes, Actinobacteria, and Proteobacteria) are able to hydrolyze *O*-glycosides from the flavonoid and nonflavonoid groups. Interestingly, several *Lactobacillus* and *Bifidobacterium* spp., highly recognized for their beneficial properties, as well as *Enterococcus* are involved in *O*-glycoside hydrolysis. Glycosylated phenolics can serve as the sole carbon and energy source of microorganisms, with the attached sugar moieties being preferentially fermented (Braune and Blaut, 2016). This could explain the microbial modulation effects of different phenolic glycosides observed for these and other bacterial groups of the gut microbiota.

Some polyphenols carry sugar moieties other than glucose, such as rutinoside (6-β-L-rhamnosyl-D-glucose) or neohesperidoside (2-β-L-rhamnosyl-D-glucose). Gut microorganisms acting on these phenolics possess α-rhamnosidases in addition to β-glucosidases (Braune and Blaut, 2016; Selma *et al.*, 2009). This is the case for some species of *Lactobacillus* such as *L.*

Table 13.1 Hydrolysis of conjugated phenolics achieved by the human intestinal microbiota

Hydrolysis	Substrate	Phyla	Species	References
O-glycosides	**Flavonoids**: Flavonols, flavanones, dihydrochalcones, isoflavones, anthocyanins **Nonflavonoids**: Ellagitannins, lignans, stilbenes	Firmicutes:	*Blautia* sp. *Clostridium sphenoides, C. saccharogumia, C. cocleatum* *Eubacterium rectale, E. ramulus, E. cellulosolvens* *Enterococcus casseliflavus, E. avium, E. faecalis* *Lactobacillus acidophilus, L. plantarum, L. casei, L. leichmanii* *Lactococcus lactis, Lactococcus* sp.	(Amaretti *et al*., 2015; Ávila *et al*., 2009a; Bang *et al*., 2015; Bokkenheuser *et al*., 1987; Braune *et al*., 2005; Braune and Blaut, 2011, 2012; Clavel *et al*., 2006; Hur *et al*., 2000; Ilett *et al*., 1990; Kim *et al*., 2014, 2015; Liu *et al*., 2003; Marotti *et al*., 2007; Miyake *et al*., 1997; Raimondi *et al*., 2009; Schneider *et al*., 1999; Schneider and Blaut, 2000; Schoefer *et al*., 2002; Shin *et al*., 2016; Tamura *et al*., 2006; Tsangalis *et al*., 2002; Zhao *et al*., 2014)
		Bacteroidetes:	*Bacteroides ovatus, B. fragilis, B. distasonis, B. uniformis* *Parabacteroides distasonis*	
		Actinobacteria:	*Bifidobacterium angulatum, B. animalis, B. catenulatum, B. pseudocatenulatum, B. longum, B. pseudolongum, B. infantis, B. breve, B. dentium, B. adolescentis, B. bifidum*	

		Proteobacteria:	*Enterobacter cloacae*	
C-glycosides	**Flavonoids**: Flavones, isoflavones	Firmicutes:	*Enterococcus* sp.	(Braune and Blaut, 2011, 2012; Jin *et al.*, 2008; Kim *et al.*, 2015; Xu *et al.*, 2014)
			Eubacterium cellulosolvens *Lactococcus* sp.	
Glucuronides	**Nonflavonoids**: Ellagitannins	Proteobacteria:	*Escherichia coli*	(Ilett *et al.*, 1990)
Ester	**Nonflavonoids**: Hydroxycinnamates	Firmicutes:	*Lactobacillus gasseri*	(Andreasen *et al.*, 2001; Couteau *et al.*, 2001)
		Actinobacteria:	*Bifidobacterium lactis*	
		Proteobacteria:	*Escherichia coli*	

Table 13.2 Enzymatic reactions of polyphenolic aglycones achieved by the human intestinal microbiota

Reaction	Substrate	Phyla	Species	References
C-ring cleavage	**Flavonoids**: Flavonols, flavanones, dihydrochalcones, flavan-3-ols, isoflavones, anthocyanins	Firmicutes:	*Butyrivibrio* sp. *Clostridium butyricum, C. orbiscindens* *Eubacterium ramulus, E. oxidoreducens* *Flavonifractor plautii* *Lactobacillus plantarum*	(Braune *et al.*, 2001; Braune and Blaut, 2011; Cheng *et al.*, 1969, 1971; Hur *et al.*, 2002; Jin and Hattori, 2012; Krumholz and Bryant, 1986; Kutschera *et al.*, 2011; Miyake *et al.*, 1997; Sanchez-Patan *et al.*, 2012; Schneider *et al.*, 1999; Schneider and Blaut, 2000; Schoefer *et al.*, 2002, 2003; Takagaki and Nanjo, 2015; Tzounis *et al.*, 2008; Wang *et al.*, 2001, 2004; Winter *et al.*, 1989, 1991; Yokoyama *et al.*, 2010)
		Actinobacteria:	*Adlercreutzia equolifaciens* *Eggerthella* spp. *SDG-2, E. lenta* *Slackia equolifaciens*	
Lactone cleavage	**Nonflavonoids**: Ellagitannins	Actinobacteria:	*Gordonibacter urolithinfaciens, G. pamelaeae* *Ellagibacter isourolithinifaciens*	(Beltrán *et al.*, 2018; Selma *et al.*, 2014a,b, 2017)
Lactonization	**Nonflavonoids**: Lignans	Firmicutes:	*Lactonifactor longoviformis*	(Bess *et al.*, 2020)

Dehydroxylation	**Flavonoids**: Flavanones, flavan-3-ols, **Nonflavonoids**: Ellagitannins, lignans, hydroxycinnamates	Firmicutes:	*Clostridium butyricum* *Eubacterium ramulus* *Flavonifractor plautii*	(Beltrán *et al.*, 2018; Bess *et al.*, 2020; Clavel *et al.*, 2005, 2006; Jin and Hattori, 2012; Selma *et al.*, 2014a,b, 2017; Takagaki and Nanjo, 2015; Wang *et al.*, 2001; Zhao *et al.*, 2014)
		Actinobacteria:	*Adlercreutzia equolifaciens* *Eggerthella lenta* *Gordonibacter urolithinfaciens, G. pamelaeae* *Ellagibacter isourolithinifaciens*	
Demethylation	**Flavonoids**: Flavanones, flavonols, flavan-3-ols, anthocyanins **Nonflavonoids**: Lignans	Firmicutes:	*Blautia producta*, *Blautia* spp. *Butyribacterium methylotrophicum* *Clostridium* *Eubacterium limosum, E. callanderi* *Lactobacillus* *Peptostreptococcus productus* *Streptococcus*	(Bess *et al.*, 2020; Clavel *et al.*, 2005, 2006; Hur and Rafii, 2000; Keppler and Humpf, 2005; Kim *et al.*, 2014; Possemiers *et al.*, 2005, 2006, 2008)

(*continued*)

Table 13.2 (Continued)

Reaction	Substrate	Phyla	Species	References
Decarboxylation	**Nonflavonoids**: Ellagitannins	Actinobacteria:	*Gordonibacter urolithinfaciens*, *G. pamelaeae* *Ellagibacter isourolithinifaciens*	(Beltrán *et al.*, 2018; Selma *et al.*, 2014a,b, 2017; Torrallardona *et al.*, 2007)
Reductions	**Flavonoids**: Isoflavones **Nonflavonoids**: Lignans, stilbenes	Actinobacteria:	*Adlercreutzia equolifaciens* *Bifidobacterium animalis*, *B.longum*, *B. pseudolongum* *Eggerthella lenta* *Slackia equolifaciens*, *S. isoflavoniconvertens*	(Bess *et al.*, 2020; Bode *et al.*, 2013; Jin *et al.*, 2008, 2010; Maruo *et al.*, 2008; Matthies *et al.*, 2009; Shimada *et al.*, 2010; Tamura *et al.*, 2006; Tsangalis *et al.*, 2002; Tsuji *et al.*, 2010; Walle *et al.*, 2004; Wang *et al.*, 2005; Yokoyama and Suzuki, 2008)

plantarum, or *Bifidobacterium* such as *B. pseudocatenulatum* (see Table 13.1). Screening approaches reveal that the rhamnosidase activity may be strain specific (Amaretti *et al.*, 2015; Ávila *et al.*, 2009b). *C*-glycoside hydrolysis has been less investigated than *O*-glycosides hydrolysis, and only some species of the Firmicutes phylum have been related (*Enterococcus* sp., *Eubacterium cellulosolvens*, and *Lactococcus* sp.). In contrast, species from different phyla are involved in ester hydrolysis, including Firmicutes, Bacteroidetes, Actinobacteria, and Proteobacteria (see Table 13.1).

After deconjugation, polyphenols are further transformed by the gut microbiota, which catalyzes reactions of *C*-ring and lactone fission, lactonization, dehydroxylation, demethylation, decarboxylation, reduction of carbon–carbon double bonds, and isomerization (Braune and Blaut, 2016; Selma *et al.*, 2009). Bacterial species or strains reported to catalyze these reactions are listed in Table 13.2. Interestingly, bacterial species of only two phyla (Firmicutes and Actinobacteria) have been related to these aglycone transformations. Furthermore, bacterial species implicated within the Actinobacteria phylum fundamentally belong to the family Eggerthellaceae although they were previously considered members of the Coriobacteriaceae family.

Thus, the class Coriobacteriia containing the family Coriobacteriaceae has been divided into the Eggerthellales ord. nov. (including the family Eggerthellaceae fam. nov.) and the emended order Coriobacteriales (including the emended families Coriobacteriaceae and Atopobiaceae fam. nov.) (Gupta *et al.*, 2013). Therefore, because of the division of the class Coriobacteriia, bacterial species described as polyphenol transformers have been regrouped within the Eggerthellales ord. nov. while phylogenetic neighbors not associated with polyphenol conversion (*Atopobium*, *Olsenella*, and *Collinsella*) are clustered in the emended order Coriobacteriales. The class Coriobacteriia and its three families are prevalent and dominant members of the human gut microbiota and can metabolize bile acids, steroid hormones, lipids, and xenobiotics such as polyphenols (Clavel *et al.*, 2016).

Further research should be performed to elucidate the potential health benefits of species from the class Coriobacteriia and especially the Eggerthellaceae family due to their ability to produce bioactive metabolites from several dietary polyphenols. As an example, two neurodegenerative disorders, Alzheimer's disease and multiple sclerosis, have recently been associated with a depletion of *Adlercreutzia equolifaciens*, an equol-producing bacterium from the Eggerthellaceae family (Chen *et al.*, 2016; Haran *et al.*, 2019). In contrast, some cardiovascular disease markers such as total cholesterol, low-density lipoprotein cholesterol (LDLc), and non-high-density lipoprotein cholesterol (non-HDLc) have been positively correlated with other Eggerthellaceae genera (*Slackia*) in overweight-obese individuals (Romo-Vaquero *et al.*, 2019). *Collinsella*, a genus also from the class Coriobacteriia but the Coriobacteriaceae family, has been positively correlated with serum cholesterol and bile acid deconjugation and impaired glucose outside and inside pregnancy (Gómez-Arango *et al.*, 2016).

As a result of the great diversity of gut bacterial genera involved in the metabolism of polyphenols and the interindividual differences in their metabolism, recent studies have given rise to the so-called polyphenol-metabolizing phenotypes or polyphenol metabotypes (Espín *et al.*, 2017; Tomás-Barberán *et al.*, 2018). As in the case of polyphenol metabotypes, enterotypes, driven by *Bacteroides*, *Prevotell*a, and *Ruminococcus*, were also proposed to cluster the population according to their microbiome and were suggested to be functional markers (Arumugam *et al.*, 2011). However, BMI, gender, or age cannot explain the observed enterotypes (Arumugam *et al.*, 2011; Zhou *et al.*, 2014). The enterotype concept is still under discussion, and the latest studies suggest the existence of microbial community gradients rather than distinct enterotypes as previously suggested (Costea *et al.*, 2017).

In contrast to enterotypes, polyphenol metabotypes (Espín *et al.*, 2017) and particularly the gut microbiota-mediated metabolism of the polyphenols ellagitannins and ellagic acid to yield urolithins (urolithin metabotypes, UMs) have been associated with the gut microbiota composition and the individual status as described below. Three UMs have been defined to date depending on the final urolithins produced;

metabotype A (UM-A) is characterized by the production of only urolithin A (Uro-A), metabotype B (UM-B) yields urolithin B (Uro-B) and IsoUro-A in addition to Uro-A, and metabotype 0 (UM-0) does not produce these final urolithins (see Figure 13.4) (Tomás-Barberán *et al.*, 2014b). Furthermore, three new urolithins (M7R, CR, and AR) have been described, which are not specific to any of the known UM-A and UM-B (García-Villalba *et al.*, 2019).

The human gut microbial ecology associated with UMs and the divergence between the enterotypes and UMs have also been identified (Romo-Vaquero *et al.*, 2019). The distribution of UMs in the healthy population and their association with aging has recently been demonstrated in a large cohort (n = 839) aged from 5 to 90 years (Cortés-Martín *et al.*, 2018). Other studies have shown that overweight-obese individuals belonging to UM-B exhibited increased cardiovascular disease risk (higher blood total cholesterol, LDL-cholesterol, and oxidized LDL levels, among others), whereas UM-A metabotype seemed to be protective against CVD risk factors (González-Sarrías *et al.*, 2018; Selma *et al.*, 2018). Preliminary observations have also associated UM-B with obesity and gastrointestinal pathologies driven by a gut microbial imbalance (dysbiosis) (Selma *et al.*, 2016; Tomás-Barberán *et al.*, 2014b). Although the association between UMs and obesity has not been unequivocally confirmed yet (Cortés-Martín *et al.*, 2018), a strong positive correlation of BMI with Coriobacteriaceae whose abundance was increased in UM-B has been shown (Romo-Vaquero *et al.*, 2019). These findings agree with the previously reported health implications of the Coriobacteriia class in lipid metabolism (Clavel *et al.*, 2016; Martínez *et al.*, 2013). Remarkably, a higher gut microbiome richness (Chao 1 index) and proinflammatory microorganisms, including methanogenic Archaea as well as the Gammaproteobacteria class, are increased in UM-B versus UM-A, which suggests that UM-B is a gut dysbiosis-prone metabotype (Romo-Vaquero *et al.*, 2019).

Therefore, variations in the gut microbiota ecology among individuals can affect polyphenol metabolism capacity, but at the same time, polyphenol gut metabotypes can indirectly reflect the individual's gut microbiome (biomarkers of gut microbiota) and health status (biomarkers of health status).

Therefore, clustering of individuals according to polyphenol metabotypes (metabotyping) could be a useful tool to predict the functionality and status of gut microbiota and to better understand the effects of bioactive compounds and the large interindividual variation observed.

The clustering of individuals according to their ability to metabolize other polyphenols such as isoflavones up to equol has also been established (equol metabotypes) (Hazim *et al.*, 2016). In contrast, no differential metabotypes have been identified in the gut metabolism of procyanidins and flavan-3-ols (Cortés-Martín *et al.*, 2019). Interestingly, with the exception of some *Bifidobacterium* (*B. animalis, B. longum, B. pseudolongum*), the rest of the bacterial genera and species described as responsible for the reduction of isoflavones to equol (*Slackia equolifaciens, S. isoflavoniconvertens, Eggerthella equolifaciens, Adlercreutzia equolifaciens*) are also capable of the dehydroxylation (*A. equolifaciens*) and *C*-ring cleavage (*A. equolifaciens, S. equolifaciens*) of flavan-3-ols (Braune and Blaut, 2016). However, other bacterial species that also have *C*-ring cleavage activity (*Eggerthella lenta* and *Lactobacillus plantarum*) are involved in the metabolism of flavan-3-ols. This divergence in the bacteria implicated in isoflavone and flavan-3-ol metabolism versus isoflavone metabolism could justify the nonexistence of procyanidin metabotypes versus equol metabotypes. Furthermore, in contrast to equol production, microbial metabolism of the flavan-3-ols has more than one potential conversion reaction (dehydroxylation, *C*-ring cleavage in three different positions and dehydroxylation of the *C*-ring cleavage product) and therefore can lead to many different metabolites. Therefore, the clustering of individuals is less feasible. Gut metabotypes of other polyphenol groups are unexplored and deserve further research.

13.3 Modulation of Gut Microbiota by Dietary Polyphenols

Gut microbiota is composed of 100 trillion microbial cells existing symbiotically with the human host. A healthy, diverse, and stable gut microbiota is essential in maintaining the homeostatic balance of barrier integrity in human metabolism, as well as regulating the gut–brain axis (Portune *et al.*, 2016; Rowland *et al.*, 2018). Recent publications have highlighted the influence of gut microbiota on different diseases such as obesity, inflammation, metabolic syndrome (Boulangé *et al.*, 2016), and neurodegenerative diseases such as Parkinson's (Sun and Shen, 2018) and Alzheimer's (Haran *et al.*, 2019). Among intestinal bacteria, bifidobacteria and lactobacilli that include many probiotic strains, besides other gut bacteria such as *Faecalibacterium prausnitzii* and *Bacteroides thethaiotamicron*, are considered beneficial in human health, helping to improve gut barrier function and host immune system, prevent diarrhea or allergies, activate provitamins, and modulate lipid metabolism (Boulangé *et al.*, 2016). However, not all microbial groups of the intestine are favorable. For this reason, it is crucial to avoid a gut microbial imbalance (dysbiosis) to maintain the host–microbiota cooperation and functional stability.

Gut microbiota composition depends on several factors such as diet, treatment with antibiotics or other medications, physical activity, type of delivery, and breastfeeding, among others (Collado *et al.*, 2015; Dueñas *et al.*, 2015). Functional foods represent an interesting option to modulate the gut microbiota. Diet modulates gut ecology via microbial fermentation. In this connection, the prebiotic effects of dietary fiber have primarily been investigated. Gut bacteria produce short-chain fatty acids (SCFA) such as acetate, propionate, and butyrate through dietary fiber fermentation. Butyrate exerts health benefits, including the preservation of the gut barrier function and integrity by stimulation of the formation of mucin, antimicrobial peptides, and tight-junction proteins (Rivière *et al.*, 2016). It also induces the differentiation of colonic regulatory T cells, which have a central role in the suppression of inflammatory responses (Rivière *et al.*, 2016).

Apart from dietary fiber, several recent studies also indicate that dietary polyphenols are relevant in the modulation of gut microbiota and that these microorganisms convert polyphenols into active and bioavailable metabolites (Dueñas *et al.*, 2015; Mosele *et al.*, 2015; Ozdal *et al.*, 2016; Selma *et al.*, 2009; Tomás-Barberán *et al.*, 2016). Consequently, the concept of prebiotics, previously restricted to carbohydrates, has been updated and a new and broader definition allows inclusion of polyphenols as potential prebiotics: "A prebiotic is a non-digestible compound that, through its metabolization by microorganisms in the gut modulates composition and/or activity of the gut microbiota, thus conferring a beneficial physiological effect on the host" (Bindels *et al.*, 2015). Prebiotic effects of different polyphenol-rich foods have been included in different review publications (Dueñas *et al.*, 2015; Mosele *et al.*, 2015; Ozdal *et al.*, 2016; Selma *et al.*, 2009; Tomás-Barberán *et al.*, 2016). As an example, polyphenols, and particularly hydrolyzable and condensed tannins, promote the growth of lactobacilli and bifidobacteria in different *in vitro* models, and in preclinical and clinical studies with polyphenol-rich foods, including cocoa, tea, grapes, wine, berries, pomegranates, and nuts (Dueñas *et al.*, 2015).

Polyphenol consumption is increasingly proposed as a strategy in the prevention and treatment of different diseases such as metabolic and neurological disorders. This is explained not only by their ability to reduce inflammation and endotoxemia, among other effects, but also by their capacity to modulate the gut microbiota, which makes them promising nutraceuticals against metabolic and brain disorders (Clavel *et al.*, 2016; Serra *et al.*, 2018). However, recent investigations have shown large interindividual differences in the health effects of dietary polyphenols because of the specific gut microbiome and bioactive microbial metabolome of each individual. Population clustering by polyphenol metabotypes is being suggested as an explanation for the interindividual variability observed in the health effects of plant bioactives (Espín *et al.*, 2017; González-Sarrías *et al.*, 2017; Manach *et al.*, 2017; Tomás-Barberán *et al.*, 2016, 2018). Indeed, volunteer metabotyping would be necessary in clinical trials because specific metabotypes harbor a particular microbial ecology, which could be related to the final health effects along with the specific bioactive metabolites produced. Metabotypes can also reflect the gut microbiota composition and metabolic status, and consequently could be biomarkers of potential polyphenol health effects. This is illustrated in a study in which UMs were associated with the host blood lipid profile but also with the interindividual variability in the improvement of cardiovascular risk biomarkers after consumption of pomegranate ellagitannins (González-Sarrías *et al.*, 2017; Selma *et al.*, 2018).

Population-level metabotyping could also be useful to elucidate the capacity of individuals to modify their microbiota and metabotype through polyphenol consumption. Indeed, in a recent study, urolithin nonproducers (UM-0) became producers (UM-A or UM-B) following ET-rich food consumption (González-Sarrías *et al.*, 2017). Therefore, the ability to produce urolithins (UM-A and UM-B) vs nonproducers (UM-0) could be partially related to the consumption of polyphenol-rich diets.

Population clustering by polyphenol metabotypes such as those from equol and urolithins suggests a personalized effect of the nutraceuticals associated with the gut microbiota inherent in the individual. This might explain the lack of consensus about health benefits associated with dietary polyphenol consumption. Therefore, intervention studies to evaluate the effect of dietary polyphenols should include microbiota-associated metabotypes together with the other factors behind the interindividual variability observed in clinical trials such as genetic and environmental factors (diet, lifestyle, medication, etc.) to allow us to move towards personalized nutrition.

Acknowledgments

This research was supported by the Projects AGL-2015-64124-R, AGL-2015-73106-EXP (MINECO, Spain), 19900/GERM/15 and 20880/PI/18 (Fundación Séneca de la Región de Murcia, Spain), and 201870E014, 201770E081 and 201870I028 (CSIC, Spain). ACM is the holder of a FPI predoctoral grant from MINECO (Spain).

References

Aguilar-Zarate, P., Cruz-Hernandez, M.A., Montañez, J.C., Belmares-Cerda, R.E., and Aguilar, C.N. (2014) Tannins structure. *Revista Mexicana de Ingeniería Química*, **13**, 63–74.

Amaretti, A., Raimondi, S., Leonardi, A., Quartieri, A., and Rossi, M. (2015) Hydrolysis of the rutinose-conjugates flavonoids rutin and hesperidin by the gut microbiota and bifidobacteria. *Nutrients*, 7, 2788–800.

Andreasen, M.F., Kroon, P.A., Williamson, G., and Garcıa-Conesa, M.T. (2001) Esterase activity able to hydrolyze dietary antioxidant hydroxycinnamates is distributed along the intestine of mammals. *Journal of Agricultural and Food Chemistry*, **49**, 5679–84.

Arumugam, M., Raes, J., Pelletier, E., *et al.* (2011) Enterotypes of the human gut microbiome. *Nature*, **473**, 174–80.

Aura, A.M. (2008) Microbial metabolism of dietary phenolic compounds in the colon. *Phytochemistry Reviews*, 7, 407–29.

Ávila, M., Hidalgo, M., Sánchez-Moreno, C., Pelaez, C., Requena, T., and de Pascual-Teresa, S. (2009a) Bioconversion of anthocyanin glycosides by Bifidobacteria and Lactobacillus. *Food Research International*, **42**, 1453–61.

Ávila, M., Jaquet, M., Moine, D., *et al.* (2009b) Physiological and biochemical characterization of the two α-L-rhamnosidases of Lactobacillus plantarum NCC245. *Microbiology*, **155**, 2739–49.

Bang, S.H., Hyun, Y.J., Shim, J., Hong, S.W., and Kim, D.H. (2015) Metabolism of rutin and poncirin by human intestinal microbiota and cloning of their metabolizing a-Lrhamnosidase from Bifidobacterium dentium. *Journal of Microbiology and Biotechnology*, **25**, 18–25.

Beltrán, D., Romo-Vaquero, M., Espín, J.C., Tomás-Barberán, F.A., and Selma, M.V. (2018) Ellagibacter isourolithinifaciens gen. nov., sp. nov., a new member of the family Eggerthellaceae, isolated from human gut. *International Journal of Systematic and Evolutionary Microbiology*, **68**, 1707–12.

Bess, E.N., Bisanz, J.E., Yarza, F., *et al.* (2020) Genetic basis for the cooperative bioactivation of plant lignans by Eggerthella lenta and other human gut bacteria. *Nature Microbiology*, **5**, 56–66.

Bindels, L.B., Delzenne, N.M., Cani, P.D., and Walter, J. (2015) Towards a more comprehensive concept for prebiotics. *Nature Reviews Gastroenterology & Hepatology*, **12**, 303.

Bode, L.M., Bunzel, D., Huch, M., *et al.* (2013) In vivo and in vitro metabolism of trans-resveratrol by human gut. *American Journal of Clinical Nutrition*, **97**, 295–309.

Bokkenheuser, V.D., Shackleton, C.H.L., and Winter, J. (1987) Hydrolysis of dietary flavonoid glycosides by strains of intestinal Bacteroides from humans. *Biochemical Journal*, **248**, 953–6.

Boulangé, C.L., Neves, A.L., Chilloux, J., Nicholson, J.K., and Dumas, M.E. (2016) Impact of the gut microbiota on inflammation, obesity, and metabolic disease. *Genome Medicine*, **8**, 1–12.

Braune, A. and Blaut, M. (2011) Deglycosylation of puerarin and other aromatic C-glucosides by a newly isolated human intestinal bacterium. *Environmental Microbiology*, **13**, 482–94.

Braune, A. and Blaut, M. (2012) Intestinal bacterium Eubacterium cellulosolvens deglycosylates flavonoid C- and O-glucosides. *Applied and Environmental Microbiology*, **78**, 8151–3.

Braune, A. and Blaut, M. (2016) Bacterial species involved in the conversion of dietary flavonoids in the human gut. *Gut Microbes*, **7**, 216–34.

Braune, A., Engst, W., and Blaut, M. (2005) Degradation of neohesperidin dihydrochalcone by human intestinal bacteria. *Journal of Agricultural and Food Chemistry*, **53**, 1782–90.

Braune, A., Gütschow, M., Engst, W., and Blaut, M. (2001) Degradation of quercetin and luteolin by Eubacterium ramulus. *Applied and Environmental Microbiology*, **67**, 5558–67.

Chaplin, A., Carp, C., and Mercader, J. (2018) Resveratrol, metabolic syndrome, and gut microbiota. *Nutrients*, **10**, e1651.

Chen, J., Chia, N., Kalari, K.R., *et al.* (2016) Multiple sclerosis patients have a distinct gut microbiota compared to healthy controls. *Scientific Reports*, **6**, 1–10.

Cheng, K.J., Jones, G.A., Simpson, F.J., and Bryant, M.P. (1969) Isolation and identification of rumen bacteria capable of anaerobic rutin degradation. *Canadian Journal of Microbiology*, **15**, 1365–71.

Cheng, K.J., Krishnamurty, H.G., Jones, G.A., and Simpson, F.J. (1971) Identification of products produced by the anaerobic degradation of naringin by Butyrivibrio sp. C3. *Canadian Journal of Microbiology*, **17**, 129–31.

Clavel, T., Henderson, G., Alpert, C.A., *et al.* (2005) Intestinal bacterial communities that produce the active estrogen-like compounds enterodiol and enterolactone in humans. *Applied and Environmental Microbiology*, **71**, 6077–85.

Clavel, T., Henderson, G., Engst, W., Dore, J., and Blaut, M. (2006) Phylogeny of human intestinal bacteria that activate the dietary lignansecoisolariciresinol diglucoside. *FEMS Microbiology Ecology*, **55**, 471–8.

Clavel, T., Desmarchelier, C., Haller, D., *et al.* (2016) Intestinal microbiota in metabolic diseases: from bacterial community structure and functions to species of pathophysiological relevance. *Gut Microbes*, **5**, 544–51.

Collado, M.C., Rautava, S., Isolauri, E., and Salminen, S. (2015) Gut microbiota: a source of novel tools to reduce the risk of human disease? *Pediatric Research*, **77**, 182–8.

Cortés-Martín, A., García-Villalba, R., González-Sarrías, A., *et al.* (2018) The gut microbiota urolithin metabotypes revisited: the human metabolism of ellagic acid is mainly determined by aging. *Food & Function*, **9**, 4100–6.

Cortés-Martín, A., Selma, M.V., Espín, J.C., and García-Villalba, R. (2019) The human metabolism of nuts proanthocyanidins does not reveal urinary metabolites consistent with distinctive gut microbiota metabotypes. *Molecular Nutrition & Food Research*, **63**, 1–10.

Costea, P.I., Hildebrand, F., Manimozhiyan, A., *et al.* (2017) Enterotypes in the landscape of gut microbial community composition. *Nature Microbiology*, **3**, 8–16.

Couteau, D., McCartney, A.L., Gibson, G.R., Williamson, G., and Faulds, C.B. (2001) Isolation and characterization of human colonic bacteria able to hydrolyse chlorogenic acid. *Journal of Applied Microbiology*, **90**, 873–81.

Dueñas, M., Muñoz-González, I., Cueva, C., *et al.* (2015) A survey of modulation of gut microbiota by dietary polyphenols. *BioMed Research International*, **2015**, 850902.

Espín, J.C., González-Sarrías, A., and Tomás-Barberán, F.A. (2017) The gut microbiota: a key factor in the therapeutic effects of (poly)phenols. *Biochemical Pharmacology*, **139**, 82–93.

García-Villalba, R., Selma, M.V., Espín, J.C., and Tomás-Barberán, F.A. (2019) Identification of novel urolithin metabolites in human feces and urine after the intake of a Pomegranate extract. *Journal of Agricultural and Food Chemistry*, **67**, 11099–107.

Gomez-Arango, L.F., Barrett, H.L., McIntyre, H.D., *et al.* (2016) Connections between the gut microbiome and metabolic hormones in early pregnancy in overweight and obese women. *Diabetes*, **65**, 2214–23.

Gonthier, M.P., Remesy, C., Scalbert, A., *et al.* (2006) Microbial metabolism of caffeic acid and its esters chlorogenic and caftaric acids by human faecal microbiota in vitro. *Biomedicine & Pharmacotherapy*, **60**, 536–40.

González-Sarrías, A., García-Villalba, R., Romo-Vaquero, M., *et al.* (2017) Clustering according to urolithin metabotype explains the interindividual variability in the improvement of cardiovascular risk biomarkers in overweight-obese individuals consuming pomegranate: a randomized clinical trial. *Molecular Nutrition & Food Research*, **61**, 1–14.

González-Sarrías, A., Núñez-Sánchez, M.A., Ávila-Gálvez, M.A., *et al.* (2018) Consumption of pomegranate decreases plasma lipopolysaccharide-binding protein levels, a marker of metabolic endotoxemia, in patients with newly diagnosed colorectal cancer: a randomized controlled clinical trial. *Food & Function*, **9**, 2617–22.

Griffiths, L.A. and Barrow, A. (1972) Metabolism of flavonoids compounds in germfree rats. *Biochemical Journal*, **130**, 1161–2.

Gupta, R.S., Chen, W.J., Adeolu, M., and Chai, Y. (2013) Molecular signatures for the class Coriobacteriia and its different clades; proposal for division of the class Coriobacteriia into the emended order Coriobacteriales, containing the emended family Coriobacteriaceae and Atopobiaceae fam. nov., and Eggerthe. *International Journal of Systematic and Evolutionary Microbiology*, **63**, 3379–97.

Haran, J.P., Bhattarai, S.K., Foley, S.E., *et al.* (2019) Alzheimer's disease microbiome is associated with dysregulation of the anti-inflammatory p-glycoprotein pathway. *MBio*, **10**, e00632–19.

Hazim, S., Curtis, P.J., Schär, M.Y., *et al.* (2016) Acute benefits of the microbial-derived isoflavone metabolite equol on arterial stiffness in men prospectively recruited according to equol producer phenotype: a double-blind randomized controlled trial. *American Journal of Clinical Nutrition*, **103**, 694–702.

Hur, H. and Rafii, F. (2000) Biotransformation of the isoflavonoids biochanin A, formononetin, and glycitein by Eubacterium limosum. *FEMS Microbiology Letters*, **192**, 21–5.

Hur, H.G., Lay, J.O. Jr., Beger, R.D., Freeman, J.P., and Rafii, F. (2000) Isolation of human intestinal bacteria metabolizing the natural isoflavone glycosides daidzin and genistin. *Archives of Microbiology*, **174**, 422–8.

Hur, H.G., Beger, R.D., Heinze, T.M., *et al.* (2002) Isolation of an anaerobic intestinal bacterium capable of cleaving the C-ring of the isoflavonoid daidzein. *Archives of Microbiology*, **178**, 8–12.

Ilett, K.F., Tee, L.B.G., Reeves, P.T., and Minchin, R.F. (1990) Metabolism of drugs and other xenobiotics in the gut lumen and wall. *Pharmacology & Therapeutics*, **46**, 67–93.

Jimenez, N., Esteban-Torres, M., Mancheno, J.M., de las Rivas, B., and Munoz, R. (2014) Tannin degradation by a novel tannase enzyme present in some lactobacillus plantarum strains. *Applied and Environmental Microbiology*, **80**, 2991–7.

Jin, J.S. and Hattori, M. (2012) Isolation and characterization of a human intestinal bacterium Eggerthella sp. CAT-1 capable of cleaving the C-ring of (+)-catechin and (-)-epicatechin, followed by p-dehydroxylation of the B-ring. *Biological & Pharmaceutical Bulletin*, **35**, 2252–6.

Jin, J.S., Nishihata, T., Kakiuchi, N., and Hattori, M. (2008) Biotransformation of C-glucosylisoflavone puerarin to estrogenic (3S)-equol in co-culture of two human intestinal bacteria. *Biological & Pharmaceutical Bulletin*, **31**, 1621–5.

Jin, J.S., Kitahara, M., Sakamoto, M., Hattori, M., and Benno, Y. (2010) Slackia equolifaciens sp. nov., a human intestinal bacterium capable of producing equol. *International Journal of Systematic and Evolutionary Microbiology*, **60**, 1721–4.

Keppler, K. and Humpf, H.U. (2005) Metabolism of anthocyanins and their phenolic degradation products by the intestinal microflora. *Bioorganic & Medicinal Chemistry*, **13**, 5195–205.

Kim, M., Kim, N., and Han, J. (2014) Metabolism of kaempferia parviflora polymethoxyflavones by human intestinal bacterium Blautia sp. MRG-PMF1. *Journal of Agricultural and Food Chemistry*, **62**, 12377–83.

Kim, M., Lee, J., and Han, J. (2015) Deglycosylation of isoflavone C-glycosides by newly isolated human intestinal bacteria. *Journal of the Science of Food and Agriculture*, **95**, 1925–31.

Krumholz, L.R. and Bryant, M.P. (1986) Eubacterium oxidoreducens sp. nov. requiring H2 or formate to degrade gallate, pyrogallol, phloroglucinol and quercetin. *Archives of Microbiology*, **144**, 8–14.

Kutschera, M., Engst, W., Blaut, M., and Braune, A. (2011) Isolation of catechin-converting human intestinal bacteria. *Journal of Applied Microbiology*, **111**, 165–75.

Larrosa, M., Yañéz-Gascón, M.J., Selma, M.V., *et al.* (2009) Effect of a low dose of dietary resveratrol on colon microbiota, inflammation and tissue damage in a DSS-induced colitis rat model. *Journal of Agricultural and Food Chemistry*, **57**, 2211–20.

Liu, Y., Dai, Y., Xun, L., and Hu, M. (2003) Enteric disposition and recycling of flavonoids and ginkgo flavonoids. *Journal of Alternative and Complementary Medicine*, **9**, 631–40.

Manach, C., Milenkovic, D., van de Wiele, T., *et al.* (2017) Addressing the inter-individual variation in response to consumption of plant food bioactives: towards a better understanding of their role in healthy aging and cardiometabolic risk reduction. *Molecular Nutrition & Food Research*, **61**, 1–16.

Marotti, I., Bonetti, A., Biavati, B., Catizone, P., and Dinelli, G. (2007) Biotransformation of common bean (Phaseolus vulgaris L.) flavonoid glycosides by Bifidobacterium species from human intestinal origin. *Journal of Agricultural and Food Chemistry*, **55**, 3913–19.

Martínez, I., Perdicaro, D.J., Brown, A.W., *et al.* (2013) Diet-induced alterations of host cholesterol metabolism are likely to affect the gut microbiota composition in hamsters. *Applied and Environmental Microbiology*, **79**, 516–24.

Maruo, T., Sakamoto, M., Ito, C., Toda, T., and Benno, Y. (2008) Adlercreutzia equolifaciens gen. nov., sp. nov., an equol-producing bacterium isolated from human faeces, and emended description of the genus Eggerthella. *International Journal of Systematic and Evolutionary Microbiology*, **58**, 1221–7.

Matthies, A., Blaut, M., and Braune, A. (2009) Isolation of a human intestinal bacterium capable of daidzein and genistein conversion. *Applied and Environmental Microbiology*, **75**, 1740–4.

Mingshu, L., Kai, Y., Qiang, H., and Dongying, J. (2006) Biodegradation of gallotannins and ellagitannins. *Journal of Basic Microbiology*, **46**, 68–84.

Miyake, Y., Yamamoto, K., and Osawa, T. (1997) Metabolism of antioxidant in lemon fruit (Citrus limon BURM. f.) by human intestinal bacteria. *Journal of Agricultural and Food Chemistry*, **45**, 10.

Mosele, J., Macià, A., and Motilva, M.-J. (2015) Metabolic and microbial modulation of the large intestine ecosystem by non-absorbed diet phenolic compounds: a review. *Molecules*, **20**, 17429–68.

Ozdal, T., Sela, D.A., Xiao, J., Boyacioglu, D., and Chen, F. (2016) The reciprocal interactions between polyphenols and gut microbiota and effects on bioaccessibility. *Nutrients*, **8**, 1–36.

Portune, K.J., Beaumont, M., Davila, A.M., Tomé, D., Blachier, F., and Sanz, Y. (2016) Gut microbiota role in dietary protein metabolism and health-related outcomes: the two sides of the coin. *Trends in Food Science & Technology*, **57**, 213–32.

Possemiers, S., Heyerick, A., Robbens, V., de Keukeleire, D., and Verstraete, W. (2005) Activation of proestrogens from hops (Humulus lupulus L.) by intestinal microbiota; conversion of isoxanthohumol into 8-prenylnaringenin. *Journal of Agricultural and Food Chemistry*, **53**, 6281–8.

Possemiers, S., Bolca, S., Grootaert, C., *et al.* (2006) The prenylflavonoid isoxanthohumol from hops (Humulus lupulus L.) is activated into the potent phytoestrogen 8-prenylnaringenin in vitro and in the human intestine. *Journal of Nutrition*, **136**, 1862–7.

Possemiers, S., Bolca, S., Eeckhaut, E., Depypere, H., and Verstraete, W. (2007) Metabolism of isoflavones, lignans and prenylflavonoids by intestinal bacteria: producer phenotyping and relation with intestinal community. *FEMS Microbiology Ecology*, **61**, 372–83.

Possemiers, S., Rabot, S., Espín, J.C., *et al.* (2008) Eubacterium limosum activates isoxanthohumol from hops (Humulus lupulus L.) into the potent phytoestrogen 8-prenylnaringenin in vitro and in rat intestine. *Journal of Nutrition*, **138**, 1310–16.

Raimondi, S., Roncaglia, L., de Lucia, M., *et al.* (2009) Bioconversion of soy isoflavones daidzin and daidzein by Bifidobacterium strains. *Applied Microbiology and Biotechnology*, **81**, 943–50.

Rivière, A., Selak, M., Lantin, D., Leroy, F., and de Vuyst, L. (2016) Bifidobacteria and butyrate-producing colon bacteria: importance and strategies for their stimulation in the human gut. *Frontiers in Microbiology*, **7**, 979.

Romo-Vaquero, M., Cortés-Martín, A., Loria-Kohen, V., *et al.* (2019) Deciphering the human gut microbiome of urolithin metabotypes: association with enterotypes and potential cardiometabolic health implications. *Molecular Nutrition & Food Research*, **63**, 1–11.

Rowland, I., Gibson, G., Heinken, A., *et al.* (2018) Gut microbiota functions : metabolism of nutrients and other food components. *European Journal of Nutrition*, **57**, 1–24.

Sanchez-Patan, F., Tabasco, R., Monagas, M., *et al.* (2012) Capability of Lactobacillus plantarum IFPL935 to catabolize flavan3-ol compounds and complex phenolic extracts. *Journal of Agricultural and Food Chemistry*, **60**, 7142–51.

Schneider, H. and Blaut, M. (2000) Anaerobic degradation of flavonoids by Eubacterium ramulus. *Archives of Microbiology*, **173**, 71–5.

Schneider, H., Schwiertz, A., Collins, M.D., and Blaut, M. (1999) Anaerobic transformation of quercetin-3-glucoside by bacteria from the human intestinal tract. *Archives of Microbiology*, **171**, 81–91.

Schoefer, L., Mohan, R., Braune, A., Birringer, M., and Blaut, M. (2002) Anaerobic C-ring cleavage of genistein and daidzein by Eubacterium ramulus. *FEMS Microbiology Letters*, **208**, 197–202.

Schoefer, L., Mohan, R., Schwiertz, A., Braune, A., and Blaut, M. (2003) Anaerobic degradation of flavonoids by Clostridium orbiscindens. *Applied and Environmental Microbiology*, **69**, 5849–54.

Selma, M.V., Espín, J.C., and Tomás-Barberán, F.A. (2009) Interaction between phenolics and gut microbiota: role in human health. *Journal of Agricultural and Food Chemistry*, **57**, 6485–501.

Selma, M.V., Tomás-Barberán, F.A., Beltran, D., Garcia-Villalba, R., and Espín, J.C. (2014a) Gordonibacter urolithinfaciens sp. nov., a urolithin-producing bacterium isolated from the human gut. *International Journal of Systematic and Evolutionary Microbiology*, **64**, 2346–52.

Selma, M.V., Beltrán, D., García-Villalba, R., Espín, J.C., and Tomás-Barberán, F.A. (2014b) Description of urolithin production capacity from ellagic acid of two human intestinal Gordonibacter species. *Food & Function*, **5**, 1779–84.

Selma, M.V., Romo-Vaquero, M., García-Villalba, R., González-Sarrías, A., Tomás-Barberán, F.A., and Espín, J.C. (2016) The human gut microbial ecology associated with overweight and obesity determines ellagic acid metabolism. *Food & Function*, **7**, 1769–74.

Selma, M.V., Beltrán, D., Luna, M.C., *et al.* (2017) Isolation of human intestinal bacteria capable of producing the bioactive metabolite isourolithin a from ellagic acid. *Frontiers in Microbiology*, **8**, 1–8.

Selma, M.V., González-Sarrías, A., Salas-Salvadó, J., *et al.* (2018) The gut microbiota metabolism of pomegranate or walnut ellagitannins yields two urolithin-metabotypes that correlate with cardiometabolic risk biomarkers: comparison between normoweight, overweight-obesity and metabolic syndrome. *Clinical Nutrition*, **37**, 897–905.

Serra, D., Almeida, L., and Dinis, T. (2018) Dietary polyphenols: a novel strategy to modulate microbiota-gut-brain axis. *Trends in Food Science & Technology*, **78**, 224–33.

Shimada, Y., Yasuda, S., Takahashi, M., *et al.* (2010) Cloning and expression of a novel NADP(H)-dependent daidzein reductase, an enzyme involved in the metabolism of daidzein, from equoL-producing Lactococcus strain 20–92. *Applied and Environmental Microbiology*, **76**, 5892–901.

Shin, N.R., Moon, J.S., Shin, S.Y., *et al.* (2016) Isolation and characterization of human intestinal Enterococcus avium EFEL009 converting rutin to quercetin. *Letters in Applied Microbiology*, **62**, 68–74.

Simons, A.L., Renouf, M., Hendrich, S., and Murphy, P.A. (2005) Metabolism of glycitein (7; 41-dihydroxy6-methoxy-isoflavone) by human gut microflora. *Journal of Agricultural and Food Chemistry*, **53**, 8519–25.

Sun, M.F. and Shen, Y.Q. (2018) Dysbiosis of gut microbiota and microbial metabolites in Parkinson's Disease. *Ageing Research Reviews*, **45**, 53–61.

Takagaki, A. and Nanjo, F. (2015) Biotransformation of (-)-epigallocatechin and (-)-gallocatechin by intestinal bacteria involved in isoflavone metabolism. *Biological & Pharmaceutical Bulletin*, **38**, 325–30.

Tamura, M., Tsushida, T., and Shinohara, K. (2006) Isolation of an isoflavone-metabolizing, Clostridium-like bacterium, strain TM-40, from human faeces. *Anaerobe*, **13**, 32–5.

Tomás-Barberán, F.A., García-Villalba, R., Quartieri, A., *et al.* (2014a) In vitro transformation of chlorogenic acid by human gut microbiota. *Molecular Nutrition & Food Research*, **58**, 1122–31.

Tomás-Barberán, F.A., García-Villalba, R., González-Sarrías, A., Selma, M.V, and Espín, J.C. (2014b) Ellagic acid metabolism by human gut microbiota: consistent observation of three urolithin phenotypes in intervention trials, independent of food source, age, and health status. *Journal of Agricultural and Food Chemistry*, **62**, 6535–8.

Tomás-Barberán, F.A., Selma, M.V., and Espín, J.C. (2016) Interactions of gut microbiota with dietary polyphenols and consequences to human health. *Current Opinion in Clinical Nutrition and Metabolic Care*, **19**, 471–6.

Tomás-Barberán, F.A., González-Sarrías, A., García-Villalba, R., *et al.* (2017) Urolithins, the rescue of "old" metabolites to understand a "new" concept: metabotypes as a nexus among phenolic metabolism, microbiota dysbiosis, and host health status. *Molecular Nutrition & Food Research*, **61**, e201500901.

Tomás-Barberan, F.A., Selma, M.V., and Espín, J.C. (2018) Polyphenols' gut microbiota metabolites: bioactives or biomarkers? *Journal of Agricultural and Food Chemistry*, **66**, 3593–4.

Tomé-Carneiro, J., Gonzálvez, M., Larrosa, M., *et al.* (2013) Resveratrol in primary and secondary prevention of cardiovascular disease: a dietary and clinical perspective. *Annals of the New York Academy of Sciences*, **1290**, 37–51.

Torrallardona, D., Badiola, I., and Broz, J. (2007) Effects of benzoic acid on performance and ecology of gastrointestinal microbiota in weanling piglets. *Livestock Science*, **108**, 210–13.

Tsangalis, D., Ashton, J.F., McGill, A.E.J., and Shah, N.P. (2002) Enzymic transformation of isoflavone phytoestrogens in soymilk by b-glucosidase-producing bifidobacteria. *Journal of Food Science*, **67**, 3104–13.

Tsuji, H., Moriyama, K., Nomoto, K., Miyanaga, N., and Akaza, H. (2010) Isolation and characterization of the equol-producing bacterium Slackia sp. strain NATTS. *Archives of Microbiology*, **192**, 279–87.

Tzounis, X., Vulevic, J., Kuhnle, G.G., *et al.* (2008) Flavanol monomer-induced changes to the human faecal microflora. *British Journal of Nutrition*, **99**, 782–92.

Walle, T., Hsieh, F., DeLegge, M.H., Oatis, J.E., and Walle, U.K. (2004) High absorption but very low bioavailability of oral resveratrol in humans. *Drug Metabolism and Disposition*, **32**, 1377–82.

Wang, D.G., Zhang, Z.W., Ju, J.F., Wang, X.Y., and Qiu, W.J. (2012) Investigation of piceid metabolites in rat by liquid chromatography tandem mass spectrometry. *Journal of Chromatography. B, Analytical Technologies in the Biomedical and Life Sciences*, **879**, 69–74.

Wang, L.Q., Meselhy, M.R., Li, Y., *et al.* (2001) The heterocyclic ring fission and dehydroxylation of catechins and related compounds by Eubacterium sp. strain SDG-2, a human intestinal bacterium. *Chemical & Pharmaceutical Bulletin*, **49**, 1640–3.

Wang, X.L., Hur, H., Lee, J.H., Kim, K.T., and Kim, S. (2005) Enantioselective synthesis of s -equol from dihydrodaidzein by a newly isolated anaerobic human intestinal bacterium. *Applied and Environmental Microbiology*, **71**, 214–19.

Wang, X.L., Kim, K.T., Lee, J.H., Hur, H.G., and Kim, S.I. (2004) C-ring cleavage of isoflavones daidzein and genistein by a newly-isolated human intestinal bacterium Eubacterium ramulus Julong 601. *Journal of Microbiology and Biotechnology*, **14**, 766–71.

Winter, J., Moore, L.H., Dowell, V.R., and Bokkenheuser, V.D. (1989) C-Ring cleavage of flavonoids by human intestinal bacteria. *Applied and Environmental Microbiology*, **55**, 1203–8.

Winter, J., Popoff, M.R., Grimont, P., and Bokkenheuser, V.D. (1991) Clostridium orbiscindens sp. nov., a human intestinal bacterium capable of cleaving the flavonoid C-ring. *International Journal of Systematic Bacteriology*, **41**, 355–7.

Xu, J., Qian, D., Jiang, S., *et al.* (2014) Application of ultra-performance liquid chromatography coupled with quadrupole time-of-flight mass spectrometry to determine the metabolites of orientin produced by human intestinal bacteria. *Journal of Chromatography. B, Analytical Technologies in the Biomedical and Life Sciences*, **944**, 123–7.

Yeh, C.T. and Yen, G.C. (2003) Effects of phenolic acids on human phenol sulfotransferase in relation to their antioxidant activity. *Journal of Agricultural and Food Chemistry*, **51**, 1474–9.

Yokoyama, S. and Suzuki, T. (2008) Isolation and characterization of a novel equol-producing bacterium from human feces. *Bioscience, Biotechnology, and Biochemistry*, **72**, 2660–6.

Yokoyama, S., Niwa, T., Osawa, T., and Suzuki, T. (2010) Characterization of an O-desmethylangolensin-producing bacterium isolated from human feces. *Archives of Microbiology*, **192**, 15–22.

Zhao, M., Du, L., Tao, J., *et al.* (2014) Determination of metabolites of diosmetin-7-o-glucoside by a newly isolated escherichia coli from human gut using UPLC-Q-TOF/MS. *Journal of Agricultural and Food Chemistry*, **62**, 11441–8.

Zhao, Z., Egashira, Y., and Sanada, H. (2004) Ferulic acid is quickly absorbed from rat stomach as the free form and then conjugated mainly in liver. *Journal of Nutrition*, **134**, 3083–8.

Zhou, Y., Mihindukulasuriya, K.A., Gao, H., *et al.* (2014) Exploration of bacterial community classes in major human habitats. *Genome Biology*, **15**, R66.

Index

Dietary Polyphenols: Metabolism and Health Effects, First Edition.
Edited by Francisco A. Tomás-Barberán, Antonio González-Sarrías, and Rocío García-Villalba.

d

e

f

Wiley The manufacturer's authorized representative according to the EU General Product Safety Regulation is Wiley-VCH GmbH, Boschstr. 12, 69469 Weinheim, Germany, e-mail: Product_Safety@wiley.com.

Printed and bound by CPI Group (UK) Ltd, Croydon, CR0 4YY

19/08/2026

02201089-0001